In Vitro Cultivation of Protozoan Parasites

Editor
James B. Jensen, Ph.D.
Associate Professor of Parasitology
Department of Microbiology and Public Health
Michigan State University
East Lansing, Michigan

CRC Press
Taylor & Francis Group
Boca Raton London New York

CRC Press is an imprint of the
Taylor & Francis Group, an **informa** business

First published 1983 by CRC Press
Taylor & Francis Group
6000 Broken Sound Parkway NW, Suite 300
Boca Raton, FL 33487-2742

Reissued 2018 by CRC Press

Library of Congress Cataloging in Publication Data
Main entry under title:

In vitro cultivation of protozoan parasites.

Bibliography: p.
Includes index.
Contents: The Coccidia / Clarence A. Speer—
Lumen dwelling Protozoa / Louis S. Diamond—
Leishmania / K. P. Chang—[etc.]
1. Protozoology—Cultures and culture media.
I. Jensen, James B., 1943- . [DNLM: 1. Protozoa—
Isolation and purification. 2. Parasitology—
Methods. QX 50 135]
QL366.I48 1983 593.1'0724 83–2769
ISBN 0-8493-5748-9

A Library of Congress record exists under LC control number: 83002769

ISBN 13: 978-1-315-89435-5 (hbk)
ISBN 13: 978-1-351-07345-5 (ebk)

Visit the Taylor & Francis Web site at http://www.taylorandfrancis.com and the
CRC Press Web site at http://www.crcpress.com

PREFACE

It has been stated more than once in mixed scientific company, that because the field of parasitology is so vast and there is so much to know about the many parasites' morphology, development epidemiology, physiology, biochemistry, pharmacology, immunology, molecular biology, etc., that parasitologists, who know where the problems are, usually are ill-equipped to solve them. It is not my purpose to lend any credence to this supposition, except to admit that there is plenty of work to do in solving the problems parasites present to man and his domestic animals. Many researchers skilled in immunology, biochemistry, molecular biology, etc., have avoided the application of such skills to problems in parasitology because of the logistical problems presented by an organism that has adopted another living organism, or, as is the case of many protozoan parasites, another living cell, as its ecologic niche. The development of techniques for in vitro cultivation of parasites has, to a great degree, reduced such difficulties and, consequently has opened the doors to many investigations that were simply not practical before. Accordingly, it is the purpose of this book to make available to parasitologists and workers in many other disciplines a review of the developments leading up to the successful cultivation of the more important protozoan parasites of man and domestic animals, a detailed description of the current state of the art of protozoan parasite cultivation, and a limited discussion of the major achievements to our understanding of parasite biology derived through experimentation using cultured parasites. I apologize to the reader for the fact that this book does not contain a discussion on one important parasite—*Trypanosoma cruzi*. The author of this chapter failed to deliver his promised contribution, despite my numerous encouragements and his avowed assurances of compliance.

James B. Jensen

THE EDITOR

James B. Jensen, Ph.D., is an Associate Professor of Parasitology in the Department of Microbiology and Public Health at Michigan State University, East Lansing, Michigan. He received his B.S. and M.S. degrees from Brigham Young University, Provo, Utah in 1970 and 1972, respectively. He began his Ph.D. studies with the late Dr. Datus M. Hammond at Utah State University where pioneering work on cultivation of mammalian coccidial parasites had been in progress for several years. After the untimely death of Dr,. Hammond, Dr. Jensen completed his Ph.D. at Auburn University, Auburn, Alabama in 1975.

Upon completion of his graduate studies, Dr. Jensen joined Dr. William Trager at the Rockefeller University in New York City where their joint efforts led to the continuous cultivation of the human malarial parasite, *Plasmodium falciparum.* This breakthrough has been heralded as the most significant advancement in research on malaria during the past 30 years. This work has led to a renewed interest in a vaccine for malaria, as well as extensive studies on malarial parasite biochemistry, pharmacology, ultrastructural development, and biology which were impossible to do before continuous cultivation was acheived.

Dr. Jensen has worked as a consultant on malaria for the World Health Organization-Special Programme for Research and Training in Tropical Diseases, and the NIAID-USAID Project on Epidemiology and Control of Arthropod-Borne Diseases in the Near East, as well as the WHO-Special Programme Scientific Working Group on Malaria Immunology and Vaccine Development. He presently is a member of the International Board of the Kuvin Centre for the Study of Infectious and Tropical Diseases of the Hadassah Medical School, Hebrew University, Jerusalem, Israel.

In addition to his duties as an Associate Professor of Parasitology at Michigan State University, Dr. Jensen maintains an active research laboratory in the Sudan where he works under the sponsorship of the NIH through the Sudanese Ministry of Health on immunology of malaria infections and other parasitic diseases. He is a member of the American Society of Tropical Medicine and Hygiene, the American Society of Parasitologists and the Society of Protozoologists. He is listed in Outstanding Young Men of America, 1979, and Who's Who in Frontier Science and Technology, 1983.

This book is dedicated to my two mentors
the late Dr. D. M. Hammond
and Dr. W. Trager

CONTRIBUTORS

C. G. D. Brown, Ph.D.
Senior Research Fellow
Centre for Tropical Veterinary Medicine
University of Edinburgh
Edinburgh, Scotland

Kwang-poo Chang, Ph.D.
Associate Professor
Rockefeller University
New York, New York

Louis S. Diamond, Ph.D.
Head
Section of Parasite Growth and Differentiation
Laboratory of Parasitic Diseases
National Institute of Allergy and Infectious Diseases
National Institutes of Health
Bethesda, Maryland

Wallace R. Fish, Ph.D.
Department of Preventive Medicine and Biometrics
Uniformed Services University of Health Sciences
Bethesda, Maryland

George C. Hill, Ph.D.
Director
Division of Biomedical Sciences
Meharry Medical College
Nashville, Tennessee

Hiroyuki Hirumi, Ph.D.
Senior Scientist
International Laboratory for Research on Animal Diseases
Nairobi, Kenya

James B. Jensen, Ph.D.
Associate Professor of Parasitology
Department of Microbiology and Public Health
Michigan State University
East Lansing, Michigan

Michael G. Levy, Ph.D.
Assistant Professor
College of Veterinary Medicine
University of Illinois
Urbana, Illinois

Miodrag Ristic, Ph.D.
Professor of Veterinary Pathology and Hygiene
College of Veterinary Medicine
University of Illinois
Urbana, Illinois

Clarence A. Speer, Ph.D.
Associate Professor of Microbiology
Department of Microbiology
University of Montana
Missoula, Montana

TABLE OF CONTENTS

Chapter 1

THE COCCIDIA

Clarence A. Speer

TABLE OF CONTENTS

I. CLASSIFICATION

Most coccidia are obligate intracellular parasites of the intestinal tracts of vertebrates. However, certain coccidia have been reported from invertebrates and in organs other than the intestinal tract of vertebrates. Although the term "coccidia" is sometimes used to include genera such as *Haemoproteus*, *Hapatocytis*, *Leucocytozoon* and *Plasmodium*, which are genera of the order Eucoccidiorida, it will be limited here to members of the suborder Eimeriorina. The most recent classification of the coccidia is that of Levine[1] and Levine et al.[1a] However, Levine[2] recently eliminated the Family Sarcocystidae and placed its genera in the Family Eimeriidae. This family now includes *Besnoitia, Eimeria, Frenkelia, Isospora, Sarcocystis*, and *Toxoplasma*, as well as other genera. In a recent taxonomic review of the genus *Toxoplasma*, Levine[3] synonymized *Hammondia* with *Toxoplasma*. However, a more recent report by Hendricks et al.[4] points out that these two genera have certain distinctly different characteristics. Therefore, *Hammondia* is considered here as a distinct genus. Dubey[5] proposed a new genus, *Levieneia*, for those isosporans *(I. felis, I. rivolta, I. canis,* and *I. ohioensis)* which are not homoxenous but rather are facultatively or obligatorily heteroxenous and form cyst-like stages in tissues of the intermediate host. For these same isosporans, Frenkel[5a] proposed the new genus *Cytoisospora*. Although the two types of *Isospora* have obvious differences in their life cycles, both types will be considered here to belong to the genus *Isospora*. Based upon the classification schemes of Levine[1,2] and Dubey,[5] Table 1 gives the classification of those genera of coccidia which are of medical and veterinary importance, and which are the subject of this chapter.

II. LIFE CYCLES AND PATHOGENESIS

Coccidia have stages which occur within the host, called endogenous stages, and exogenous stages which occur outside the host. Certain species of coccidia are homoxenous in which all endogenous stages of the parasite occur within a single host. Certain other species are heteroxenous in which certain stages of the endogenous life cycle occur within an intermediate host and other or similar stages occur in the definitive host. Regardless, whether the coccidian is homoxenous or heteroxenous, the coccidia are known to progress sequentially through certain stages of their life cycle.

The exogenous stages are oocysts which are usually shed in the feces by the definitive host. Oocysts may undergo sporulation within the host or outside the host. In those species which sporulate outside the host, oxygen, moisture, and proper temperatures cause the oocyst to undergo sporulation. During sporulation of those coccidia which are of medical or veterinary importance, the protoplasm (sporont) of the oocyst condenses away from the inner layer of the oocyst wall, and then cleaves to form two or four sporoblasts which then develop into sporocysts, each of which contain two or four sporozoites. Oocysts which contain sporocysts with sporozoites are infective to the definitive and/or intermediate host.

When sporulated oocysts of those coccidian species which infect mammals are ingested by a definitive or intermediate host, CO_2, trypsin and/or bile cause the sporozoites to become

Table 1
CLASSIFICATION OF THE COCCIDIA OF MEDICAL AND VETERINARY IMPORTANCE

Phylym Apicomplexa Levine, 1970

Apical complex, generally consisting of polar ring(s), micronemes, rhoptries, subpellicular microtubules, and conoid, present at some stage; micropore(s) generally present; single type of nucleus; cilia and flagella absent, except for flagellated microgametes in some groups; sexuality, when present, syngamy; cysts often present; all species parasitic.

Class Sporozoea Leuckart, 1879

Conoid, if present, forming complete cone; reproduction generally both sexual and asexual; oocysts generally containing infective sporozoites which result from sporogony; locomotion of mature organisms by body flexion, gliding, or undulation or longitudinal ridges; flagella present only in microgametes of some groups; pseudopods ordinarily absent, if present, used for feeding, not locomotion; homoxenous or heteroxenous.

Subclass Coccidia Leuckart, 1879

Mature gamonts small, typically intracellular without mucron or epimerite; syzygy generally absent, if present, involves markedly anisogamous gametes; life cycle characteristically consisting of merogony, gametogony, and sporogony; most species in vertebrates.

Order Eucoccidida Leger and Duboscq, 1910

Merogony present; invertebrates and/or invertebrates.

Suborder Eimeriorina Leger, 1911

Macrogamont and microgamont develop independently; no syzygy; microgamont typically producing many microgametes; zygote not motile; sporozoites typically enclosed in sporocysts within oocyst; homoxenous or heteroxenous.

Family Eimeriidae Minchin, 1903

Development in host cell proper; oocysts and meronts (schizonts) without attachment organelle; oocysts with 0, 1, 2, 4, or many sporocysts, each with one or more sporozoites; homoxenous or heteroxenous; merogony and gametogony within the host, sporogony typically (but not necessarily) outside; microgametes with two or three flagella; in vertebrates or invertebrates.

Genus *Besnoitia* Henry, 1913

Oocysts with two sporocysts, each with four sporozoites; facultatively or obligatorily heteroxenous with definitive host a felid so far as is known; merogony only in intermediate host so far as is known; meronts especially in skin, fibroblasts and presumably other cells; metrocysts not formed; parenteral meront wall thick, with hyperplastic, hypertrophic host cell nuclei; meronts and gamonts in enteron cells of definitive host; parenteral meronts usually rounded, large, and aseptate; sporogony outside of host.

Genus *Eimeria* Schneider, 1875

Oocysts with four sporocysts, each with two sporozoites; homoxenous; spotogony outside of host.

Genus *Frenkelia* Biocca, 1968

Oocysts with two sporocysts, each with four sporozoites, presumably obligatorily heteroxenous, with intermediate host a prey animal and definitive host a predator; merogony in intermediate host; parenteral meront wall thin; meronts in brain and spinal cord; parenteral meronts usually rounded quite small and sometimes septate; metrocytes formed; gamonts in enteron cells of definitive host; sporogony within definitive host.

Genus *Hammondia* Frenkel, 1974

Oocysts with two sporocysts, each with four sporozoites; obligatory heteroxenous, with intermediate host a prey animal and definitive host a predator so far as is known; merogony in intermediate host; merogony and gametogony in enterons of definitive host so far as is known; thin-walled cysts lack septae and metrocytes; sporogony outside of host.

Genus *Isospora* Schneider, 1881

Oocysts with two sporocysts, each with four sporozoites; in homoxenous meronts and gamonts usually in enteron cells; in facultatively or obligatorily heteroxenous organisms asexual cycle not restricted to intestine; cysts

Table 1 (continued)
CLASSIFICATION OF THE COCCIDIA OF MEDICAL AND VETERINARY IMPORTANCE

in many organs of intermediate host, typically unizoite; asexual stages preceding sexual development in carnivores; tachyzoites and bradyzoites nontransmissible to intermediate hosts; oocysts infectious to definitive and intermediate hosts; sporogony inside or outside of host.

Genus *Sarcocystis* Lankester, 1882

Oocysts with two sporocysts, each with four sporozoites; obligatory heteroxenous, with intermediate host a prey animal and definitive host a predator so far as is known; merogony only in intermediate host so far as is known; first generation meronts in vascular endothelial cells; there is possibly at least one additional intermediate meront generation in the intermediate host; final generation meronts in striated muscle cells; metrocytes formed; wall of final meront generation without noticeable host cell nuclei when mature; final generation meronts usually elongate, large and often septate; gamonts but not meronts in enteron cells of definitive host; sporogony within definitive host.

Genus *Toxoplasma* Nicolle and Manceaux, 1909

Oocysts with two sporocysts, each with four sporozoites; facultatively or obligatorily heteroxenous, with definitive host a felid so far as is known; merogony in both intermediate and definitive hosts; merozoites and oocysts can cause infection in intermediate and definitive hosts; meronts in brain and almost any other cell of intermediate host; parenteral mecont wall thin, without noticeable host cell nuclei when mature; metrocysts not formed; meronts and gamonts in enteron cells of definitive host; sporogony outside of host.

motile, excyst from the sporocyst and leave the oocyst via the micropyle. In those *Eimeria* species which infect avian hosts, the oocyst wall is broken by the mechanical grinding action of the gizzard which liberates the sporocysts from which sporozoites excyst in response to chymotrypsin and bile.[6,7] Sporozoites actively penetrate cells of the intestinal tract, usually intestinal epithelial cells or cells of the lamina propria of the small intestine. Intracellular sporozoites transform into spheroid stages called trophozoites or uninucleate schizonts or meronts.[8] Several or many nuclear divisions without cytoplasmic division occur in each meront. Eventually, budding occurs at the margin of the meront to form elongate uninucleate merozoites. These merozoites usually leave their host cell and enter other cells to form one or more additional meront generations. Each coccidian species usually forms a characteristic number of merozoites in each meront generation, as well as a characteristic number (usually two to four) of asexual generations. During part of their asexual cycle, certain coccidia multiply asexually by endodyogeny in addition to merogony. In endodyogeny, two daughter organisms are formed within a mother cell, which is ultimately destroyed. Zoites which result from such forms and proliferate relatively rapidly are called tachyzoites, whereas those which multiply more slowly are called bradyzoites. Tachyzoites form colonies or pseudocysts, bradyzoites occur in cysts.

After a certain number of generations, merozoites enter the sexual portion of the endogenous cycle by developing into male or female gamonts, called microgamonts and macrogamonts, respectively. Certain merozoites form microgamonts, whereas others form macrogamonts. In the microgamont, several nuclear divisions occur without cytoplasmic division. After a certain number of nuclei have been formed, several to many uninucleate microgametes, each with two or three flagella, are formed at the margin of the gamont. The macrogamont undergoes no nuclear division, but increases several times in size with the proliferation of various cytoplasmic organelles. A microgamete fertilizes a macrogamont to form a zygote, which forms at its margin a resistant oocyst wall consisting of one or more layers. The oocyst causes the host cell to rupture and is shed from the host into the external environment.

In *Eimeria*[9] and certain *Isospora*[10-12] species, the asexual and sexual stages of the life cycle occur within a single host. In species of certain other *Isospora*,[13-17] *Besnoitia*[18] and

Toxoplasma,[19-21] asexual stages only occur in the intermediate host, and asexual and sexual stages occur in the definitive host. In these genera, oocysts or bradyzoites and/or tachyzoites can cause infection in intermediate and definitive hosts. So far as is known in *Frenkelia* and *Sarcocystis*, merogony and/or endodyogeny occur only in the intermediate host and gametogony occurs only in the definitive host.[2] Intermediate hosts become infected by ingesting oocysts, or sporocysts, whereas definitive hosts become infected by ingesting tissue cysts which contain bradyzoites or tachyzoites.

The coccidia are known to be capable of causing disease in humans and various domestic and wild animals. The term coccidiosis usually used to designate disease caused by infections by *Eimeria* species only. *Toxoplasma* causes toxoplasmosis, *Sarcocystis* causes sarcocystosis or sarcosporidiosis and so on.

A. *Besnoitia*

So far as is known *Besnoitia* uses felids only as definitive hosts, in which occur merogony, gametogony, and oogony.[18,22] In the intermediate host, *Besnoitia* grows intracellularly and destroys parasitized cells of various connective tissue, and produces cysts with characteristically thick walls.[23-26] Depending upon the virulence of the strain of *Besnoitia*, a generalized infection may occur in the intermediate host which may die within a week.[5] Usually the host recovers from the acute infection and cysts are formed primarily in connective tissue. Cysts which rupture may cause a granulomatous reaction in tissues of the intermediate host.[27] Asexual forms of *B. darlingi* and *B. jellisoni* have been shown to be highly pathogenic for marmosets.[28] *Besnoitia besnoiti* can cause severe economic losses in cattle due to weight loss, hide condemnations, and death.[29]

B. *Eimeria*

In many domestic and wild animals *Eimeria* species cause the disease known as coccidiosis. Coccidiosis is transmitted from animal to animal by means of sporulated oocysts which are ingested in contaminated water or feed. Coccidiosis occurs primarily in young animals and especially in those animals housed or penned under crowded conditions. Economic losses due to coccidiosis are difficult to determine because of the lack of adequate records as well as the complex nature of the disease. The disease causes mortality, morbidity, and weight loss in various domestic animals, but probably has its greatest economic impact in the poultry and cattle industries around the world. Fitzgerald[30] estimated that bovine coccidiosis alone caused a worldwide economic loss of $472 million.

Coccidiosis is still the most economically important parasitic disease of chickens.[6] Because of the widespread use of anticoccidials by chicken growers, little economic loss due to mortality occurs in the U.S. Losses due to coccidiosis in chickens are primarily due to morbidity which results in reduced weight gains, egg production, and decreased food conversion.[6] Ruff and Reid[6] estimated $60 to $120 million are lost each year in the U.S. due to weight loss plus another $35 million which are spent on anticoccidials.

Some *Eimeria* species are highly pathogenic, whereas other species are not, and certain strains of a given *Eimeria* species may be more pathogenic than others. During an *Eimeria* infection, the asexual stages cause little tissue damage to their host, whereas the gamonts and oocysts may cause extensive tissue destruction. Host cells infected with meronts are probably destroyed as the meront ruptures, releasing merozoites into the lumen of the intestinal tract or into adjacent tissue. Gamonts and oocysts cause a marked alteration of their host cells which may become distorted, rupture, separate from adjacent cells, and slough from the intestinal epithelium, exposing the deeper tissues. Since gamonts and oocysts are more numerous than the asexual stages, they cause more host cell destruction and, therefore, more tissue damage which may lead to a denuding of the intestinal epithelium.

In severe infections, the clinical signs usually include diarrhea with blood and portions of intestinal tissue in the feces. An enteritis develops which is complicated by secondary bacterial infection in denuded epithelial areas of the intestinal tract. Infected animals experience poor weight gain or weight loss, emaciation, and weakness and severely infected animals may die.

C. *Frenkelia*

Little is known about the life cycle and pathogenesis of *Frenkelia*. *Frankelia* species appear to be structurally and biologically similar to *Sarcocystis*.[5] *Frenkelia* is known to form thin-walled cysts in the brain of various intermediate hosts such as field voles, meadow mice, chinchillas, muskrats, and bank voles.[5] The complete life cycle has been determined for only *F. buteonis* (synonyms *I. buteonis* and *F. clethrionomyobuteonis*.[31,32] Gamonts and oocysts develop in the lamina propria of the small intestine of the buzzard after ingestion of cysts from the brains of mice. In the intermediate host, meronts appear in the liver as early as 7 days after ingesting *F. buteonis* sporocysts and cysts appear in the brain beginning at 18 days.[33,34]

D. *Hammondia*

Hammondia hammondi, *H. heydorni*, and *H. pardalis* are three recently described species of coccidia which use domestic cats or wild felids as definitive hosts[4,32,35] and various other animals such as monkeys, mice, rats, guinea pigs, deer mice, dogs, and hamsters as intermediate hosts.[35-39] The rat, *Rattus rattus*, appears to be a natural intermediate host for *H. hammondi*.[38] *Hammondia hammondi* is structurally and antigenically similar to *Toxoplasma gondii*.[35,39] However, *H. hammondi* differs from *T. gondii* by having an obligatory two-host cycle, by not being transmitted transplacentally, by not infecting extraintestinal organs of cats, and by being relatively nonpathogenic for the intermediate host.[35-37,39,40] Intermediate hosts become infected with *H. hammondi* by ingesting sporulated oocysts. Sporozoites excyst in the small intestine and develop into tachyzoites within the intestinal lamina propria, Peyer's patches, and mesenteric lymph nodes, causing death of infected cells. At 2 weeks after infection, cysts appear in other tissues, particularly skeletal muscle.[35,40] Cats become infected by ingesting cysts in tissues of intermediate hosts. Bradyzoites escape from the cysts in the small intestine of the cat and develop into meronts. Merozoites develop into gamonts and unsporulated oocysts are shed in the feces at 5 to 10 days after ingestion of cysts by the cat.

E. *Isospora*

The genus *Isospora* contains two structurally and biologically distinct types of organisms. The first type contains organisms with a direct life cycle without intermediate hosts. Infection with this type of organism occurs primarily in intestinal or other epithelial cells of the host. Organisms of the second type have a facultative or an obligatory two-host life cycle, and cyst-like stages are formed in tissues of the intermediate host. Dubey[5] and Frenkel[5a] described new genera, *Levineia* and *Cytoisospora*, respectively, for organisms of the second type, and suggested that those isosporans of the first type which have a direct life cycle and no cyst-like stages should remain in the genus *Isospora*. Both types will be considered here to be in the genus *Isospora*.

Isospora belli is an uncommon parasite of man but can cause acute and chronic clinical symptoms characterized by fever, malaise, persistent diarrhea, weight loss, steatorrhea, and even death.[11,41] After ingesting sporulated oocysts, merogony and gametogomy occur primarily in the intestinal epithelial cells of the upper small intestine.[11,12] Oocysts may sporulate and excyst in the duodenum, thus causing continued merogony and gametogony, and therefore prolonged chronic symptoms.

Isospora felis and *I. rivolta* use cats as definitive hosts, and *I. canis* and *I. ohioensis* use dogs as definitive hosts. Suitable intermediate hosts are mice, rats, and hamsters. Infections in intermediate and definitive hosts by these *Isospora* species are asymptomatic.[5] Oocysts are infective to both the intermediate and definitive hosts. After ingestion of oocysts, cysts which are typically unizoite are formed in many organs of the intermediate host. Merogony precedes gametogony in the intestinal epithelium of the definitive host.

F. *Sarcocystis*

Sarcocystis is an obligate two-host parasite which infects humans and various animals. In mammals, *Sarcocystis* uses carnivores or omnivores as final hosts and herbivores or omnivores as intermediate hosts. *Sarcocystis* is known to be capable of causing disease in humans[42] as well as livestock such as cattle, swine, and sheep.[43] *Sarcocystis* probably causes an annual economic loss of several million dollars in cattle and sheep due to condemnations at slaughter of meat carcasses or trimmings thereof, abortion, mortality, and morbidity, including poor feed conversion and general unthriftiness.[44-49]

The impact of *Sarcocystis* infections in humans and domestic and free-ranging animals is still not well understood. A relatively high prevalence of infections by *Sarcocystis* has been reported in various free-ranging herbivores and carnivores.[50-59] *Sarcocystis hemionilatrantis* obtained from coyotes has been found to be capable of killing mule deer fawns.[52]

Evidently, there is little or no damage caused by *Sarcocystis* infections in the carnivore host. Heydorn and Rommel[60] and Fayer[61] found that dogs could be reinfected with *S. bovicanis* in which there were little or no microscopic lesions or cellular reaction to the parasite. Speer et al.[59] found that coyotes could be reinfected with *S. hemionilatrantis* by ingesting mule deer muscle tissue. However, unlike *S. bovicanis* infections in dogs, 14-day primary infection and especially reinfection with *S. hemionilatrantis* caused a marked mononuclear cellular infiltration, hyperemia, and vascular wall changes in the lamina propria of the small intestine of coyotes.

In intermediate hosts such as cattle and sheep, the pathological changes due to *Sarcocystis* are most severe during acute infection between about 24 to 33 days after inoculation,[51,62,63] Before and during this period, there are numerous meronts within endothelial or subendothelial cells[63a,b] throughout the body which cause a severe inflammatory response including myositis, lymphadenopathy, and petechial hemorrhages. Severely infected animals may die from such an acute infection. During chronic infections, cysts of *Sarcocystis* occur in various muscular tissue, in which they cause little or no pathology.[43] However, when muscle cysts rupture, a granulocytic infiltration followed by mononuclear infiltration and ultimately fibrosis may occur.[64]

G. *Toxoplasma*

Toxoplasma gondii is the etiological agent of toxoplasmosis. *Toxoplasma* was originally discovered in a North African rodent by Nicolle and Manceaux.[65] It is an unusual coccidian parasite because it has little specificity for intermediate hosts. Evidently, it can infect any warm blooded animal, bird or mammal, as an intermediate host, but uses wild and domestic felids only as definitive hosts. *Toxoplasma* can be transmitted congenitally, by carnivorism or fecal contamination.[20] The resistant oocysts are the key to the epidemiology of this disease and assure the dissemination of the parasite. Felids and intermediate hosts become infected by ingesting tachyzoites in pseudocysts, bradyzoites in cysts, or sporozoites in oocysts. Merogony, gametogony, and oocyst formation occur in the intestinal epithelium of felids that previously ingested oocysts, tachyzoites, or cysts. Extraintestinal merogony also occurs in felids which may be lethal depending upon the number of organisms ingested as well as the age and immune status of the host. Only asexual reproduction by endodyogeny occurs in various tissues in intermediate hosts which have ingested *Toxoplasma* pseudocysts, cysts, or oocysts.

Worldwide, the prevalence of *Toxoplasma* infection in man and various wild and domestic animals, as determined by antibody tests, has been shown to vary between 7 to 94%.[66] Even though infection is usually very common, clinically evident infections are relatively rare. Concomitant infections, extremes of age, exposure, poor nutrition, pregnancy, lactation, blood diseases, concomitant malignancy, and immunosuppressive drugs may cause a *Toxoplasma* infection to become sufficiently high to cause clinical symptoms. In order for congenital toxoplasmosis to occur, a woman must experience an acute infection during pregnancy. The severity of congenital toxoplasmosis depends upon the time during gestation that infection occurred as well as the infecting strain of *Toxoplasma*. The earlier during pregnancy that maternal infection occurs, the greater the likelihood of severe disease in the fetus. The parasite may cause death of the fetus or varying degrees of central nervous system damage which may be manifested by reintochoroiditis, epilepsy, or mental deficiency associated with cerebral necrosis and calcification.[66]

Toxoplasma also causes serious disease in domestic animals such as sheep,[67-69] swine,[70,71] dogs,[72] and cats.[73] It can cause abortion in sheep which is of considerable economic importance in New Zealand and Great Britain.[67-69] Although the prevalence of *Toxoplasma* infections in cattle is relatively common,[74] the importance of toxoplasmosis in cattle and the significance of infected beef as a means of human infection are still not well understood.

III. HISTORY OF CULTIVATION

Early attempts to cultivate various coccidia used liquid culture medium free or relatively free of host cells or tissue. MacGowen[75] was one of the first researchers to attempt cultivation of a coccidian species. He described changes in sarcocysts of *S. tenella* cultured in a 1% glucose solution. Triffitt[76] was the first to attempt in vitro cultivation of an *Eimeria* species. She inoculated immature and mature meronts of *E. gilruthi* (synonym *Gastrocystis giltruthi*) obtained from sheep into various nutrient liquid media in which the meronts disintegrated after 8 or 9 days. Levaditi et al.[77] were the first to report attempts to culture *T. gondii*. Cook and Jacobs[78] reviewed the literature on the in vitro cultivation of *Toxoplasma* between 1929 and 1958. Scott[79] discussed his attempts as well as those of other investigators from 1872 to 1943 to culture sarcocysts of *Sarcocystis*. Spindler and Zimmerman[80] reported attempts to culture *Sarcocystis* in a sterile dextrose solution.

These early attempts to cultivate coccidia prior to about 1950 were, of course, unsuccessful because of the lack of a suitable host cell in the in vitro systems used. During this period techniques for isolating and cultivating cells and tissues were being developed elsewhere.[81] In 1948, investigators at the National Institutes of Health discovered that in order to induce single cells to divide in vitro, they must be given a medium similar to their immediate surroundings within a tissue. Also, by the late 1940s antibiotics became available to protect large-scale, long-term cell cultures from otherwise certain contamination. By 1950 a system of maintaining cultured cells obtained from various animals, particularly birds and mammals, became available for use by coccidian researchers.

Probably one of the first clues that coccidia could be cultured in vitro was provided by Frenkel[82] in 1953 when he briefly mentioned that subinoculation of cyst forms of *B. jellisoni* into chick embryos resulted in acute infection similar to toxoplasmosis. Frenkel[83] again briefly mentioned that *B. jellisoni* multiplied in various chick embryos and several tissue culture cell lines. In 1962, Bigalke[84] reported successful cultivation and proliferation of *B. besnoiti* in cultured lamb and calf kidney cells.

The use of cultured cells for studying coccidian species did not receive much attention until Patton[85] reported the development of sporozoites of *E. tenella*, a parasite of the cecum of chickens, to first generation meronts in monolayers of various avian and mammalian cell types. Between 1965 and 1972 numerous reports on the cultivation of *Toxoplasma* and

Table 2
DEVELOPMENT OF *BESNOITIA* ZOITES IN CULTURED CELLS

Species	Intermediate host	Cell[a] type	Farthest development	Original descriptions
B. besnoiti	Rabbit	BK	?	84
		LK		
B. jellisoni	Mouse	ECF	Pairs of zoites, rosettes	209
		PEK		
		REK		
		TK		
	White mice or kangaroo rat	BES	Rosettes	174
		BETr		
		MDBK		

[a] Abbreviations for cell type: BK, bovine kidney; BES, embryonic bovine spleen; BETr, embryonic bovine trachea; ECF, embryonic chicken fibroblasts; PEK, embryonic pig kidney; REK, embryonic rabbit kidney; TK, turtle kidney.

various *Eimeria* species from avian and mammalian hosts appeared in the literature. During this period, certain species of *Besnoitia, Isospora*, and *Sarcocystis* were also grown in cultured cells (see review by Doran[86]). Before 1972, most of the reports were descriptive, pertaining to structural development and mechanics of in vitro cultivation. After 1972, the research emphasis changed to the use of in vitro cultivation to study anticoccidial drugs, cytochemistry and biochemistry, ultrastructure, genetics and immunity. Tables 2 through 8 list those species of coccidia which have been grown in cultured cells. Table 9 lists some species of coccidia which are of medical or veterinary importance which have not yet been grown in cultured cells.

Long[87,88] and Long and Tanielian[89] described the development of sporozoites of *E. Brunetti, E. mivati*, and *E. tenella* to oocysts in embryonated chickens. Fitzgerald[90] described the development of *E. stiedai* sporozoites to oocysts in avian embryos. Between 1966 and 1973, several reports of the development of *Eimeria* species in embryonated chicken eggs appeared in the literature (see review by Doran[86]). Since 1973 there have been only a few such reports.[91-95]

IV. TECHNIQUES

A. Isolation, Purification, and Preservation of Parasites

Coccidian parasites must first be isolated from the feces or tissues of the host, concentrated, and purified before they can be inoculated into cultured cells. Many useful techniques for isolating and purifying parasites were reviewed by Davis.[96] Techniques which are particularly useful for isolating, purifying, and inoculating coccidian parasites into cultured cells, and initiating and maintaining cell cultures are presented below.

1. Oocysts and Sporocysts

Since one species of animal can serve as a suitable host for several different species of coccidia and since one host is often simultaneously naturally infected by two or more species, it is necessary to establish pure infections by a single species. Edgar and Seibold[97] described a method for obtaining oocysts from animals infected with a single oocyst which was obtained by serial dilution of an oocyst suspension. Since in vitro cultivation usually requires a relatively large number of parasites to be used as inoculum, oocysts resulting from single oocyst inoculations should be inoculated into several animals in order to obtain large numbers of oocysts.

Table 3
DEVELOPMENT OF *EIMERIA* SPOROZOITES IN CULTURED CELLS

Species	Host	Cell[a] type	Farthest[b] development	Original report
E. acervulina	Chicken	CK	MM (4)	232
E. adenoeides	Turkey	BEK	MM (1)	134
E. alabamensis	Cow	MDBK	UM (2)	222
E. auburnensis	Cow	BES	MM (1)	173
		BETr		
		MDBK		
E. bilamellata	Ground squirrel	BEK	MM (1)	155
		BETr		
		BHK		
		MDBK		
E. bovis	Cow	BETr	IM (2)	387
E. brunetti	Chicken	CK	MM (3)	213
E. callospermophili	Ground squirrel	BEK	UM (2)	155
		EGS		
		BHK		
		BES		
		BESy		
		BET		
		BETh		
		BETr		
		Int 407		
		MDBK		
E. canadensis	Cow	BEK	MM (1)	184
		BEL		
		BK		
E. contorta	Rat	BEK	MM (2)	167
		MDBK		
E. crandallis	Sheep	BEL	MM (1)	214
		OETh		
		OETr		
E. diminuta	Jungle fowl	CTr	MM (2)	95
E. dispersa	Turkey	TK	MM (2)	228
		CK		
E. ellipsoidallis	Cow	MDBK	MM (1)	215
		BES		
		BETr		
E. gallopavonis	Turkey	TK	MA	228
E. larimerensis	Ground squirrel	MDBK	IM (2)	153
		EGS		
		BEL		
		OEK		
		MDBK	MM (2)	
		OETh		
E. magna	Rabbit	MDBK	MM (2)	154
		BEL		
E. meleagrimitis	Turkey	TK	MA	140
E. mivati	Chicken	CK	MM (2)	91
E. necatrix	Chicken	CK	MM (2)	241
E. nieschulzi	Rat	BEI	MM (1)	168
		BEK		
		BEL		

Table 3 (continued)
DEVELOPMENT OF *EIMERIA* SPOROZOITES IN CULTURED CELLS

Species	Host	Cell[a] type	Farthest[b] development	Original report
E. ninakohlyakimovae	Sheep	BEK	MM (1)	182
		BK		
		OEK		
		OET		
		OETh		
		OETr		
E. stiedai	Rabbit	GF	MM (?)	388
		RL		
E. tenella	Chicken	CK	OO	138
		GL		252
		PK		
		QK		
		TK		
		PhK		
E. tsunodae	Quail	CE	MM (1)	389
		CK		
		HK		
		QE		
E. uzura	Quail	CE	MM (1)	389
		CK		
		HK		
		QE		
E. vermiformis	Mouse	ME	MM (1)	219
		BEK		
E. zuernii	Cow	BEK	MM (1)	216
		MDBK		
		BETr		

[a] Abbreviations for cell types: BEI, embryonic bovine intestine; BEK, embryonic bovine kidney; BEL, embryonic bovine liver; BES, embryonic bovine spleen; BESy, embryonic bovine syvoiral; BET, embryonic bovine thymus; BETh, embryonic bovine thyroid; BETr, embryonic bovine trachea; BHK, Syrian hamster kidney; BK, bovine kidney; CE, chicken embryo; CE, chicken embryo; CK, chicken kidney; CTr, chicken trachea; EGS, embryonic ground squirrel; GF, goat fibroblasts; GL, guinea fowl; HK, hamster kidney; Int 407, embryonic human intestine; MDBK, Madin-Darby bovine kidney; ME, mouse embryo; OEK, embryonic ovine kidney; OET, embryonic ovine thymus; OETh, embryonic ovine thyroid; OETr, embryonic ovine trachea; PhK, pheasant kidney; PK, partridge kidney; QE, quail embryo; QK, quail kidney; RL, rabbit liver; TK, turkey kidney.

[b] Abbreviations for farthest developmental stage of parasite in culture: IM, immature meront; MA, macrogamont; MM, mature meront; 00, accept; UM, uninucleate meront (= trophozoite). Number of asexual generation listed in parentheses.

Oocysts or sporocysts collected in feces must be separated from fecal debris, which would have a deleterious effect upon cultured cells. This can be accomplished by passing the fecal material containing oocysts or sporocysts through a graded series of three sieves beginning with a 100-mesh, then a 200-mesh, and finally a 300 or 325-mesh.[98] After passage through the sieves, the oocysts in tap water should be allowed to sediment for 12 to 24 hr. The supernatant should be discarded and the sediment which contains the oocysts or sporocysts retained.

Oocysts of most coccidia are usually stored under refrigeration in water alone or in 2.5% potassium dichromate. Leek and Fayer[99] found that sporocysts of *S. bovicanis* stored in

Table 4
DEVELOPMENT OF *EIMERIA* MEROZOITES IN CULTURED CELLS

Species	Host	Cell[a] type	Merozoite[b] generation	Farthest[c] development	Original report
E. acervulina	Chicken	CK	4	OO	232
E. bovis	Cow	BEK	1	OO	244
E. brunetti	Chicken	ECF	2	GA	246
		EQF			
E. larimerensis	Ground squirrel	BEK	1	GA	218
E. magna	Rabbit	MDBK	?	OO	142
		MDBK	?	OO	169
		RK	?	OO	
E. meleagrimitis	Turkey	TK	1	MM (2)	140
		TK	2	OO	
		TK	3	OO	
E. stiedai	Rabbit	RK	?	MM (4)	247
E. tenella	Chicken	ECF	2	OO	245
		CC	2	OO	141
		CE	2	OO	
		CL	2	OO	

[a] Abbreviations for cell types: BEK, embryonic bovine kidney; CC, chicken cecum; CE, chick embryo; CK, chicken kidney; CL, chicken liver; ECF, embryonic chicken fibroblast; EQF, embryonic quail fibroblast; MDBK, Madin-Darby bovine kidney; RK, rabbit kidney; TK, turkey kidney.

[b] Merozoite generation obtained from infected host.

[c] Abbreviations for farthest development stage of parasite in culture: GA, gamonts; MM, mature meront, asexual generation in culture listed in parentheses; OO, oocyst.

Table 5
DEVELOPMENT OF *HAMMONDIA* SPOROZOITES IN CULTURED CELLS

Species	Definitive host	Cell[a] type	Farthest development	Original description
H. hammondi (strain WC1170)	Cat	MK	Pairs of zoites	257
		ME		
		WI-28		
H. heydorni (syn. *I. begemina* small form; *I. wallacei*)	Cat	ECF	Pairs of zoites	258
		HA		
		MAK		
		MOK		
		QF		
		RK		
		SEK		

[a] Abbreviations for cell types: ECF, embryonic chicken fibroblasts; HA, human amnion; MAK, marmoset kidney; ME, embryonic mouse fibroblasts; MK, monkey kidney; MOK, monkey kidney, QF, quail fibroblasts; RK, rabbit kidney; SEK, embryonic sheep kidney; WI-38, human diploid fibroblasts.

distilled water or Hank's Balanced Salt Solution (HBSS) (pH 7.0—7.4), containing 10,000 units penicillin/mℓ, 0.01 g streptomycin/mℓ, 0.05 mg Fungizone/mℓ, and 500 units mycostatin/mℓ (PSFM), for 173 to 302 days had 47 to 74% excystation, whereas only 1 to 26% excystation occurred after storage for the same time period in 1% sodium hypochlorite, 2% sulfuric acid or 2.5% potassium dichromate. Sporocysts stored in water had fungal,

Table 6
DEVELOPMENT OF *ISOSPORA* SPOROZOITES IN CULTURED CELLS

Species	Host	Cell type[a]	Farthest development	Original report
I. canis	Dog	BEK BETr	Pairs of zoites	262
I. felis	Cat	BEK BETr CK F6 HeLa Int 407 MDCK Minn-EE	Pairs of zoites	263
I. lacazei and/or *I. chloridis*	Sparrow	CF CRK	Mature 2nd-generation meront	208
I. rivolta		BEK CAK	Pairs of zoites	261

[a] Abbreviations for cell types: BEK, embryonic bovine kidney; BETr, embryonic bovine trachea; CaEK, embryonic canary fibroblasts; CAK, canine kidney intestine; CEF, embryonic chicken fibroblasts; CEI, embryonic canine; CEK, embryonic canine kidney; CK, chick kidney; FL, human amnion; HeLa, human epitheloid cervical carcinoma; Int 407, embryonic human intestine; MDCK, Madin-Darby canine kidney; Minn-EE, human esophageal epithelium WI-38, human diploid lung.

bacterial, and protozoal growths, whereas no such microbial growths occurred in the HBSS-PSFM storage medium. Thus, HBSS-PSFM appears to be the best storage medium in which to maintain previously cleaned *Sarcocystis* sporocysts. This medium should also prove to be superior to conventional media for the storage of other coccidian oocysts and sporocysts.

Except for *Frenkelia, Sarcocystis,* and certain *Isospora* species, oocysts of most coccidia are passed unsporulated and must be sporulated in order to obtain sporozoites for inoculation of cultured cells. Usually, a higher percent sporulation occurs if the oocysts are sporulated at room temperature prior to being refrigerated. Oocysts can be sporulated in tap water[100] or 2.5% potassium dichromate[101] at room temperature by gently stirring or bubbling air through the aqueous suspension for 1 to 7 days. During this period, samples of the oocyst suspension should be examined periodically by light microscopy to determine the extent of sporulation. Seventy to eighty-five percent or more sporulation is usually considered excellent. Attempts to achieve higher percents of sporulation by continuing the sporulation process beyond 3 or 4 days may cause a reduction in the number of viable oocysts.

Once the oocysts have undergone sporulation, it is necessary to further clean the oocysts of fecal debris not eliminated by the sieves. Various techniques have been devised for this purpose such as coverslip flotation,[102] gravity pan flotation,[103] linear gradients,[104] zonal gradients,[105,106] continuous flow differential density flotation,[107] and centrifugal elutriation.[108] Probably the most simple and convenient method is that of Jackson[103] as modified by Hammond et al.[109] This technique which has been extensively used to clean *Eimeria* oocysts has also been used to separate oocysts of *Hammondia*[4] and *Isospora*[13] and sporocysts of *Sarcocystis*.[110] The procedure is as follows:

Table 7
DEVELOPMENT OF *SARCOCYSTIS* ZOITES IN CULTURED CELLS

Species	Definitive host/ intermediate host	Cell type	Farthest development	Original descriptions
Sarcocystis sp.	Unknown	BEK	GA	156,253
		BETr		
		MDCK		
		ECK		
		ECM		
		ETK		
		BEK	OO	267
		BETr	OO	319
Sarcocystis sp. (Fischer, 1978)	Dog/goat	MDCK	OO	266
Sarcocystis sp.	Unknown/monkey	CL	ND	266
		MDCK		
		HSF		
		SK		
S. muris	Cat/mouse	CL	GA	266
S. ovicanis	Dog/sheep	MDCK	OO	266
S. suihominis	Human/pig	HSF	OO	149
		Int 407		266
S. tenella	Unknown/sheep	SEK	MA	265

[a] Abbreviations for cell types: BEK, embryonic bovine kidney; BETr, embryonic bovine trachea; CL, cat lung; ECK, embryonic chicken kidney; ECM, embryonic chicken muscle; ETK, embryonic turkey kidney; HSF, human skin fibroblasts; Int 407, human intestine; MDCK, Madin-Darby canine kidney; SEK, embryonic sheep kidney; SK, swine kidney (PK 15).

[b] Abbreviations for farthest developmental stage in culture: GA, gamonts, MA, macrogamonts; ND, no development; OO, oocysts.

- Mix 25 mℓ oocyst suspension with 25 mℓ Sheather's sugar solution 500 g sucrose, 320 g distilled water, 6.5 g liquid phenol (80% solution, J. T. Baker Chemical Co., Phillipsburg, N.J.) in a glass petri dish, 90 mm in diameter.
- After 5 to 10 min, gently place the bottom surface of the bottom portion of a plastic or glass petri dish which barely fits inside of the 90 mm petri dish on the surface of the oocyst-Sheather's sugar solution.
- Allow to stand 1 hr.
- Gently lift the plastic or glass petri dish from the surface of the oocyst-Sheather's sugar solution and wash the adhering oocysts and liquid from the bottom surface of the dish into a beaker by using a jet of water from a plastic squeeze bottle.
- Repeat the process 1 or 2 times.
- Centrifuge the contents of the beaker containing the oocysts 3 or more times to remove the sugar solution and resuspend the oocysts in water, a saline solution, HBSS-PSFM or 2.5% potassium dichromate.
- Store the oocysts at refrigerated temperature (−4°C) until they are to be used for experimentation.

Oocysts of *Toxoplasma* should be handled somewhat differently than those of *Eimeria* species. Dubey et al.[74,111,112] suggest the following:

- Float oocysts of *Toxoplasma* from cat feces by using a 40% sucrose (w/v) solution (1.15 sp. gr.) with 0.8% phenol. Sheather's sugar solution floats too much fecal debris.
- Sporulate and store *Toxoplasma* oocysts in 2% H_2SO_4 instead of 2.5% K_2CrO_7. Infectivity of chromated oocysts is lower than those preserved in H_2SO_4.

Table 8
DEVELOPMENT OF *TOXOPLASMA GONDII* IN CULTURED CELLS

Inoculum	Cell type[a]	Farthest development	Some recent reports
Sporozoite	MK	Zoites in pairs or rosettes	264
Sporozoite (Beverley strain)	LC	Zoites in pairs or rosettes	270
Tachyzoites (RH strain)	HEL	Zoites in pairs or rosettes, merogony	271
Tachyzoites (SF strain)	DK	Zoites in pairs or rosettes	145
Tachyzoites (RH strain)	MC Vero	No development zoites in pairs or rosettes	390
Tachyzoites (RH strain)	HLC	Zoites in pairs or rosettes	146
Tachyzoites (RH and Beverley strains)	LC HC HU WI38	Zoites in pairs or rosettes	352
Tachyzoites (RH strain)	PM	Zoites in pairs or rosettes	391
Lagrave strain	PM	Parasites disintegrated	391

[a] Abbreviations for cell types: DK, dog kidney; HC, human conjunctiviva; HEL, human embryonic lung; HLC, human larynx carcinoma; HU, human U cells; LC, L-cells; MC, mosquito cells; MK, monkey kidney; PM, peritoneal macrophages; WI38, human embryonic lung.

- Aerate on a shaker (150 rpm, Laboratory Rotator, Model G2, New Brunswick Scientific Co., Inc., New Brunswick, New Jersey) at room temperature for 3 to 7 days to cause sporulation of oocysts.
- Oocysts should be sporulated in capped bottles and separated from cat feces before sporulation in order to minimize risk to humans.
- Release sporocysts from Clorox®-treated oocysts by shaking with the help of alundum or sand grinders. Scattering parasites probably occurs when tissue-grinders are used.

Most species of *Sarcocystis* form oocysts within the lamina propria of the small intestine of the final host. Because of this location the oocysts and/or sporocysts are slowly released into the lumen of the intestinal tract as indicated by prolonged patent periods of 10 to 40 or more days (see review by Markus[43]). Thus much time and effort are required in order to recover most of the oocysts or sporocysts of *Sarcocystis* in the host feces. Recently, Dubey[113] (personal communication) devised a method of improving the recovery rate of *Sarcocystis* oocysts and sporocysts. This technique is as follows:

- Kill the carnivore host when relatively high numbers of oocysts and sporocysts are present, and remove the small intestine.
- Cut the small intestine into 5 or 6 segments, lay each segment flat on a paper towel and open lengthwise with scissors, and spread flat with mucosa surface upward.
- Use a glass microscope slide to scrap the intestinal mucosa and submucosa, leaving the muscularis externa intact.
- Add 20 volumes of distilled H_2O to 1 part mucosal scraping and gently homogenize in a Waring® blender for 60 sec.

Table 9
VARIOUS COCCIDIAN PARASITES OF HUMANS AND DOMESTIC AND WILD ANIMALS WHICH HAVE NOT YET BEEN GROWN IN CULTURED CELLS

Parasite species	Final hosts or prey/predator hosts
Besnoitia	
B. bennett Babudieri, 1932	Horse, ass
B. darlingi (Brumpt, 1913) Schneider, 1967	Lizards, opossum, marmoset
B. tarandi (Hadwen, 1922) Levine, 1961	Reindeer, caribou
Cryptosporidium	
C. meleagridis Slavin, 1955	Domestic turkey
Eimeria	
E. ahsata Honess, 1942	Rocky Mountain bighorn sheep, domestic sheep, mouflon
E. anseris Kotlán, 1932	Domestic goose, blue goose, Richardson's Canada goose
E. arloingi (Marotel, 1905) Martin, 1909	Domestic goat
E. balphae Ernst, Chobotar and Anderson, 1967	Ord Kangaroo rat
E. bukidnonensis Tubangui, 1931	Ox, zebu, water buffalo
E. chinchillae de Vos and van der Westhuizen, 1968	Chinchilla
E. cylindrica Wilson, 1931	Ox, zebu, water buffalo
E. debliecki Douwes, 1921	Domestic and wild pigs
E. falciformis (Eimer, 1870) Haberkorn, 1970	Mouse
E. falciformis var. pragensis Cerná, Mehlhorn and Scholtyseck, 1974	Mouse
E. faurei (Moussu and Marotel, 1902) Martin, 1909	Domestic sheep, Rocky Mountain bighorn sheep, domestic goat and others
E. ferrisi Levine and Ivens, 1965	Mouse
E. illinoisensis Levine and Ivens, 1967	Ox
E. intestinalis Kheisin, 1948	Domestic rabbit
E. intricata Spiegl, 1925	Domestic sheep, Rocky Mountain Bighorn sheep, domestic goat
E. irresidua Kessel and Jankiewicz, 1931	Domestic rabbit
E. kotlani Gräfner and Graubmann, 1964	Domestic goose
E. labbeana Pinto, 1928	Domestic pigeon, doves
E. leukarti (Flesch, 1883) Reichenow, 1940	Horse, ass
E. matsubayashii Tsunoda, 1952	Domestic rabbit
E. maxima Tyzzer, 1929	Chicken
E. media Kessel, 1929	Domestic rabbit
E. meleagridis Tyzzer, 1927	Domestic and wild turkey
E. mitis Tyzzer, 1929	Chicken
E. neoleporis Carvalho, 1942	Cottontail and domestic rabbits
E. nocens Kotlan, 1933	Domestic and blue goose
E. ovina Levine and Ivens, 1970	Domestic sheep, bighorn sheep
E. papillata Ernst, Chobotar and Hammond, 1971	Mouse
E. pellita Supperer, 1952	Ox
E. perforans (Leuckart, 1879) Sluiter and Swellengrebel, 1912	Domestic rabbit, cottontail
E. scabra Henry, 1931	Domestic and wild pigs
E. sigmodontis Barnard, Ernst and Dixon, 1974	Cotton rat
E. spinosa Henry, 1931	Pig
E. subspherica Christensen, 1941	Ox, zebu, water buffalo
E. truncata (Railliet and Lucet, 1891) Wasielewski, 1904	Domestic goose, Canada goose, domestic ducks and others
E. utahensis Ernst, Hammond and Chobotar, 1968	Kangaroo rats
E. wyomingensis Huizinga and Winger, 1942	Ox, zebu, water buffalo
Frenkelia	
F. buteonis (synonyms *I. buteonis* Rommel and Krampitz, 1975, *F. clethrionomyobuteonis* Tadros and Laarman, 1976)	Mice/buzzard

Table 9 (continued)
VARIOUS COCCIDIAN PARASITES OF HUMANS AND DOMESTIC AND WILD ANIMALS WHICH HAVE NOT YET BEEN GROWN IN CULTURED CELLS

Parasite species	Final hosts or prey/predator hosts
Hammondia	
H. pardalis Hendricks, Ernst, Courtney and Speer, 1979	Mice/ocelot and other felids
Isospora	
I. arctopitheci Rodhain, 1933	Capuchin, marmoset
I. belli Wenyon, 1923	Man
I. bigemina (Stiles, 1891) Lüke, 1906 (large form)	Dog, cat, fox, polecat, mink, man (?)
I. neorivolta Dubey and Mahrt, 1978	Dog
I. ohioensis Dubey, 1975	Dog
I. suis Biester, 1934	Pig
I. vulpina Nieschulz and Bos, 1933	Dog
Sarcocystis	
S. betrami Doflein, 1901	Horse/dog
S. fayeri Dubey, Streitel, Stromberg and Toussant, 1977	Horse/dog
S. hemionilatrantis Hudkins and Kistner, 1977	Mule deer/coyote
S. hirsuta Moulé, 1888	Cattle/dog
S. hominis (Railliet and Lucet, 1891) Dubey, 1976	Cattle/man
S. leporum Crawley, 1914	Cottontail rabbit/cat
S. lindemanni (Lindemann, 1868) Rivolta, 1878	Man/?
S. miescheriana (Kuhn, 1865) Lankester, 1882	Pig/dog
S. porcifelis Dubey, 1976	Pig/cat
Tyzzeria	
T. anseris Nieschulz, 1947	Domestic goose, blue and snow goose
T. perniciosa Allen, 1936	Domestic duck
Wenyonella	
W. gallinae Ray, 1945	Chicken

- Centrifuge the mixture at 400 g for 10 min, rehomogenize and sediment in distilled H_2O and recentrifuge.
- Repeat the process until the supernatant becomes clear.
- Suspend sporocysts in 10 to 100 mℓ distilled water or HBSS-PSFM and count the number of sporocysts by a hemacytometer.

2. Sporozoites

Most coccidian oocysts can be treated with Clorox® for 20 to 30 min in order to kill other microorganisms which would be certain to contaminate the in vitro cultures. In certain *Eimeria* and *Isospora* species, the outer layer but not the inner layer of the oocyst wall is removed by the Clorox® treatment.[103,114-117] However, such Clorox®-treated oocysts sporulate normally, are capable of undergoing excystation and are infectious to susceptible hosts.[118]

After sporulated oocysts have been treated with Clorox®, the oocysts should be washed several times by centrifuging and resuspending the oocysts in physiological saline or culture medium until the odor of Clorox® can no longer be detected or until the color of the culture medium turns from yellow to orange.

In vitro techniques for causing sporozoites to excyst from oocysts or sporocysts has been reported by several investigators (see review by Long and Speer[119]). Oocysts of *Eimeria* species from mammalian hosts require CO_2 in order to initiate excystation. The CO_2 acts upon the micropyle of the oocyst causing it to become permeable to trypsin and bile salts, which cause the sporozoites to excyst from the sporocysts and leave the oocyst via the micropyle.[120,121] In the avian host, eimerian oocysts are broken by the mechanical action of the gizzard, liberating the sporocysts from which sporozoites excyst in response to chymotrypsin.[7]

In preparing sporozoites for in vitro cultivation, the CO_2 requirement can be eliminated by mechanically rupturing the oocyst wall to release the sporocysts. Sporocysts can then be exposed to a trypsin and/or bile salt solution to cause sporozoite excystation. Sporozoites of *T. gondii* will excyst from intact oocysts and free sporocysts in response to 5% bovine bile.[122] Sporocysts of *S. bovihominis* will excyst in the presence of bovine bile but not in canine bile, trypsin-canine bile, or trypsin alone.[110] Preincubation of *S. bovihominis* sporocysts in cysteine hydrochloride and an atmosphere of 50% CO_2-50% air decreased the time required for excystation in the trypsin-bovine bile solution. Sporocysts of *I. serini,*[123] *I. arctopitheci,*[124] and *I. endocallimici,*[125] will excyst in a trypsin-bile salt solution. Patton and Brigman[126] obtained far better excystation of *E. tenella* in a trypsin solution containing sodium taurodeoxycholate than in trypsin solutions containing other bile salts such as chicken bile, Oxgall, taurocholate, and glycocholate. Sporozoites will excyst from sporocysts of *E. callospermophili* and *E. larimerensis* upon exposure to trypsin only but not to bile salts only,[116] whereas those of *I. canis*[117] and *H. pardalis* (unpublished data) will excyst considerably faster from sporocysts exposed to a solution containing both trypsin and bile salt.[116,117]

A method for preparing oocysts, excysting sporozoites and inoculating cultures with sporozoites is as follows:

- Determine the number of sporulated oocysts/mℓ in the cleaned oocyst suspension by counting with a hemocytometer and remove the appropriate quantity of oocysts needed for the experiment. These oocysts should have been previously cleaned by the method of Jackson[103] as modified by Hammond et al.[109]
- Centrifuge the oocysts to form a pellet, resuspend them in undiluted Clorox®, and incubate them at room temperature for at least 20 to 30 min.
- Remove the Clorox® by centrifuging and resuspending the oocysts 2 to 3 times in a physiological saline solution.
- Suspend the oocysts in 2 to 3 mℓ physiological saline and grind in a Teflon®-coated tissue grinder for 1 to 10 min or longer to release intact sporocysts.
- Dilute suspension to 30 to 40 mℓ in physiological saline and centrifuge at 200 to 300 g for 8 min.
- Resuspend pellet of sporocysts in appropriate excysting fluid, which has been sterilized by membrane filtration. Most species will excyst in physiological saline containing 0.25% trypsin (1:250 or 300, DIFCO Laboratories, Detroit, Michigan) and 0.75% sodium taurocholate (bile salt, DIFCO Laboratories).
- Incubate at 37°C until most of the sporozoites have excysted. Usually this requires 5 to 30 min or longer depending upon the excystation rate of the coccidian species used.
- Wash excysted sporozoites twice in physiological saline to remove the trypsin-bile salt solution and resuspend in 10 mℓ of the appropriate serum-free culture medium for the cell type used.
- Determine the concentration of sporozoites by using a hemocytometer and adjust the volume to a concentration of about 300 to 600 sporozoites/mm^2 surface area of culture vessel. A Leighton tube with a 9 × 35 mm coverslip should be inoculated with 10^5 to 2×10^5 sporozoites. Strout et al.[127] found that more asexual development occurred in cultures inoculated with 10^4 sporozoites of *E. tenella* than in those inoculated with 10^6 sporozoites. By using inocula ranging from 2.5×10^4 to 10^6 sporozoites Doran[128] found that the greatest number of oocysts developed in cultures inoculated with 10^5 sporozoites of *E. tenella*.
- Inoculate culture vessels with the sporozoite suspension, and incubate at the appropriate temperature (37°C for most mammalian species, 38 to 39°C for bovine and 39 to 41°C for avian species) for about 2 to 6 hr.
- Remove the inoculum by aspiration and inoculate cultures with fresh culture medium containing 3 to 10% serum and antibiotics.

- Incubate cultures at the appropriate temperatures in stoppered vessels or in covered vessels in an atmosphere of 5% CO_2, 95% air.

The inoculum prepared by the above method often contains empty sporocysts and oocyst walls in addition to sporozoites. Although it is not absolutely necessary, sporozoites may be further purified by using glass bead columns[129,130] or by centrifugal elutriation.[108]

Sharma and Foster[131] found that fluid extracts of oocysts and sediments containing oocysts of *E. tenella* were dermally toxic when injected into rabbits. Fayer and Hammond[132] reported that oocyst fluid released by grinding *E. bovis* oocysts killed cultured cells within a few hours after inoculation. Doran[129] found that a greater portion of *E. adenoeides* sporozoites, which were cleaned of debris by passage through two glass columns (240 by 15 mm) of packed glass beads (100 μm), developed to first generation meronts than did uncleaned sporozoites. This indicates that the oocyst and sporocyst walls or residual oocyst fluid in uncleaned sporozoite inocula have a slight deleterious effect upon parasite development within cultured cells. Much of the toxic oocyst fluid is probably removed during the routine washing of sporozoites after excystation in culture medium.

Stotish et al.[108] used centrifugal elutriation to separate mixtures of sporozoites, sporocysts, oocysts, and debris into highly purified components. The method is relatively simple and is superior to the glass bead column technique because it can be used to rapidly prepare large numbers of highly purified parasites. The conditions used in centrifugal elutriation are also very mild; the parasites are subjected to relatively low centrifugal forces (about 400 g). This technique should prove quite valuable in preparing sporozoites and merozoites for inoculation of cell cultures, in quantitative biochemical investigations of the coccidia as well as in separating a mixed population of coccidial oocysts or merozoites from infected intestinal epithelial cells or from cell culture.

3. *Cryopreservation*

Cryopreservation may be used in order to simplify the techniques used in preparing coccidian parasites for in vitro cultivation. Norton and Joyner[133] described a technique of freeze-storage of sporozoites within sporocysts. Doran,[134,135] and Doran and Vetterling[136,137] described techniques for freeze-storing excysted sporozoites, which can be used immediately after thawing without the manipulations necessary for excysting sporozoites from sporocysts or oocysts. Doran[129] found that although fewer sporozoites were intracellular at 5 hr after inoculation of previously frozen specimens than fresh specimens, there was no apparent effect of freezing upon parasite development. He also found a lower percentage of developmental stages from frozen, older sporozoites than from frozen, younger sporozoites.

A method based upon that of Doran[134,135] for freezing and thawing excysted sporozoites is as follows:

- Release sporocysts from oocysts by grinding.
- Excyst sporozoites by suspending sporocysts in a 0.25% trypsin-0.75% bile salt solution at 37 to 43°C (depending upon body temperature of host) until most of the sporozoites have excysted.
- Concentrate sporozoites by centrifugation and resuspend in 10 to 12.5% dimethylsulfoxide (DMSO). Doran[135] found that 10% DMSO was an optimal concentration for the subsequent freezing of *E. tenella* sporozoites, whereas 12.5% DMSO was optimal for *E. adenoeides* and *E. mivati*.
- Adjust sporozoite-DMSO suspension to pH 7.0 to 7.2 with 0.5 *N* NaOH or HCL.
- Adjust concentration to 500,000 sporozoites/mℓ and place 3 mℓ amount in 10 mℓ round bottom, screw-top tubes.
- Freeze specimens beginning with a cooling rate of 1°/min from room temperature to −30°C, and then 10°/min from −30° to −80°C.

- Immediately after reaching −80°C, place the tubes containing the sporozoites in liquid nitrogen vapor for storage.
- When frozen sporozoites are going to be used for experimentation, thaw the frozen specimens by vigorously rotating the tube for 1 min in a 37 to 43°C water bath.
- Dilute and rinse out the DMSO suspension with culture medium and centrifugation.
- Resuspend sporozoites in culture medium and adjust volume to desired concentration of sporozoites for inoculation of cultured cells.

4. Merozoites and Tachyzoites

Eimeria tenella is the only coccidian to have been grown from sporozoites to oocysts in cultured cells.[138,139] Within cultured cells, most coccidian sporozoites have been found to develop only to first generation meronts, a few form second or third generation meronts and a couple of species other than *E. tenella* have developed to gamonts. In order to achieve further asexual development, gametogony and oogony, merozoites can be removed from the intestine of the infected host and inoculated into cultured cells. Techniques for recovering *Eimeria* merozoites from the intestinal tract of the final host have been described by Augustine and Doran,[140] Bedrnik,[141] and Speer and Hammond.[142] By using these methods the generation of merozoite obtained from the host is often difficult or impossible to determine and the inoculum may contain a mixture of merozoites from more than one generation. A workable technique based upon that of Speer and Hammond[142] for the recovery of merozoites is as follows:

- Inoculate several hosts with a relatively large number of oocysts usually 100,000 to 200,000 of the coccidian species to be studied.
- At various intervals after inoculation at which time certain meront generations are present, kill one host and remove that portion of the intestinal tract which contains the parasites.
- Lay the intestine on paper toweling and open the intestine lengthwise with a pair of scissors.
- Spread the intestine flat with the epithelial surface upward.
- Scrape the intestinal mucosa with a glass microscope slide or spatula, leaving the muscularis externa and most of the submucosa intact.
- Place the mucosal scrapings in a 90 mm petri dish and gently stir with a glass rod.
- Rinse the mucosal scrapings in saline A or culture medium containing 5,000 units penicillin G/mℓ and 5,000 μg dihydrostreptomycin/mℓ.
- Incubate the scrapings in this highly concentrated antibiotic solution for about 30 min.
- Centrifuge the suspension at 200 g for 1 min and remove the supernatant fluid which contains the merozoites.
- Resuspend the merozoites in culture medium with 50 units penicillin G/mℓ and 50 μg dihydrostreptomycin/mℓ.
- Merozoites should be inoculated immediately into vessels containing cultured cells. Determine the number of merozoites by counting in a hemocytometer. Usually, culture vessels are inoculated at the rate of 10^2 to 10^3 merozoites per mm^2 of culture vessel surface.

Techniques for isolating merozoites or tachyzoites of those coccidia other than *Eimeria* species have not yet been reported. However, only slight modification of the techniques used to isolate *Eimeria* merozoites should have to be made to isolate merozoites or tachyzoites of those coccidia which have asexual stages in the intestinal tract of the host. Different and new techniques may have to be developed for the isolation of merozoites of those coccidia which have extraintestinal meronts (i.e., *Sarcocystis*). Extraintestinal stages of *Isospora* species multiply at a slow rate and are difficult to demonstrate histologically.[5] Thus, techniques to isolate zoites of *Isospora* from extra intestinal tissues in sufficient quantities for in vitro cultivation may prove to be inordinately difficult.

Tachyzoites of *Toxoplasma* can be easily isolated from the peritoneal cavities of mice or from tissue culture.[5] Tachyzoites will grow in the peritoneal cavity as well as most other tissues of mice after intraperitoneal inoculation of oocysts, tachyzoites, or bradyzoites of *T. gondii*. Tachyzoites of virulent strains usually kill the host within 1 to 2 weeks, whereas those of avirulent strains do not kill the host, and therefore, can be aspirated periodically from the peritoneal cavity of infected animals during light anesthesia. Tachyzoites of virulent strains need to be passed at least weekly into the peritoneal cavity of other mice. Relatively few tachyzoites occur in the peritoneal cavities of mice infected with avirulent strains. Increased numbers of tachyzoites of avirulent strains can be obtained from the peritoneal cavities of mice immunosuppressed with corticosteroids (cortisone acetate, 2.5 mg, injected subcutaneously twice weekly).[5]

Toxoplasma tachyzoites will multiply in numerous types of cultured cells and can be maintained indefinitely in cell culture by serial passage without causing any apparent alteration in parasite morphology or virulence.[143-145] Recently, Braveny et al.[146] described a technique for the mass cultivation of *T. gondii* tachyzoites in cell culture.

Tryon et al.[147] used a hemolysin, produced by *Pseudomonas aeruginosa*,[148] to isolate *Toxoplasma* tachyzoites from peritoneal exudate and infected HeLa cell suspensions. Hemolysin effectively lysed all of the host cells such as white and red blood cells and tissue cells, but left the tachyzoites intact and infectious. Hemolysin is a glycolipid surfactant which solubilizes phospholipids, and serum proteins suppress its activity. The method described by Tryon et al.[147] to isolate *Toxoplasma* tachyzoites is as follows:

- Centrifuge parasites and peritoneal exudate cells or cultured cells to form a pellet.
- Remove supernatant fluid and add 1 mℓ culture medium containing 8 units purified, heat-stable hemolysin to the pellet, and agitate for 45 sec.
- Prevent further action of hemolysin by adding 2 mℓ 50% fetal calf serum.
- Centrifuge suspension for 3 min at 1000 g and resuspend pellet in culture medium.

5. Bradyzoites

Bradyzoites occur within cysts in intermediate hosts of the tissue-cyst-forming coccidia (i.e., *Besnoitia, Frenkelia, Hammondia,* certain *Isospora, Sarcocystis,* and *Toxoplasma).* Cysts of *Besnoitia* occur primarily in connective tissue cells; those of *Frenkelia* occur in the brain; *Hammondia* in skeletal and cardiac muscle; *Isospora* in many organs (*I. felis* is found mainly in mesenteric lymph nodes); *Sarcocystis* primarily in muscle tissue, occasionally in the brain; *Toxoplasma* occurs in visceral organs, but persistent cysts occur mainly in neural and muscular tissues such as the brain, eye and skeletal and cardiac muscle.[5]

Bradyzoites of the tissue cyst-forming coccidia can be obtained from infected hosts by crushing cysts in aseptically excised pieces of tissue with a glass tissue grinder. Bradyzoites of *Isospora* species might be difficult to obtain because they form few cysts, each of which contain only a few organisms. In order to obtain cysts from mice infected with a virulent strain of *Toxoplasma*, it is necessary to treat animals with an anti-*Toxoplasma* drug in order to prevent death from acute toxoplasmosis.[5] Sulfadiazine can be used to control acute toxoplasmosis by giving sodium sulfadiazine in the drinking water (15 mg/100 mℓ drinking water from day 4 to 12 of infection with M-7741 strain of *T. gondii*; 125 mℓ/100 mℓ drinking water from day 1 to 21 after infection with RH strain).[5]

A technique for isolating bradyzoites is as follows:

- Aseptically excise pieces of tissues containing cysts from infected animals.

- Crush tissue and cysts in a glass tissue grinder. Mehlhorn and Heydorn[149] used trypsinization to separate zoites of *S. suihominis* cysts from pigs.
- Suspend tissue-parasite mixture in 2 to 3 mℓ of culture medium plus 5 to 10% fetal calf serum.
- Pour suspension through several layers of culture medium and centrifuge filtrate at 250 to 300 g for 5 min.
- Resuspend parasites in tissue culture medium, count number of zoites by a hemocytometer and adjust volume of suspension to 10^5 to 2.5×10^5 zoites/mℓ.
- Inoculate culture vessels at the rate of 6.5×10^{12} to 1.6×10^3 zoites per mm^2 of culture vessel surface area.

Sarcocystis infections by microscopic cysts can be detected by the technique of Jacobs et al.[150] as modified by Box and McGuinness.[151] To my knowledge, zoites of *Sarcocystis* isolated by this technique, which uses a digestion fluid of 0.75% pepsin (w/v), 0.86% NaCl (w/v) and 1% v/v HCl, have not been examined for their ability to penetrate and develop within cultured cells. Jacobs et al.[150] reported that after digestion the *Sarcocystis* organisms were motile, whereas Box and McGuinness[151] detected no motility.

B. Cell Culture Techniques

Various types of animal cell lines can be obtained from the American Type Culture Collection (12301 Parklawn Drive, Rockville Md. 20852, USA). Primary cell cultures, cell lines, and established cell lines can be initiated from embryonic or juvenile tissue and maintained in one's laboratory. A short presentation of the techniques for initiating, maintaining, and transferring cell lines is as follows:

1. Cell Culture Media

When cells are removed from the parent organism and placed in culture, they must be exposed to conditions similar to those within tissues of the parent organism. These conditions are supplied by a culture medium containing serum, the pH of which is maintained by balanced salt solutions (Table 10). Antibiotics are usually added to the medium to suppress microbial contamination. Chemically defined media, sera, and balanced salt solutions can be obtained from Grand Island Biological Company (GIBCO, 3175 Stanley Road, Grand Island, New York, N.Y. 14072, USA). Some of the media routinely used for culturing the coccidia are minimal essentials medium with Earle's salts, medium 199 with Earle's salts, and RPMI 1640 medium with HEPES buffer.

2. Transfer and Maintenance of Cultured Cells

Most cultured cells cannot be maintained in the same container indefinitely without being subcultured. Cells which are not contact inhibited will continue to grow until overcrowding causes the monolayer to detach from the growing surface. Even cells adapted to suspension culture need to be diluted periodically. The interval of growth period at which subculturing or serial passage becomes necessary varies depending upon the generation time of the cell type, the concentration of cells originally inoculated into the culture vessel, the availability of nutrients (e.g., serum concentration) and the temperature of incubation.

Usually, culture vessels are inoculated at the rate of 2×10^5 cells/mℓ culture medium. The serum concentration may vary between 5% to 20% depending upon the cell type. Once a monolayer has formed, most cell types can be maintained for 1 to 3 weeks in medium containing 2 % serum, which reduces mitotic activity. The frequency at which the medium needs to be changed is dependent upon the metabolic rate of the cell type and the serum concentration.

Cell cultures need to be passaged or transferred when a confluent monolayer of cells has

Table 10
BALANCED SALT SOLUTIONS FOR CULTURE MEDIA[a]

	Earle's g/ℓ	Hank's (g/ℓ)
NaCl	6.8	8.0
KCl	0.4	0.4
$CaCl_2$	0.2	0.14
$MgSO_4 \cdot 7H_2O$	0.2	0.2
$NaH_2PO_4 \cdot H_2O$	0.14	—
Na_2HPO_4	—	0.06
KH_2PO_4	—	0.06
$NaHCO_3$	1.1 — 2.2	0.35
Glucose	1.0	1.0
Phenol red (optional)	0.01	0.01

[a] Calcium chloride ($CaCl_2$) should be dissolved separately or added slowly to the solution of other salts in order to avoid precipitation. Hank's balanced salt solution mixed with medium needs no gassing with CO_2, whereas media prepared with Earle's balanced salt solution should be gassed with a mixture of 5% CO_2-95% air. Earle's balanced salt solution has a more stable pH than that of Hank's.

been achieved. Overgrowth of cells will cause cells to pile up and detach, and a depletion of nutrients may cause atypical cell behavior and eventually cell death. Cells which are slow to detach from growing surfaces in a trypsin solution, will readily detach in a trypsin solution containing a chelating agent such as EDTA or versene (Table 11). A procedure for passage of cultured cells and to set up cultures for experimentation is as follows:

- Remove culture medium from culture flask containing a confluent monolayer of cells and add enough trypsin-versene solution to just cover the monolayer.
- Incubate at 37°C and at about 3 to 5 min intervals gently agitate the medium back and forth across the monolayer until nearly all of the cells have detached from the substrate.
- Decant the trypsin-versene solution containing suspended cells into a centrifuge tube and add culture medium containing 5 to 20% serum.
- Centrifuge at about 250 g for 8 min and resuspend pellet of cells in culture medium.
- Determine cell density by counting cells in a hemocytometer and adjust volume by adding culture medium to contain a suitable number of cells for inoculation into sterile culture vessels. Usually a concentration of 2×10^5 cells/mℓ is sufficient. Immediately after inoculation of culture vessels with cells, each vessel should be gassed with 5% CO_2-95% air. Table 12 gives the appropriate volume of medium, number of cells and gassing time according to culture vessel size.
- Cultures to be used for experimentation are usually incubated for 1 to 2 days at 37 to 41°C before the inoculation of coccidia.
- When cultures are ready to be inoculated with parasites, remove the culture medium by aspiration and inoculate culture medium containing parasites.
- Incubate cultures at 37 to 41°C.

3. Primary Cell Cultures

Primary cell cultures can be started from cells, tissues, or organs taken directly from embryonic or juvenile organisms. A procedure for preparing primary cell cultures is as follows:

Table 11
TRYPSIN-VERSENE (EDTA) SOLUTION[a]

	g/ℓ TDH_2O
NaCl	8.0
KCl	0.4
$NaHCO_3$	0.6
Glucose	1.0
Trypsin	1.0
Versene or EDTA	0.2

[a] Mix all ingredients except versene or EDTA and $NaHCO_3$ in triple glass distilled water (TDH_2O) for 1 hr, than add versene or EDTA and $NaHCO_3$. Filter sterilize trypsin-versene solution.

Table 12
RELATIVE AMOUNTS OF MEDIUM, CELLS AND CO_2 GASSING FOR VARIOUS SIZES OF CULTURE VESSELS

Vessel size	Volume of medium	Total cells	Gassing time 5% CO — 95% air
Leighton tube with 9 × 35mm coverslip	2 mℓ	4×10^5	1 sec
25 mℓ	5 mℓ	10^6	3 sec
75 mℓ	20 mℓ	4×10^6	6 sec
150 mℓ	30 to 50 mℓ	6×10^6 to 10^7	12 sec

- Aseptically remove the organ or tissue from the organism and place it in culture medium containing 1000 units/mℓ penicillin G and 500 μg/mℓ dihydrostreptomycin.
- Mince tissue into 1 mm^3 pieces.
- Place tissue pieces in a trypsinization flask with a magnetic stirring bar and add enough trypsin-versene solution to immerse tissues completely.
- Trypsinize by agitating the tissue pieces with the magnetic stirrer for 10 min, discard supernatant, add fresh trypinizing solution and trypsinize for about 10 min. Save supernatant and repeat once.
- Decant suspended cells through two layers of sterile gauze into centrifuge tubes, centrifuge at 250 g for 8 min and resuspend pelleted cells in culture medium.
- Count the number of cells by a hemocytometer and adjust concentration to about 10^6 cells/mℓ.
- Inoculate culture vessels and incubate at 37 to 41°C for 2 days without disturbing the cultures.

V. IN VITRO CULTIVATION

A. Motility and Host Cell Penetration

Motility is characteristically present in various stages of the coccidia and related Apicomplexa including ookinetes, sporozoites, merozoites, tachyzoites, bradyzoites, and microgametes. Motility has also been reported in sporozoite-shaped meronts[152-155] and macrogamonts.[156,157] Microgametes move by two or three flagella and macrogamonts of *E.*

magna move by 'pseudopodia'.[157] Apicomplexan ookinetes and zoites have no apparent specialized organelles which might provide locomotion. They have a somewhat similar shape, undergo a similar type of helical gliding movement, and also have an apical complex, multilayered pellicle and subpellicular microtubules.[158-160] Thus, movement in these stages probably occurs by a similar process. Forward movement may occur by helical gliding in which the body of the parasite rotates upon its longitudinal axis[161] or by annular waves of contraction which originate at the anterior end and then move posteriorly.[159,162-165] Flexing of the bodies of sporozoites and merozoites is probably caused by the subpellicular microtubules (Figure 4).[164] Mechanisms involved in pivoting movements characteristic of sporozoites and merozoites, which attach to the substrate by their posterior ends and rotate upon their longitudinal axis, and the occasional movement in a posterior direction by sporozoites and merozoites (personal observation) are not known. Vetterling and Doran[166] showed that amylopectin (Figures 5 and 8) is probably the energy source used by eimerian sporozoites for the motility required to undergo excystation and penetrate cells.

Bile and bile salts are known to stimulate motility in sporozoites[116,117] and merozoites[167,168] of certain *Eimeria* and *Isospora* species. When bile solution was added after excystation had occurred in the presence of trypsin only, free sporozoites of *E. larimerensis* became considerably more motile.[116] After adding bile salt solutions, sporozoites of *I. canis* began to glide about within sporocysts in Clorox®-treated oocysts.[117] Merozoites of five different *Eimeria* species from mammalian hosts that had developed in cultured cells increased in degree of motility when exposed to solutions containing bile or bile salts.[168] Microgametes of *E. magna* also exhibited an enhanced motility and escaped from cultured cells upon exposure to bile or bile salt solutions.[169] In vivo, bile may stimulate merozoite and microgamete motility causing these stages to leave their host cells and enter other cells to form meronts or gamonts or to fertilize macrogamonts.

The mechanism by which bile and bile salts stimulate motility in sporozoites, merozoites, and microgametes is not known. Motility is probably caused by subpellicular microtubules in sporozoites and merozoites and by flagellar microtubules in microgametes. Cilia and flagella are known to move by the sliding of microtubular doublets of the axoneme past each other. A similar mechanism of movement in sporozoites and merozoites may occur in which a sliding action occurs between the inner membrane complex of the pellicle and the subpellicular microtubules. The force needed to cause displacement between adjacent doublets and hence the bending of cilia and flagella is provided by hydrolysis of ATP.[170] Calcium ions are known to facilitate microtubular sliding. The level of Ca^{2+} is controlled by the cell membrane and membrane stimulation causes Ca^{2+} influx.[170] Bile or bile salts used to enhance motility of sporozoites, and merozoites and microgametes which developed in cultured cells, may have altered their pellicular membranes causing an influx of Ca^{2+} which in turn resulted in increased microtubular sliding and, thus, enhanced motility.

Each motile stage of coccidia is capable of penetrating and/or leaving cells in order to develop further or, as in the case of microgametes, to fertilize macrogamonts. The process of penetration of cultured cells by sporozoites, sporozoite-shaped meronts, merozoites, tachyzoites, and macrogamonts have been described for various species of coccidia.[78,132,152-155,157,164,168,171-189]

Zoite penetration of cultured cells appears to occur by a similar process among the coccidia (Figure 1). Coccidian sporozoites are highly motile and appear to enter cells actively as demonstrated by cinemicrography.[187,188] In fact, Jensen and Edgar[178] found that motility was required for sporozoites of *E. magna* to penetrate cultured cells. Probably the most detailed description of cell penetration by live sporozoites is that of Roberts et al.[164] Sporozoites of *E. larimerensis* were found to undergo gliding or helical movements immediately before entering cells. Sporozoites first thrust a slender anterior protuberance into the host cell to a depth of 1 to 2 μm (Figure 1). Sometimes this protuberance was repeatedly retracted and

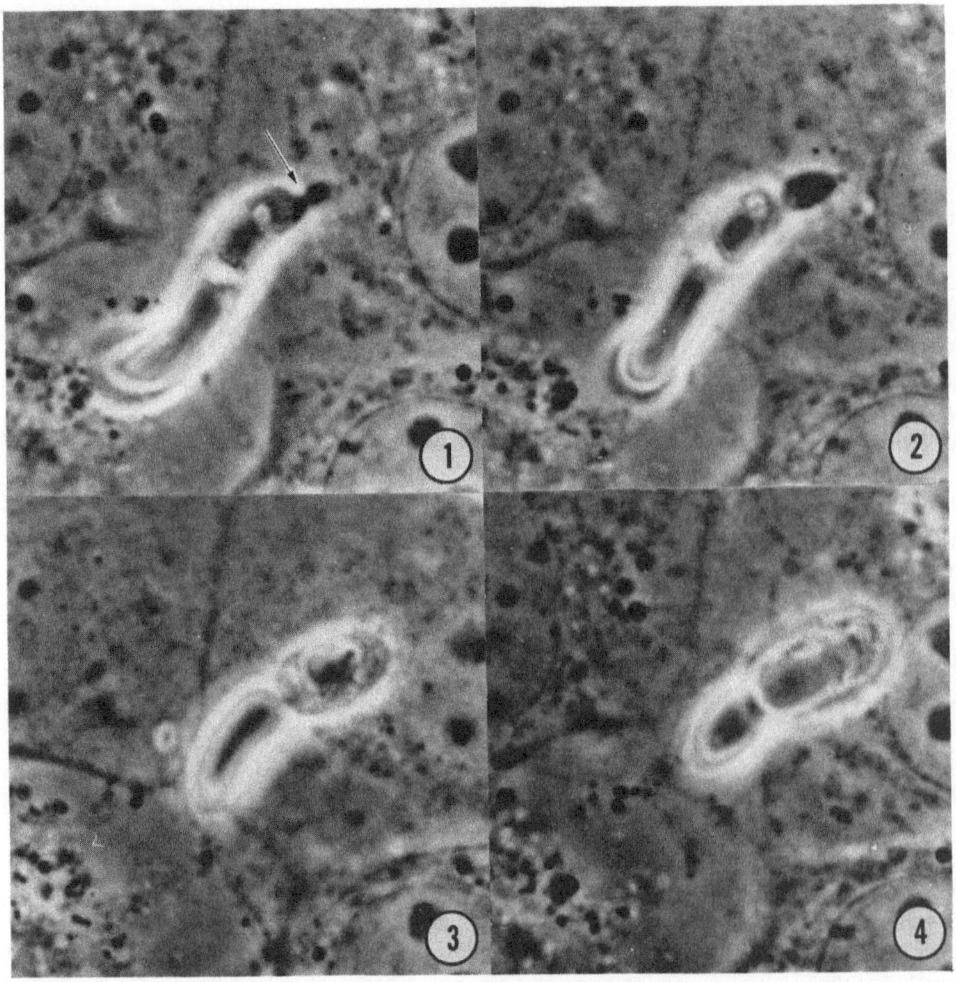

FIGURE 1. Phase-contrast photomicrographs of a living *Eimeria larimerensis* sporozoite in the process of penetrating a Madin-Darby bovine kidney cell immediately after inoculation, × 1600. (1) Sporozoite penetrating host cell with constriction (arrow) at anterior tip of parasite, which is intracellular. (2) Sporozoite has moved slightly forward with its anterior end further within the host cell. (3) Sporozoite is about half way into cell as indicated by the constriction in the middle of its body. (4) About 2/3 of the sporozoite body is intracellular.

inserted into the cell at different locations in the immediate vicinity. Usually the protuberance moved laterally for 1 to 2 sec within the host cell cytoplasm, and then appeared to swell, increasing in width and occasionally decreasing in length. Within 3 sec, the remainder of the sporozoite entered the host cell. During penetration, the sporozoite body was constricted to approximately 1/3 to 1/2 its normal width at the site of entry (Figure 1). Similar constrictions were also observed to pass along the bodies of some sporozoites which passed through host cell cytoplasm, penetrated the host cell nucleus and/or left the host cell. Such constrictions may be caused by sporozoites passing through the host cell plasmalemma, nuclear envelope, endoplasmic reticulum, and/or microfilaments. [164,180] Immediately after penetration, sporozoites occasionally leave a clear pathway approximately the width of the sporozoite in the host cell cytoplasm.[164,187] No host cell cytoplasm has been observed to escape during penetration of cultured cells by sporozoites[164,187] or merozoites,[168] whereas loss of host cell cytoplasm or nucleoplasm frequently occurs when sporozoites leave cells or host cell nuclei.[164,187]

The ultrastructure of sporozoite or tachyzoite penetration of cultured cells has been reported for *E. larimerensis*,[164,186] *E. magna*,[178-181] *I. canis*,[180] and *T. gondii*.[190,191] The role that each component of the apical complex (i.e., conoid, rhoptries, micronemes, polar rings, apical rings, subpellicular microtubules) (Figure 4) plays in cell penetration by coccidian zoites is not well understood. The idea that rhoptries may form secretory substances which aid in host cell penetration has been proposed by numerous investigators (see reviews by Scholtyseck and Mehlhorn[192] and Vivier and Petitprez[193]). Jensen and Edgar[179] found empty saccules believed to be rhoptry remnants in the anterior end of *E. magna* sporozoites which were fixed in the process of penetrating cultured cells. They suggested that rhoptries are indeed secretory and probably function as an aid in penetration.

A host cell penetration enhancing factor (PEF) has been isolated from zoites of *T. gondii* which appears to be enzymatic in nature, probably of lysosomal origin.[194-196] Evidently, it is spent during penetration of cultured cells and is again synthesized by the parasite while intracellular.[197] The PEF can be isolated from the fluid of parasite suspensions and, if added to parasite inocula, it increases the virulence of various *T. gondii* strains to mice. Immunofluorescence indicated that PEF was located in the anterior end of *Toxoplasma* zoites, perhaps the rhoptries and/or micronemes.[198] *Toxoplasma* zoites also require Ca^{++} and Mg^{++} in order to penetrate cells, whether in the presence or absence of PEF.[191] Penetration enhancing factor does not cause an increase in phagocytic activity of cultured cells or improved motility of *Toxoplasma* zoites. High concentrations of PEF can cause disruption of cellular membranes.[191]

The conoid is also believed to participate in cell penetration but its role still remains uncertain. During penetration, the conoids of invading sporozoites of *I. canis* were fully extended,[180] those of *E. magna* were partially or completely retracted[181] and those of *E. larimerensis* were completely retracted.[164]

Stadtsbaeder et al.[199] found a greater number of live than killed *Toxoplasma* zoites in cultured macrophages, and macrophages treated with cytochalasin, an antiphagocytic substance, still contained *Toxoplasma*. This and other studies which used cinemicrography or electron microscopy[172,190,191,200] indicate that *Toxoplasma* zoites enter most 'nonphagocytic' cells by active penetration, whereas they may actively penetrate phagocytic cells or be phagocytosed by them.

During penetration by *E. larimerensis* sporozoites, the plasmalemma of the host cell is indented for a short distance and then interrupted.[164] In vivo penetration of host cells by *E. ferrisi* merozoites occasionally causes similar interruptions in the host cell plasmalemma.[201] Evidently, after penetration the host cell then forms new membranes which completely surround the parasite to form the parasitophorous vacuole (Figure 4). Jensen and Hammond[181] suggested that the invaginated host cell plasmalemma which initially surrounds intracellular sporozoites of *E. magna* breaks down and is later replaced by another membrane of host cell origin which becomes the parasitophorous vacuole membrane. Sporozoites of *E. larimerensis*[164] and multinucleate merozoites of *E. magna* (Figure 5) have been observed to penetrate and then leave several cultured cells, often passing from one cell into an adjacent cell. Such organisms carry with them portions of the host cell plasmalemma (Figure 15) and cytoplasm,[164] which evidently does not interfere with their ability to penetrate other cells. During penetration of cultured cells by *I. canis* sporozoites, Jensen and Edgar[180] considered the plasmalemma of the host cell to be mechanically stretched as indicated by a single line rather than the typical trilaminar appearance.[180] They were unable to determine the fate of the indented host cell plasmalemma.[180] Pulvertaft et al.[185] and Klainer et al.[183] found that *T. gondii* zoites entered cultured cells and mouse peritoneal macrophages by active penetration or by phagocytosis. Klainer et al.[183] and Jones and Hirsch[200] reported that *Toxoplasma* zoites could be phagocytosed by cells normally considered to be nonphagocytic (e.g., HeLa cells, L cells, and fibroblasts). However, Zaman and Colley[189] found that

Toxoplasma zoites actively penetrated macrophages within the peritoneal cavities of mice. The macrophage plasmalemma was interrupted by the zoite at the site of penetration. Later the parasite became enclosed in a parasitophorous vacuole within a newly formed host cell membrane.

Lycke et al.[202] found that hyaluronidase enhanced penetration of cultured cells by *Toxoplasma* zoites, whereas hyaluronic acid inhibited the effects of hyaluronidase. Hsu[203] found an increase in the number of intracellular zoites of *Toxoplasma* after adding lysozyme or hyaluronidase to HeLa cell cultures. Fayer et al.[204] showed that hyaluronidase did not enhance penetration of cultured cells by *E. adenoeides* sporozoites and that chondroitin sulfate and hyaluronic acid did not reduce sporozoite penetration. Fayer[205] found that quinine inhibited penetration of cultured cells by *E. meleagrimitis* and *E. tenella* sporozoites, but had no effect on sporozoite motility. Colchicine, colcemide, and vinblastine, which interfere with the assembly of microtubules, and antiphagocytic agents, such as sodium fluoride, iodoacetate, and 2-deoxyglucose, have no effect upon penetration of cultured cells by *E. magna* sporozoites, which were found by Jensen and Edgar[178] to have retained their motility. Cytocholasin B, which disrupts microfilaments, caused a loss of motility and, therefore, cell penetration by *E. magna* sporozoites.[178] However, sporozoites of coccidia contain few or no microfilaments. Thus, microtubules may not function in locomotion or they may be atypical microtubules which are sensitive to the inhibitory actions of cytochalasin B.[178] Another possibility is that microfilament-like connections which are sensitive to cytochalasin B occur between the subpellicular microtubules and the inner surface of the pellicular inner membrane complex.

Treatment of *Eimeria* sporozoites with immune serum or treatment of cultured cells with immune serum prevents cell penetration.[206] This indicates that antibodies may attach to cell surfaces as well as the pellicle of sporozoites which prevents sporozoite penetration. Treatment of cultured cells with interferon or an interferon inducer (polyinosinic-polycytidylic acid) has been shown to reduce the number of developing stages of *E. tenella*, but has no effect upon sporozoite penetration.[207] Fewer sporozoites of *I. lacazei* and/or *chloridis* from sparrows penetrated cultured cells when suspended in culture medium containing fetal calf serum (FCS) than when FCS was omitted.[208] More *E. tenella* sporozoites penetrated cells when suspended in medium containing FCS than in medium containing chicken serum.[128]

B. Development

1. Besnoitia

The in vitro development of bradyzoites obtained from tissue cysts has been described for certain *Besnoitia* species[84,174,209-211] (Table 2). Bradyzoites of *B. jellisoni* are capable of actively penetrating cultured cells and host cell nuclei,[174] and multiply asexually by an internal budding process, called endodyogeny,[212] within the host cell cytoplasm or nucleus.[210] In vitro development of sporozoites, merozoites, and tachyzoites of *Besnoitia* has not yet been reported.

2. Eimeria

The development of *Eimeria* species in cultured cells has been more extensively studied than any other coccidia. To date, those *Eimeria* species studied in cultured cells have been either parasites of avian or mammalian hosts. Most reports concerning avian *Eimeria* species describe the occurrence of first or subsequent generations of meronts after inoculation of sporozoites, whereas most reports of *Eimeria* species from mammalian hosts describe in detail the process of asexual development to first- and/or second-generation meronts.

Various cell types have been used to support in vitro development of *Eimeria* species (Tables 3 and 4). In general, better development of *Eimeria* species occurs in cell types obtained from the natural host than from unnatural hosts. Kidney cells which are relatively easy to grow in vitro have been more extensively used than any other single cell type and appear to be the most suitable cell type for development of *Eimeria* species.

Within a few hours or perhaps as long as several days after host cell penetration and before nuclear division begins, sporozoites of *E. bovis*,[132] *E. brunetti*,[213] *E. contorta*,[167] *E. crandallis*,[214] *E. ellipsoidalis*,[215] *E. ninakohlyakimovae*,[182] *E. tenella*,[85] and *E. zuernii*[216] undergo transformation to spheroid stages called trophozoites or uninucleat meronts. In *E. bilamellata*,[155] *E. callospermophili*,[155,217] *E. canadensis*,[184] *E. larimerensis*,[153,188,218] and *E. vermiformis*[219] one or more nuclear divisions occur before the parasites transform to spheroid stages. Such multinucleated organisms which are still elongate have been called sporozoite-shaped schizonts or meronts (Figure 4). Sporozoite-shaped meronts of *E. callospermophili*,[155,187] *E. larimerensis*,[153,188,218] and *E. magna*[154] have a motility similar to that of sporozoites and are capable of leaving and penetrating cultured cells. Such meronts also move about within the host cell cytoplasm causing enlargement of the parasitophorous vacuole (Figure 4). In *E. alabamensis*, *E. auburnensis*, and *E. meleagrimitis*, transformation to a spheroid stage may occur before or after nuclear division.[140,173,220,222] Sporozoite-shaped meronts of *E. auburnensis* also occur in in vivo infections.[223]

Transformation of uninucleate or multinucleated organisms to spheroid stages (Figure 2) occurs by an increase in width and/or by means of a lateral outpocketing which gradually progresses anteriorly as well as posteriorly[153,155,182,187,218,224]

Ultrastructural studies of organisms undergoing such transformation within cultured cells have shown that the inner membrane complex of the pellicle and the subpellicular microtubules disappear at various sites over the surface of the parasite, except for the anterior[224] and posterior regions.[225] After transformation to a spheroid stage has been completed (Figure 2), the inner membrane complex of the pellicle, subpellicular microtubules, and other organelles associated with the apical complex or the posterior end of the original sporozoite also disappear.[224-226]

Most *Eimeria* sporozoites studied thus far have been found to have an anterior as well as a posterior refractile body and a single nucleus (Figures 2 and 4).[227] After sporozoites enter cultured çells changes begin to occur in the anterior (if present) and posterior refractile bodies, and the parasite nucleus.[132,153,155,173,182,217,222,228] At about 6 hr after inoculation of cultured cells, the nucleus and nucleolus of *E. callospermophili* sporozoites change from a spheroid to an ellipsoid shape and markedly increase in size.[217] At 8 to 10 hr, the first nuclear division occurs in which the nucleolus elongates, becomes dumbbell shaped, and then divides into two separate nucleoli. Two daughter nuclei form by an infolding of the nuclear envelope between the two nucleoli. Additional nuclear divisions occur similarly except that the nuclei gradually decrease in size, eventually becoming indistinct.[217] The anterior and posterior refractile bodies gradually decrease in size, usually by forming small granules at their margins.[153,217]

In those *Eimeria* species which have been studied in vitro and have large first generation meronts, the refractile body material gradually disappears as development progresses.[132,173,182,184,214,216] In some of the *Eimeria* species which form small meronts with relatively few merozoites in cultured cells, portions of the refractile body material of the original sporozoite are incorporated into each merozoite as an anterior and posterior refractile body (Figures 2 and 5).[40,153,155,187,188,225,229] The refractile body material present in first-generation merozoites of *E. magna* and *E. larimerensis* may or may not be carried over into the second-generation merozoite.[153,154] The refractile body material may be incorporated into merozoites of the second-generation only if two or four merozoites are formed.[153,154] If second-generation meronts of *E. magna* or *E. larimerensis* contain six or more merozoites, the refractile body material is not present in specimens examined by light or electron microscopy (Figure 5). Merozoites of *E. magna* beyond the second asexual generation also lack refractile body material (Figure 2). Refractile bodies have also been observed in vivo

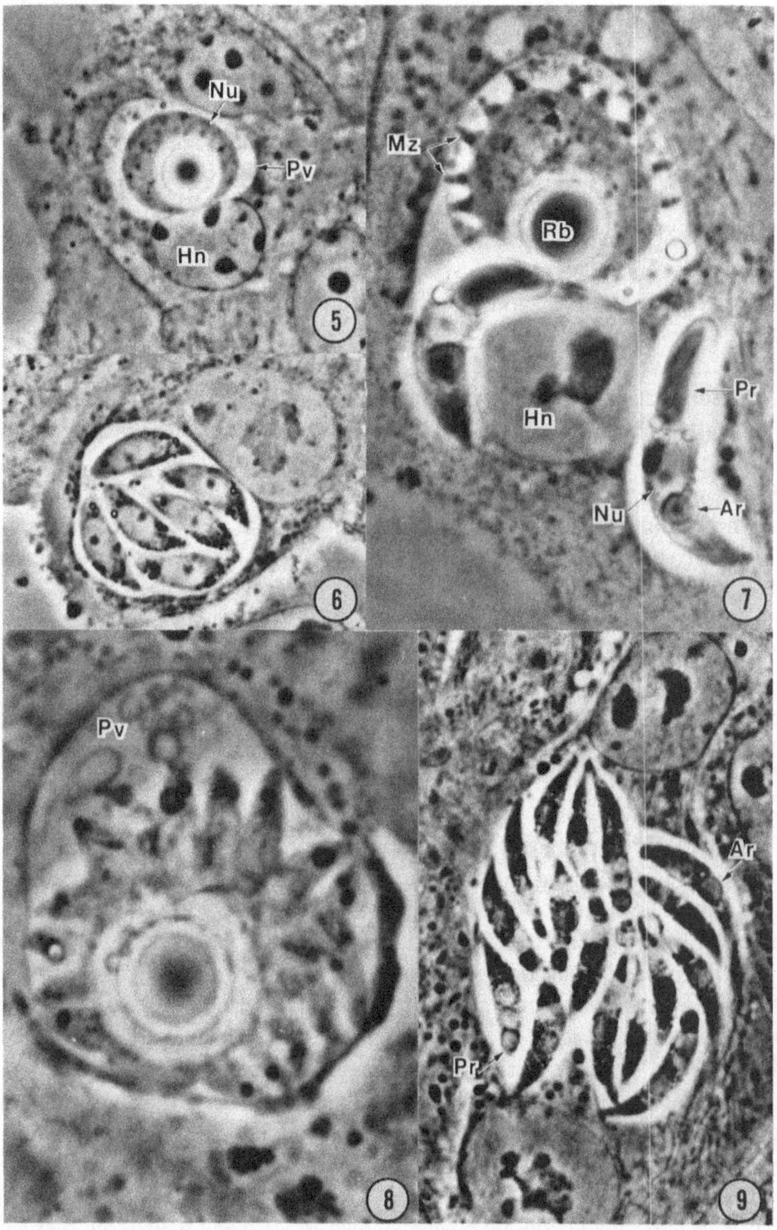

FIGURE 2. Phase-contrast photomicrographs of various asexual stages of *Eimeria larimerensis* (Figure 2) and *E. magna* (Figure 2) in Madin-Darby bovine kidney cell cultures; interval after inoculation of sporozoites or merozoites and magnification listed in parentheses. Abbreviations: Ar, anterior refractile body; Hn, host cell nucleus; Mz, developing merozoites; Nu, nucleus of parasite; Pr, posterior refractile body; Pv, parasitophorous vacuole; Rb, refractile body. (5) Spheroid first-generation meront; note eccentrically located refractile body and relatively large size of nuclei (24, × 1200). (6) Mature meront with six merozoites, each with an elongated nucleus and prominent nucleolus; cultures inoculated with merozoites obtained from rabbits (24, × 1350).[235] (7) Host cell containing two sporozoites and a budding first-generation meront (24, × 2000). (8) Stage of merozoite formation in first-generation meront; note small refractile granules at tip of each forming merozoite (48, × 3200).[153] (9) Mature first-generation meront; note that each merozoite contains an anterior and posterior refractile body (60, × 2000). (From Speer et al., *J. Parasitol.*, 59, 613, 1973; Speer, C. A. and Hammond, O. M., *Z. Parasitenkd.*, 35, 105, 1970. With permission.)

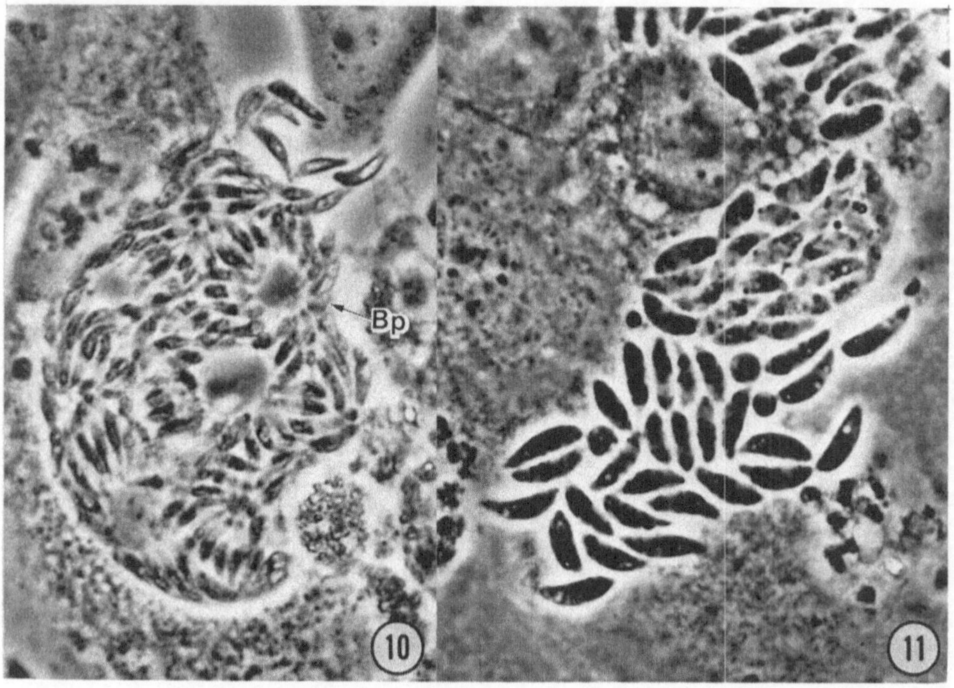

FIGURE 3. Phase-contrast photomicrographs of *Eimeria magna* in primary rabbit kidney and Madin-Darby bovine kidney cells at 72 hr after inoculation with merozoites obtained from infected rabbits. (10) Composite second-generation meront with merozoites budding from blastophores (Bp) (× 900). (11) Merozoites of a second-generation meront in culture escaping from a host cell (× 1000).

in first generation merozoites of *E. callospermophili, E. bilamellata* (L. C. Anderson, personal communication) and *E. larimerensis* (personal observation). The significance of refractile bodies in sporozoites and merozoites is not known.

First generation merozoites of all *Eimeria* species studied in cell culture form by a radial budding process at the margin of immature meronts *(E. alabamensis, E. auburnensis, E. bilamellata, E. bovis, E. callospermophili, E. canadensis, E. contorta, E. crandallis, E. larimerensis, E. magna, E. meleagrimitis, E. ninakohlyakimovae, E. tenella, and E. zuernii)* (Figures 2 and 3). Certain of these Eimeria species (*E. auburnensis, E. bovis, E. canadensis, E. crandallis, E. ninakohlyakimovae, E. tenella, and E. zuernii*) also form merozoites at the margins of blastophores. Fine structural studies of the development of first-generation merozoites in cell culture have been reported for *E. alabamensis,*[221] *E. bovis,*[230] *E. callospermophili,*[225] *E. ninakohlyakimovae,*[224,231] and *E. tenella.*[226] In all of these species except *E. callospermophili* merozoite formation begins at the margin of the meront. In *E. callospermophili,* merozoite formation begins internally with merozoites eventually budding at the margin of the meront.[225]

Development and rate of development of *Eimeria* sporozoites to first-generation meronts in cell culture appears to be similar to that which occurs in the host, except that meronts developing in cell culture are usually smaller and contain fewer merozoites. Those *Eimeria* species which form either small or large composite first-generation meronts within the host also form small or relatively large meronts in cultured cells. All four merogonous generations of *E. acervulina,* however, are slightly larger than those in the chicken.[232] Speer et al.[216] found that first-generation meronts of *E. zuernii* in cell culture were relatively large containing 500 to 1000 merozoites, whereas Davis and Bowman[233] found only small meronts with 24 to 36 merozoites in calves. Recently, Stockdale[233a] confirmed the findings of Speer et al.[216]

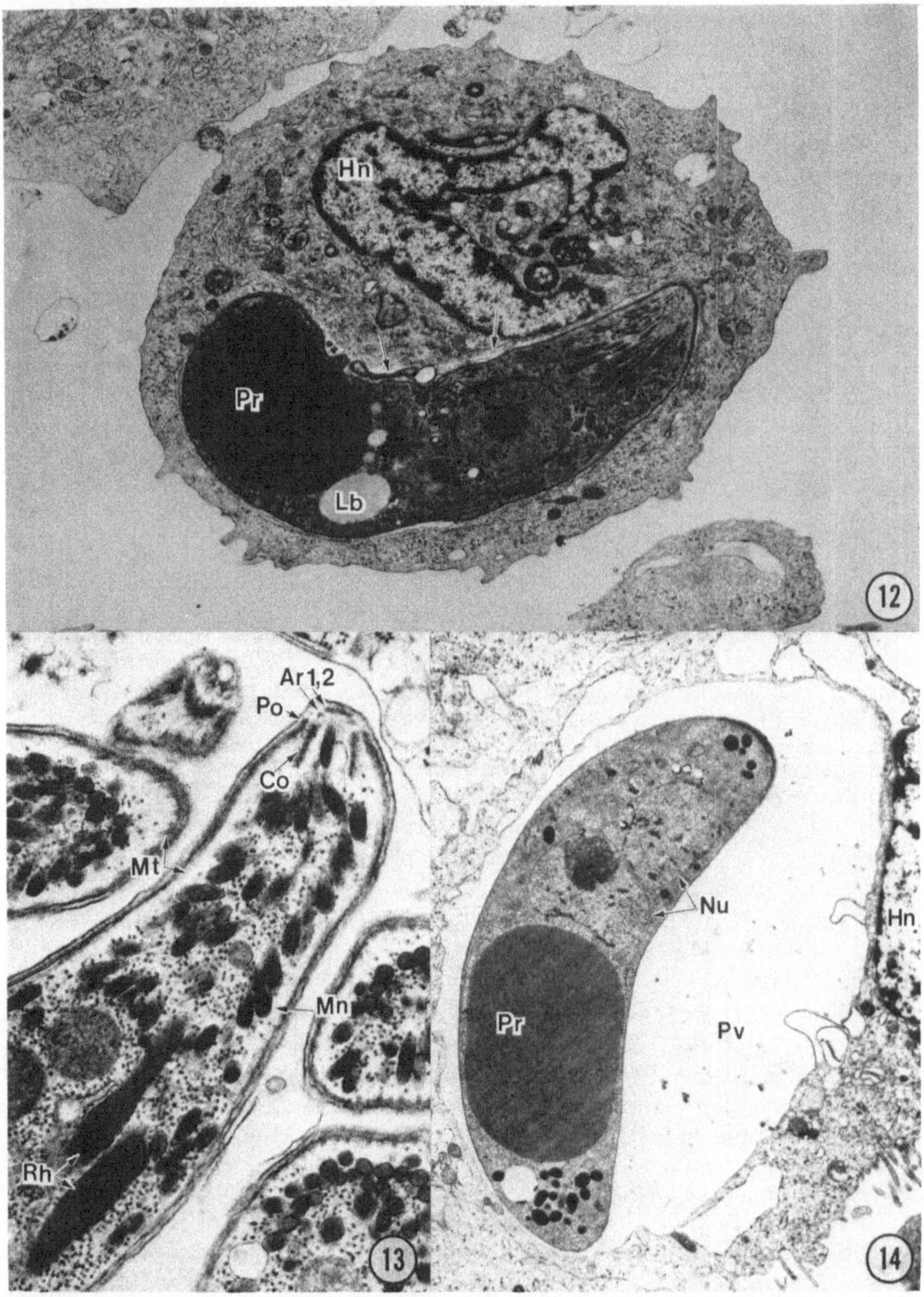

FIGURE 4. Transmission electron micrographs of *Eimeria larimerensis* in Madin-Darby bovine kidney cells and *E. magna* in a primary rabbit intestinal cell. Abbreviations: Ar 1, 2, apical rings 1 and 2; Co, conoid; Hn, host cell nucleus; Lb, lipid body; Mn, microneme; Mt, subpellicular microtubule; Nu, nucleus of parasite; Po, polar ring; Pr, posterior refractile body; Pv, parasitophorous vacuole; Rh, rhoptry. (12) Intracellular sporozoite at 5 hr after inoculation; note that the parasite is within a parasitophorous vacuole, which also has several layers of host cell membranes (arrows) near the convex surface of the sporozoite (×8750). (13) Apical complex of second-generation merozoite 60 hr after inoculation of cultures with merozoites from rabbits (×32,000). (14) Binucleate sporozoite-shaped meront at 72 hr after inoculation (×7200).

by discovering that *E. zuernii* does indeed form large first-generation meronts in the small intestine of calves.

Muller et al.[167] observed large meronts with more numerous merozoites in bovine cell cultures which had not been treated with bile salts than in bile salt-treated cultures. They suggested that in untreated cultures the larger meronts resulted from first-generation merozoites developing to second-generation merozoites within the same parasitophorous vacuole in which they had originated. Such development is known to occur in vivo or in vitro in *E. magna*,[169,234,235] *E. tenella*,[236] *E. stiedae*,[237] *E. larimerensis*,[218] *I. felis*,[238] and *I. canis*,[239]

Eimerian merozoites escape from cultured cells (Figure 3 by destroying the host cell membrane causing cell lysis (Figure 5).[219,222] Merozoites of *E. callospermophili* are slightly constricted as they escape from the parasitophorous vacuole and host cell.[225] Usually, merozoites enter cells in the immediate vicinity and develop into subsequent stages (Figure 6).

Within cell cultures, most *Eimeria* species from mammals develop only to first-generation meronts or immature second-generation meronts. *Eimeria contorta*,[167] *E. larimerensis*,[218] and *E. magna*[154] are the only *Eimeria* species from mammals which develop from sporozoites to mature second-generation meronts. Much further development in cultured cells has been obtained with *Eimeria* species from avian hosts. Sporozoites of *E. acervulina* will develop to mature fourth-generation meronts;[232] those of *E. brunetti* develop to mature third-generation meronts;[213] *E. diminuta*,[95] *E. dispersa*,[228] *E. mivati*,[240] and *E. necatrix*[241] develop to mature second-generation meronts; *E. gallopavonis* [228] and *E. meleagrimitis*[140] develop to macrogamonts; and *E. tenella* develops to gamonts and oocysts.[138,139,242,243]

In order to achieve further asexual development, gametogony and oogony, certain investigators have inoculated cell cultures with merozoites obtained from the intestines of infected animals (Table 4). Such studies have shown that merozoites of certain *Eimeria* species taken from the host are more likely to develop advanced endogenous stages in cell culture than are sporozoites. Merozoites of *E. acervulina*,[232] *E. bovis*,[244] *E. magna*,[142,157,169] *E. meleagrimitis*,[140] and *E. tenella*[141,245] develop to meronts, gamonts, and oocysts in cell culture, whereas merozoites of *E. brunetti*, [246] *E. larimerensis*,[218] and *E. stiedae*[247] develop to meronts and/or gamonts. Since sporozoites of several *Eimeria* species develop only to mature first- or second-generation meronts in cell culture, whereas merozoites obtained from infected animals develop to gamonts or oocysts, substances or conditions for development of the entire endogenous portion of the life cycle in cell cultures are lacking for the species previously studied except for *E. tenella*.

Merozoites of *E. magna* obtained from infected rabbits develop to meronts or gamonts and oocysts within cultured cells (Figures 6 to 8). Certain uninucleate merozoites of *E. magna* remain within the parasitophorous vacuole in which they originated and develop into multinucleate merozoites (Figure 5) or microgamonts.[142,157,169]

Multinucleate merozoites have been described in vitro or in vivo for *E. stiedai*,[237,248,249] *E. magna*,[142,154,157,169,234,250,251] *E. piriformis*,[9] and *E. callospermophili* [155] but their significance remains obscure. Pellérdy and Dürr[248] considered multinucleate merozoites of *E. stiedai* to be formed in response to an immune reaction of the host, or to an inadequate supply of nutrients which resulted in degeneration and fragmentation of nuclei. However, fine structural studies have shown that merozoite anlagen are present in some multinucleate merozoites of *E. magna* developing in vivo[234] as well as in vitro (Figure 5), indicating that certain multinucleate merozoites represent a stage in the formation of new merozoites. Speer et al.[235] and Speer[169] found that most *E. magna* multinucleated merozoites within cultured cells transformed to spheroid stages, each of which underwent additional nuclear divisions and gave rise to uninucleate merozoites. This resulted in the formation of large composite meronts with 40 to 500 merozoites[235] (Figure 3). A few multinucleate merozoites formed microgamonts.[169,235] Thus, such in vitro studies have shown that multinucleate merozoites are an important part in the life cycle of *E. magna*. Whether such stages are as equally important in the life cycle of other coccidia in which they occur remains to be determined.

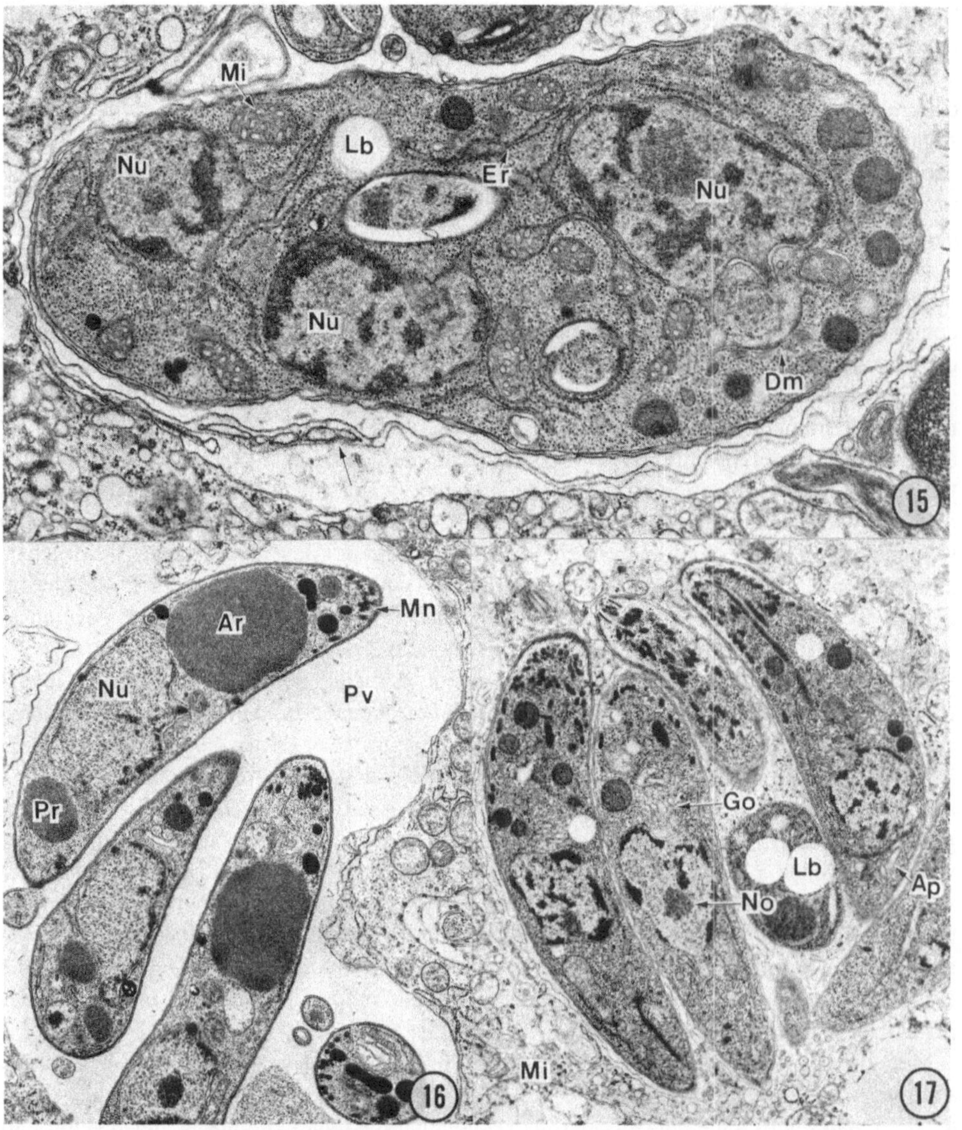

FIGURE 5. Transmission electron micrographs of *Eimeria magna* in primary rabbit kidney cells and *E. larimerensis* in Madin-Darby bovine kidney cell. Abbreviations: Ap, amylopectin; Ar, anterior refractile body; Dm, developing merozoite; Er, endoplasmic reticulum; Go, Golgi complex; Lb, lipid body; Mi, mitochondrion; Mn, microneme; No, nucleolus of parasite; Nu, nucleus of parasite; Pr, posterior refractile body; Pv, parasitophorous vacuole. (15) Multinucleate merozoite at 48 hr after inoculation of sporozoites; note that parasite is partially surrounded by several layers of host cell membranes (arrow) (×15,000). (16) Portion of first-generation meront at 100 hr after sporozoite inoculation (×9500). (17) Second-generation meront at 72 hr after inoculation of sporozoites; note that the parasitophorous vacuole membrane is discontinuous particularly near the anterior ends of the merozoites (×9500).

Microgamonts of *E. magna* (Figures 7 and 8) arise from uninucleate or multinucleate merozoites.[169,235] Speer and Danforth[278] found no differences in ultrastructure between microgamonts of *E. magna* which developed in cell culture and those in tissues of infected rabbits. They also found that microgametogony of *E. magna* was similar to that described previously for various *Eimeria* species, *Toxoplasma* and *Sarcocystis*.[212,267] Several nuclear divisions occur in microgamonts before microgamete formation. Nuclei usually become

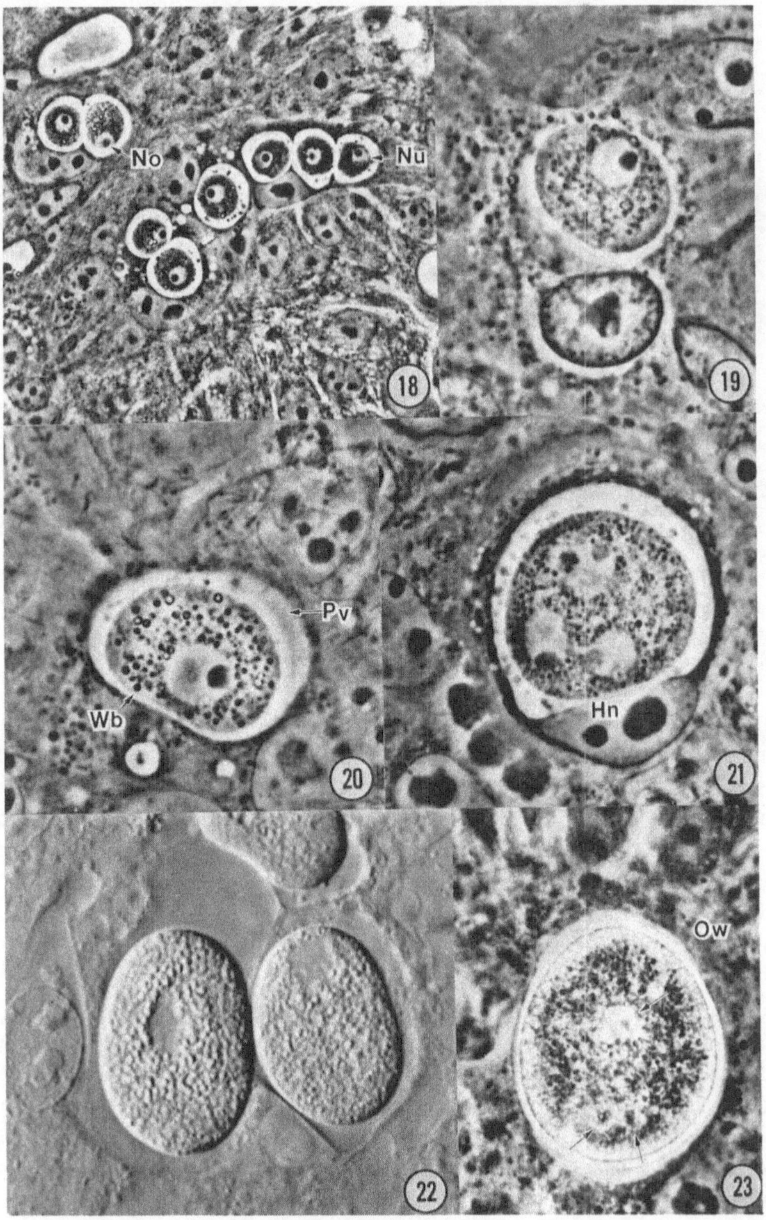

FIGURE 6. Photomicrographs of macrogamonts and an oocyst of *Eimeria magna* in Madin-Darby bovine kidney (MDBK) cells inoculated with merozoites obtained from rabbits, and of a macrogamont of *Isospora canis* in MDBK cells inoculated with sporozoites. Phase-contrast (18-21,23); interference-contrast microscopy (22). Interval after parasite inoculation of cultures and magnification are listed in parentheses. Abbreviations: Hn, host cell nucleus; No, nucleolus of parasite; Nu, nucleus of parasite; Ow, oocyst wall; Pv, parasitophorous vacuole; Wb, wall forming body. (18) Several young macrogamonts, each within separate parasitophorous vacuoles; note large nucleus and nucleolus in each gamont (84, ×650). (19) Macrogamont in degenerate host cell; note lack of prominent parasitophorous vacuole (96, ×1400). (20) Intermediate macrogamont (48, ×1400). (21) Multinucleate macrogamont; note prominent parasitophorous vacuole and apparent loss of host cell cytoplasm at lower left (60, ×1400). (22) Two mature macrogamonts (60, ×1400). (23) Relatively large oocyst with three nuclei (arrows), one of which is centrally located (72, ×1000).

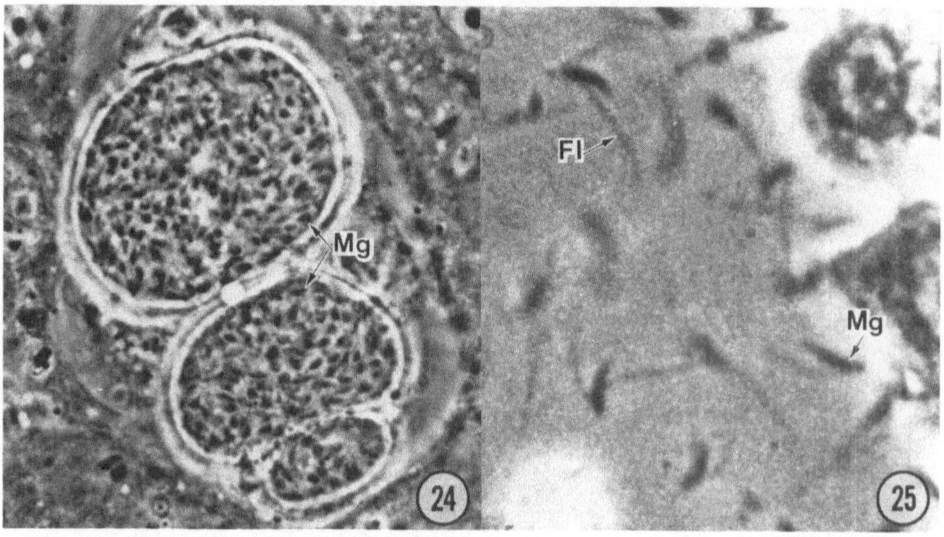

FIGURE 7. Phase-contrast photomicrographs of *Eimeria magna* in Madin-Darby bovine kidney cells. Interval after inoculation of merozoites obtained from rabbits into cultures and magnification are listed in parentheses. Abbreviations: Fl, flagellum of microgamete; Mg, body of microgamete. (24) Three mature microgamonts in the same host cell (72, × 1200). (25) Several biflagellate microgametes (72, × 3200).

elongate in shape, compact, and electron dense at the end oriented toward the periphery and lucent centrally. Usually, only the dense portion of the nucleus is incorporated into the gamete as the gamete develops by protruding from the gamont surface (Figure 8). The electron-lucent portion of the nucleus remains in the residual body of the gamont (Figure 8). Fully formed microgametes (Figure 7) have two or three flagella and associated basal bodies, and contain a nucleus, mitochondrion, and several microtubules.[212,267]

The occurrence of multinucleate macrogamonts (Figure 6) in cultured cells has been reported for *E. magna*[169] and *E. tenella*.[252] Those of *E. magna* form by fusion of several gamonts within the same parasitophorous vacuole.[169] Multinucleate macromonts also occur in marmosets infected with *I. arctopitheci* (unpublished data). As suggested by Doran and Augustine,[252] multinucleate macrogamonts are probably abnormalities caused by in vitro cultivation and have no significance in the life cycle of these parasites. However, recently I observed unusually large multinucleate oocysts of *E. magna* in cultured cells. One nucleus was always centrally located whereas the other nuclei were more peripherally located and appeared degenerate (Figure 6). Thus, some multinucleate macrogamonts of *E. magna* appear to be capable of developing to oocysts.

Macrogamonts of *E. magna* were observed by Speer and Hammond[157] and Speer[169] to move about by means of a pseudopodium-like extension of lobopodium type within the parasitophorous vacuole of cultured cells, to leave cells and to pass from one cell into an adjacent cell. Fayer[156,253] found that macrogamonts of *Sarcocystis* in cell culture rapidly rotated clockwise and counterclockwise on their longitudinal axis while within the parasitophorous vacuole. Since axis while within the parasitophorous vacuole. Since macrogamonts of the coccidia typically complete development into the oocyst without leaving the host cells, the significance of their movement is difficult to explain.

McDougald and Jeffers[254,255] found that a precocious strain (Wis-F) of *E. tenella* underwent gametogony and oogony in cultured cells after only one asexual generation, whereas the Wisconsin strain underwent two asexual generations. Previously, the life cycle of *E. tenella* had been considered to be fixed, with a minimum of two asexual generations before gametogony. Such a finding is particularly significant because it indicates that control of the

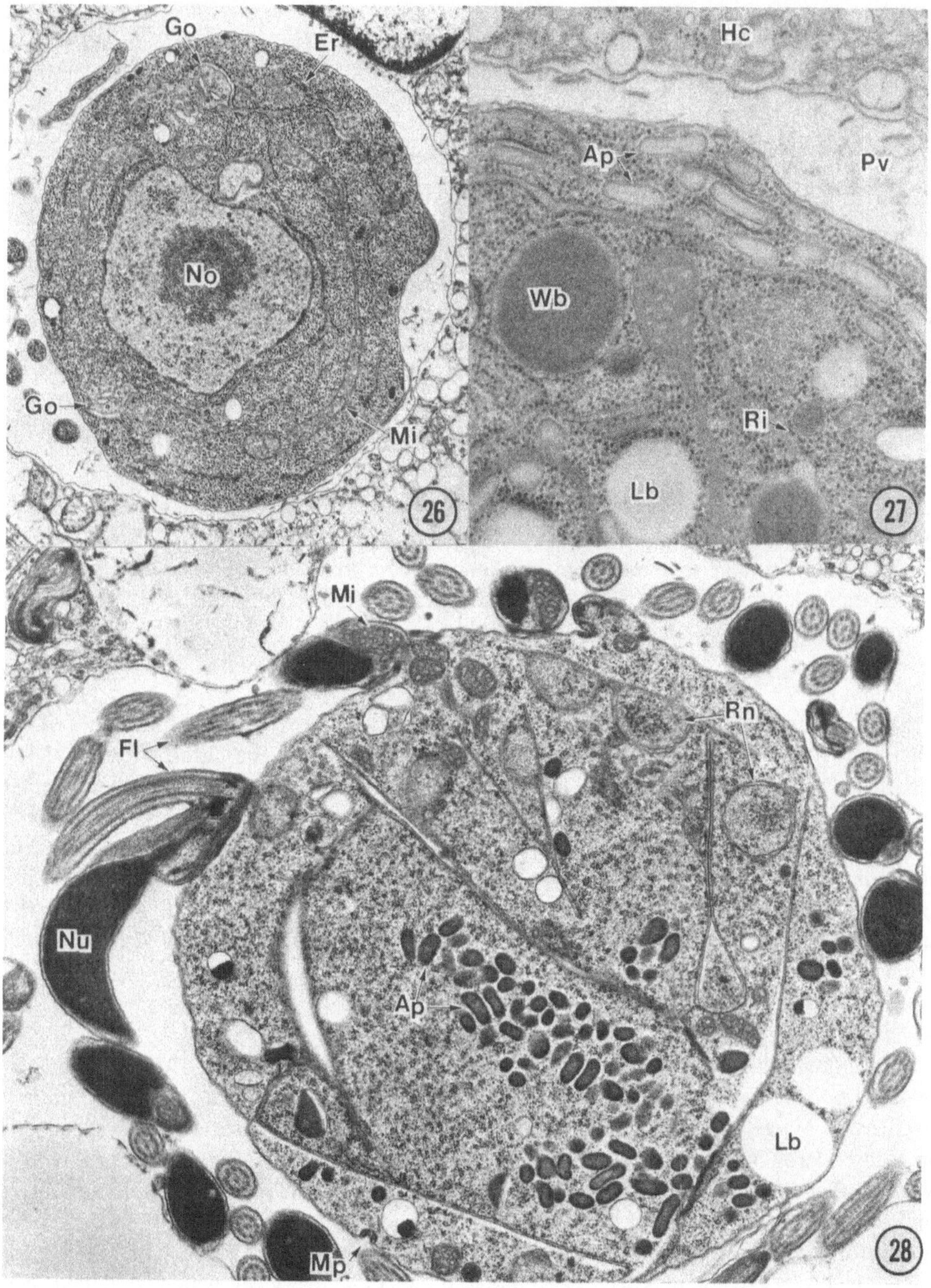

FIGURE 8. Transmission electron micrographs of *Eimeria magna* gamonts in Madin-Darby bovine kidney cells. Abbreviations: Ap, amylopectin; Er, endoplasmic reticulum; Fl, flagellum of microgamete; Go, Golgi complex; Hc, host cell cytoplasm; Lb, lipid body; Mi, mitochondrion; Mp, micropore; No, nucleolus of macrogamont; Nu, nucleus of microgamete; Pv, parasitophorous vacuole; Ri, ribosome; Rn, residual nucleus. (26) Young macrogamont (× 10,000). (27) Peripheral portion of nearly mature macrogamont (× 30,400). (28) Microgamont with nearly completely formed microgamonts at its periphery (× 11,300).

transition from schizogony to gametogony is a genetic trait, largely independent of host influence.[254] McDougald and Jeffers[254,255] also found that several microgamonts and macrogamonts occurred in clusters, indicating that haploid sporozoites may be bisexual. Doran[138] and Strout and Ouellette[243] found that strains of *E. tenella,* which evidently were different than the Wis-F and Wisconsin strains, underwent three asexual generations in cell culture before gametogony and oogony. Klimes et al.[256] studied sexual differentiation of *Eimeria* in cell cultures and found that Periodic-Acid-Schiff (PAS)-positive merozoites gave rise to macrogamonts, whereas PAS-negative merozoites formed microgamonts.

Eimeria magna and *E. tenella* oocysts produced in culture sporulate at room temperature and are infectious to their natural host.[138,140,169,254,255] Sporozoites of *E. magna,* obtained from oocysts produced in culture or from oocysts collected from rabbits that had been inoculated with in vitro-produced oocysts, developed to first- and second-generation meronts in cell culture[169] similar to that described previously.[154]

Oocysts of *E. magna* produced in cell culture were similar in size, shape, and appearance to those obtained from rabbits as described by Pellerdy[9] and Ryley and Robinson,[250] except that they were slightly smaller and usually lacked the prominent collar at the margin of the micropyle.[169] Rabbits inoculated with in vitro-produced oocysts shed typical *E. magna* oocysts which had the prominent collar.[169] In vitro-produced oocysts of *E. acervulina* and *E. tenella* are similar to those from chickens, except that a few *E. tenella* oocysts are considerably longer,[138,232] and in certain cell types they are slightly smaller.[252] Oocysts of *E. meleagrimitis* are smaller than those from the turkey.[140]

3. Hammondia

Sporozoites of *H. hammondi* and *H. heydorni* actively penetrate and form pairs of organisms by endodyogeny within cultured cells[257,258] (Table 5). The ultrastructural characteristics of intracellular sporozoites and of organisms undergoing endodyogeny were similar to those of *T. gondii,*[259] except that sporozoites and early intracellular stages of *H. hammondi* contain a crystalloid body which is not present in *T. gondii.* Similar crystalloid bodies occur in *I. canis* sporozoites.[260] Attempts to obtain long term cultivation by Sheffield et al.[257] were unsuccessful; organisms were no longer present in cultured cells at 4 to 6 weeks after inoculation. Attempts to obtain development of tachyzoites and bradyzoites of *Hammondia* species have not yet been reported.

4. Isospora

In cultured cells sporozoites of *I. lacazei* and/or *chloridis* develop to mature second-generation meronts,[208] whereas those of *I. canis, I. felis,* and *I. rivolta* appear to multiply exclusively by endodyogeny.[261-263] However, recently this author observed a few specimens in Madin-Darby bovine kidney cells at 96 hr after inoculation of *I. canis* sporozoites which were similar in appearance to macrogamonts (Figure 6). Such specimens were similar in appearance to intermediate macrogamonts of *E. magna* (Figure 6), except that they were not surrounded by a prominent parasitophorous vacuole (Figure 6). The lack of a prominent parasitophorous vacuole may have been due to the fact that every cell which contained such a specimen was degenerate (Figure 6). Stages similar to microgamonts or oocysts were not observed.

Based upon differences between the development of sporozoites of *I. felis* in cell culture and that which occurs in the host, Fayer and Thompson[263] suggested that sporozoites of *I. felis* may react to various intracellular environments by undergoing multiplication in different ways. For example, merogony and gametogony occur in intestinal epithelial cells of the feline host,[238] whereas in cultured cells and in extraintestinal tissue cells of the feline parasite multiplication occurs by endodyogeny.[16,263] A similar phenomenon may occur with *I. rivolta* from cats.[14,261]

5. Sarcocystis

The development of *Sarcocystis* in cultured cells has been described for only a few species and all reports thus far describe the development of tissue cyst zoites obtained from intermediate hosts to gamonts and/or oocysts (Table 7). The development of *Sarcocystis* sporozoites and merozoites in cultured cells has not yet been reported.

Fayer[156,253] was the first to describe the development of *Sarcocystis* zoites in cell culture to gamonts and oocysts, which was also the first description of such stages in the life cycle of *Sarcocystis*. Based upon his findings as well as those of others which demonstrated that *Toxoplasma* developed typical coccidian stages similar to *Eimeria* in the intestine of cats,[20,21,264] Fayer[156] suggested that *Sarcocystis* species may also have similar intestinal stages. *Sarcocystis* was later discovered to be an obligate two host parasite, tissue cysts of which were found to cause intestinal infections in carnivores.[60]

Dubremetz et al.[265] found that *S. tenella* zoites obtained from the esophagus of sheep developed in cell cultures to stages similar to macrogamonts as described by Fayer.[156,253] Fayer[253] was unsuccessful in obtaining sporulation of in vitro-produced oocysts of *Sarcocystis*, whereas Becker et al.[266] describe the sporulation of four species of *Sarcocystis* within various mammalian cell types.

The ultrastructure of gametogony, oogony, and/or sporogony has been described for several species of *Sarcocystis* developing in cell cultures.[149,266,267] Becker et al.[266] found that the ultrastructural features of gamonts and oocysts of five different species of *Sarcocystis* formed in cultured cells were essentially identical to those formed in the intestine of the natural host. Tissue cyst zoites develop directly into microgamonts and macrogamonts. They also found that each species of *Sarcocystis* rapidly underwent gametogony in vitro and in vivo, usually within 18 hr after inoculation. Some degree of cell type specificity was also found in that gametogony of each *Sarcocystis* species studied occurred only in cell cultures obtained from the final host. Each microgamont had a large irregularly shaped nucleus which gave rise simultaneously to 20 to 30 microgametes.[266] Only an electron dense portion of the large nucleus was incorporated as a nucleus into each microgamete. Electron lucent areas of the large nucleus remained within the residual body of the microgamont. Microgametes were 4 to 5 μm × 1 μm in size and had two or three flagella. Becker et al.[266] observed a microgamete flattened against the surface of a macrogamont which probably represented an early stage in fertilization. At 22 hr after inoculation, oocysts began to undergo sporulation to form two sporocysts. The ultrastructure of the sporocysts and sporocyst walls of *S. muris, S. ovicanis, S. suihominis*, and *Sarcocystis* species from a goat-dog cycle and from a monkey[266] was similar to that described for *Toxoplasma*,[122] *I. canis*,[268] and *I. endocallimici*.[269]

6. Toxoplasma

Toxoplasma gondii can be grown in cell culture more easily than any other coccidian parasite (see review by Doran[8 6]) (Table 8). However, to date only asexual reproduction of *T. gondii* has been obtained in cultured cells. Sporozoites as well as tachyzoites obtained from the peritoneal cavities of infected animals undergo multiplication in cell cultures by endodyogeny.[259,264,270] A single report by Azab and Beverley[271] describes the development of tachyzoites of the RH strain of *Toxoplasma* in human embryonic lung cells by means of merogony as well as endodyogeny. There seems to be some question as to whether the meronts observed by Azab and Beverley[271] were those of *Toxoplasma* or of some other coccidian, because the culture forms do not resemble the stages in the gut of the cat.

Bommer[171] used cinemicrography to demonstrate that *Toxoplasma* tachyzoites actively penetrate cultured cells. Within two to five hr after entering cells, toxoplasms of the RH and Beverley strains undergo a relatively rapid multiplication by endodyogeny.[171,172,264,270,272] Usually, the more virulent the strain of *Toxoplasma* the more rapid the rate of asexual reproduction in cultured cells.

Sheffield and Melton[259] described the ultrastructure of endodyogeny of *Toxoplasma gondii* in monkey kidney cells. After entering the host cell, *T. gondii* transforms from a crescent to an ovoid shape. The Golgi divides into two separate complexes, near which the apical complex of each daughter zoite begins to form. The inner membrane complexes and associated subpellicular microtubules of each daughter zoite extend posteriorly around each end of the nucleus which becomes horseshoe-shaped. The nucleus then divides with each portion becoming enclosed by the forming daughter membranes. The daughter organisms increase in size within the mother cell, the inner membrane complex of the mother cell disappears and the outer membrane of the mother cell forms most of the outer membrane of each daughter zoite. Rosettes of zoites are formed by successive divisions without separation of the posterior ends of the offspring.[259]

By using cinemicrography, Meyer and Musacchio[273] found that tachyzoites of *T. gondii* within cultured mouse macrophages may divide by binary fission, rather than endodyogeny. A similar type of 'binary fission' has been observed in first-generation meronts of *E. dispersa*, second-generation meronts of *E. magna* and *E. larimerensis*, and third-generation meronts of *E. meleagrimitis*[140,154,218,228,235].

Toxoplasma can be maintained indefinitely by serial passage of infected cell cultures without alteration of parasite structure or virulence.[143-145] By maintaining cultures at 22°C and 4°C rather than at 37 to 39°C, the interval between culture transfer can be prolonged so that only one transfer is necessary every 3 or 4 weeks.[78,274,275]

The effects of pH, culture media, temperature of incubation, cell type, age and passage number of cell types, etc., on the development of various coccidia in cell culture have been reviewed by Doran.[86] To date, there have been no reports of the in vitro cultivation of *Frenkelia* species.

C. Nutrition and Biochemistry

There is little information concerning the type of nutrition in the coccidia. Most investigators believe that coccidian nutrition may occur via micropores (Figure 9), which are ultrastructural organelles present in the pellicle of all stages of apicomplexan protozoans, except microgametes. Micropores in the Haemosporidia have been shown to ingest host cell cytoplasm which results in the formation of food vacuoles whose contents undergo digestion in the presence of acid phosphatase to form hemazoin, a residual granular substance (see review by Aikawa[158]). Thus, in this group micropores function as feeding structures. However, in other apicomplexans the role of the micropore has not been well defined. Micropores have been observed in certain stages of various coccidia developing in cell cultures, including sporozoites, and/or meronts of various *Eimeria* species,[221,224-226,230,276-278] zoites of *H. hammondi*[257] and *T. gondii*[259] and microgamonts of *E. magna*,[278] a *Sarcocystis* sp. from grackles,[267] and *S. suihominis*.[149] Becker et al.[266] briefly described the ultrastructure of gametogony and sporogony of five species of *Sarcocystis (S. suihominis, S. ovicanis, S. muris, S. ovifelis*, and a *Sarcocystis* species from monkeys) and did not mention the presence or absence of micropores. Most of the micropores observed in coccidia within cultured cells have been inactive (Figure 9) which is similar to observations of coccidia developing within cells of the natural host (see review by Senaud et al.[279]). Although most micropores appear inactive, food vacuoles are frequently observed near micropores in the parasite cytoplasm (Figure 9). Active micropores usually contain substances similar in appearance to that which occurs in the parasitophorous vacuole and parasite food vacuoles.[279] In microgamonts and macrogamonts of *E. nieschulzi*, Sibert and Speer[280,281] observed active micropores which contained material similar to that of the crescent body which was present in the parasitophorous vacuole. Such a finding indicates that crescent bodies, which are commonly seen in the parasitophorous vacuole surrounding eimerian parasites, may actually represent partially digested host cell material. *Eimeria alabamensis* is the only coccidian studied thus far

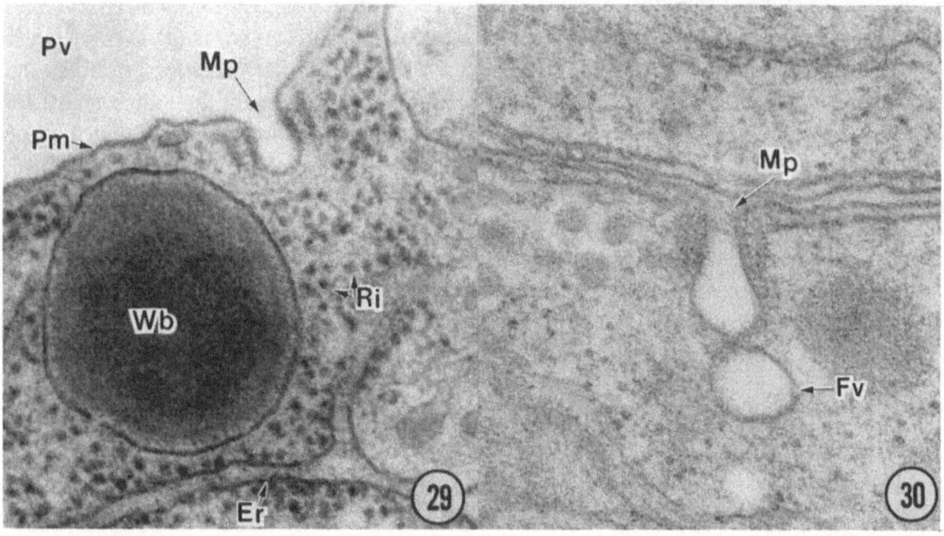

FIGURE 9. Transmission electron micrographs of *Eimeria magna* in a Madin-Darby bovine kidney cell and *E. stiedai* from rabbit bile duct epithelium. Abbreviations: Er, endoplasmic reticulum; Fv, food vacuole; Mp, micropore; Pm, macrogamont pellicle membrane; Pv, parasitophorous vacuole; Ri, ribosome; Wb, wall forming body. (29) Peripheral portion of a nearly mature macrogamont with an inactive micropore (×40,000). (30) Portions of two adjacent merozoites, one of which has an active micropore (×38,000).

which appears to directly ingest unaltered host cell cytoplasm.[276] Sénaud et al.[279] suggested that in the coccidia the host cell cytoplasm becomes altered before entering the micropores as indicated by the dissimilarity between host cell cytoplasm and material within the parasitophorous vacuole, the micropore cavity and parasite food vacuoles. The mechanisms responsible for the production and delivery of nutritive elements necessary for parasite maintenance and growth need to be studied further by using cytochemical and physiological methods.

Little is known about the nutritional requirements of the coccidia in vivo or in vitro (see review by Ryley[282]). This is primarily due to the fact that coccidian parasites are obligate intracellular organisms which depend upon their host cells for chemical stimuli and nutritional factors in order to undergo growth and development. McDougald[283] suggested that coccidian sporozoites may need a certain stimulus to activate the parasite genome in order for the parasite to undergo further development, which may be largely nutritional after initiation. Once these stimuli have been identified or supplied in an in vitro system, then coccidia might be able to undergo development in cell-free culture media.

There are only a few reports concerning nutritional requirements for the development of coccidia in cell cultures.[284-286] Amino acid requirements of *E. tenella* were studied by Sofield and Strout.[286] They found that certain levels of glutamine, isoleucine, methionine, tryptophan, and tyrosine are essential for optimum growth of *E. tenella* in embryonic chick kidney cells. Sporozoites of *E. tenella* failed to develop to second-generation meronts when glutamine was omitted from the culture medium, whereas omission of isoleucine, methionine, tryptophan, or tyrosine had little adverse effect on parasite development. Amino acids present in serum had little effect upon parasite development.[286] Because coccidian parasites develop only within host cells, such findings cannot be interpreted as absolute parasite requirements. Deletion of certain amino acids or other substances from the culture medium may have an adverse effect upon the host cell which indirectly affects the parasite.

Thirteen amino acids are required for growth and survival of most animal cells in culture.[287,288] Six or seven "nonessential" amino acids are often used in cell culture media because they may be stimulatory for growth or actually be required for growth.[289] Requirements for certain amino acids are frequently dependent upon cell density. For example, cultures with high cell populations have drastically reduced requirements for certain amino acids such as cystine, glutamine, serine, and asparagine.[289] Studies on the requirements for "nonessential" amino acids by cultured cells with or without parasites are complicated by the fact that animal cells are able to synthesize substantial amounts of these amino acids.[289] Uptake of "nonessential" amino acids during one phase of cell growth may be obscured by later excretion. Therefore, such parameters must be considered in designing experiments to investigate the nutritional requirements of coccidia within cultured cells.

Similar problems are inherently present in studying the vitamin requirements of coccidia in cell cultures. Ryley and Wilson[285] found that *p*-aminobenzoic acid (PABA), nicotinamide, pyridoxine, and thiamin may be necessary for the development of *E. tenella* sporozoites to second-generation meronts. Doran and Augustine[284] found that vitamin A, biotin, PABA, folic acid, nicotinomide, Ca pantothenate, pyridoxine, pyridoxol, riboflavin, thiamin, ascorbic acid, calciferol, α-tocopherol, and menadione were all required for first- and second-generation merogony of *E. tenella*, whereas all such vitamins except calciferol and folic acid were required for gametogony. Vitamin requirements by coccidia within the host may be different than requirements within cell cultures. Warren[290] found a reduction in the number of *E. tenella* oocysts produced by chickens on diets lacking ascorbic acid, biotin, Ca pantothenate, choline, folic acid, nicotinamide, PABA, pyridoxine, riboflavin, and thiamin, whereas diets lacking vitamins A, D, E, and K had no apparent effect on oocyst production. Biotin, choline, folic acid, inositol, nicotinamide, Ca pantothenate, pyridoxol, riboflavin, and thiamin have been shown to be the minimal vitamin requirements of most mammalian cell lines[291,292] and deleting such vitamins from the culture medium would be expected to have an indirect adverse effect upon parasite development.

During development, coccidia are known to consume considerable amounts of carbohydrates, primarily glucose, some of which are degraded to lactic acid.[282] Glucose is also the principal carbohydrate in cell metabolism although most cultured cells can utilize fructose, galactose, and mannose.[289]

In addition to amino acids, vitamins and carbohydrates, various other substances such as lipids, inorganic ions, keto-acids, hormones, and purine and pyrimidines need to be studied in order to determine if they are necessary or stimulatory for growth of coccidia within cultured animal cells. Exogenous lipids are derived from serum in the medium and are well used by animal cells in culture.[293] Keto-acids such as α-ketoglutarate, oxaloacetate, and pyruvate and hormones such as insulin, 1-thyroxine, serotonin, and hydrocortisone are known to stimulate growth of certain cultured cells.[294-298]

Little is known about the requirements of the coccidia for nucleic acid synthesis precursors. *Toxoplasma gondii* incorporates pyrimidine nucleosides as well as their precursors.[299,300] Zoites of *Toxoplasma* appear to possess ribosomal and transfer RNAs which are distinctly different than those of cultured mammalian host cells.[300] Morgan and Canning[301] found that ^{3}H-adenosine was readily incorporated into developing first-generation meronts of *E. tenella* in chick embryos, whereas ^{3}H-thymidine was not. Similar observations were reported by Mayberry and Marquardt[302] concerning *E. nieschulzi* in rats. Eimeria tenella and *E. callospermophili* do not incorporate ^{3}H-thymidine during asexual development in cell culture.[303-306] By using a thymidine analogue, 5-bromo-2′- deoxyuridine, and an inhibitor of *de novo* thymidylic acid synthesis, 5-fluoro-2′-deoxyuridine, Ouellette et al.[305] showed that *E. tenella* meronts were unable to utilize either exogenous thymidine or 5-bromo-2′-deoxyuridine but were dependent upon the methylation of deoxyuridylic acid via the thymidylate synthetase reaction to supply thymidine for DNA synthesis. Some meronts but not sporozoites of *E. tenella* incorporate a limited amount of ^{3}H-cytidine and some sporozoites and all meronts incorporate ^{3}H-uridine.[304] Tritiated cytidine appeared to come via the host cell cytoplasm from label originally bound to RNA in the host cell nuclei.[304] Tritiated uridine was accumulated more readily by meronts of *E. tenella* than by host cells, indicating that ^{3}H-uridine was incorporated directly into parasite RNA or the RNA was being synthesized more rapidly in the parasites than in the cultured cells.[304]

Nuclear hypertrophy of infected cells occurs in vitro (Figures 2 and 6) as well as in vivo.[86,307] By using microspectrophotometry and autoradiography, Fernando et al.[308] studied DNA synthesis in nuclei of intestinal cells of chicks infected with *E. maxima* or *E. necatrix*. *Eimeria maxima* did not affect the nuclear DNA of villous epithelial cells, whereas the nuclei of crypt cells infected with *E. necatrix* underwent enlargement and increased DNA synthesis. Browning et al.[303] reported a parasite-induced increase in thymidine incorporation in nuclei of host cell cultures infected with asexual stages of *E. tenella*. They suggested that such an increased synthesis of host cell DNA might be a host cell defense mechanism or an adaptation which facilitates parasite development. Speer et al.[235] observed a relatively greater number of asexual stages of *E. magna* in cultured cells containing two nuclei than in those with a single nucleus. They pointed out that this effect was similar to the effects of cytochalasins A and B which do not interfere with mitosis but inhibit cytokinesis in cultured mammalian cells.[309] Patton[310] used cytochalasin B to induce binucleate cells and found more asexual stages of *E. tenella* in binucleate cells than in mononucleate cells.

Little is known about the various types of enzymes which coccidian parasites use to catalyze biochemical reactions (see reviews by Frandsen[311] and Ryley[282]). Biochemical assays are nearly impossible because of the difficulty in separating enzymes of intracellular parasites from host cell enzymes. Enzyme cytochemistry has been used to a limited extent to study the site of enzyme activity within certain coccidia[312-316] and some such studies were of parasites within cultured cells.[314,316] Akao[313,314] found that mitochondria of *T. gondii* were peculiar in that they contained peroxidase but not aspartate aminotransferase. Aspartate aminotransferase was present in the inner membrane complex of the parasite pellicle, whereas peroxidase was not present in any other parasite organelle. In mammalian cells, aspartate aminotransferase occurs in mitochondria, limiting membranes of microbodies and nuclear membranes,[317] and mitochondria usually lack peroxidase.[318] Vetterling and Waldrop[316] studied the effects of various chemical fixatives, fixative concentrations and dehydrating agents on the preservation of enzyme activity and site of enzyme activity within *E. tenella* and cultured host cells. In general, formaldehyde fixation eliminated less activity than glutaraldehyde. Reaction products of glucose-6-phosphatase activity were found as granules in the cytoplasm of merozoites, and as fine deposits in meronts and macrogamonts. Adenosine triphosphatiase occurred in the parasite cytoplasm and adjacent to parasite nucleoli. Reaction products of 5′-nucleotidase were found in the cytoplasm and adjacent to nucleoli of all endogenous stages except for microgamonts. Thiamin pyrophosphatase was scattered and sparse. Alkaline phosphatase was not found in any of the endogenous stages, whereas acid phosphatase occurred in the parasite cytoplasm.

Fayer and Thompson[263,319] and Hare and Strout[320] reported cytochemical and cytological observations on *I. felis* from cats, a *Sarcocystis* sp. from grackles and *E. tenella* from chickens propagated in cell cultures. Beyer et al.[321] used cytophotometrics to study the dynamics of cytoplasmic RNA during merogonous development of *E. tenella* in vivo, in cell culture and in chick embryos. In *E. tenella*, Hare and Strout[320] observed chromatin rings composed of DNA in sporozoites, meronts, and microgamonts, but not in macrogamonts. Ribonucleic acid occurred in the nucleoli of sporozoites and macrogamonts, in acentric granules within merozoite nuclei and in the cytoplasm of all meronts. Carbohydrates disappeared during early sporozoite development, but accumulated during merogony and gametogony. Lipid droplets (Figures 5 and 8) were present in all stages of *E. tenella* but were particularly concentrated in residual bodies or unsegmented areas of meronts and microgamonts. Acid mucopolysaccharides occurred as cytoplasmic granules in all developmental stages of *E. tenella*.[320] Similar observations were reported by Fayer and Thompson[263,319] for asexual stages of *I. felis* and sexual stages and oocysts of a *Sarcocystis* species in cell culture. Fayer and Thompson[319] also studied the presence of protein and found that protein-positive granules in macrogamonts appeared to participate in the formation of the oocyst wall of *Sarcocystis*. Such bodies probably correspond to the wall-forming bodies described for *Sarcocystis*[266] and *Eimeria* species.[212]

D. In Vitro Activity of Anticoccidial Agents

Various anticoccidial agents have been tested for their effectiveness to inhibit in vitro development of *T. gondii*[322-333] and of certain avian species of *Eimeria*.[95,205,285,334-346]

Current in vivo methods for evaluating the effectiveness of various anticoccidial agents are expensive and involve elaborate facilities. Because of this, several investigators have attempted to establish workable methods for screening chemotherapeutic compounds for their anticoccidial activity in cell cultures. Most of these studies have used *E. tenella* because it is one of the more common pathogenic species of avian coccidia and commercially acceptable anticoccidial agents, therefore, must show activity against it. In addition, *E. tenella* is the only coccidian which will develop from sporozoites to oocysts within cultured cells.[138,139] This provides a particularly useful in vitro system for drug screening since different drugs are known to inhibit different stages of parasite development. For example, methyl benzoquate inhibits development of intracellular sporozoites of *E. tenella* whereas robenidine allows development in immature first-generation meronts.[341,342]

Screening of anticoccidial drugs in cell culture is presently not a satisfactory or reliable alternative to screening in chickens, although such in vitro studies may play an important role in the final evaluation of an active drug.[343] In vitro studies can be particularly useful in determining the stage in the life cycle which is effected by the drug, determining whether the drug is coccidiostatic or coccidiocidal, and describing the structural changes caused in the parasite by the drug.[338,341,343,345] Uneven drug distribution and metabolism, and drug toxicity in the host are important aspects of drug screening which cannot be adequately examined in an in vitro system. However, if a drug is toxic to cultured cells, it is likely to be toxic to the host. Also, different species of avian coccidia develop at different sites in the intestinal tract, indicating that they may also differ significantly in their basic metabolism and, therefore, drug sensitivity as well as in degree of drug exposure.

Ryley and Wilson[343] compared various aspects of screening or prescreening anticoccidial drugs in vitro to in vivo drug screening. A summary of their comparison is as follows: (1) Relatively small quantities of the drug are needed; in vivo screens require about 700 mg of drug for each eimerian species, whereas 10 mg may be sufficient for in vitro screening; (2) In vitro screening is less expensive than in vivo screening, but the amounts of effort are nearly equal. Nevertheless, in vivo screening must ultimately be performed in the final evaluation of the drug; (3) Cell toxicity may not be expressed in vitro, whereas in the host the drug may be metabolized to a toxic agent. Also, a drug without activity in vitro may be metabolized to a highly effective anticoccidial agent within the host; (4) However, activity in vitro provides no guarantee of detectable activity in the chick. Even if a drug is active in vitro and in vivo, the level of in vitro activity gives no reliable indication of in vivo activity; (5) In vivo and in vitro screens should examine drug activity against other coccidian species as well as to *E. tenella*.

Although Fungizone, which contains the active ingredient Amphotericin B, was found to lyse sporozoites extracellularly[335] and quinine inhibited sporozoite penetration of cultured cells,[205] most coccidiostats, such as amprolium, decoquinate, monensin, and other, have no apparent effect upon extracellular sporozoites or sporozoite penetration.[336,339] Most coccidiostats in vitro inhibit sporozoite development beyond first-generation meronts.

To date, approximately 33 chemicals have been examined for their in vitro activity against avian coccidia, particularly *E. tenella*. Nearly all of these chemicals have been found to be coccidiostatic, whereas monensin only proved to be coccidiocidal.[346] After incubating sporozoites of *E. tenella* in cultured cells for 24 hr in the presence of amprolium, lasalocid, methyl benzoquate, monensin, nicarbazin, or robenidine, Wilson[346] removed the chemical and found that all sporozoites proceeded with development except for those exposed to monensin. However, Ryley and Wilson[343] pointed out that most chemicals which are normally coccidiostatic would probably become coccidiocidal if the parasites were exposed for longer periods to higher concentrations of the drug.

Coccidiostats act as growth factor antagonists or they interfere in some other way with the normal metabolism of the parasite. Ryley and Wilson[285] studied antagonism between coccidiostats and various growth factors in cell cultures infected with *E. tenella*. Such an in vitro system minimizes the complications of coccidiostat distribution and metabolism in the host, being reduced to the simple combination of host cell, parasite and external medium. Ryley and Wilson[343] found that the anticoccidial activity of amprolium on *E. tenella* was significantly or completely prevented by adding thiamin to the culture medium. Similar results were obtained with sulphonamides and *p*-amino-benzoic acid, 6-aminonicotinamide and nicotinomide, and 4-desoxypyridoxine and pyridoxine. They observed no interactions between various antagonists and biotin, riboflavin, purines, or pyrimidines.

Different *Eimeria* species and strains vary in their degree of susceptibility to various coccidiostats.[342,343] Ryley and Wilson[343] found that buquinolate, decoquinate, and methyl benzoquate had no activity in vitro against the Chile strain or *E. tenella* at 2, 5, and 20 ppm, respectively, whereas 0.0005 to 0.1 ppm of these same chemicals were effective against six other strains.

The effects of coccidiostats on parasite ultrastructure, and differences in ultrastructure and enzyme content between sensitive and insensitive strains may not be readily apparent. Quinolone anticoccidial drugs stop growth of *E. tenella* by blocking mitochondrial respiration. In an ultrastructure study, Wang[347] detected no differences in mitochondrial physical properties, cytochrome content, or respiratory activity between a quinolone-resistant strain and a quinolone-sensitive strain of *E. tenella*.

Various chemicals have been examined for their ability to inhibit *T. gondii* penetration and development within cultured cells[300,322,324,332,333] as well as to obtain drug-resistant or temperature-sensitive parasite mutants.[325-330] Remington et al.[331] found that rifampin in concentrations more than 100 times that required to inhibit RNA synthesis in bacterial cells did not inhibit RNA synthesis in cell-free toxoplasms. They found that rifampin inhibited multiplication of *T. gondii* within cultured L-cells, but suggested that this may have been due to the toxic effects of the drug on the cultured host cells. Bedrnik[322] examined the activity of five coccidiostatic drugs used in the poultry industry to control *Eimeria* infections in chickens on *T. gondii* in HeLa cell cultures. Buquinolate, clopidol, and VUFB 6207 had no effect on *T. gondii*, whereas nicarbazin and robenziden suppressed its intracellular multiplication. Roberts et al.[332] used the plaque assay in cell cultures to evaluate the activity of various chemotherapeutic agents of *T. gondii*. Relatively high doses of sulfisoxale, cleocin, and aureomycin were required in order to inactivate *Toxoplasma*, whereas a relatively low dose of lincocin inhibited plaque formation by 100%. Melton and Sheffield[324] found that ormetoprim, sulfadimethoxine, and a combination of the two drugs were less effective in inhibiting penetration and multiplication of *T. gondii* in cultured cells than lasalocid or monensin. Sheffield and Melton[333] reported that pyrimethamine inhibited *T. gondii* multiplication in cultured cells and caused considerable structural changes. A synergistic effect occurred when sulfadiazine and pyrimethamine were combined, whereas sulfadiazine alone had no effect. *Toxoplasma* organisms became rounded, usually had fragmented nuclei and had abnormal daughter membrane formation during endodyogeny.

Pfefferkorn[325] and Pfefferkorn and Pfefferkorn[326-330] have used nitroso-guanidine or 5-

fluorodeoxyuridine to obtain drug-resistant and temperature-sensitive mutants of *T. gondii* in order to facilitate biochemical and genetic studies. By using a single treatment of *N*-methyl-*N*-nitro-*N*-nitrosoguanidine, Pfefferkorn and Pfefferkorn[326]obtained a mutant of *T. gondii* which was 50 times more resistant to adenine arabinoside, which inhibits synthesis of DNA by intracellular *T. gondii*. In a later study, Pfefferkorn and Pfefferkorn[329] found that the mutant and wild type had similar activities of adenosine deaminase, deoxyadenosine deaminase, and deoxyadenosine kinase, whereas the mutant had less than 0.1% of the adenosine kinase activity present in the wild type *T. gondii*. Based upon these findings, they suggested that the mutant was resistant to adenine arabinoside because without adenosine kinase to phosphorylate adenine arabinoside it could not convert the analogue to the active form, adenine arabinoside triphosphate.

Mutagenesis with *N*-methyl-*N*'-nitro-*N*-nitrosoguanidine was found by Pfefferkorn and Pfefferkorn[327] to yield seven temperature sensitive mutants *T. gondii*, all of which lost the ability to form plaques in cultured cells at 40°C but still grew at 33°C. Mutants were also markedly less virulent for mice than the wild type RH strain of *T. gondii*. Pfefferhorn[325] studied the enzymic defect of a mutant resistant to 5-fluorodeoxyuridine. The resistant mutant and the wild type RH strain had little or no uridine kinase; however, the mutant had about 100 times less uracil phosphoribosyltransferase than the wild type. A back mutant of the resistant strain regained a normal amount of uracil phosphoribosyltransferase and sensitivity to fluorodeoxyuridine. Such findings indicated that the wild type RH strain of *T. gondii* incorporates deoxyuridine into nucleic acids by phosphorolysis to produce uracil which is converted to uridylic acid by uracil phosphoribosyltransferase.[325]

E. Immunology

Cell cultures have been used to obtain parasite antigens for serologic tests or immunization, and to evaluate the roles which specific and nonspecific serum factors, lymphocytes, and phagocytic cells may play in host immune responses to coccidia.

1. Antigens and Immunization

Braveny et al.[140] described a technique for obtaining large quantities of *T. gondii* in human larynx carcinoma cells maintained in minimal essentials medium with 5% fetal calf serum 200,000 IU/ℓ penicillin and 20 mg/ℓ sulmycin. The greatest number of in vitro-produced zoites were obtained when an inoculum of 2.5 to 5 $\times$ 10^6 *T. gondii* zoites/Roux flask was used. At 2 to 3 days after inoculation, 150 to 200 times the amount of *Toxoplasma* inoculated could be harvested from the cultures.

Toxoplasma antigens obtained from cell cultures are more sensitive as determined by the complement fixation test than are antigens from the peritoneal exudate of mice.[348,349] Also, in contrast to peritoneal exudate antigen, it is easier to prepare larger quantities of *T. gondii* antigen from cell cultures and such antigens have been found to be more stable when stored at 4°C.[348]

There are only a few reports of attempts to immunize experimental animals against *Toxoplasma* with parasites obtained from in vitro cultivation. Cysts of *Toxoplasma* can be maintained in the brain and other organs of animals experimentally infected with avirulent strains of *T. gondii*. Organisms obtained from such animals can be used to immunize other animals, but the procedures are laborious. Zoites of an avirulent strain from cell cultures are capable of protecting mice and guinea pigs against lethal challenge with a virulent *Toxoplasma* strain.[350] Mice immunized with fractions obtained from disrupted *T. gondii* zoites as well as 200 μg of *Toxoplasma* RNA were found by Araujo and Remington[351] to resist challenge with *Toxoplasma* 30 days later.

2. Serum Factors

Various interferon preparations appear to be inconsistent in their ability to inhibit intracellular multiplication of *Toxoplasma* and *Eimeria* species. A crude rabbit interferon preparation has been found to inhibit *Toxoplasma* replication in rabbit kidney cells, whereas a similar mouse interferon preparation does not inhibit *Toxoplasma* multiplication in mouse L cells.[352] Treatment of cultured cells with interferon or polyinosinic-polycytidylic acid (In-Cn), an interferon inducer of cultured mammalian cells, significantly inhibited in vitro development of *E. tenella* sporozoites.[207] Chick or mouse monolayers treated with interferon were significantly protected from destruction by *Toxoplasma*.[353] Pretreatment of certain mammalian cell cultures with In-Cn will induce the formation of interferon within 24 hr. Even so, inhibition of *T. gondii* within such cultures did not begin until 72 hr.[352] Thus, these studies with *E. tenella* and *Toxoplasma* suggest that In-Cn and interferon may inhibit parasite multiplication by a mechanism other than that directly attributed to interferon.

Specific antiserum has been found to inhibit sporozoite or tachyzoite penetration as well as multiplication of *Eimeria* and *Toxoplasma* species within cultured cells.[183,203,354-358] Kambara et al.[356] obtained cyst-like bodies of *Toxoplasma* by overlaying HeLa cell cultures with rat plasma in the presence of an appropriate amount of specific antiserum. High titer antiserum caused a suppression in the formation of the cyst-like bodies. Shimada et al.[358a] found that antibody would enhance the formation of cysts containing bradyzoites in vitro as shown by a daily progression in the number of cysts in rabbit corneal cells. However, recently, Hoff et al.[358b] found that antibody was not necessary for the initiation of cyst formation. Thus, even though specific antibody appears not to be necessary for the initiation of cyst formation, it will enhance the number of *Toxoplasma* cysts formed in cultured cells.

After treatment in specific antiserum, *Toxoplasma* tachyzoites lose their ability to penetrate cultured cells[356] and many of those treated with immune serum plus complement collapse, have irregular vacuoles on their surfaces, and extrude their cytoplasmic contents.[183]

Davis and Porter[359] used the ability of *E. tenella* sporozoites to penetrate and develop normally in cultured chick kidney cells as an indicator of the anticoccidial effects of sera and extracts of cecal contents or mucosal tissue. Pretreatment of *E. tenella* sporozoites with normal serum enhanced the frequency of intracellular development, whereas pretreatment in balanced salt solution without protein did not interfere with the ability of the parasite to penetrate cells but did damage sporozoites so that most of them lost the ability to differentiate. Similar results of inhibition of intracellular multiplication of *E. tenella* were obtained by incubating sporozoites in extracts of cecal contents, whereas no inhibition occurred in the presence of a mucosal extract. Thus, sporozoites can evidently lose the ability to differentiate even before penetrating host cells. Davis and Porter[359] suggested that there may be a possible synergism between nonspecific factors and secretory antibodies in anticoccidial immunity.

Based upon their studies of secretory IgA in anticoccidial immunity in the chicken[360] and their observations of inhibition of multiplication of *E. tenella* sporozoites in cultured cells by immune serum and cecal extracts, Davis and Porter[359] proposed the following mechanism of secretory IgA-mediated anticoccidial immunity in the chicken. Supepithelial plasma cells secrete IgA, anti-sporozoite antibodies, which enters the epithelial cell. Within the epithelial cell secretory IgA (sIgA) is assembled from IgA and secretory component, which is synthesized in the cytoplasm of the epithelial cell.[361] Secretory IgA then leaves the epithelial cell and is present in the mucin on the mucosal surface of the cecum. Sporozoites become attached to the sIgA antibodies on the mucosal surface, during which time they are held stationary making them susceptible to damage by proteolytic enzymes or other substances in the cecal contents. Most of these sporozoites would presumably be unable to penetrate host cells and most of those which did enter cells would probably have lost their ability to differentiate. If such a mechanism does indeed exist, then IgA appears to have gained more significance in anticoccidial immunity than previously considered.

3. Cellular Mechanisms

a. Lymphokines

Various soluble mediators of cellular immunity (lymphokines) have been studied in in vitro and in vivo immune responses to *Toxoplasma* and *Besnoitia*. These include growth inhibiting factor (GIF),[362,363] *Toxoplasma* growth inhibiting factor (Toxo-GIF)[364] and anti-*Toxoplasma* arming factor (ATAF).[365] Chinchilla and Frenkel[362] and Frenkel and Chinchilla[363] found that GIF from antigen-treated lymphocytes from immune hamsters would specifically protect not only cultured peritoneal macrophages, but also cultured fibroblasts and kidney cells infected with *T. gondii* or *B. jellisoni*. Cultured kidney cells infected with *Besnoitia* or *Toxoplasma* and treated with homospecific GIF were better protected than with heterospecific GIF.[362] Intracellular multiplication of both parasites was inhibited by GIF. Macrophages were not required for protection of cultured cells or for the elaboration of the GIF in lymphocyte cultures. Immune lymphocytes needed only to be exposed to parasite antigen in order to produce GIF. Growth inhibiting factor afforded better protection to hamsters injected with *Besnoitia* than did specific antiserum. Chinchilla and Frenkel[362] suggested that GIF might be useful for clinical treatment of toxoplasmosis.

Sethi and Brandis[365] found that nonimmune thioglycollate stimulated mouse peritoneal macrophages could be armed to exert anti-toxoplasma activity with a soluble factor (ATAF) obtained from cultures of specifically stimulated T-cells from animals immune to *Toxoplasma*. This lymphokine appears to be different than GIF. Growth inhibiting factor (GIF) has a molecular weight of 4,000 to 5,000 and is resistant to RNase and DNase and to heat at 56°C for 30 min, but is sensitive to chymotrypsin.[362,363] Antitoxoplasma arming factor (ATAF) has a molecular weight of 50,000 to 100,000, is resistant to RNase and DNase, but can be inactivated by trypsin or heating at 58°C for 30 min.[365]

Shirahata et al.[364] found that Toxo-GIF was evidently produced by specifically sensitized T-cells and that concanavalin A and phytohemagglutinin P increased production of Toxo-GIF by T-lymphocytes, whereas bacterial lipopolysaccharide had no effect. Nonimmune mouse peritoneal mouse macrophages treated with Toxo-GIF inhibited intracellular multiplication of *T. gondii*. Further study may prove that Toxo-GIF is different or identical to GIF or ATAF.

Bedrnik[354] found that spleen lymphocytes obtained from chickens immune to *E. tenella* were unable to prevent development of *E. tenella* sporozoites in chicken kidney cell cultures.

b. Macrophages

Normal, nonspecifically and specifically activated macrophages from various hosts have been examined in vitro for their ability to phagocytose and destroy various species of coccidia. Many coccidia are known to be phagocytosed by macrophages within tissues of the host[366-368] but the roles that macrophages play in anticoccidian immunity are not well understood.

The specificity of macrophage-related immunity is difficult to assess because nonspecifically activated macrophages may be appreciably more inhibitory to parasites than normal macrophages.[369] However, specifically armed (or activated) macrophages usually show significantly greater inhibition than nonspecifically activated ones.[199,370] Macrophages from immune animals have been shown to be more effective in killing *T. gondii* zoites than those from nonimmune animals.[199,370-372] Peritoneal macrophages from guinea pigs chronically infected with *T. gondii* became significantly more resistant to *Listeria monocytogenes* when such macrophage cultures were incubated with sensitized spleen cells and *T. gondii* antigen.[373] Normal macrophage cultures infected with mycoplasmas have increased microbicidal activity for *T. gondii* as well as *L. monocytogenes* by intracellular killing.[374] Peritoneal macrophages from mice chronically infected with *Toxoplasma* have the capacity to nonspecifically destroy tumor cells in vitro by a nonphagocytic mechanism.[375,376] Such cytotoxicity may be due to lysosomal enzymes on the surfaces of activated macrophages. Macrophages are known to secrete lysosomal enzymes and then recapture them upon their surfaces.[377] Macrophages from animals infected with *Toxoplasma* may be armed with more lysosomal enzymes than macrophages from normal mice.

Hoff and Frenkel[371] found that peritoneal macrophages from *Besnoitia*-immune hamsters delayed growth of homologous parasites about 3 times longer than growth of *Besnoitia* in *Toxoplasma*-immune or nonimmune macrophages. However, *Besnoitia*-immune cultures reduced *Toxoplasma* growth by one-half. Specifically committed lymphocytes could instruct macrophages through close lymphocyte-macrophage proximity to reduce the homologous organisms 10-fold, whereas heterologous organisms were reduced twofold. Lymphocyte supernatants could initiate hypersensitivity as indicated by macrophage activation and giant cell formation in culture, indicating that lymphokines might participate in hypersensitivity. Nguyen and Stadtsbaeder[378] found that lymphocytes and syngeneic macrophages spontaneously form rosettes in culture which are not caused by immunoglobulin receptors or "cytophilic antibody". Percentages of rosette formation were the same irrespective whether thymic lymphocytes came from normal or *Toxoplasma*-immune mice. Berman and Johnson[379] reported that phagocytosis and intracellular multiplication of *T. gondii* were identical in normal placental and adult monocytes. However, *Toxoplasma* multiplication was significantly inhibited by both types of monocytes preincubated with supernatants from sensitized lymphocytes or *Toxoplasma* antigen.

Various sera have been examined for their effects on phagocytosis and intracellular fate of *Toxoplasma* and certain *Eimeria* species within macrophages. Rose[380] found that immune serum from chickens infected with *E. tenella* or *E. maxima* had both "cytophilic" and opsonic adherence properties and that the opsonic property was species specific. Peak activity of immunity coincided with the onset of complete immunity to infection in the host. Although the uptake of *E. tenella* sporozoites by peritoneal exudate cells (heterophils and macrophages) from immunized birds was greater than by cells from susceptible birds, no differences in appearance of phagocytosed parasites were evident by electron microscopy.[381]

Macrophages activated with new born calf serum have a markedly enhanced capacity to phagocytose *T. gondii* zoites as compared to macrophages cultured in fetal calf serum.[382] However, their ability to restrict intracellular multiplication of *T. gondii* was not markedly increased.

Alveolar and peritoneal macrophages treated with specific antiserum have been found to have an increased rate of phagocytosis and intracellular killing of *Toxoplasma* zoites.[199,355,383] Shirahata et al.[384] found that antitoxoplasma antibodies did not enhance phagocytic activity of *T. gondii* by cultured macrophages; however, activated and normal macrophages treated with fresh or heat-inactivated immune serum significantly inhibited intracellular multiplication of *T. gondii*.

The intracellular fate of *Toxoplasma* has been shown by Akinshina et al.[385] to depend also on the virulence of the *T. gondii* strain. The Lagrave strain of *T. gondii* has low virulence and undergoes disintegration soon after entering macrophages and are completely digested by 24 to 48 hr after inoculation. In contrast, zoites of the highly virulent RH strain of *Toxoplasma* undergo multiplication and cause considerable alteration in nuclei of infected macrophages.[384]

Jones and Hirsch[200] studied the intracellular fate of live and glutaraldehyde-killed *Toxoplasma* zoites in peritoneal macrophages, the lysosomes of which were labeled with thorium dioxide or ferritin. They found that killed *Toxoplasma* zoites were destroyed within phagolysosomes, as indicated by the presence of thorium dioxide or ferritin within the vacuole surrounding the parasite. Live parasites were present within phagosomes but escaped destruction by somehow interfering with the formation of phagolysosomes. The phagocytic vacuolar membrane appeared altered by the appearance of microvillus protrusions into the phagosome. Mitochondria and strips of endoplasmic reticulum were located in the cytoplasm immediately outside of the phagosome. Jones and Hirsch[200] suggested that the above cellular alterations may have interfered with the formation of phagolysosomes.

McLeod and Remington[386] devised a method to determine the extent of intracellular inhibition or killing of *T. gondii* zoites by mononuclear phagocytes by measuring the incorporation of tritium-labeled deoxyuridine ((6-^{3}H) UdR) into nucleic acids of intracellular *Toxoplasma*. Extracellular and inhibited or killed intracellular zoites of *T. gondii* do not incorporate significant amounts of (6-^{3}H) UdR.

VI. ADDENDUM

Antibodies labeled with visual markers such as ferritin, horseradish peroxidase, viruses, and hemocyanin have recently been used extensively to help understand a variety of biological phenomena. In the field of immunoparasitology, labeled antibodies for electron microscopy are now being used to answer questions about the arrangement of antigenic sites on the surfaces of parasites as well as to characterize the interactions of antibodies including monoclonal antibodies with parasites.[392-394] Monoclonal antibodies may be used to isolate protective parasite antigens which may be eventually incorporated into vaccines to combat parasitic disease or to identify portions of the parasite genome responsible for the production of protective antigens.

Recently, Danforth[395] reported the successful establishment of 10 cloned antibody-secreting cell lines. The hybridoma cell lines were produced by fusing P3-X63-Ag8 myeloma cells with splenocytes from BALB/cByJ mice which had been previously immunized with *E. tenella* or *E. mitis* sporozoites. Monoclonal antibodies exhibited eight different binding patterns on or in air-dried sporozoites as determined by indirect immunofluorescence. Binding patterns included a general internal fluorescence, fluorescence of the apical end, pellicle, and/or refractile refractile body of the sporozoite. Most of the monoclonal antibodies developed against *E. tenella* sporozoites were found to be stage specific and reacted with only the sporozoite or early trophozoite, whereas only two clones produced antibodies that cross reacted with mature merozoites.

Ionophorous antibiotics, such as monensin, lasalocid, and narasin, are effective anticoccidial compounds which appear to be capable of killing intracellular parasites without injuring host cells.[396] These antibiotics possess unique physical properties making them capable of forming complexes with cations and transporting them into and through biological membranes.[397] McDougald and Galloway[339] found that treatment of extracellular sporozoites of *E. tenella* with monensin had no effect on subsequent intracellular development of the parasite in vivo. However, Smith and Strout[396,398] and Smith et al.[399] showed that pretreatment of extracellular sporozoites of *E. tenella* with lasalocid, monensin, narasin, or salinomycin severely curtailed subsequent intracellular asexual development of the parasite in cell culture. Smith et al.[400] studied monensin-treated sporozoites by scanning and transmission electron microscopy and found that sporozoites had irregular surfaces and often exhibited a gross swelling. Although the specific lethal mechanism of anticoccidial ionophorous antibiotics has not yet been determined, a theory has been proposed that only intracellular stages are susceptible to their actions.[396] However, Smith et al.[399] showed that extracellular *E. tenella* sporozoites may incorporate a potentially lethal concentration of ionophorous antibiotics and that a coccidiocidal activity may be expressed whether or not penetration of the host cell occurs.

Augustine[400] and Augustine and Danforth[401] studied the effects of various substances on the penetration of cultured cells by sporozoites of *E. meleagrimitis*. Cell cultures treated with poly-L-histidine, poly-L-lysine, Ca^{++} or chmotrypsin contained significantly fewer intracellular sporozoites than untreated control cultures, whereas heparin, dextran, sulfate, or trypsin did not significantly affect the number of intracellular parasites.[400] The finding that fewer sporozoites occurred in cells treated with chymotrypsin, whereas trypsin had no effect, indicates that chymotrypsin-sensitive membrane receptors may play an important role in sporozoite penetration. Miller et al.[402] found that chymotrypsin but not trypsin inhibited invasion of human erythrocytes by *Plasmodium knowlesi* merozoites and suggested the presence of erythrocyte membrane receptors for *Plasmodium* species. These receptors were later found by Miller et al.[403] to belong to the positive Duffy blood group determinants. Augustine and Danforth[401] studied the effects of four lectins on penetration of cultured cells by *E. meleagrimitis* sporozoites. Pokeweed mitogen and castor bean agglutinin had no effect, whereas phytohemagglutinin-treated cultures contained significantly greater numbers of intracellular sporozoites than control cultures and wheat germ agglutinin significantly reduced penetration.

ACKNOWLEDGMENTS

I thank Ms. Diane Stanich, Mr. Andrew Olcott, and Ms. Judy Sanderson for technical assistance in preparation of the manuscript, and Drs. J. P. Dubey and Steven Reed for their suggestions concerning the content of this chapter.

REFERENCES

1. **Levine, N. D.,** Introduction, history, and taxonomy, in *The Coccidia,* Hammond, D. M. and Long, P. L., Eds., University Park Press, Baltimore, 1973, 1.

1a. **Levine, N. D., Corliss, J. O., Cox, F. E. G., Deroux, G., Grain, J., Honigberg, B. M., Leedale, G. F., Loeblich, III, A. R., Lom, J., Lynn, D., Merinfeld, E. G., Page, F. C., Poljansky, G., Sprague, V., Vavra, J., and Wallace, F. G.,** A newly revised classification of the Protozoa, *J. Protozool.*, 27, 37, 1980.

2. **Levine, N. D.,** Nomenclature of *Sarcocystis* in the ox and sheep and of fecal coccidia of the dog and cat, *J. Parasitol.*, 63, 36, 1977.

3. **Levine, N. D.,** Taxonomy of *Toxoplasma, J. Protozool.*, 24, 36, 1977.

4. **Hendricks, L. D., Ernst, J. V., Courtney, C. H., and Speer, C. A.,** *Hammondia pardalis* sp. n. (Sarcocystidae) from the ocelot, *Felis paradalis,* and experimental infection in other felines, *J. Protozool.*, 26, 39, 1979.

5. **Dubey, J. P.,** *Toxoplasma, Hammondia, Besnoitia, Sarcocystis,* and other tissue cyst-forming coccidia of man and animals, in *Parasitic Protozoa, Vol. 3, Kreier, J. P., Ed., Academic Press, New York, 1977, 101.*

5a. **Frenkel, J. K.,** *Besnoitia wallacei* of cats and rodents: With a reclassification of other cyst-forming isosporoid coccidia, *J. Parasitol.*, 63, 611, 1977.

6. **Ruff, M. D. and Reid, W. M.,** Avian coccidia, in *Parasitic Protozoa,* Vol. 3, Academic Press, New York, 1977, 34.

7. **Wang, C. C. and Stotish, R. L.,** Pancreatic chymotrypsin as the essential enzyme for excystation of *Eimeria tenella, J. Parasitol.*, 61, 923, 1975.

8. **Hammond, D. M.,** Ultrastructure and Development of Coccidia, in *Proc. Symp. Coccidia and Related Organisms,* Guelph, Ontario, Canada, 1973, 11.

9. **Pellérdy, L. P.,** *Coccidia and Coccidiosis,* Verlag Paul Parey, Berlin, 1974.

10. **Box, E. D.,** Exogenous stages of *Isospora serini* (Aragão) and *Isospora canana* sp. n. in the canary *(Serinus canarius* Linnaeus), *J. Protozool.*, 22, 165, 1975.

11. **Brandborg, L. L., Goldberg, S. B., and Breidenback, W. C.,** Human coccidiosis — a possible cause of malabsorption, *N. Engl. J. Med.*, 283, 1306, 1970.

12. **Trier, J. S., Moxey, P. C., Schimmel, E. M., and Robles, E.,** Chronic intestinal coccidiosis in man: intestinal morphology and response to treatment, *Gastroenterology,* 66, 923, 1974.

13. **Dubey, J. P.,** Experimental *Isospora canis* and *Isospora felis* infection in mice, cats, and dogs, *J. Protozool.*, 22, 416, 1975.

14. **Dubey, J. P.,** Life cycle of *Isospora rivolta* (Grassi, 1879) in cats and mice, *J. Protozool.*, 26, 433, 1979.

15. **Dubey, J. P. and Mehlhorn, H.,** Extraintestinal stages of *Isospora ohioensis* from dogs in mice, *J. Parasitol.*, 64, 689, 1978.
16. **Frenkel, J. K. and Dubey, J. P.,** Rodents as vectors of feline coccidia *Isospora felis* and *Isospora rivolta*, *J. Infect. Dis.*, 125, 69, 1972.
17. **Mehlhorn, H. and Markus, M. B.,** Electron microscopy of stages of *Isospora felis* of the cat in the mesenteric lymph node of the mouse, *Z. Parasitenkd.*, 51, 15, 1976.
18. **Wallace, G. D. and Frenkel, J. K.,** *Besnoitia* species (Protozoa, Sporozoa, Toxoplasmatidae): recognition of cyclic transmission by cats, *Science*, 188, 369, 1975.
19. **Dubey, J. P. and Frenkel, J. K.,** Cyst-induced toxoplasmosis in cats, *J. Protozool.*, 19, 155, 1972.
20. **Frenkel, J. K., Dubey, J. P., and Miller, N. L.,** *Toxoplasma gondii* in cats: fecal stages identified as coccidian oocysts, *Science*, 167, 893, 1970.
21. **Hutchison, W. M., Dunachie, J. F., Siim, Chr., and Work, K.,** Coccidian-like nature of *Toxoplasma gondii*, *Br. Med. J.*, 1, 142, 1970.
22. **Rommel, M.,** Vergleichende Dorstellung der Entwicklungsbiologie der Gattungen *Sarcocystis, Frenkelia, Isospora, Cystoisospora, Hammondia, Toxoplasma*, and *Besnoitia*, *Z. Parasitenkd.*, 57, 269, 1978.
23. **Basson, P. A., McCully, R. M., and Bigalke, R. D.,** Observations on the pathogenesis of bovine and antelope strains of *Besnoitia besnoiti* (Marotel, 1912) infection in cattle and rabbits, *Onderstepoort J. Vet. Res.*, 37, 105, 1970.
24. **Chobotar, B., Anderson, L. C., Ernst, J. V., and Hammond, D. M.,** Pathogenicity of *Besnoitia jellisoni* in naturally infected kangaroo rats *(Dipodomys ordii)* in northwestern Utah, *J. Parasitol.*, 56, 192, 1970.
25. **Ernst, J. V., Chobotar, B., Oaks, E. C., and Hammond, D. M.,** *Besnoitia jellisoni* (Sporozoa: Toxoplasmea) in rodents from Utah and California, *J. Parasitol.*, 54, 545, 1968.
26. **Webster, G.,** Besnoitiosis in a woodland caribou, *J. Wildl. Dis.*, 12, 566, 1976.
27. **Frenkel, J. K.,** Pathogenesis of toxoplasmosis and of infections with organisms resembling *Toxoplasma*, *Ann. N.Y. Acad. Sci.*, 64, 215, 1956.
28. **Schneider, C. R.,** Susceptibility of the marmoset, *Saguinus geoffroyi* Pucheran, to interperitoneal and oral infections with *Besnoitia* (Protozoa: Toxoplasmea), *J. Parasitol.*, 53, 1135, 1967.
29. **Bigalke, R. D.,** New concepts on the epidemiological features of bovine besnoitiosis as determined by laboratory and field investigations, *Onderstepoort J. Vet. Res.*, 35, 3, 1968.
30. **Fitzgerald, P. R.,** The economics of bovine coccidiosis, *Feedstuffs*, 44, 28, 1972.
31. **Rommel, M. and Krampitz, H. E.,** Beitrage zum Lebenszyklus der Frenkelian I. Die Identitat von *Isospora buteonis* aus dem Mausebussard mit einer Frenkelienart (*F. clethrionomyobuteonis* spec. n.) aus der Rotelmaus, *Berl. Munch. Tierarztl. Wochenenschr.*, 88, 338, 1975.
32. **Tadros, W. and Laarman, J. J.,** *Sarcocystis* and related coccidian parasites: a brief general review, together with a discussion on some biological aspects of their life cycles and a new proposal for their classification, *Acta Leidensia*, 44, 1, 1976.
33. **Krampitz, H. E., Rommel, M., Geisel, O., and Kaiser, E.,** Beitrage zum Lebenszyklus der Frenkelien. II. die ungeschlechtliche entwicklung von *Frenkelia clethrionomyobuteonis* in der Rotelmaus, *Z. Parasitenkd.*, 51, 7, 1976.
34. **Rommel, M., Krampitz, H. E., Gobel, E., Geisel, O., and Kaiser, E.,** Untersuchungen Uber den Lebenszylus von *Frenkelia clethrionomyobuteonis*, *Z. Parasitenkd.*, 50, 204, 1976.
35. **Frenkel, J. K. and Dubey, J. P.,** *Hammondia hammondi* gen. nov., sp. nov., from domestic cats, a new coccidian related to *Toxoplasma* and *Sarcocystis*, *Z. Parasitenkd.*, 46, 3, 1975.
36. **Dubey, J. P.,** Experimental *Hammondia hammondi* infection in dogs, *Br. Vet. J.*, 131, 741, 1975.
37. **Dubey, J. P. and Wong, M. M.,** Experimental *Hammondia hammondi* infection in monkeys, *J. Parasitol.*, 64, 551, 1978.
38. **Mason, R. W.,** The detection of *Hammondia hammondi* in Australia and the identification of a free-living intermediate host, *Z. Parasitenkd.*, 57, 101, 1978.
39. **Wallace, G. D.,** Observations on a feline coccidian with some characteristics of *Toxoplasma* and *Sarcocystis*, *Z. Parasitenkd.*, 46, 167, 1975.
40. **Dubey, J. P. and Streitel, R. H.,** Further studies on the transmission of *Hammondia hammondi* in cats, *J. Parasitol.*, 62, 548, 1976.
41. **Jarpa, G. A.,** Coccidiosis humana, *Biologica*, 39, 3, 1966.
42. **Jeffery, H. C.,** Sarcosporidiosis in man, *Trans. R. Soc. Trop. Med. Hyg.*, 68, 17, 1974.
43. **Markus, M. B.,** *Sarcocystis* and sarcocystosis in domestic animals and man, *Adv. Vet. Sci. Comp. Med.*, 22, 159, 1978.
44. **Fayer, R.,** Economic losses to *Sarcocystis*, *Natl. Wool Grower*, 66, 22, 1976.
45. **Fayer, R., Johnson, A. J., and Lunde, M.,** Abortion and other signs of disease in cows experimentally infected with *Sarcocystis fusiformis* from dogs, *J. Infect. Dis.*, 134, 624, 1976.
46. **Frelier, P., Mayhew, I. G., Fayer, R., and Lunde, M. N.,** *Sarcocystis:* a clinical outbreak in dairy calves, *Science*, 195, 1341, 1976.

47. **Leek, R. G. and Fayer, R.**, Sheep experimentally infected with *Sarcocystis* from dogs. II. Abortion and disease in ewes, *Cornell Vet.*, 68, 108, 1978.
48. **Leek, R. G., Fayer, R., and Johnson, A. J.**, Sheep experimentally infected with *Sarcocystis* from dogs. I. Disease in young lambs, *J. Parasitol.*, 63, 642, 1977.
49. **Munday, B. L., Barker, I. K., and Rickard, M. D.**, The developmental cycle of a species of *Sarcocystis* occurring in dogs and sheep, with observations on pathogenicity in the intermediate host, *Z. Parasitenkd.*, 46, 111, 1975.
50. **Dubey, J. P., Fayer, R., and Seesee, F. M.**, *Sarcocystis* in coyotes from Montana: prevalence and experimental transmission to sheep and cattle, *J. Am. Vet. Med. Assoc.*, 173, 1167, 1978.
51. **Fayer, R. and Johnson, A. J.**, *Sarcocystis fusiformis* infection in the coyote *(Canis latrans)*, *J. Infect. Dis.*, 131, 189, 1975.
52. **Hudkins, G. and Kistner, T. P.**, *Sarcocystis hemionilatrantis* (sp. n.) life cycle in mule deer and coyotes, *J. Wildl. Dis.*, 13, 80, 1977.
53. **Hudkins, G., Kistner, T. P., and Fayer, R.**, Possible species differences between *Sarcocystis* from mule deer and cattle, *J. Wildl. Dis.*, 12, 86, 1976.
54. **Karstad, L. and Trainer, D. O.**, *Sarcocystis* in white-tailed deer, *Bull. Wildl. Dis. Assoc.*, 5, 25, 1969.
55. **Kelley, A. L., Penner, L. R., and Pickard, R. J.**, *Sarcocystis* in the mouse, *J. Mammal.*, 31, 462, 1950.
56. **Mills, H. B.**, Observations on Yellowstone elk, *J. Mammal.*, 7, 250, 1936.
57. **Pond, D. B. and Speer, C. A.**, *Sarcocystis* in free-ranging herbivores on the National Bison Range, *J. Wildl. Dis.*, 15, 51, 1979.
58. **Scott, J. W.**, Economic importance of Sarcosporidia, with special reference to *Sarcocystis tenella*, *Univ. Wyo. Ag. Exp. Sta. Bull.*, 262, 1943.
59. **Speer, C. A., Pond, D. B., and Ernst, J. V.**, Development of *Sarcocystis hemionilatrantis* Hudkins and Kistner, 1977 in the small intestine of coyotes, *Proc. Helm. Soc. Wash.*, 47, 106, 1980.
60. **Heydorn, A. O. and Rommel, M.**, Beitrage zum Lebenzyklus der Sarkosporidien. II. Hund and Katze als Ubertrager der Sarkosporidien des Rindes, *Berl. Munch. Tieraerztl. Wochenschr.*, 85, 333, 1972.
61. **Fayer, R.**, Development of *Sarcocystis fusiformis* in the small intestine of the dog, *J. Parasitol.*, 60, 660, 1974.
62. **Gestrich, R., Heydorn, A. O., and Baysu, N.**, Beitrage zum Lebenszyklus der Sarkosporidien. II. Untersuchungen zur Artendifferenzierung bei *Sarcocystis fusiformis* und *Sarcocystis tenella*, *Berl. Munch. Tierarztl. Wochenschr.*, 88, 191, 1975.
63. **Johnson, A. J., Hildebrandt, P. K., and Fayer, R.**, Experimentally induced *Sarcocystis* infection in calves: pathology, *Am. J. Vet. Res.*, 36, 995, 1975.

63a. **Dubey, J. P., Weisbrode, S. E., Speer, C. A., and Sharma, S. P.**, Sarcocystosis in goats: Clinical signs and pathologic and hematologic findings, *J. Am. Vet. Med. Assoc.*, 178, 683, 1981.

63b. **Speer, C. A. and Dubey, J. P.**, Scanning transmission electron microscopy of ovine mesenteric arteries infected with first-generation meronts of *Sarcocystis tenella*, *Can. J. Zool.*, 60, 203, 1982.

64. **Ruiz, A. and Frenkel, J. K.**, Recognition of cyclic transmission of *Sarcocystis muris* by cats, *J. Infect. Dis.*, 133, 409, 1976.
65. **Nicolle, C. and Manceaux, L.**, Sur une infection á corps de Leishman (ou organisms voisins) du gondi, *C. R. Hebd. Seances Acad. Sci.*, 147, 763, 1908.
66. **Beverley, J. K. A.**, Toxoplasmosis, *Br. Med. J.*, 2, 475, 1973.
67. **Beverley, J. K. A. and Watson, W. A.**, Ovine abortion and toxoplasmosis in Yorkshire, *Vet. Rec.*, 73, 6, 1961.
68. **Beverley, J. K. A. and Watson, W. A.**, Prevention of experimental and of naturally occurring ovine abortion due to toxoplasmosis, *Vet. Rec.*, 88, 39, 1971.
69. **Hartley, W. J. and Moyle, G.**, Observations on an outbreak of ovine congenital toxoplasmosis, *Aust. Vet. J.*, 44, 105, 1968.
70. **Harding, J. D. J., Beverley, J. K. A., Shaw, I. G., Edwards, B. L., and Bennet, G. H.**, *Toxoplasma* in English pigs, *Vet. Rec.*, 73, 3, 1961.
71. **Work, K., Eriksen, L., Fennestad, K. L., Moller, T., and Siim, J. C.**, Experimental toxoplasmosis in pregnant sows. I. Clinical, pathological and serological observations, *Acta Pathol. Microbial. Scand.*, B78, 129, 1970.
72. **Koestner, A. and Cole, C. R.**, Neuropathology of canine toxoplasmosis, *Am. J. Vet. Res.*, 21, 831, 1960.
73. **Frenkel, J. K.**, Pursuing toxoplasma, *J. Infect. Dis.*, 122, 553, 1970.
74. **Dubey, J. P., Miller, N. L., and Frenkel, J. K.**, Characterization of the new fecal form of *Toxoplasma gondii*, *J. Parasitol.*, 56, 447, 1970.
75. **MacGowen, J. P.**, Some points relating to the morphology and development of *Sarcocystis tenella*, *Parasitology*, XV, 139, 1923.
76. **Triffitt, M. J.**, Observations on *Gastrocystis gilruthi*, a parasite of sheep in Britain, *Protozoology*, 1, 7, 1925.

77. **Levaditi, C., Sanchis-Bayarri, V., Lepine, P., and Schoen, R.,** Etude sur l'encephalomyelite provoquee par le *Toxoplasma caniculi, Ann. Inst. Pasteur, Paris,* 43, 673, 1929.
78. **Cook, M. K. and Jacobs, L.,** Cultivation of *Toxoplasma gondii* in tissue cultures of various derivations, *J. Parasitol.,* 44, 172, 1958.
79. **Scott, J. W.,** Life history of Sarcosporidia, with particular reference to *Sarcocystis tenella, Univ. Wyo. Agr. Exp. Sta. Bull.,* 259, 1, 1943.
80. **Spindler, L. A. and Zimmerman, H. F.,** The biological status of *Sarcocystis, J. Parasitol.,* 31 (Suppl.), 13, 1945.
81. **Pollack, R., Ed.,** *Readings in Mammalian Cell Culture,* Cold Spring Harbor Laboratory, New York, 1973, 3.
82. **Frenkel, J. K.,** Infections with organisms resembling *Toxoplasma,* together with the description of a new organism: *Besnoitia jellisoni, Atti Congr. Int. Microbiol.,* 5, 426, 1953.
83. **Frenkel, J. K.,** The development of the cyst of *Besnoitia jellisoni,* usefulness of this infection as a biological model, *Excerpta Med. Int. Congr. Ser.,* 91, 125, 1965.
84. **Bigalke, R. D.,** Preliminary communication on the cultivation of *Besnoitia besnoiti* (Morotel, 1912) in tissue culture and embryonated eggs, *J. S. Afr. Vet. Med. Assoc.,* 23, 523, 1962.
85. **Patton, W. H.,** *Eimeria tenella:* cultivation of the asexual stages in cultured animal cells, *Science,* 150, 767, 1965.
86. **Doran, D. J.,** Cultivation of coccidia in avian embryos and cell culture, in *The Coccidia: Eimeria, Isospora, Toxoplasma* and Related Genera, Hammond, D. M., and Long, P. L., Eds., University Park Press, Baltimore, 1973, 183.
87. **Long, P. L.,** Development of *Eimeria tenella* in avian embryos, *Nature (London),* 208, 509, 1965.
88. **Long, P. L.,** The growth of some species of Eimeria in avian embryos, *J. Parasitol.,* 56, 575, 1966.
89. **Long, P. L. and Tanielian, Z.,** The isolation of *Eimeria mivati* in Lebanon during the course of a survey of *Eimeria* spp. in chickens, *Lebanon Agr. Res. Inst.,* 6, 1, 1965.
90. **Fitzgerald, P. R.,** Development of *Eimeria stiedae* in avian embryos, *J. Parasitol.,* 56, 1252, 1970.
91. **Long, P. L.,** The growth of *Eimeria* in cultured cells and a chicken embryos: a review, in *Proc. Symp. Coccidia and Related Organisms,* Guelph, Ontario, Canada, University Press, 1973, 57.
92. **Long, P. L.,** Further studies on the pathogenicity and immunogenicity of an embryo-adapted strain of *Eimeria tenella, Avian Pathol.,* 3, 255, 1974.
93. **Long, P. L. and Millard, B. J.,** Pathogenicity, immunogenicity and site specificity of an attenuated strain of *Eimeria tenella, J. Protozool.,* 22, 53A, 1975.
94. **Rollinson, D.,** Development of *Eimeria tenella* in quail embryos, *Trans. R. Soc. Trop. Med. Hyg.,* 70, 21, 1976.
95. **Shirley, M. W., Millard, B. J., and Long, P. L.,** Studies on the growth, chemotherapy and enzyme variation of *Eimeria acervulina* var. *diminuta* and *E. acervulina* var. *mivati, Parasitology, 75, 165, 1977.*
96. **Davis, L. R.,** Techniques, in *The Coccidia: Eimeria, Isospora, Toxoplasma* and Related Genera, Hammond, D. M. and Long, P. L., Eds., University Park Press, Baltimore, 1973, 411.
97. **Edgar, S. A. and Seibold, C. T.,** A new coccidium of chickens, *Eimeria mivati* sp. n. (Protozoa: Eimeriidae) with deals on its life history, *J. Parasitol.,* 50, 193, 1964.
98. **Goff, O. E.,** Flowers of sulfur and charcoal in the prevention of experimentally produced coccidiosis, *Poult. Sci.,* 21, 23, 1942.
99. **Leek, R. G. and Fayer, R.,** Survival of sporocysts of *Sarcocystis* in various media, *Proc. Helm. Soc. Wash.,* 46, 151, 1979.
100. **Lotze, J. C. and Leek, R. G.,** A practical method for culturing coccidial oocysts in tap water, *J. Parasitol.,* 47, 588, 1961.
101. **Hammond, D. M. and Davis, L. R.,** An improved method for sporulating oocysts in bovine fecal material, *Am. J. Vet. Res.,* 5, 70, 1944.
102. **Lane, C.,** The mass diagnosis of ankylostome infestation, *Trans. R. Soc. Trop. Med. Hyg.,* 17L407, 1924.
103. **Jackson, A. R. B.,** The isolation of viable coccidial sporozoites, *Parasitology,* 54, 87, 1964.
104. **Sharma, M. N. and Reid, W. M.,** A cleaning method for coccidial oocysts using density gradient sedimentation, *J. Parasitol.,* 49, 159, 1963.
105. **Marquardt, W. C.,** Separation of nematode eggs from fecal debris by gradient centrifugation, *J. Parasitol.,* 47, 248, 1961.
106. **Patnaik, B.,** A technique of obtaining oocysts of coccidia in pure state from chicken feces by modified Marquardt's method, *Ind. Vet. J.,* 43, 414, 1966.
107. **Vetterling, J. M.,** Continuous-flow differential density flotation of coccidial oocysts and a comparison with other methods, *J. Parasitol.,* 55, 412, 1969.
108. **Stotish, R. L., Simashkevich, P. M., and Wang, C. C.,** Separation of sporozoites, sporocysts, and oocysts of *Eimeria tenella* by centrifugal elutriation, *J. Parasitol.,* 63, 1124, 1977.
109. **Hammond, D. M., Chobotar, B., and Ernst, J. V.,** Cytological observations on sporozoites of *Eimeria bovis* and *E. auburnensis,* and an *Eimeria* species from the Ord Kangaroo Rat, *J. Parasitol.,* 54, 550, 1968.

110. **Fayer, R. and Leek, R. G.,** Excystation of *Sarcocystis fusiformis* sporocysts from dogs, *Proc. Helminthol. Soc. Wash.*, 40, 294, 1973.
111. **Dubey, J. P., Miller, N. L., and Frenkel, J. K.,** The *Toxoplasma gondii* oocysts from cat feces, *J. Exp. Med.*, 132, 636, 1970.
112. **Dubey, J. P., Swan, G. V., and Frenkel, J. K.,** A simplified method for isolation of *Toxoplasma gondii* from the feces of cats, *J. Parasitol.*, 58, 1005, 1972.
113. **Dubey, J. P.,** Coyote a final host for *Sarcocystis* species of goats, sheep, cattle, elk, bison and moose in Montana, *Am. J. Vet. Res.*, in press.
114. **Monné, L. and Honig, G.,** On the properties of the shells of the coccidian oocysts, *Arkiv. Zool.*, 7, 251, 1954.
115. **Nyberg, P. A. and Knapp, S. E.,** Effect of sodium hypochlorite on the oocyst wall of *Eimeria tenella* as shown by electron microscopy, *Proc. Helminthol. Soc. Wash.*, 37, 32, 1970.
116. **Roberts, W. L., Speer, C. A. and Hammond, D. M.,** Electron and light microscope studies of the oocyst walls, sporocysts, and excysting sporozoites of *Eimeria callospermophili* and *E. larimerensis*, *J. Parasitol.*, 56, 918, 1970.
117. **Speer, C. A., Hammond, D. M., Mahrt, J. L., and Roberts, W. L.,** Structure of the oocyst and sporocyst walls and excystation of sporozoites of *Isospora canis*, *J. Parasitol.*, 59, 35, 1973.
118. **Wagenbach, G. E., Challey, J. R., and Burns, W. C.,** A method for purifying coccidian oocysts employing clorox and sulfuric acid-dichromate solution, *J. Parasitol.*, 52, 1222, 1966.
119. **Long, P. L. and Speer, C. A.,** Invasion of host cells by coccidia, in *Parasite Invasion*, Blackwell Scientific, Oxford, 1977.
120. **Bunch, T. D. and Nyberg, P. A.,** Effects of carbon dioxide on coccidian oocysts from 6 host species, *J. Protozool.*, 17, 364, 1970.
121. **Jensen, J. B., Nyberg, P. A., Burton, S. D., and Jolley, W. R.,** The effects of selected gases on excystation of coccidian oocysts, *J. Parasitol.*, 62, 195, 1976.
122. **Christie, E., Pappas, P. W., and Dubey, J. P.,** Ultrastructure of excystment of *Toxoplasma gondii* oocysts, *J. Protozool.*, 25, 438, 1978.
123. **Speer, C. A. and Duszynski, D. W.,** Fine structure of the oocyst walls of *Isospora serini* and *Isospora canaria* and excystation of *Isospora serini* from the canary, *Serinus canarius* L., *J. Protozool.*, 22, 476, 1975.
124. **Duszynski, D. W. and Speer, C. A.,** Excystation of *Isospora arctopitheci* Rodhain, 1933 with notes on a similar process in *Isospora bigemina* (Stiles, 1891) Luke, 1906, *Z. Parasitenkd.*, 48, 191, 1976.
125. **Speer, C. A., Marchiondo, A. A., Duszynski, D. W., and File, S. K.,** Ultrastructure of the sporocyst wall during excystation of *Isospora endocallimici*, *J. Parasitol.*, 62, 984, 1976.
126. **Patton, W. H. and Brigman, G. P.,** The use of sodium taurodeoxycholate for excystation of *Eimeria tenella* sporozoites, *J. Parasitol.*, 65, 526, 1979.
127. **Strout, R. G., Ouellette, C. A., and Gangi, D. P.,** Effect of inoculum size on development of *Eimeria tenella* in cell cultures, *J. Parasitol.*, 55, 406, 1969.
128. **Doran, D. J.,** Increasing the yield of *Eimeria tenella* oocysts in cell culture, *J. Parasitol.*, 57, 891, 1971.
129. **Doran, D. J.,** Effect of age and freezing on development of *Eimeria adenoeides* and *E. tenella* sporozoites in cell culture, *J. Parasitol.*, 56, 27, 1970.
130. **Wagenbach, G. E.,** Purification of *Eimeria tenella* sporozoites with glass bead columns, *J. Parasitol.*, 55, 833, 1969.
131. **Sharma, N. N. and Foster, J. W.,** Toxic substance in various constituents *Eimeria tenella* oocysts, *Am. J. Vet. Res.*, 25, 211, 1964.
132. **Fayer, R. and Hammond, D. M.,** Development of first-generation schizonts of *Eimeria bovis* in cultured bovine cells, *J. Protozool.*, 14, 764, 1967.
133. **Norton, C. C. and Joyner, L. P.,** The freeze preservation of coccidia, *Res. Vet. Sci.*, 9, 598, 1968.
134. **Doran, D. J.,** Cultivation and freezing of poultry coccidia, *Acta Vet. (Brno)*, 38, 25, 1969.
135. **Doran, D. J.,** Freezing excysted coccidial sporozoites, *J. Parasitol.*, 55, 1229, 1969.
136. **Doran, D. J. and Vetterling, J. M.,** Preservation of coccidial sporozoites by freezing, *Nature (London)*, 217, 1262, 1968.
137. **Doran, D. J. and Vetterling, J. M.,** Infectivity of two species of poultry coccidia after freezing and storage in liquid nitrogen vapor, *Proc. Helminthol. Soc. Wash.*, 36, 30, 1969.
138. **Doran, D. J.,** *Eimeria tenella*: from sporozoites to oocysts in cell culture, *Proc. Helminthol. Soc. Wash.*, 37, 84, 1970.
139. **Strout, R. G. and Ouellette, C. A.,** Gametogony of *Eimeria tenella* (coccidia) in cell cultures, *Science*, 163, 696, 1969.
140. **Augustine, P. C. and Doran, D. J.,** Development of *Eimeria meleagrimitis* Tyzzer from sporozoites and merozoites in turkey kidney cell cultures, *J. Protozool.*, 25, 82, 1978.
141. **Bedrnik, P.,** Cultivation of *Eimeria tenella* in tissue cultures. I. Further development of second generation merozoites in tissue cultures, *Acta Protozool. Warszawa*, 7, 87, 1969.

142. **Speer, C. A. and Hammond, D. M.,** Development of gametocytes and oocysts of *Eimeria magna* from rabbits in cell culture, *Proc. Helminthol. Soc. Wash.*, 39, 114, 1972.
143. **Chernin, E. and Weller, T. H.,** Further observations on the growth of *Toxoplasma gondii* in roller tube cultures of mouse and primate tissues, *J. Parasitol.*, 43, 33, 1957.
144. **Csóka, R. and Kulcsár, G.,** Maintenance of *Toxoplasma gondii* in primary human amniotic cell culture, *Acta Microbiol. Acad. Sci. Hung.*, 15, 357, 1968.
145. **Wang, C.,** Multiplication of *Toxoplasma gondii* in tissue culture and its subsequent virulence in mice, *Chin. J. Micro.*, 3, 111, 1970.
146. **Braveny, I., Winter, W., and Disko, R.,** A method of mass cultivation of *Toxoplasma gondii* in cell culture, *Tropenmed. Parasitol.*, 29, 432, 1978.
147. **Tryon, J. C., Weidner, E., Larson, A. D., and Hart, L. T.,** A rapid method for isolating purified *Toxoplasma* tachyzoites from peritoneal exudate and cell culture, *J. Parasitol.*, 64, 1127, 1978.
148. **Sledge, W. E., Larson, A. D., and Hart, L. T.,** Costae of *Tritrichomonas foetus:* purification and chemical composition, *Science*, 199, 186, 1978.
149. **Mehlhorn, H. and Heydorn, A. O.,** Electron microscopical study on gametogony of *Sarcocystis suihominis* in human tissue cultures, *Z. Parasitenkd.*, 58, 97, 1979.
150. **Jacobs, L., Remington, J. S., and Melton, M. L.,** A survey of meat samples from swine, cattle, and sheep for the presence of encysted *Toxoplasma*, *J. Parasitol.*, 46, 23, 1960.
151. **Box, E. D. and McGuinness, T. B.,** *Sarcocystis* in beef from retail outlets demonstrated by digestion technique, *J. Parasitol.*, 64, 161, 1978.
152. **Mueller, B. E. G. and Speer, C. A.,** *Eimeria canadensis:* intracellular motility of a sporozoite-shaped meront in vitro, *Proc. Helminthol. Soc. Wash.*, 48, 87, 1981.
153. **Speer, C. A. and Hammond, D. M.,** Development of *Eimeria larimerensis* from the Uinta ground squirrel in cell cultures, *Z. Parasitenkd.*, 35, 105, 1970.
154. **Speer, C. A. and Hammond, D. M.,** Development of first- and second-generation schizonts of *Eimeria magna* from rabbits in cell cultures, *Z. Parasitenkd*, 37, 336, 1971.
155. **Speer, C. A., Hammond, D. M., and Anderson, L. C.,** Development of *Eimeria callospermophili* and *E. bilamellata* from the Uinta ground squirrel *Spermophilus armatus* in cultured cells, *J. Protozool.*, 17, 274, 1970.
156. **Fayer, R.,** *Sarcocystis:* development in cultured avian and mammalian cells, *Science*, 168, 1104, 1970.
157. **Speer, C. A. and Hammond, D. M.,** Motility of macrogamonts of *Eimeria magna* (Coccidia) in cell culture, *Science*, 178, 763, 1972.
158. **Aikawa, M.,** *Plasmodium:* the fine structure of malarial parasites, *Exp. Parasitol.*, 30, 285, 1971.
159. **Garnham, P. C. C., Bird, R. G., Baker, J. R., Desser, S. S. and El-Nahal, H. M. S.,** Electron microscope studies on motile stages of malaria parasites VI. The ookinete of *Plasmodium berghei yoelii* and its transformation into the early oocyst, *Trans. R. Soc. Trop. Med. Hyg.*, 63, 187, 1969.
160. **Roberts, W. L., Hammond, D. M., and Speer, C. A.,** Ultrastructural study of the intra- and extracellular sporozoites of *Eimeria callospermophili*, *J. Parasitol.*, 56, 907, 1970.
161. **Speer, C. A., Rosales-Ronquillo, M. C., and Silverman, P. H.,** Motility of *Plasmodium berghei* ookinetes in vitro, *J. Invertebr. Pathol.*, 25, 73, 1975.
162. **Freyvogel, R. A.,** Shape, movement in situ and locomotion of plasmodial ookinetes, *Acta Tropica*, 23, 201, 1966.
163. **Roberts, W. L., Mahrt, J. L., and Hammond, D. M.,** The fine structure of the sporozoites of *Isospora canis*, *Z. Parasitenkd.*, 40, 183, 1972.
164. **Roberts, W. L., Speer, C. A. and Hammond, D. M.,** Penetration of *Eimeria larimerensis* sporozoites into cultured cells as observed with the light and electron microscopes, *J. Parasitol.*, 57, 615, 1971.
165. **Speer, C. A., Rosales-Ronquillo, M. C., and Silverman, P. H.,** Scanning electron microscope observations of *Plasmodium berghei* ookinetes in primary mosquito cell cultures, *J. Invertebr. Pathol.*, 24, 179, 1974.
166. **Vetterling, J. M. and Doran, D. J.,** Storage polysaccharide in coccidial sporozoites after excystation and penetration of cells, *J. Protozool.*, 16, 772, 1969.
167. **Muller, B. E. G., Hammond, D. M., and Scholtyseck, E.,** In vitro development of first- and second-generation schizonts of *Eimeria contorta* Haberkorn, 1971 (Coccidia, Sporozoa), *Z. Parasitenkd.*, 41, 173, 1973.
168. **Speer, C. A., Hammond, D. M., and Kelley, G. L.,** Stimulation of motility in merozoites of five *Eimeria* species by bile salts, *J. Parasitol.*, 56, 927, 1970.
169. **Speer, C. A.,** Further studies on the development of gamonts and oocysts of *Eimeria magna* in cultured cells, *J. Parasitol.*, 65, 591, 1979.
170. **Dyson, R. D.,** *Cell Biology: A Molecular Approach*, 2nd ed., Allyn and Bacon, Inc., Boston, 1978.
171. **Bommer, W.,** The life cycle of virulent *Toxoplasma* in cell culture, *Aust. J. Exp. Biol. Med. Sci.*, 47, 505, 1969.

172. **Bommer, W., Heurnert, H. H., and Milthaler, B.,** Kinematographische Studien über die Eigenbewegung von *Toxoplasma gondii, Z. Tropenmed. Parasitol.*, 20, 450, 1969.
173. **Clark, W. N. and Hammond, D. M.,** Development of *Eimeria auburnensis* in cell cultures, *J. Protozool.*, 16, 646, 1969.
174. **Fayer, R., Hammond, D. M., Chobotar, B., and Elsner, Y. Y.,** Cultivation of *Besnoitia jellisoni* in bovine cell cultures, *J. Parasitol.*, 55, 645, 1969.
175. **Fayer, R., Romanowski, R. D., and Vetterling, J. M.,** The influence of hyaluronidase and hyaluronidase substrates on penetration of cultured cells by eimerian sporozoites, *J. Protozool.*, 17, 432, 1970.
176. **Hirai, K., Hirato, K., and Yanagawa, R.,** A cinematographic study of the penetration of cultured cells by *Toxoplasma gondii, Jpn. J. Vet. Res.*, 14, 81, 1966.
177. **Itagaki, K., Hirayama, N., Tsubokura, M., and Otsuki, K.,** Development of *Eimeria tenella, E. brunetti* and *E. acervulina* in cell cultures, *Jpn. J. Vet. Sci.*, 36, 467, 1974.
178. **Jensen, J. B. and Edgar, S. A.,** Effects of antiphagocytic agents on penetration of *Eimeria magna* sporozoites into cultured cells, *J. Parasitol.*, 62, 203, 1976.
179. **Jensen, J. B. and Edgar, S. A.,** Possible secretory function of the rhoptries of *Eimeria magna* during penetration of cultured cells, *J. Parasitol.*, 62, 988, 1976.
180. **Jensen, J. B. and Edgar, S. A.,** Fine structure of penetration of cultured cells by *Isospora canis* sporozoites, *J. Protozool.*, 25, 169, 1978.
181. **Jensen, J. B. and Hammond, D. M.,** Ultrastructure of the invasion of *Eimeria magna* sporozoites into cultured cells, *J. Protozool.*, 22, 411, 1975.
182. **Kelley, G. L. and Hammond, D. M.,** Development of *Eimeria ninakohlyakimovae* from sheep in cell cultures, *J. Protozool.*, 17, 340, 1970.
183. **Klainer, A. S., Krahenbuhl, J. L., and Remington, J. S.,** Scanning electron microscopy of *Toxoplasma gondii, J. Gen. Microbiol.*, 75, 111, 1973.
184. **Muller, B. E. G., deVos, A. J., and Hammond, D. M.,** In vitro development of first- generation schizonts of *Eimeria canadensis* (Bruce, 1921), *J. Protozool.*, 20, 293, 1973.
185. **Pulvertaft, R. J. V., Valentine, J. C., and Lane, W. F.,** The behavior of *Toxoplasma gondii* on serum agar culture, *Parasitology*, 44, 478, 1954.
186. **Roberts, W. L. and Hammond, D. M.,** Scanning electron microscope study of invasion of host cells by *Eimeria larimerensis* sporozoites, *Proc. Helminthol. Soc. Wash.*, 40, 118, 1973.
187. **Speer, C. A. and Hammond, D. M.,** Cinemicrographic observations on the development of *Eimeria callospermophili* in cultured cells, *J. Protozool.*, 16 (Suppl.), 16, 1969.
188. **Speer, C. A., Davis, L. R., and Hammond, D. M.,** Cinemicrographic observations on the development of *Eimeria larimerensis* in cultured bovine cells, *J. Protozool.*, 18 (Suppl.), 19, 1971.
189. **Zaman, V. and Colley, F. C.,** Ultrastructural study of penetration of macrophages by *Toxoplasma gondii, R. Soc. Trop. Med. Hyg.*, 66, 781, 1972.
190. **Aikawa, M., Komata, Y., Asai, T., and Midorikawa, O.,** Transmission and scanning electron microscopy of host cell entry by *Toxoplasma gondii, Am. J. Pathol.*, 87, 285, 1977.
191. **Lycke, E., Carlberg, K., and Norrby, R.,** Interactions between *Toxoplasma gondii* and its host cells: function of the penetration-enhancing factor of *Toxoplasma, Infect. Immun.*, 11, 853, 1975.
192. **Scholtyseck, E. and Mehlhorn, H.,** Ultrastructural study of characteristic organelles (paired organelles, micronemes, micropores) of Sporozoa and related organisms, *Z. Parasitenkd.*, 34, 97, 1970.
193. **Vivier, E. and Petitprez, A.,** Données ultrastructurates complémentaires, morphologiques et cytochimiques, sur *Toxoplasma gondii, Protistologica*, 8, 199, 1972.
194. **Lycke, E. and Noorby, R.,** Demonstration of a factor of *Toxoplasma gondii* enhancing the penetration of Toxoplasma parasites into cultured host cells, *Br. J. Exp. Pathol.*, 47, 248, 1966.
195. **Norrby, R., Lindholm, L., and Lycke, E.,** Lysosomes of *Toxoplasma gondii* and their possible relation to the host cell penetration of Toxoplasma, *J. Bacteriol.*, 96, 916, 1968.
196. **Norrby, R. and Lycke, E.,** Factors enhancing the host cell penetration of *Toxoplasma gondii, J. Bacteriol.*, 95, 53, 1967.
197. **Norby, R.,** Host cell penetration of *Toxoplasma gondii, Infect. Immun.*, 2, 250, 1970.
198. **Norrby, R.,** Immunological study on the host cell penetration factor of *Toxoplasma gondii, Infect. Immun.*, 3, 278, 1971.
199. **Stadtsbaeder, S., Nguyen, B. T., and Calvin-Preval, M. C.,** Respective role of antibodies and immune macrophages during acquired immunity against Toxoplasmosis in mice, *Ann. Immunol. (Inst. Pasteur)*, 126 C, 461, 1975.
200. **Jones, T. C. and Hirsch, J. G.,** The interaction between *Toxoplasma gondii* and mammalian cells. II. the absence of lysosomal fusion with phagocytic vacuoles containing living parasites, *J. Exp. Med.*, 136, 1173, 1972.
201. **Scholtyseck, E. and Chobotar, B.,** Electron microscope observations concerning the penetration of a host cell by *Eimeria ferrisi in vivo, Z. Parasitenkd.*, 49, 91, 1975.

202. **Lycke, E., Lund, E., and Strannegard, D.,** Enhancement by lysosome and hyaluronidase of the penetration of toxoplasma parasites into cultured host cells, *Br. J. Exp. Pathol.*, 46, 189, 1965.
203. **Hsu, K. Y.,** Studies on the tissue culture of *Toxoplasma gondii* with HeLa cells, *Trop. Med.*, 13, 73, 1971.
204. **Fayer, R., Romanowski, R. D., and Vetterling, J. M.,** The influence of hyaluronidase and hyaluronidase substrates on penetration of cultured cells by eimerian sporozoites, *J. Protozool.*, 17, 432, 1970.
205. **Fayer, R.,** Quinine inhibition of host cell penetration by eimerian sporozoites in vitro, *J. Parasitol.*, 57, 901, 1971.
206. **Long, P. L. and Rose, M. E.,** Immunity to coccidiosis: effect of serum antibodies on cell invasion by sporozoites of *Eimeria* in vitro, *Parasitology*, 65, 437, 1972.
207. **Fayer, R. and Baron, S.,** Activity of interferon and an inducer against development of *Eimeria tenella* in cell culture, *J. Protozool.*, 18 (Suppl.), 12, 1971.
208. **Turner, M. B. F. and Box, E. D.,** Cell culture of *Isospora* from the English sparrow, *Passer domesticus domesticus*, *J. Parasitol.*, 56, 1218, 1970.
209. **Akinshina, G. T. and Doby J. M.,** Multiplication de *Besnoitia jellisoni* Frenkel 1953 (Protozoaires Toxoplasmatea) en cultures de cellules de tissus d'origines differentes, *Protistologica*, 5, 249, 1969.
210. **Akinchina, G. T. and Zasukhin, D. N.,** Intranuclear localization of *Toxoplasma gondii* and *Besnoitia jellisoni* in conditions of tissue culture and some remarks on the intranuclear parasitism, *Acta Protozool.*, 8, 341, 1971.
211. **Doby, J. M. and Akinshina, G. T.,** Possibilites de developpement de *Besnoitia jellisoni* (Protozaire parasite Toxoplasmatea) en culture de cellules. Quelques aspects de son comportement en fibroblastes d'embryon de Poulet, *C. R. Seances Soc. Biologie*, 162, 1207, 1968.
212. **Scholtyseck, E.,** Ultrastructure, in *The Coccidia: Eimeria, Isospora, Toxoplasma* and Related Genera, University Park Press, Baltimore, 1973, 81.
213. **Ryley, J. F. and Wilson, R. G.,** The development of *Eimeria brunetti* in tissue culture, *J. Parasitol.*, 58, 660, 1972.
214. **deVos, A. J., Hammond, D. M., and Speer, C. A.,** Development of *Eimeria crandallis* Honess from sheep in cultured cells, *J. Protozool.*, 19, 335, 1972.
215. **Speer, C. A. and Hammond, D. M.,** Development of *Eimeria ellipsoidalis* from cattle in cultured bovine cells, *J. Parasitol.*, 57, 675, 1971.
216. **Speer, C. A., deVos, A. J., and Hammond, D. M.,** Development of *Eimeria zuernii* in cell cultures, *Proc. Helminthol. Soc. Wash.*, 40, 160, 1973.
217. **Speer, C. A. and Hammond, D. M.,** Nuclear divisions and refractile body changes in sporozoites and schizonts of *Eimeria callospermophili* in cultured cells, *J. Parasitol.*, 56, 461, 1970.
218. **Speer, C. A., Hammond, D. M., and Elsner, Y. Y.,** Development of second-generation schizonts and immature gamonts of *Eimeria larimerensis* in cultured cells, *Proc. Helminthol. Soc. Wash.*, 40, 147, 1973.
219. **Kelley, G. L. and Youssef, N. N.,** Development in cell cultures of *Eimeria vermiformis* Ernst, Chobotar and Hammond, 1971, *Z. Parasitenkd.*, 53, 23, 1977.
220. **Doran, D. J., and Vetterling, J. M.,** Cultivation of the turkey coccidian, *Eimeria meleagrimitis* Tyzzer, 1929, in mammalian kidney cell cultures, *Proc. Helminthol. Soc. Wash.*, 34, 59, 1967.
221. **Sampson, J. R. and Hammond, D. M.,** Fine structural aspects of development of *Eimeria alabamensis* schizonts in cell cultures, *J. Parasitol.*, 58, 311, 1972.
222. **Sampson, J. R., Hammond, D. M., and Ernst, J. V.,** Development of *Eimeria alabamensis* from cattle in mammalian cell cultures, *J. Protozool.*, 18, 120, 1971.
223. **Chobotar, B., Hammond, D. M., and Miner, M. L.,** Development of the first-generation schizonts of *Eimeria auburnensis*, *J. Parasitol.*, 55, 385, 1969.
224. **Kelley, G. L. and Hammond, D. M.,** Fine structural aspects of early development of *Eimeria ninakohlyakimovae* in cultured cells, *Z. Parasitenkd.*, 38, 271, 1972.
225. **Roberts, W. L., Hammond, D. M., Anderson, L. C., and Speer, C. A.,** Ultrastructural study of schizogony in *Eimeria callospermophili*, *J. Protozool.*, 17, 584, 1970.
226. **Pacheco, N. D., Vetterling, J. M., and Doran, D. J.,** Ultrastructure of cytoplasmic and nuclear changes in *Eimeria tenella* during first-generation schizogony in cell culture, *J. Parasitol.*, 61, 31, 1975.
227. **Roberts, W. L. and Hammond, D. M.,** Ultrastructural and cytological studies of the sporozoites of four *Eimeria* species, *J. Protozool.*, 17, 76, 1970.
228. **Doran, D. J. and Augustine, P. C.,** *Eimeria dispersa* and *Eimeria gallopavonis:* infectivity, survival, and development in primary chicken and turkey kidney cell cultures, *J. Protozool.*, 24, 172, 1977.
229. **Hammond, D. M., Speer, C. A., and Roberts, W. L.,** Occurrence of refractile bodies in merozoites of *Eimeria* species, *J. Parasitol.*, 56, 189, 1970.
230. **Dubremetz, J. F. and Elsner, Y. Y.,** Ultrastructural study of schizogony of *Eimeria bovis* in cell cultures, *J. Protozool.*, 26, 367, 1979.
231. **Kelley, G. L. and Hammond, D. M.,** Fine structural aspects of nuclear division and merogony of *Eimeria ninakohlyakimovae* in cultured cells, *J. Parasitol.*, 59, 1071, 1973.

232. **Naciri-Bontemps, M.,** Reproduction of the cycle of coccidia *Eimeria acervulina* (Tyzzer, 1929) in cell cultures of chicken kidneys, *Ann. Rech. Veter.*, 7, 223, 1976.
233. **Davis, L. R. and Bowman, G. W.,** The endogenous development of *Eimeria zuernii*, a pathogenic coccidium of cattle, *Am. J. Vet. Res.*, 18, 569, 1957.
233a. **Stockdale, P. H. G.,** Schizogony and gametogony of *Eimeria zuernii* (Rivolta, 1878) Martin 1909, *Vet. Parasitol.*, 1, 367, 1976.
234. **Danforth, H. D. and Hammond, D. M.,** Stages of merogony in multinucleate merozoites of *Eimeria magna* Pérard, 1925, *J. Protozool.*, 19, 454, 1972.
235. **Speer, C. A., Hammond, D. M., and Elsner, Y. Y.,** Further asexual development of *Eimeria magna* merozoites in cell cultures, *J. Parasitol.*, 59, 613, 1973.
236. **Long, P. L.,** In vitro culture of *Eimeria tenella*. Second Internat. Congr. Parasitol., *J. Parasitol.*, 56 (Suppl.), 214, 1970.
237. **Heller, G.,** Elektronemikroskopische Untersuchungen zur Schizogonie in den sog. kleinen Schizonten von *Eimeria steidae* (Sporozoa, Coccidia), *Protistologica*, 7, 461, 1971.
238. **Shah, H. L.,** The life cycle of *Isospora felis* Wenyon, 1923, a coccidium of the cat, *J. Protozool.*, 18, 3, 1971.
239. **Lepp, D. L. and Todd, K. S., Jr.,** Life cycle of *Isospora canis* Nemeséri, 1959 in the dog, *J. Protozool.*, 21, 199, 1974.
240. **Long, P. L.,** The growth of embryo-adapted strains of *Eimeria* in cell cultures, *Proc. Int. Congr. Parasitol.*, 1, 121, 1974.
241. **Doran, D. J.,** Survival and development of five species of chicken coccidia in primary chicken kidney cell cultures, *J. Parasitol.*, 57, 1135, 1971.
242. **Doran, D. J.,** Comparative development of *Eimeria tenella* in primary cultures of kidney cells from the chicken, pheasant, partridge and turkey, *J. Parasitol.*, 57, 1376, 1971c.
243. **Strout, R. G. and Ouellette, C. A.,** Schizogony and gametogony of *Eimeria tenella* in cell culture, *Am. J. Vet. Res.*, 31, 911, 1970.
244. **Speer, C. A. and Hammond, D. M.,** Development of second-generation schizonts, gamonts and oocysts of *Eimeria bovis* in bovine kidney cells, *Z. Parasitenkd.*, 42, 105, 1973.
245. **Bedrnik, P.,** Development of sexual stages and oocysts from the 2nd generation of *Eimeria tenella* merozoites in tissue cultures, *Folia Parasitol.*, 14, 364, 1967.
246. **Shibalova, T. A.,** Cultivation of the endogenous stages of chicken coccidia in embryos and tissue cultures, *J. Parasitol.*, 56, 315, 1970.
247. **Coudert, P. and Provot, F.,** Developpement interne et schizogonie en cultures cellulares *d'Eimeria stiedae* (Lindmann, 1865) Kissalt et Hartmann, 1907, *C. R. Acad. Sci., Paris*, 279, 911, 1974.
248. **Pellérdy, L. P. and Dürr, U.,** Zum endogenen Entwicklungszyklus von *Eimeria stiedai* (Lindemann, 1865) Kisskalt and Hartmann, 1907, *Acta Vet. Acad. Sci. Hung.*, 20, 227, 1970.
249. **Cerná, Z. and Sénaud, J.,** Some peculiarities of the fine structure of merozoites of *Eimeria stiedai, Folia Parasitol.*, 18, 177, 1971.
250. **Ryley, J. F. and Robinson, T. E.,** Life cycle studies of *Eimeria magna* Pérard, 1925, *Z. Parasitenkd.*, 50, 257, 1976.
251. **Sénaud, J., and Cerná, Z.,** Etude ultrastructurale des merozoites et de la schizogonie des Coccidies (Eimeriina): *Eimeria magna* (Pérard, 1925) de l'intestin des lapins et *E. tenella* (Raillet et Lucet, 1891) des coecums des poulets, *J. Protozool.*, 16, 155, 1969.
252. **Doran, D. J. and Augustine, P. C.,** Comparative development of *Eimeria tenella* from sporozoites to oocysts in primary kidney cell cultures from gallinaceous birds, *J. Protozool.*, 20, 658, 1973.
253. **Fayer, R.,** Gametogony of *Sarcocystis* sp. in cell culture, *Science*, 175, 65, 1972.
254. **McDougald, L. R. and Jeffers, T. K.,** *Eimeria tenella* (Sporozoa, Coccidia): gametogony following a single asexual generation, *Science*, 192; 258, 1976.
255. **McDougald, L. R. and Jeffers, J. K.,** Comparative in vitro development of precocious and normal strains of *Eimeria tenella* (Coccidia), *J. Protozool.*, 23, 530, 1976.
256. **Klimes, B., Rootes, D. G., and Tanielian, Z.,** Sexual differentiation of merozoites of *Eimeria tenella*, *Parasitology*, 65, 131, 1972.
257. **Sheffield, H. G., Melton, M. L., and Neva, F. A.,** Development of *Hammondia hammondi* in cell cultures, *Proc. Helminthol. Soc. Wash.*, 43, 217, 1976.
258. **Shibalova, T. A. and Petrenko, V. I.,** *Isospora bigemina:* Sporozoite-induced development of the life cycle in tissue culture, (in Russian; English summary), *Akad. Nauk. USSR*, 6, 201, 1972.
259. **Sheffield, H. G. and Melton, M. L.,** The fine structure and reproduction of *Toxoplasma gondii, J. Parasitol.*, 54, 209, 1968.
260. **Roberts, W. L., Mahrt, J. L., and Hammond, D. M.,** The fine structure of the sporozoites of *Isospora canis, Z. Parasitenkd.*, 40, 183, 1972.
261. **Fayer, R.,** Cultivation of feline *Isospora rivolta* in mammalian cells, *J. Parasitol.*, 58, 1207, 1972.

262. **Fayer, R. and Mahrt, J. L.,** Development of *Isospora canis* (Protozoa; Sporozo) in cell culture, *Z. Parasitenkd.*, 38, 313, 1972.
263. **Fayer, R. and Thompson, D. E.,** *Isospora felis:* development in cultured cells with some cytological observations, *J. Parasitol.*, 60, 160, 1974.
264. **Sheffield, H. G. and Melton, M. L.,** *Toxoplasma gondii:* the oocyst, sporozoite, and infection of cultured cells, *Science*, 167, 892, 1970.
265. **Dubremetz, J. F., Porchet-Hennére, E., and Parenty, M. D.,** Croissance de *Sarcocystis tenella* en culture cellulaire, *C. R. Acad. Sc., Paris*, 280, 1793, 1975.
266. **Becker, B., Mehlhorn, H., and Heydorn, A. O.,** Light and electron microscopic studies on gamogony and sporogony of 5 *Sarcocystis* species in vivo and in tissue cultures, *Zbl. Bakt. Hyg., I. Abt. Ong.*, A244, 394, 1979.
267. **Vetterling, J. M., Pacheco, N. D., and Fayer, R.,** Fine structure of gametogony and oocyst formation in *Sarcocystis* sp. in cell culture, *J. Protozool.*, 20, 613, 1973.
268. **Speer, C. A., Hammond, D. M., Mahrt, J. L., and Roberts, W. L.,** Structure of the oocyst and sporocyst walls and excystation of sporozoites of *Isospora canis*, *J. Parasitol.*, 59, 35, 1973.
269. **Speer, C. A., Marchiondo, A. A., Duszynski, D. W., and File, S. K.,** Ultrastructure of the sporocyst wall during excystation of *Isospora endocallimici J. Parasitol.*, 62, 984, 1976.
270. **Kusunoki, Y.,** In vitro study on the invasion and multiplication of *Toxoplasma gondii* sporozoites in cultured cells, *Jap. J. Parasitol.*, 26, 6, 1977.
271. **Azab, M. and Beverley, J. K. A.,** Schizogony of *Toxoplasma gondii* in tissue cultures, *Z. Parasitenkd.*, 44, 33, 1974.
272. **Lund, E., Lycke, E., and Sourander, P.,** A cinematographic study of *Toxoplasma gondii* in cell cultures, *Br. J. Exp. Pathol.*, 42, 357, 1961.
273. **Meyer, H. and deOliveira Musacchio, M.,** *Toxoplasma gondii* in tissue cultures: a microcinematographic study in phase contrast, *Rev. Inst. Med. Trop. Sao Paulo*, 16, 7, 1974.
274. **Akinshina, G. T.,** Prolonged preservation of *Toxoplasma* in tissue cultures, *Byull. Eksp. Biol. Med.*, 58, 98, 1964.
275. **Akinshina, G. T.,** Use of the methods of tissue cultures for diagnostics of toxoplasmosis and for maintenance of Toxoplasma's strains, in *Toxoplasmosis of Animals*, Academy of Science Kazakh SSR, Almaata, 1965, 386.
276. **Sampson, J. R. and Hammond, D. M.,** Ingestion of host cell cytoplasm by micropores in *Eimeria alabamensis*, *J. Parasitol.*, 57, 1133, 1971.
277. **Strout, R. G. and Scholtyseck, E.,** The ultrastructure of first generation development of *Eimeria tenella* (Railliet and Lucet, 1891) Fanthan, 1909 in cell cultures, *Z. Parasitenkd.*, 35, 87, 1970.
278. **Speer, C. A. and Danforth, H. D.,** Fine-structural aspects of microgametogenesis of *Eimeria magna* in rabbits and in kidney cell cultures, *J. Protozool.*, 23, 109, 1976.
279. **Senaud, J., Chobotar, B., and Scholtyseck, E.,** Role of the micropore in nutrition of the Sporozoa. Ultrastructural study of *Plasmodium cathemurium, Eimeria ferrisi, E. Stiedai, Besnoitia jellisoni*, and *Frenkelia* sp., *Tropenmed. Parasitol.*, 27, 145, 1976.
280. **Sibert, G. L. and Speer, C. A.,** Fine structure of zygotes and oocysts of *Eimeria nieschulzi*, *J. Protozool.*, 27, 374, 1980.
281. **Sibert, G. L. and Speer, C. A.,** Fine structure of nuclear division and microgametogony of *Eimeria nieschulzi*, Dieben, 1924, *Z. Parasitenkd.*, 66, 179, 1981.
282. **Ryley, J. F.,** Cytochemistry, physiology and biochemistry, in *The Coccidia: Eimeria, Isospora, Toxoplasma*, and Related Genera, Hammond, D. M. and Long, P., Eds., University Park Press, Baltimore, 1973, 145.
283. **McDougald, L. R.,** The growth of avian *Eimeria* in vitro, in *Avian Coccidiosis*, Long, P. L., Boorman, K. N., and Freeman, B. M., Eds., British Polymer Science, Ltd., Edinburgh, 1978, 185.
284. **Doran, D. J. and Augustine, P. C.,** *Eimeria tenella:* vitamin requirements for development in primary cultures of chicken kidney cells, *J. Protozool.*, 25, 544, 1978.
285. **Ryley, J. F. and Wilson, R. G.,** Growth factor antagonism studies with coccidia in tissue culture, *Z. Parasitenkd.*, 40, 31, 1972.
286. **Sofield, W. L. and Strout, R. G.,** Amino acids essential for in vitro cultivation of *Eimeria tenella*, *J. Protozool.*, 21 (Suppl.), 434, 1974.
287. **Eagle, H.,** The specific amino acid requirements of a mammalian cell (strain L) in tissue culture, *J. Biol. Chem.*, 214, 839, 1955.
288. **Eagle, H.,** The specific amino acid requirements of a human carcinoma cell (strain HeLa) in tissue culture, *J. Exp. Med.*, 102, 37, 1955.
289. **Higuchi, K.,** Qualitative requirements and utilization of nutrients: mammalian and avian cells in culture, in *CRC Handbook Series in Nutrition and Food*, Section D: Nutritional Requirements, Vol. I., CRC Press, Inc., Boca Raton, Fla., 1977, 501.

290. **Warren, E. W.,** Vitamin requirements of the coccidia of the chicken, *Parasitology*, 58, 137, 1968.
291. **Eagle, H.,** The minimum vitamin requirements of the L and HeLa cells in tissue culture, the production of specific vitamin deficiencies, *J. Exp. Med.*, 102, 595, 1955.
292. **Eagle, H., Oyama, V. I., Levy, M., and Freeman, A. E.,** Myo-inositol as an essential growth factor for normal and malignant human cells in tissue culture, *J. Biol. Chem.*, 226, 191, 1957.
293. **Bailey, J. M.,** Lipid metabolism in cultured cells. V. Comparative lipid nutrition in serum and in lipid-free chemically defined medium, *Proc. Soc. Exp. Biol. Med.*, 115, 747, 1964.
294. **Boucek, R. J. and Alvarez, T. R.,** 5-hydroxytryptamine: a cytospecific growth stimulator of cultured fibroblasts, *Science*, 167, 898, 1970.
295. **Bartfeld, H. and Siegel, S. M.,** Antioxidant activity of thyroxine and related substances: effect on in vitro cell growth, *Exp. Cell Res.*, 49, 25, 1968.
296. **Higuchi, K. and Robinson, R. C.,** Studies on the cultivation of mammalian cell lines in a serum-free, chemically defined medium, *In Vitro*, 9, 114, 1973.
297. **Neuman, R. E. and McCoy, T. A.,** growth-promoting properties of pyruvate, oxalacetate, and α-ketoglutarate for isolated Walker carcinosarcoma 256 cells, *Proc. Soc. Exp. Biol. Med.*, 98, 303, 1958.
298. **Siegel, E. and Tobias, C. A.,** Actions of thyroid hormones on cultured human cells, *Nature (London)*, 212, 1318, 1961.
299. **Perrotto, J., Keister, D. B., and Gelderman, A. H.,** Incorporation of precursors into *Toxoplasma* DNA, *J. Protozool.*, 18, 470, 1971.
300. **Remington, J. S., Bloomfield, M. B., Russell, E., Jr., and Robinson, W. S.,** The RNA of *Toxoplasma gondii*, *Proc. Soc. Exp. Biol. Med.*, 133, 623, 1970.
301. **Morgan, K. and Canning, E. U.,** Incorporation of ^{3}H-thymidine and ^{3}H-adenosine by *Eimeria tenella* grown in chick embryos, *J. Parasitol.*, 60, 364, 1974.
302. **Mayberry, L. F. and Marquardt, W. C.,** Nucleic acid precursor incorporation by *Eimeria nieschulzi* (Protozoa: Apicomplexa) and jejunal villus epithelium, *J. Protozool.*, 21, 599, 1974.
303. **Browning, R. F., Patton, W. H., and Lytle, C. F.,** *Eimeria tenella:* stimulation of DNA synthesis in infected cultured animal cells, *Exp. Parasitol.*, 39, 195, 1976.
304. **Ouellette, C. A., Strout, R. G., and McDougald, L. R.,** Incorporation of radioactive pyrimidine nucleosides into DNA and RNA of *Eimeria tenella* (Coccidia) cultured in vitro, *J. Protozool.*, 20, 150, 1973.
305. **Ouellette, C. A., Strout, R. G., and McDougald, L. R.,** Thymidylic acid synthesis in *Eimeria tenella* (Coccidia) cultured in vitro, *J. Protozool.*, 21, 398, 1974.
306. **Roberts, W. L., Elsner, Y. Y., Shigematsu, A., and Hammond, D. M.,** Lack of incorporation of ^{3}H-thymidine in *Eimeria callospermophili* in cell cultures, *J. Parasitol.*, 56, 833, 1970.
307. **Hammond, D. M.,** Life cycles and development of Coccidia, in *The Coccidia: Eimeria, Isospora, Toxoplasma*, and Related Genera, Hammond, D. M. and Long, P. L., Eds., University Park Press, Baltimore, 1973, 45.
308. **Fernando, M. A., Pasternak, J., Barrell, R., and Stockdale, P. H. G.,** Induction of host nuclear DNA synthesis in coccidia-infected chicken intestinal cells, *Int. J. Parasitol.*, 4, 267, 1974.
309. **Carter, S. B.,** Effects of cytochalasins on mammalian cells, *Nature (London)*, 213, 261, 1967.
310. **Patton, W. H.,** *Eimeria tenella:* comparative entry and development in mixed cultured populations of mononucleated and binucleated mammalian cells, Proc. 49th Ann. Meeting N.E. Conf. Avian Dis. (unpaged), 1977.
311. **Frandsen, J. C.,** *Eimeria stiedae:* cytochemical identification of acid and alkaline phosphatases, carboxylic ester hydrolases, and succinate, lactate, and glucose-6-phosphate dehydrogenases in endogenous stages from rabbit tissues, *Exp. Parasitol.*, 23, 398, 1968.
312. **Akao, S.,** Ultramicroscope studies of the localization of adenosine triphosphatase activity and ^{3}H-glucose transport in *Toxoplasma gondii*, *Jap. J. Parasitol.*, 18, 488, 1969.
313. **Akao, S.,** *Toxoplasma gondii:* aspartate aminotransferase in cell membrane, *Exp. Parasitol.*, 29, 26, 1971.
314. **Akao, S.,** *Toxoplasma gondii:* localization of peroxidase activity, *Exp. Parasitol.*, 29, 250, 1971.
315. **Sénaud, J., Mehlhorn, H., and Scholtyseck, E.,** Cytochemische untersuchungen an Cystenstadien von *Besnoitia jellisoni*. Elektronenmikroskopischer Nachweis von saurer Phosphatase und Polysacchariden, *Z. Parasitenkd.*, 40, 165, 1972.
316. **Vetterling, J. M. and Waldrop, H. R.,** Effect of fixation on demonstration of phosphatases of *Eimeria tenella* grown in chick kidney cultures, *J. Protozool.*, 23, 397, 1976.
317. **Lee, S. H. and Torack, R. M.,** Electron microscope studies of glutamic oxalacetic transaminase in rat liver cell, *J. Cell Biol.*, 39, 716, 1968.
318. **Becker, N. H., Novidkoff, A. H., and Zimmerman, H. M.,** Fine structure observations of the uptake of intravenously injected peroxidase by the rat choroid plexus, *J. Histochem. Cytochem.*, 15, 160, 1967.
319. **Fayer, R. and Thompson, D. E.,** Cytochemical and cytological observations on *Sarcocystis* sp. propagated in cell culture, *J. Parasitol.*, 61, 466, 1975.
320. **Hare, J. F. and Strout, R. G.,** Cytochemical observations on *Eimeria tenella* (Coccidia) propagated in cell culture, *J. Parasitol.*, 58, 567, 1972.

321. **Beyer, T. V., Shibalova, T. A., and Ovchinnikova, L. P.,** A comparative cytometrical investigation of the dynamics of cytoplasmic RNA in the endogenous development of *Eimeria tenella* grown in vivo, in tissue culture and in chick embryo. I. Schizogonous development, 9, 339, 1972.
322. **Bedrnik, P.,** Antitoxoplasma activity of coccidiostats, *Folia Parasitol. (Praha)*, 19, 129, 1972.
323. **Cook, M. K. and Jacobs, L.,** In vitro investigations on the action of pyrimethamine against *Toxoplasma gondii*, *J. Parasitol.*, 44, 280, 1958.
324. **Melton, M. L. and Sheffield, H. G.,** Activity of the anticoccidial compound, Lasalocid, against *Toxoplasma gondii* in cultured cells, *J. Parasitol.*, 61, 713, 1975.
325. **Pfefferkorn, E. R.,** *Toxoplasma gondii:* the enzymic defect of a mutant resistant to 5-fluorodeoxyuridine, *Exp. Parasitol.*, 44, 26, 1978.
326. **Pfefferkorn, E. R. and Pfefferkorn, L. C.,** Arabinosyl nucleosides inhibit *Toxoplasma gondii* and allow the selection of resistant mutants, *J. Parasitol.*, 62, 993, 1976.
327. **Pfefferkorn, E. R. and Pfefferkorn, L. C.,** *Toxoplasma gondii:* isolation and preliminary characterization of temperature-sensitive mutants, *Exp. Parasitol.*, 39, 365, 1976.
328. **Pfefferkorn, E. R. and Pfefferkorn, L. C.,** *Toxoplasma gondii:* characterization of a mutant resistant to 5-fluorodeoxyuridine, *Exp. Parasitol.*, 42, 44, 1977.
329. **Pfefferkorn, E. R. and Pfefferkorn, L. C.,** The biochemical basis for resistance to adenine arabinoside in a mutant to *Toxoplasma gondii*, *J. Parasitol.*, 64, 486, 1978.
330. **Pfefferkorn, E. R. and Pfefferkorn, L. C.,** Quantitative studies on the mutagenesis of *Toxoplasma gondii*, *J. Parasitol.*, 65, 364, 1979.
331. **Remington, J. S., Yagura, T., and Robinson, W. S.,** The effect of Rifampin on *Toxoplasma gondii*, *Proc. Soc. Exp. Biol. Med.*, 135, 167, 1970.
332. **Roberts, C. O., Chaparas, S. D., and McLaughlin, D.,** The use of the plaque assay in chemotherapeutic and dermal hypersensitivity studies on *Toxoplasma gondii*, *Trans. Am. Micros. Soc.*, 95, 470, 1976.
333. **Sheffield, H. G. and Melton, M. L.,** Effect of pyrimethamine and sulfadiazine on the fine structure and multiplication of *Toxoplasma gondii* in cell cultures, *J. Parasitol.*, 61, 704, 1975.
334. **Augustine, P. C., Vetterling, J. M., and Doran, D. J.,** *Eimeria tenella:* growth characteristics of drug-resistant strains in chicks and cell culture, *Proc. Helminthol. Soc. Wash.*, 44, 147, 1977.
335. **Bedrnik, P.,** The sensitivity of *Eimeria tenella* sporozoites, maintained in tissue culture, to Fungizone, *Acta Vet. Brno.*, 39, 473, 1970.
336. **Fayer, R. and McLoughlin, D. K.,** Effect of Amprolium on *Eimeria adenoides* in cell culture, *J. Parasitol.*, 56, 388, 1970.
337. **Itagaki, K., Tsubokura, M., and Otsuki, K.,** Studies on methods for evaluation of anticoccidial drugs in vitro, *Jap. J. Vet. Sci.*, 36, 195, 1974.
338. **McDougald, L. R. and Galloway, R. B.,** *Eimeria tenella*: anticoccidial drug activity in cell cultures, *Exp. Parasitol.*, 34, 189, 1973.
339. **McDougald, L. R. and Galloway, R. B.,** Anticoccidial drugs: effects on infectivity and survival intracellularly of *Eimeria tenella* sporozoites, *Exp. Parasitol.*, 40, 314, 1976.
340. **McDougald, L. R. and Galloway, R. B.,** *Eimeria tenella:* inhibition of development in cell culture by serum from chickens fed anticoccidial drugs, *Z. Parasitenkd.*, 54, 95, 1977.
341. **Ryley, J. F. and Wilson, R. G.,** Studies on the mode of action of the coccidiostat robenidine, *Z. Parasitenkd.*, 37, 85, 1971.
342. **Ryley, J. F. and Wilson, R. G.,** Comparative studies with anticoccidials and three species of chicken coccidia in vivo and in vitro, *J. Parasitol.*, 58, 664, 1972.
343. **Ryley, J. F. and Wilson, R. G.,** Drug screening in cell culture for the detection of anticoccidial activity, *Parasitology*, 73, 137, 1976.
344. **Ryley, J. F., Wilson, R. G., and Betts, M. J.,** Anticoccidial activity of an azauracil derivative, *Parasitology*, 68, 69, 1974.
345. **Strout, R. G. and Ouellette, C. A.,** *Eimeria tenella:* screening of chemotherapeutic compounds in cell cultures, *Exp. Parasitol.*, 33, 477, 1973.
346. **Wilson, R. G.,** Studies on short-term application of drugs to *Eimeria tenella* in tissue culture, *Parasitology*, 73, 283, 1976.
347. **Wang, C. C.,** Studies of the mitochondria from *Eimeria tenella* and inhibition of the electron transport by quinolone coccidiostats, *Biochim. Biophys. Acta*, 396, 210, 1975.
348. **Kumar, P. S., Mohapatra, L. N., Balaya, S., and Kumar, R.,** Haemagglutination antigen from *Toxoplasma gondii* grown in tissue culture, *Indian J. Med. Res.*, 66, 756, 1977.
349. **Tós-Luty, S.,** Investigation on the *Toxoplasma* antigen, prepared from tissue culture, in the complement fixation test, *Wiad. Parazytol.*, 13, 41, 1967.
350. **Krahenbuhl, J. L., Blazkovec, A. A., and Lysenko, M. G.,** The use of tissue culture-grown trophozoites of an avirulent strain of *Toxoplasma gondii* for the immunization of mice and guinea pigs, *J. Parasitol.*, 57, 386, 1971.

351. **Araujo, F. G. and Remington, J. S.,** Protection against *Toxoplasma gondii* in mice immunized with *Toxoplasma* cell fractions, RNA and synthetic polyribonucleotides, *Immunology,* 27, 711, 1974.
352. **Schmunis, G., Weissenbacher, M., Chowchuvech, E., Sawicki, L., Galin, M. A., and Baron, S.,** Growth of *Toxoplasma gondii* in various tissue cultures treated with In·Cn or interferon, *Proc. Soc. Exp. Biol. Med.,* 143, 1153, 1973.
353. **Remington, J. S. and Merigan, T. C.,** Interferon: protection of cells infected with an intracellular protozoan *(Toxoplasma gondii), Science,* 161, 804, 1968.
354. **Bedrnik, P.,** Attempts at transfer of immunity against *Eimeria tenella* in vitro and in vivo, *Folia Parasitol. (Praha),* 24, 277, 1977.
355. **Jones, T. C., Len, L., and Hirsch, J. G.,** Assessment in vitro of immunity against *Toxoplasma gondii, J. Exp. Med.,* 141, 466, 1975.
356. **Kambara, H., Enriquez, G. L., and Takayanagi, T.,** The effects of antiserum on the cyst formation of *Toxoplasma gondii* in tissue culture, *Jap. J. Parasitol.,* 20, 91, 1971.
357. **Long, P. L. and Rose, M. E.,** Immunity to coccidiosis: effect of serum antibodies on cell invasion by sporozoites of *Eimeria* in vitro, *Parasitology,* 65, 437, 1972.
358. **Tós-Luty, S., Dutkiewicz, J., and Uminski, J.,** Influence of different sera on the in vitro reproduction of *Toxoplasma gondii, Acta Parasitol. Pol.,* 19, 227, 1971.
358a. **Shimada, K., O'Conner, G. R., and Yoneda, C.,** Cyst formation by *Toxoplasma gondii* (RH strain) in vitro. The role of immunologic mechanisms, *Arch. Ophthalmol.,* 92, 496, 1974.
358b. **Hoff, R. L., Dubey, J. P., Behbehani, A. M., and Frenkel, J. K.,** *Toxoplasma gondii* cysts in cell culture: new biologic evidence, *J. Parasitol.,* 63, 1121, 1977.
359. **Davis, P. J. and Porter, P.,** A mechanism for secretory IgA-mediated inhibition of the cell penetration and intracellular development of *Eimeria tenella, Immunology,* 36, 471, 1979.
360. **Davis, P. J., Parry, S. H., and Porter, P.,** The role of secretory IgA in anti-coccidial immunity in the chicken, *Immunology,* 34, 879, 1978.
361. **Poger, M. E. and Lamm, M. E.,** Localization of free and bound secretory component in human intestinal epithelial cells. A model for the assembly of secretory IgA, *J. Exp. Med.,* 139, 629, 1974.
362. **Chinchilla, M., and Frenkel, J. K.,** Mediation of immunity to intracellular infection *(Toxoplasma* and *Besnoitia)* with somatic cells, *Infect. Immun.,* 19, 999, 1978.
363. **Frenkel, J. K. and Chinchilla, M.,** Mediation of immunity to *Toxoplasma* and *Besnoitia* within somatic cells, *Fed. Proc.,* 37, 711, 1978.
364. **Shirahata, T., Shimizu, K., Noda, S., and Suzuki, N.,** Studies on production of biologically active substance which inhibits the intracellular multiplication of *Toxoplasma* within mouse macrophages, *Z. Parasitenkd.,* 53, 31, 1977.
365. **Sethi, K. K. and Brandis, H.,** Characteristics of soluble T-cell derived factor(s) which can induce non-immune murine macrophages to exert antitoxoplasma activity, *Z. Immun. Forsch.,* 154, 226, 1978.
366. **Scholtyseck, E., Strout, R. G., and Haberkorn, A.,** Schizonten und merozoiten von *Eimeria tenella* in Makrophagen, *Z. Parasitenkd.,* 32, 284, 1969.
367. **Sénaud, J., Mehlhorn, H., and Scholtyseck, E.,** *Besnoitia jellisoni* in macrophages and cysts from experimentally infected laboratory mice, *J. Protozool.,* 21, 715, 1974.
368. **Wacha, R. S., Hammond, D. M., and Miner, M. L.,** The development of the endogenous stages of *Eimeria ninakohlyakimovae* (Yakimoff and Rastegaieff, 1930) in domestic sheep, *Proc. Helminthol. Soc. Wash.,* 38, 167, 1971.
369. **Ruskin, J., McIntosh, J., and Remington, J. S.,** Studies on the mechanisms of resistance to phylogenetically diverse intracellular organisms, *J. Immunol.,* 103, 252, 1969.
370. **Hoff, R. L.,** In vitro expression of immunity by macrophages from hamsters chronically infected with *Besnoitia jellisoni, Fed. Proc.,* 32, 3549, 1973.
371. **Hoff, R. and Frenkel, J. K.,** Cell-mediated immunity against *Besnoitia* and *Toxoplasma* in specifically and cross-immunized hamsters and in cultures, *J. Exp. Med.,* 139, 560, 1974.
372. **Pelster, B.,** Zellulare Immunreaktionen bei der weissen Maus nach Superinfektion mit *Toxoplasma gondii, Z. Parasitenkd.,* 48, 95, 1975.
373. **Krahenbuhl, J. L. and Remington, J. S.,** In vitro induction of nonspecific resistance on macrophages by specifically sensitized lymphocytes, *Infect. Immun.,* 4, 337, 1971.
374. **Hibbs, J. B.,** Evidence for activation of macrophages and modification of target cells by in vitro association with mycoplasmas, *Clin. Res.,* 21, 212, 1973.
375. **Hibbs, J. B., Lambert, L. H., and Remington, J. S.,** Possible role of macrophage mediated nonspecific cytotoxicity in tumour resistance, *Nature (London) New Biol.,* 235, 48, 1972.
376. **Hibbs, J. B., Taintor, R. R., Chapman, Jr., H. A., and Weinberg, J. B.,** Macrophage tumor killing: influence of the local environment, *Science,* 197, 279, 1977.
377. **Beeck, H., Ullrich, K., and von Figura, K.,** Effect of lectins on endocytosis and secretion of lysosomal enzymes by cultured fibroblasts, *Biochem. Biophys. Acta,* 583, 179, 1979.

378. **Nguyen, B. T. and Stadtsbaeder, S.**, Spontaneous interaction in vitro between lymphocytes and syngeneic peritoneal macrophages of mice, *Infect. Immun.*, 13, 884, 1976.
379. **Berman, J. D. and Johnson, Jr., W. D.**, Monocyte function in human neonates, *Infect. Immun.*, 19, 898, 1978.
380. **Rose, M. E.**, Immune responses in infections with Coccidia: macrophage activity, *Infect. Imm.*, 10, 862, 1974.
381. **Rose, M. E. and Lee, D. L.**, Interactions in vitro between sporozoites of *Eimeria tenella* and host peritoneal exudate cells: electron microscopical observations, *Z. Parasitenkd.*, 54, 1, 1977.
382. **Reikvam, A. and Hoiby, E. A.**, Phagocytosis and microbicidal capacity of mouse macrophages non-specifically activated in vitro, *Acta Path. Microbiol. Scand.*, 83, 121, 1975.
383. **Ryning, F. W. and Remington, J. S.**, Effect of alveolar macrophages on *Toxoplasma gondii*, *Infect. Immun.*, 18, 746, 1977.
384. **Shirahata, T., Shimizu, K., and Suzuki, N.**, Effects of immune lymphocyte products and serum antibody on the multiplication of *Toxoplasma* in murine peritoneal macrophages, *Z. Parasitenkd.*, 49, 11, 1976.
385. **Akinshina, G. T., Abakarova, E. G., and Kirillova, F. M.**, Fate of *Toxoplasma gondii* in macrophages in vitro depending on virulence of the parasites, *Byull. Eksp. Biol. Med.*, 80, 60, 1976.
386. **McLeod, R. and Remington, J. S.**, A new method for evaluation of intracellular inhibition of multiplication or killing by mononuclear phagocytes, *J. Immunol.*, 119, 1894, 1977.
387. **Hammond, D. M. and Fayer, R.**, Cultivation of *Eimeria bovis* in three established cell lines and in bovine tracheal cell line cultures, *J. Parasitol.*, 54, 559, 1968.
388. **Doerr, U. and Hohn, H.**, Zur in vitro-Kultur von *Eimeria stiedae*, *Z. Parasitenkd.*, 3, 56, 1972.
389. **Ogimoto, K., Komatsu, M., Tsunoda, K., and Tanaka, Y.**, Development of *Eimeria* from Japanese quail in cell cultures, *Jap. J. Parasitol.*, 25 (Suppl.), 23, 1976.
390. **Buckley, S. M.**, Survival of *Toxoplasma gondii* in mosquito cell lines and establishment of continuous infection in Vero cell cultures, *Exp. Parasitol.*, 33, 23, 1973.
391. **Akinshina, G. T., Abakarova, E. G., and Kirillova, F. M.**, Fate of *Toxoplasma gondii* in macrophages in vitro depending on virulence of the parasites, *Byull' Eksp. Biol. Med.*, 80, 60, 1975.
392. **Freeman, R. R., Trejdosiewicz, A. J., and Cross, G. A. M.**, Protective monoclonal antibodies recognizing stage specific merozoite antigens of a rodent malaria parasite, *Nature (London)*, 284, 366, 1980.
393. **Rener, J., Carter, R., Rosenberg, Y., and Miller, L. H.**, Anti-gamete monoclonal antibodies synergistically block transmission of malaria by preventing fertilization in the mosquito, *Proc. Natl. Acad. Sci. USA*, 77, 6797, 1980.
394. **Aikawa, M., Rener, J., Carter, R., and Miller, L. H.**, An electron microscopical study of the interaction of monoclonal antibodies with gametes of the malarial parasite *Plasmodium gallinaceum*, *J. Protozool.*, 28, 383, 1981.
395. **Danforth, H. D.**, Development of Monoclonal Antibodies Against Avian Coccidia, 17th Annual Coccidiosis Conf., Montreal, Canada, 1981.
396. **Smith, C. K., II., and Strout, R. G.**, *Eimeria tenella*: accumulation and retention of anticoccidial ionophores by extracellular sporozoites, *Exp. Parasitol.*, 48, 325, 1979.
397. **Pressman, B. C.**, Biological applications of ionophores, *Ann. Rev. Biochem.*, 45, 501, 1976.
398. **Smith, C. K., II., and Strout, R. G.**, *Eimeria tenella:* effect of narasin, a polyether antibiotic on the ultrastructure of intracellular sporozoites, *Exp. Parasitol.*, 50, 426, 1980.
399. **Smith, C. K., II., Galloway, R. B., and White, S. L.**, Effect of ionophores on survival, penetration, and development of *Eimeria tenella* sporozoites in vitro, *J. Parasitol.*, 67, 511, 1981.
400. **Augustine, P. C.**, Effect of polyions, Ca^{++}, and enzymes on penetration of cultured cells by *Eimeria meleagrimitis* sporozoites, *J. Parasitol.*, 66, 498, 1980.
401. **Augustine, P. C. and Danforth, H. D.**, Effect of lectins on penetration of cultured cells by *Eimeria meleagrimitis* sporozoites, *Ann. Meet. Am. Soc. Parasitol.*, 151, 1981.
402. **Miller, L. H., Dvorak, J. A., Shiroishi, T., and Durocher, R.**, Influence of erythrocyte membrane components on malaria merozoite invasion, *J. Exp. Med.*, 138, 1597, 1973.
403. **Miller, L. H., Mason, S. J., Dvorak, J. A., McGinnis, M. H., and Rothman, I. K.**, Erythrocytic receptors for *Plasmodium knowlesi* malaria. Duffy blood group determinants, *Science*, 189, 561, 1975.

Chapter 2

LUMEN DWELLING PROTOZOA: *ENTAMOEBA*, TRICHOMONADS, AND *GIARDIA*

Louis S. Diamond

TABLE OF CONTENTS

I. INTRODUCTION

The following protozoa are considered in this chapter: *Entamoeba histolytica, E. histolytica*-like amebae (Laredo type), *E. hartmanni, E. coli, E. gingivalis* and *E. invadens; Trichomonas vaginalis, T. tenax, T. gallinae, Tritrichomonas foetus* and *Pentatrichomonas hominis; Giardia lamblia,* and *G. duodenalis.*

No attempt has been made to conduct an exhaustive coverage of all known media for cultivating these protozoa. Instead, I have restricted my presentation to media which have had wide acceptance, especially those with which I have had personal experience. More comprehensive coverage can be found in the works of Taylor and Baker.[1,2] Additional useful information will be found in the *Biology Data Book*.[3,4]

In vitro cultivation of protozoa, free-living or parasitic, is presently more an art than a science, and whether the reverse will be true some day is a moot point. In my own experience, which dates back some 45 years, first as an amateur and later as a professional, I have seen little indication of this change. Perhaps, science will have replaced the art when the addition of totally defined nutrients, the removal of metabolic wastes, monitoring of physical and chemical conditions of growth, and harvesting the crop have become automated.

To be successful in cultivating parasitic protozoa, or any cell or organism, one must have infinite patience, persistence and a willingness to devote countless hours and intensive effort to what may be a very elusive goal. The cultivator must be observant, sensitive to the slightest alteration in behavior of the organisms and able to respond appropriately. Since these responses are gained principally through experience, the novice is urged to begin study under supervision of more experienced individuals. Last, one must find excitement and challenge in separating a protozoan from its habitat in the wild and inducing it to take up a new existence in the culture tube.

The terminology used to designate culture systems follows the scheme of Dougherty.[5] Cultures in which the parasite is grown in the absence of all other metabolizing cells are referred to as axenic, with one known associate as monoxenic, two known associates as dixenic, several known associates as polyxenic. The term xenic is applied to cultures in which the associates are unknown. The latter have also been referred to as conventional or crude cultures.

II. *ENTAMOEBA*

Entamoeba histolytica, E. hartmanni, E. coli, and *E. gingivalis* are parasites primarily of man, apes, and monkeys. Their presence has also been recorded in dogs, cats, rats, and pigs.[6] *Entamoeba histolytica*-like amebae, characterized by their ability to grow optimally at 24°C and to survive in vitro under extreme hypotonic conditions, have been recognized only in man.[7] *Entamoeba invadens* is a parasite of reptiles. All inhabit the alimentary tract of their hosts, but only two *E. histolytica* and *E. invadens* are known to be pathogenic. The latter is a commensal in reptilian herbivores, and a pathogen in carnivores.

A. Classification

The taxonomic scheme followed throughout this chapter is one recently adopted by the Committee on Systematics and Evolution of the Society of Protozoologists.[8] The *Entamoeba* are grouped in the family Endamoebidae, order Amebida, class Lobosea, phylum Sarcomastigophora.

B. Geographical Distribution and Incidence

The *Entamoeba* under discussion, with the possible exception of the *E. histolytica*-like amebae, are distributed world-wide. *Entamoeba histolytica*-like amebae have been encountered only in temperate climates.

The incidence of *E. histolytica* in man is difficult to assess despite the numerous surveys conducted. Until recently, most workers considered *E. hartmanni* to be a race of *E. histolytica* and did not differentiate it from the latter. The existence of the *E. histolytica*-like amebae has been known only since 1961. Levine discusses the problem at length.[6]

Reported incidences of *E. histolytica* range from less than 1% to more than 50%. The lower incidences usually being reported in developed countries located in temperate zones, the higher in developing countries situated in warmer climates. The world incidence is probably between 10 and 15%. Relatively few infected persons present symptoms of disease. However, the morbidity and mortality associated with *E. histolytica* infection is of sufficient magnitude to make amebiasis a major public health problem in many areas of the world.

Entamoeba coli is the most common ameba of man. It has been found in 28% of individuals surveyed throughout the world.[6]

C. Life Cycles and Disease

The *Entamoeba* discussed, with the exception of *E. gingivalis*, have two stages in their life cycle: the trophozoite — a motile feeding form, and the cyst — a nonmotile relatively resistant form involved in transmission of the parasite. Both stages have reproductive capacity. *Entamoeba gingivalis* which inhabits the oral cavity, especially the area between the teeth and gingivae, does not have a cyst stage.

The life cycle of *E. histolytica* is as follows: ingested mature cysts excyst in the lower ileum. Colonization occurs throughout the colon. Encystment takes place in the lower colon and the cysts leave the host via feces. In some hosts, for reasons not understood, the trophozoites invade the intestinal wall causing lesions of varying severity and the symptoms of amebic colitis. Amebae spread hematogenously may reach and invade other organs of the body, most frequently the liver, causing severe disease often leading to death if untreated.

D. Cultivation

The cultivation of *E. histolytica* and *E. invadens* has received more attention than any of the other *Entamoeba*, undoubtedly because both are proven pathogens, and the following discussion will center about both species. The latter, although neither a parasite of humans nor of domestic animals is included because its study has, and predictably will continue to broaden our knowledge of the genus.

1. Historical Review

The history of the cultivation of *Entamoeba* is dealt with in three sections: xenic, monoxenic, and axenic culture systems. Those studies considered influential in advancing the technology will receive most attention. Additional references can be found in the review of Neal.[9] Cultivation in the presence of fresh tissue slices or mince is not covered (see References 1, 2).

a. Xenic Cultures

Cultivation of *E. histolytica* was achieved first by Boeck and Drbohlav in 1925, 50 years after discovery of the parasite.[10] The accomplishment was serendipitous. The original cultures had been inoculated with feces as part of a study on intestinal flagellates, a special interest of the senior author. Several media were described and tested, two were recommended: LES and LEA. Both were diphasic consisting of a slant of inspissated whole egg overlayed with Locke's solution and serum (LES) or egg albumin (LEA). Amebic growth was obtained in the presence of a mixed bacterial flora introduced with the fecal inoculum. These media as originally formulated are seldomly used today. However, modifications are in use and one commonly employed is presented in the section on media preparation.

Dobell and Laidlaw made major contributions to xenic cultivation of *E. histolytica* and other entozoa.[11] Recombining the basic ingredients used by Boeck and Drbohlav, they devised several new diphasic media. One, HSre + S, in which serum replaced the inspissated egg slant of LES and LEA is still widely employed. Most important, they introduced the use of rice starch as a source of carbohydrate. In its presence amebic growth was markedly enhanced and encystment was obtained. To this day rice starch or flour is routinely incorporated in all media in which growth of the amebae is dependent upon actively multiplying bacteria. In xenic cultures soluble carbohydrate is utilized so rapidly by the bacteria that little is left for the amebae. Dobell and Laidlaw also reported the cultivation of *E. coli, E. gingivalis,* and *Endolimax nana* in their media.

The inspissated egg or serum slants of these workers were replaced with a liver infusion-peptone agar slant by Cleveland and Sanders,[12] and Cleveland and Collier.[13] Nelson incorporated alcoholic extracts of chicken egg or mammalian tissue in an agar base and overlayed the slant with a phosphate buffered saline.[14]

Several monophasic liquid media have been devised. A widely employed medium is the egg yolk infusion devised by Balamuth and Sandza,[15] later modified by Balamuth.[16] Diamond recently described a medium consisting essentially of Trypticase, yeast extract, gastric mucin, and serum.[17]

The media mentioned can be used for both isolation of *E. histolytica* from the host and long-term maintenance (serial subculture). Some, such as Boeck and Drbohlav's LES, and Dobell and Laidlaw's HSre + S, are useful for cultivation of several other lumen dwelling protozoa.

b. Monoxenic Cultures

The cultivation of *E. histolytica* with single microbial associates was achieved in most cases during pursuit of the goal of axenic cultivation. Except possibly for Robinson's medium,[18] the techniques and media used for monoxenic cultivation are not suitable for isolation of the parasite. Their primary usefulness is in the study of interactions between the ameba and a single microbial species. They have had limited use as a source of amebic antigen for serodiagnosis of amebic disease,[19] and in vitro testing of amebicides.[20]

Cleveland and Sanders were the first to cultivate *E. histolytica* monoxenically.[21] Trophozoites isolated from bacteriologically sterile hepatic abscesses, produced experimentally in kittens, were cultured in the presence of 6 of 13 bacterial species tested. Their diphasic liver infusion-peptone agar medium was used for these cultures.[13]

Nothing further appeared in the literature on this subject until Rees and his associates reported monoxenic cultivation of *E. histolytica* with 12 additional bacterial species in 1941 and 1942.[22,23] Their cultures were initiated with amebic cysts derived from xenic cultures by micro-isolation procedures, and freed of associated bacteria by chemical treatment, a more reliable source of amebae than experimentally induced liver abscesses. Best growth of amebae was obtained with *Clostridium perfringens* or a Gram-negative, micro-aerophilic rod, organism 't'.[24] Cultures of the ameba and organism 't' were studied exhaustively in an effort to define the role played by bacteria in supporting in vitro growth of *E. histolytica*. Hopefully, these investigations would supply the information crucial to development of axenic cultures. These and related studies were summarized by Rees.[25]

Jacobs introduced the use of antibiotics (penicillin) to inhibit multiplication and eventually eliminate a sensitive bacterium from an amebic culture.[26] Further, he showed that *E. histolytica* could be maintained by serial subculture in a medium preconditioned by growth of a bacterium, to which heat-inactivated bacteria were added immediately before inoculation with the ameba. As noted by Jacobs, these cultures were extremely difficult to establish and growth of the ameba was poor. The longest time these cultures were maintained was 6 months.[27]

Jacobs' investigations were extended by Shaffer and his co-workers in studies leading eventually to development of Shaffer-Frye (S-F) medium.[28-31] In the S-F culture system a Gram-negative anaerobic bacterium, *Fusobacterium symbiosum* (= *Bacteroides symbiosus*) was used to precondition a fluid thioglycollate medium. After removal of most of the bacteria and other extraneous particulate matter by centrifugation, multiplication of the remaining organisms was inhibited with penicillin and the medium inoculated with amebae. Later modifications simplified the technique of cultivation and made the system more versatile. In the modified Shaffer-Frye (MS-F) technique of Reeves, Meleney, and Frye, preconditioning was circumvented by introduction of washed bacterial cells into the medium immediately before inoculation with amebae.[32] In the CLG and CTG media of McDade and Shaffer, Trypticase was replaced with other peptones.[33] The S-F medium, and particularly the above modifications are probably the most widely employed for monoxenic cultivation of *E. histolytica*. *Entamoeba histolytica*-like amebae, *E. invadens*, *E. coli*, *Balantidium coli*, *Trichomonas vaginalis*, and *Pentatrichomonas hominis* have been grown in these media.[30,34]

A novel approach to monoxenic cultivation was Phillips' use of the protozoan, *Trypanosoma cruzi* as the associate of the amebae.[35,36] The first cultures were derived from a prototype of S-F medium.[28] The trypanosome was substituted for the bacterial associate which was gradually eliminated with penicillin. Later, the ameba-trypanosome (A-T) cultures were initiated with cysts micro-isolated from xenic ameba-bacteria cultures.[37] Pan showed that several other trypanosomatids (*Trypanosoma lewisi*, *Leishmania donovani*, *L. brasiliensis*, *L. enriettii*, *Endotrypanum schaudinni*) could also support monoxenic cultivation of *E. histolytica*.[38] Diamond employed amebae grown with a trypanosomatid (*Crithidia* sp.) isolated from the cog-wheel bug (*Arilus cristatus*) to initiate the first axenic cultures of *E. histolytica*,[39] and devised improved media for cultivation of *Entamoeba* with trypanosomatids.[40]

Entamoeba invadens was grown monoxenically the first time by McConnachie.[41] She used HSre + S medium and *Staphylococcus albus* or *Escherichia coli* as the monobacterial associate. The ameba has been grown monoxenically in S-F type media by Reeves and Bischoff,[34] and by Balamuth in egg yolk infusion medium with *Clostridium perfringens*.[42]

Entamoeba hartmanni appears to have been cultivated monoxenically one time. Reardon and Bartgis grew it with organism 't'.[43] *Entamoeba coli*, as far as I am aware, has not been grown monoxenically.

c. Axenic Cultures

The first of the lumen dwelling amebae to be cultivated axenically was *E. invadens*. Stoll accomplished this in 1957 using a complex monophasic liquid medium consisting of a heat-labile (raw liver extract) and a heat-stable (liver infusion-gastric mucin-peptone) portion.[44,45] Axenic cultures were initiated with amebae from Miller-type cultures in which the parasites were grown bacteria-free in the presence of fresh liver.[46] Diamond cultivated *E. invadens* axenically in TYM, a liquid medium devised for cultivation of trichomonads.[47,48] It consisted of an autoclavable broth of Trypticase, yeast extract and maltose, and a filter sterilized serum supplement. Advantages of TYM over Stoll's were its clarity, simplicity of preparation, and ability to support greater amebic growth. The cultures were initiated with amebae grown in Miller-type cultures,[46] or monoxenically with *Trypanosoma cruzi* (technique of Phillips).[36] McConnachie reported axenic growth with a simple, autoclavable liquid medium of Trypticase, Panmede (enzymatic digest of ox-liver), glucose, and saline.[49] Later she showed serum supplemented medium enhanced amebic growth approximately 10-fold.[50] Diamond reported TP-S-1, a medium devised primarily for axenic cultivation of *E. histolytica*, supported luxuriant growth of the reptilian parasite.[51] Van Vliet, Op Den Kamp, and Van Deenen replaced the serum supplement of this medium with a mixture of defined ingredients: albumin, oleic acid, cholesterol, and Tween 80.[52] Unfortunately, this serum substitute does not support axenic growth of *E. histolytica* (Diamond, unpublished data). Rengpien and Bailey devised a medium for mass encystation of *E. invadens* under axenic conditions.[53]

Axenic cultivation of *E. histolytica* was reported in 1961 by Diamond.[39] A diphasic medium was used consisting of a serum enriched nutrient agar slant overlayed with a broth supplemented with a cell-free extract of chick embryos (CCE) and vitamins. A small quantity of nutrient agar included in the overlay aided in maintenance of a low redox potential and provided nutrient binding sites which could be readily engulfed by the phagocytic amebae. The extracts were prepared from quick-frozen or boiled embryos and freed of intact cells by centrifugation. Equivalent growth was obtained with either extract. Results with the boiled embryos demonstrated conclusively that growth was not dependent on viable cells. The cultures were initiated with trophozoites obtained from monoxenic ameba-crithidia (A-C) cultures.[40] Later studies showed that CEE although necessary for axenization was not required for subsequent maintenance of cultures. Moreover, when Panmede, a ox-liver digest was substituted for yeast extract, an ingredient of the original medium, axenization could be attained in the absence of CEE.[51] Following introduction of Panmede, a clear liquid medium, TP-S-1, was devised for both initiation and maintenance of axenic cultures, and a technique for mass cultivation developed. Problems encountered later in obtaining useable lots of Panmede led to formulation of a more efficient medium, TYI-S-33, in which the liver digest was replaced with yeast extract supplemented with iron, vitamin B_{12}, thioctic acid, and Tween 80.[54] In addition to *E. histolytica* and *E. invadens* the following have been cultivated in both TP-S-1 and TYI-S-33: *E. histolytica*-like (Laredo type) amebae, *E. terrapinae, E. barreti,* and the free-living *E. moshkovskii*. Gillin and Diamond described a method for cloning axenized *Entamoeba* in deep agar suspensions.[55,56] Diamond published a detailed account of the development of axenic cultures of *E. histolytica*.[57]

2. Media Preparation and Techniques of Cultivation

Glass containers of borosilicate are preferred for cultures. They are essential for axenic cultivation of *E. histolytica*. My attempts to cultivate the parasite in soda-lime glass available in the U.S. have usually failed. Sterile plasticware is useable, but in my hands has never provided amebic yields obtainable with borosilicate glassware. Culture containers fitted with screw-capped, rubber lined enclosures are preferred over cotton-plugged or rubber-stoppered containers. Henceforth unless otherwise stated culture tube refers to a 16 × 125 mm, screw-capped tube, and Erlenmeyer flask to a 250 or 300 mℓ capacity flask.

Double distilled water is used exclusively in my laboratory, the second distillation always being performed in glass apparatus. Demineralized water has been used by some workers.

To transfer organisms for subculture, I routinely use serological pipets in place of the more commonly employed Pasteur pipets, the former permit more precise control of inoculum size.

Sterilization of media by autoclave is most commonly carried out for 15 min, 121°C, 15 pounds pressure. For sterilization by filtration, I prefer membrane filters (e.g., Millipore, Nucleopore, pore size 0.45 μM). I have also used porcelain filters (Selas #3).

Customarily, media are tested for sterility by incubating them for 24 to 48 hr at 37°C. In the case of media used for axenic cultivation this is not recommended, since as little as 24 hr incubation will reduce their ability to support growth. One must rely entirely on aseptic procedures.

Most media are safely stored in the refrigerator at 4 to 5 °C. Storage life varies from a few days to several months. A medium may have a storage-life of 2 or 3 months when used for maintaining established cultures, but only 1 or 2 weeks when used for isolation and initiation of cultures. In general, media devised primarily for xenic cultivation are storable for several weeks or months, while those devised for monoxenic or axenic cultivation are storable a week or two.

Sera used in preparation of culture media, unless otherwise stated, have been inactivated at 56°C, 30 min.

Frequent reference will be made to media and techniques used in initiation of cultures and to those employed for maintenance of cultures. This is a necessary distinction since the requirements of the one are not necessarily the requirements of the other. Furthermore, an established culture is defined as one which has survived ten subcultures, and yields consistent harvests from a specified inoculum within a designated period of incubation.

a. Xenic Cultivation

1. Media

a. LE (Locke-Egg) Medium

(Readron and Rees' Modification of Boeck and Drbohlav's LES Medium)[10,58,59]

Although this modification was first reported in 1939,[58] details were not presented until 1943.[59] LE medium differs from LES in the use of Stone's modification of Locke's solution,[60] and the omission of serum in the overlay. The medium as presented below is from the files of the Media Section of the National Institutes of Health, Bethesda, Maryland. It is frequently referred to in the literature as the NIH modification.

Locke's Solution

Sodium chloride	8.00 g
Calcium chloride	0.20 g
Potassium chloride	0.20 g
Magnesium chloride	0.01 g
Sodium phosphate, dibasic	2.00 g
Sodium bicarbonate	0.40 g
Potassium phosphate, monobasic	0.30 g
Distilled water	1000 mℓ

Add ingredients to water, dissolve by stirring. Heat 15 min in flowing steam (Arnold sterilizer or autoclave). Precipitate will form. Cool overnight at room temperature (RT). Remove precipitate by filtration (Whatman #1 paper). Dispense into containers, sterilize by autoclaving.

Egg slant — Select fresh hen's eggs. Scrub with sterile distilled water, dip in 70% ethanol and flame. Break eggs into graduated cylinder. Measure. Pour into Erlenmeyer flask containing glass beads, or preferably into electric blender of the Waring® type. Add Locke's solution (12.5 to 45 mℓ egg) and thoroughly emulsify eggs. Filter through gauze into flask. Place under vacuum to draw out air bubbles. Dispense in 5 mℓ amounts to culture tubes. Inspissate egg in flowing steam (Arnold sterilizer) or autoclave. In either case tubes are positioned to produce a short butt and 40 to 50 mm slant. Inspissation in flowing steam is carried out at a temperature of 80°C, 60 to 70 min. In using the autoclave for inspissation the following procedure will ensure smooth surfaced slants free of bubbles. Allow pressure in autoclave jacket to reach 15 pounds. Position tubes properly in chamber. Close all exhaust valves and the chamber door to entrap air in chamber. Allow steam from jacket to enter chamber as fast as possible until an internal pressure of 15 pounds is reached. Hold pressure 15 min. Shut off steam supply, keep all valves and door tightly closed, allow pressure to drop gradually. Remove slants from inspissator, cool to RT.

Complete medium — Overlay slants with 6 mℓ of Locke's solution. Sterilize in autoclave. When cool add rice starch or flour (see item f below, Rice Supplement). Incubate 24 to 48 hr to test sterility. Store at 4°C. Use within 2 months.

b. HSre + S (Horse Serum-Rice-Egg-Serum) Medium
(Dobell and Laidlaw)[11]

Ringer's solution

Sodium chloride	9.00 g
Potassium chloride	0.20 g
Calcium chloride	0.20 g
Distilled water	1000 mℓ

Sterilize in autoclave.

Serum slant — Dispense sterile inactivated horse serum in 5 mℓ amounts to culture tubes. Inspissate at 80°C, 70 min, in slanted position to produce a slant 40 mm in length without a butt. Incubate 24 to 48 hr to check sterility.

Overlay — Select fresh hen's eggs. Scrub with sterile distilled water, dip into 70% ethanol, remove and flame. Cut a small hole with sterile scissors in blunt end of egg. Invert over Erlenmeyer flask containing Ringer's solution. Puncture other end of egg and allow albumin to run out of large hole. Mix thoroughly. Proportion of albumin to Ringer's solution is 4 eggs/ℓ.

Complete medium — Add to each slant sufficient egg albumin-Ringer mixture to cover. Add rice starch. Incubate 24 to 48 hr to check for sterility. Store at 4°C. Use within 1 week.

c. AE (Alcoholic Extract) Medium
(Nelson)[14]

Buffered saline (pH 7.4)

Sodium phosphate, dibasic, M/30, in 0.5% saline	82.0 mℓ
Potassium phosphate, monobasic, M/30 in 0.5% saline	18.0 mℓ

Sterilize in autoclave.

Alcoholic extract (10%) — Fresh liver (human, calf, beef, guinea pig, or cat) cut into small pieces; intestine (cat) split open, contents removed, rinsed in saline, cut into pieces; or fresh hen's egg yolk can be used. Nine hundred milliliters of 95% ethanol are placed in a bottle and the tissue or egg yolk added to bring volume to 1000 mℓ. In using egg yolk I prefer to place it in a beaker, beat it slightly for a few seconds, then add it to the ethanol. The bottle with its contents is shaken occasionally. The extract is ready to use in 48 hr, but it is better to wait a few days until the supernate clears. It is drawn off when needed. Nelson reported using the extracts for more than 2 years with little loss of potency.

Agar slant — Place 50 mℓ of extract in Erlenmeyer flask and set flask in boiling water bath to evaporate alcohol. Add 100 mℓ of sterile molten 2% (w/v) agar prepared in buffered saline. Agitate flask to redissolve sediment. Some soluble oil and waxy material may remain adherent to glass, but is of no consequence. Dispense while hot in 2 mℓ amounts to culture tubes. Slant tubes in almost horizontal position to obtain steep slants.

Complete medium — Overlay slants with 8 mℓ of sterile buffered saline containing 0.025% (w/v) Bacto agar. Add rice starch. Incubate 24 to 48 hr, to test sterility. Store at 4°C. Use within a few weeks.

d. EYI (Egg Yolk Infusion) Medium
(Balamuth and Sandza, Balamuth)[15,16]

Egg yolk infusion — Boil four fresh hen's eggs 15 min. Discard whites and place yolks in Waring® blender containing 125 mℓ of 0.8% (w/v) saline solution. Blend 2 to 3 min, or blend 36 g dehydrated egg yolk with equal volume of distilled water and 125 mℓ of 0.8% saline. Heat either egg preparation to 80°C in double boiler. Maintain constant heat 20 min. Add 20 mℓ distilled water to compensate for evaporation. Strain through two layers of muslin or four layers of gauze previously soaked in distilled water. Measure filtrate and restore to original volume with 0.8% saline. Autoclave 20 min, 15 pounds, 121°C to precipitate protein. Remove from autoclave taking special care not to agitate. Cool to RT. Store overnight at 4°C. Carefully siphon off supernate and pass through filter paper (Whatman #1) held in a Buchner funnel. Measure filtrate and add equal amount of M/15 phosphate buffer (pH 7.5) prepared from monobasic and dibasic potassium phosphate salts. Dispense in 25 mℓ amounts to tall tubes. Sterilize by autoclaving. Final pH 7.3. Store at 4°C. Can be used for several months after preparation.

Complete medium — Just prior to use bring egg yolk infusion to RT. Dispense clear supernate in 8 mℓ amounts to culture tubes. Add rice starch. Balamuth suggested supplementing the medium with liver extract when it was used for diagnostic purposes.[16] The liver preparations recommended are no longer available. Acceptable substitutes are Panmede (Paines and Byrne), Oxoid liver infusion (L-25), Oxoid liver digest neutralized (L-27). These can be conveniently added in a final concentration of 0.2% w/v to the egg yolk infusion medium just before autoclaving.

e. TYSGM-9 (Trypticase-Yeast Extract-Serum-Gastric Mucin) Medium
(Diamond)[17]

Nutrient broth

Potassium phosphate, dibasic		2.8 g
Potassium phosphate, monobasic		0.4 g
Sodium chloride		7.5 g
Trypticase (BBL)*		2.0 g
Yeast extract (BBL)*		1.0 g
Distilled water	q.s.	970 mℓ

*Biosate (BBL), 3 g, can be substituted for Trypticase and yeast extract.

Complete medium — Place 200 mg gastric mucin in dry, sterile, screw-capped Erlenmeyer flasks or bottles (125 mℓ capacity). Add 97 mℓ of broth to each flask or bottle. Autoclave to solubilize gastric mucin and sterilize medium. Cool to RT. Add aseptically 3 mℓ bovine serum and 0.5 mℓ Tween 80 (10% v/v in absolute ethanol). The medium can be stored in bulk at this point or dispensed as follows. Swirl vigorously to keep insoluble particulates of gastric mucin in suspension and distribute in 8 mℓ amounts to culture tubes. Add 2 to 3 mg rice starch or 4 to 6 mg unpolished rice flour per mℓ medium (see below, Rice supplement). Incubate 24 to 48 hr to test sterility. Store at 4°C. Can be used a month or more. The pH of the medium is 7.2, the osmolality 320 mOsm/kg. The gastric mucin used is a product derived from hog stomach prepared according to New and Non-Official Remedies (N.N.R.) 1953. Gastric mucin #16030, U.S. Biochemical Corp., meets the requirement.

f. Rice Supplement

As noted earlier, rice has been added to all media for xenic cultivation of *Entamoeba* since Dobell and Laidlaw's discovery of its importance in enhancing growth. These investigators used purified rice starch, others used unpolished rice flour, and some have used commercial rice starch. Many workers fail to indicate the rice product employed. Often no distinction is made between flour and starch. Amebic yields are clearly related to the quantity and quality of rice employed.[9,61,62] Nevertheless, most workers are content to use a loopful (bacteriological) or a pinch of rice. A few use weighed amounts. For a more thorough discussion of rice and the various forms used in xenic cultures consult Nelson and Jones.[62]

I have used unpolished rich flour (e.g., Bacto Rice Powder), custom milled whole brown rice, and purified rice starch prepared by a modification of Griffin and McCarten's method which is presented below.[63] Of these, purified rice starch provides the highest amebic yields. Starch grains are discrete and free of debris present in most rice preparations, a distinct advantage when examining a culture.

I use varying amounts of rice depending on the preparation, the type of medium, and whether the medium is to be used for isolation and initiation of a culture or maintenance of established cultures. To initiate a culture in liquid medium, I add 2 to 3 mg of purified rice starch per mℓ of medium, and for diphasic medium the same amount per mℓ of overlay. For established cultures half of these amounts are used. They are doubled for both initiation and maintenance when unpolished rice or milled whole rice is used.

The following is a convenient method for sterilizing rice and assuring uniform distribution. Introduce 500 mg of rice into each of several culture tubes. Do not tighten caps. Place in dry heat sterilizer in horizontal position, making sure that rice is distributed loosely over lower surface of tube. Heat treat 2.5 hr at 150°C. Allow to cool. Tighten caps. Store at RT. To use, add 9.5 mℓ of 0.8% (w/v) sterile saline or phosphate buffered saline (pH 7.4). Suspend rice by vigorous shaking, preferably with a vortex-type mixer. Dispense in required volumes to tubes of culture medium. To do this, thoroughly mix suspension each time before withdrawing it from stock tube and take up in a pipet a volume sufficient for five culture tubes.

Most workers add rice immediately before inoculating medium. Since adding rice earlier is not detrimental, a more convenient method is to add it when the medium is placed in tubes. The medium and rice supplement are then tested together for sterility by incubating 24 to 48 hr.

g. Preparation of Purified Rice Starch
(Griffin and McCarten,[63] as Modified by McCarten)

(1) Place 300 g rice (unpolished or milled whole brown kernel) in Waring® blender. Add distilled water to make a thick slurry and permit grinding. Grind for 2 min. (2) Add 0.5 g Difco Trypsin (1:250) per 100 g rice. Adjust pH to 7.5 with 10% (w/v) sodium bicarbonate. (3) Incubate at 37°C, readjusting pH to 7.5 until it remains constant (2 to 4 hr). (4) Transfer to 1 ℓ graduated cylinders. Add distilled water bringing total volume to 1000 mℓ. Mix by inverting several times. (5) Allow to settle until a sharp line of demarcation appears. Siphon off supernate and save. (6) Repeat steps 4 and 5 until only a few rice particles remain in the sediment. Discard sediment. (7) Pool the rice-rich supernates and centrifuge. Discard the newly formed supernate. Scrape the brownish layer off of sediment and discard. (8) Repeat steps 2 through 7. (9) Suspend rice in water. Add 1/3 volume of toluene, shake vigorously. (10) Centrifuge — Discard supernate. Scrape brownish layer off of sediment and discard. (11) Repeat steps 9 and 10 until the toluene-starch interface is practically clear and the brownish light component eliminated. (12) Suspend in an equal volume of acetone. Centrifuge and discard supernate. Repeat twice to remove all water. (13) Remove the acetone from the packed rice starch in vacuo. (14) Dry to a constant weight by heating at 110°C.

2. Techniques of Cultivation

Each of the media described above has its advocates. The novice, in selecting a medium, will do well to heed Craig's advice.[64] "A medium used by one thoroughly trained in its use, and having a long experience with it, will give better results than one used by an untrained observer with little practical experience, and it is best to select a certain medium and use it indefinitely rather than try one medium after another without learning how to use any to the best advantage."

Insofar as diagnosis of *E. histolytica* infection is concerned cultures serve as a supplemental procedure. They do not replace diagnosis by microscopic examination of feces, biopsy material, exudates, etc. They are most useful in detecting cases of intestinal infection.

It is relatively easy in the initial culture to grow out amebae in sufficient numbers to make a diagnosis. It is more difficult to nurture and maintain them in serial subculture, and this accounts, partially, for the existence of so few established cultures. In either case, control of the associated bacterial flora is the key to successful cultivation. Bacteria favorable to growth must be conserved and those which are harmful eliminated, especially those capable of rapidly degrading rice. In some cases a whole flora or a species known to favor amebic growth is introduced into a culture. Bacterial overgrowth must be avoided. This is especially important in initiating cultures.

The method commonly employed to control bacterial growth is to introduce dyes with bacteriostatic properties (e.g., acriflavine and gentian violet) and/or antibiotics, especially penicillin and streptomycin. Spingarn and Edelman used 1000 to 2000 units penicillin G, or 1 to 2 mg streptomycin sulfate/mℓ of overlay to isolate *E. histolytica* from feces.[65] Norman and Brooke used a combination of both of these antibiotics and in much lower concentrations.[66] Dobell and Laidlaw were the first to use acriflavine. Das and Singh used a combination of acriflavine and gentian violet.[67] These dyes because of their availability, relative low cost, and stability are recommended for use in areas of the world in which antibiotics are not readily available. After a culture has been established it may or may not require continued use of the antibacterial substances.

Mention has been made of introducing a whole flora into cultures to improve and stabilize amebic growth. I have had unusual success with the NRS flora. This flora was derived from the NRS strain of *E. histolytica* isolated from feces of a *Macaca* monkey by Dobell in 1929.[68] In 1940, Dr. C. W. Rees then at the NIH, received this strain from Dr. H. E. Meleney. The flora was separated from the amebae and has been maintained at the NIH for the last 40 years. A recent analysis of the flora performed by Dr. V. J. Gill, Clinical Pathology Department, Clinical Center, NIH, showed the presence of the following: *Escherichia coli, Streptococcus faecium* (Group D), *Clostridium butyricum, Clostridium* sp., *Eubacterium* sp., and *Fusobacterium symbiosum*. Of special interest, the *Fusobacterium symbiosum* used in S-F medium was derived from this flora.[28]

The NRS flora has an uncommon ability to induce cyst formation in *E. histolytica*. I have also used it to initiate cultures from bacteria-free microisolated cysts. Added as a supplement to existing flora it has improved growth, and in some cases has permitted establishment of strains whose associated bacteria would not support the amebae. I have employed the NRS flora in cultivation of *E. histolytica, E. coli, E. gingivalis*, and an *Entamoeba* resembling *E. gingivalis* isolated from the uterus of an IUD user.[69] *Entamoeba gingivalis* is difficult to maintain in vitro, apparently because the oral bacteria isolated with the amebae are frequently unfavorable for their growth. Recently, using the NRS flora as a supplement I established four isolates in culture which could not be maintained with the patient's flora.

a. Diphasic Media

Bring two tubes of medium to RT. To the first add 200 to 300 units penicillin G sodium, and 200 to 300 μg streptomycin sulfate/mℓ of overlay. No antibiotics are added to the second tube; the patients' flora may be compatible and will not need manipulation. Select from the surface of the stool specimen a bit of feces the size of a small pea and emulsify in 1 or 2 mℓ of sterile saline. Include blood and/or mucus if present. Divide emulsion into two equal portions and introduce each into a tube. Incubate at 35.5°C with screw-cap tightly closed. After 48 hr withdraw a small sample of sediment from the interface of the overlay and base of the slant, place on slide, and examine with microscope. If no amebae are found make a blind passage as follows. Gently agitate sediment with a pipet and withdraw 1 mℓ. Transfer to tube of fresh medium. Add antibiotics only to that tube receiving an inoculum from an antibiotic containing culture. Restore total volume of the parent culture by addition of fresh overlay fluid. Place parent and newly initiated culture in incubator. At end of two successive 48 hr periods examine for amebae. If none are found after second examination discard cultures.

In the event amebae grow out in any culture and it is desired to perpetuate the isolate in vitro, subculturing at regular intervals is initiated.

To subculture, most workers transfer a portion of the old culture directly to tubes of fresh medium. To give me more precise control of the inoculum, I use the following method. Chill old culture 5 min in ice-water. With a 2 mℓ pipet take up all of the amebae-bearing sediment located between the interface of the overlay and the base of the slant. Place in a clean sterile tube and mix thoroughly. Examine a drop with the microscope to determine vigor and density of the amebae. The number of amebae transferred can only be determined by trial and error. In the early stages of establishing a culture transfer one-half of the material taken from the old culture to each of two tubes of fresh medium. Later as amebic growth improves transfer smaller portions, e.g., one-third or one-fourth. If only a few amebae are present transfer all the sediment to a single fresh tube. Subculture three times per week.

In establishing cultures, concentrations of antibiotics may have to be increased or decreased to control bacterial populations. Their use may be continuous or intermittent. Rice-splitting bacteria can be eliminated frequently through use of acriflavine (50 to 100 μg/mℓ of overlay) for one or two subcultures. Ridding the cultures of these bacteria also facilitates elimination of *Blastocystis hominis*, a common and annoying protozoan contaminant.

b. Liquid Media

These media have a distinct advantage over diphasic media. Their relative clarity permits examination of the amebae *in situ* without the necessity of sampling the culture between transfers in order to monitor them.

Cultures are initiated by inoculating them in the same manner as are the diphasic media. The concentrations of antibiotics used are 100 to 200 units penicillin G, potassium, and 100 to 200 μg streptomycin sulfate/mℓ of medium. Incubate with screw-cap tightly closed, 48 hr. To examine a culture, gently invert tube several times to disperse its contents uniformly. Place in incubator at a slight angle (5°) for 1 hr. Remove from incubator and examine with inverted microscope fitted with wide-angle occulars and an Abbe type condenser. Avoid disturbing contents of tube. Search entire length of tube using a magnification of 100×. Amebae when present will be attached to the glass walls, a few to clumps of bacteria. If no amebae are found, or only a few are observed, centrifuge culture 3 min, 275 g. Carefully remove supernate with pipet and discard. Replenish culture with fresh medium, add rice if depleted. Incubate another 48 hr. Examine. Amebae are usually present in sufficient numbers at this time to allow subculture. Occasionally a second refeeding may be required. If no amebae are seen after an additional 48 hr incubation, the culture can be discarded.

To subculture freshly isolated cultures, chill tubes 10 min in ice-water, invert 10 times to loosen attached amebae, centrifuge 3 min, 275 × g. Decant and discard supernate. Resuspend sedimented amebae in 1.0 mℓ of fresh medium and transfer 0.5 mℓ to each of two tubes of medium. As growth improves with successive subcultures the size of the inoculum can be decreased. Subculture three times per week.

In TYSGM-9 medium well established cultures (15 or more subcultures) can be incubated in a slanted position (15°). This maneuver is accompanied by increased amebic yields which in turn permits simplification of subculturing. The heavy growth of amebae circumvents the need to concentrate them by centrifugation and transfer can be made directly. The cultures are chilled 10 min and inverted as in the previously described technique. Aliquants are transferred directly with a pipet to fresh medium. The size of the inoculum can vary from 0.3 to 1.5 mℓ of the old culture depending on the species and/or strain of ameba. The need to use more than 1.5 mℓ to maintain the culture signifies that the amebae are not present in sufficient number to use the direct method of transfer.

Acriflavine treatment of liquid medium is performed with lower concentrations (2.5 to 10 μg/mℓ) than used with diphasic medium.

b. Monoxenic Cultivation

1. Media for Growth with Bacteria

a. MS-F (Modified Shaffer-Frye) Medium
(Reeves, Meleney, and Frye)[32]

Nutrient broth

Trypticase (BBL)		20.0 g
Glucose		10.0 g
Sodium chloride		2.5 g
Potassium phosphate, dibasic.$3H_2O$		2.0 g
Thiomalic acid		1.5 g
Distilled water	q.s	1000 mℓ

Dissolve Thiomalic acid in water and neutralize with 1 *N* NaOH. Add salts, Trypticase, and glucose, dissolve by stirring. Dispense in 12 mℓ portions to culture tubes. Autoclave. Store at 4°C. Use within 1 week. A 10 × concentrated solution of Thiomalate and salts can be stored for several weeks.

Medium for bacterial associate: Fusobacterium symbiosum (=Bacteroides symbiosus) **ATCC #14940 —** Prepare basal medium, add 2% (w/v) yeast extract. Autoclave. Store at 4°C. Use within 72 hr.

b. CTG (Cysteine-Tryptone-Glucose), CLG (Cysteine-Lactalbumin-Glucose) Media
(McDade and Shaffer)[33]

Nutrient broth

Tryptone (Difco) or		20.0 g
Lactalbumin hydrolysate (ICN-NBC)		
Yeast extract (Difco)		10.0 g
Glucose		5.0 g
Sodium phosphate, dibasic		4.0 g
L-Cysteine hydrochloride		0.5 g
Distilled water	q.s.	1000 mℓ

Add ingredients to distilled water, dissolve by stirring. Autoclave. Dispense in 40 mℓ portions to large culture tubes (25 × 200 mm).

2. Media for Growth with Trypanosomatids

c. TTY-SB (Tryptose-Trypticase-Yeast Extract-Serum-Blood) Medium*
(Diamond)[40]

* For cultivation of amebae with *Crithidia* sp. or *Trypanosoma cruzi*, and maintenance of stock cultures of *Crithidia* sp. *Crithidia* sp. (Strain ReF-1:PRR) isolated from feces of cog-wheel bug, *Arilus cristatus*.[40] *Trypanosoma cruzi* (Culbertson strain ATCC #30013) isolated from a human.[40]

Nutrient broth TTY

Tryptose (Difco)		10.0 g
Trypticase (BBL)		10.0 g
Yeast extract (BBL)		10.0 g
Glucose		5.0 g
L-Cysteine hydrochloride		1.0 g
Ascorbic acid		0.2 g
Sodium chloride		5.0 g
Potassium phosphate, dibasic		0.8 g
Potassium phosphate, monobasic		0.8 g
Distilled water	q.s.	950 mℓ

Dissolve salts in 600 mℓ distilled water. Add Tryptose and dissolve with heat (50°C) if necessary. Add remaining ingredients dissolving each by stirring. Bring total volume to 950 mℓ with distilled water. Adjust pH to 7.0 with 1 *N* NaOH. Dispense in 190 mℓ portions to Erlenmeyer flasks. Autoclave. Cool to RT.

Complete medium TTY-SB — Add aseptically to each flask 10 mℓ bovine serum and 0.5 mℓ defibrinated rabbit blood. Dispense in 25 mℓ portions to 125 mℓ Erlenmeyer flasks for maintenance of stock crithidia cultures, and in 8 mℓ portions to culture tubes for ameba-crithidia (A-C) cultures. Store at 4°C. Use within 10 days. Blood is frozen and thawed to lyse erythrocytes. Can be stored at −22°C for 1 year.

d. TTY-6-SB (Tryptose-Trypticase-Yeast Extract-Serum-Blood) Medium* (Diamond)[40]

Nutrient broth TTY-6

Tryptose (Difco)		5.0 g
Trypticase (BBL)		5.0 g
Yeast extract (BBL)		5.0 g
Sodium chloride		5.0 g
Potassium phosphate, dibasic		0.8 g
Potassium phosphate, monobasic		0.8 g
Distilled water	q.s.	800 mℓ

Prepare as in TTY. Adjust pH to 7.0. Dispense in 160 mℓ portions in Erlenmeyer flasks. Autoclave. Cool to RT.

Complete medium TTY-6-SB — Add aseptically to each flask of broth 38 mℓ of inactivated rabbit serum and 2 mℓ frozen and thawed defibrinated rabbit blood. Dispense in 12 mℓ portions to 125 mℓ Erlenmeyer flasks. Store at 4°C. Use within 2 weeks.

3. Techniques of Cultivation

A few monoxenic cultures have been initiated with amebae derived directly from host material including human liver aspirates, pleural exudates, and pus from a cutaneous lesion.[57] Most, however, have been initiated with amebic trophozoites or cysts from sources heavily contaminated with bacteria, e.g., xenic cultures or feces. Elimination of these unwanted bacteria is a major problem. Various means have been used to accomplish this including microisolation,[22] treatment of the cultures with chemical disinfectants and/or bacteriostatic dyes,[41,70] and antibiotics.[26,30,42,72,73]

In my early studies I combined Rees' microisolation procedure with antibiotic treatment to free cysts of associated bacteria, and antibiotics alone to free trophozoites.[40] In subsequent

* For maintenance of *T. cruzi* stock cultures.

studies I have tended to initiate these cultures with trophozoites and have relied entirely on mixtures of antibiotics. The following has proved most useful: penicillin G, potassium, 1000 units; streptomycin sulfate, 1000 μg; gentamycin, 30 μg; and oxytetracycline, 30 μg/mℓ of medium. Occasionally, I have added kanamycin sulfate, 30 μg. The mixture is prepared so that 1 volume added to 24 volumes of medium provides the required concentrations of antibiotics. Usually, the offending bacteria are eliminated within four to eight successive treatments.

In monoxenic culture systems the microbial associate is usually maintained independent of the amebae and is only added at the time of subculture of the latter.

a. MS-F Technique

The associate, *Fusobacterium symbiosum*, may be used either in the fresh state or after storage. The choice is dictated by the needs of the investigator. Freshly grown bacteria are obtained from stock cultures transferred daily. Six-tenths mℓ of the old culture is transferred to tubes of yeast supplemented basal medium (Section II.D.2.b) and incubated at 37°C. Bacteria to be stored are grown in flasks containing 400 to 800 mℓ medium. Inocula are taken from 24 hr tube cultures after several daily transfers and are equal to 2 to 5% of the volume of medium in the flask. These flasks are incubated between 16 and 18 hr (never more than 24 hr). To harvest, centrifuge cultures in large sterile bottles, 45 min, 2000 g. Transfer sediment to 50 mℓ centrifuge tubes. Wash twice with normal saline containing 0.02 *M* sodium thiomalate. Store under 0.1 *M* thiomalate at 4°C. Preparations can be used up to 3 months.

M-SF cultures are maintained as follows. To each screw-capped tube containing 12 mℓ basal medium add 0.1 to 0.5 mℓ sterile fresh horse serum, 5000 units penicillin G, potassium, and 2 mℓ freshly grown *F. symbiosum* or its equivalent of washed and stored cells. Immediately afterward transfer per mℓ of medium 500 to 1000 amebae from the old culture. In established cultures amebic yields range between 10 and 50,000 amebae/mℓ after 48 to 72 hr incubation.

b. CTG, CLG Technique

Place 40 mℓ of basal medium (Section II.D.2.b) in large tubes (25 × 200 mm) and inoculate with 0.3 mℓ of a 48-hr culture of *F. symbiosum*. Incubate 24 hr, 37°C. Decant supernate, collecting it in 50 mℓ centrifuge tubes. Overlay with melted petrolatum. Centrifuge 45 min, 900 g. Collect supernate, it should contain between 90 and 140 million bacteria and have a pH of 6.2 to 6.4. Place 2.5 mℓ in a culture tube. Add 0.1 mℓ of normal horse serum and 0.7 mℓ of 0.85% saline containing 1000 units penicillin G, potassium. Inoculate with 0.5 mℓ of a 48 to 72-hr culture of amebae. Inocula of 500 amebae/mℓ of medium should yield 15 to 25,000 cells/mℓ in 72 hr.

c. Trypanosomatid Stock Culture Maintenance

For *Crithidia* sp. inoculate flasks of TTY-SB medium (Section II.D.2.b) with 10 to 20 million flagellates, and for *Trypanosoma cruzi* inoculate flasks of TTY-6-SB (Secton II.D. 2.b) medium with 100 million trypanosomes. Incubate at 24°C with container caps loosened to admit air. Transfer crithidia cultures alternately at 72 and 96 hr, and the trypanosomes at weekly intervals.

d. Monoxenic Amebae Cultures: Initiation and Maintenance

To initiate ameba-crithidia (A-C) cultures or ameba-trypanosome (A-T) cultures inoculate tubes of TTY-SB first with 50 to 100,000 amebic trophozoites or 100 to 200 cysts, then with the appropriate trypanosomatid. For A-C cultures use per mℓ of medium 2 to 3 million crithidia from 72-hr stock cultures, and for A-T cultures 1.5 to 3 million trypanosomes from 7-day stock cultures. Incubate at 35.5°C with screw-cap tightened. Subculture trophozoite-initiated cultures after 48 hr and cyst-initiated cultures after 96 hr. Subsequently, transfer both culture types at alternate intervals of 48 and 72 hr, and extend to 72 and 96 hr after cultures have become established.

To subculture, chill old A-C or A-T cultures 5 min in ice-water bath. Invert tubes gently several times to disperse amebae and loosen those adhering to glass walls. Centrifuge 5 min, 275 g. Decant supernate and discard. Add 1 mℓ fresh medium to tube and resuspend sedimented amebae by repeated, but gentle pipetting. Transfer aliquants to medium freshly inoculated with trypanosomatids. For A-C cultures of *E. histolytica* add 3 to 4 million crithidia/mℓ medium taken from 72 or 96-hr stock cultures, and one-half the numbers for *E. histolytica*-like amebae. For all A-T cultures add 2 million trypanosomes/mℓ medium from 7-day stock cultures.

During the initial stages of establishing a culture, amebic harvests may be scanty and the entire yield of a single tube will be required to seed the subculture. Yields eventually increase. In A-C cultures, depending on the strain, 10 to 20-fold increases can be expected in 72 hr from inocula of 10,000 amebae/mℓ medium. Half of this number can be expected with *E. histolytica*-like amebae. The A-T cultures produce between 4 and 6-fold increases and no differences in yields are observed between *E. histolytica* and *E. histolytica*-like amebae.

Entamoeba invadens can be monoxenized with *T. cruzi* by following the same general scheme. For these cultures I prefer an incubation temperature of 24°C and extend the interval between transfers to 7 and 10 days. *Crithidia* sp. cannot be used because of its tendency to overgrow the amebae at this temperature.

Cultures are monitored by observing the amebae *in situ* with an inverted microscope in much the same manner described for xenic cultures grown in liquid media. The cultures which are incubated in the upright position are inverted gently several times to disperse the amebae, then positioned at an angle of 5° for 10 min to allow settling of the amebae in a uniform layer. The cultures are removed from the incubator and examined with the microscope. Healthy amebae display active pseudopod formation and are attached to the glass surfaces of the tube or to each other in clumps of various size.

c. *Axenic Cultivation*

1. Media

a. TP-S-1 (Trypticase-Panmede Liver Digest-Serum) Medium (Diamond)[51]

Nutrient broth TP

Trypticase (BBL)		10.0 g
Panmede liver digest (P & B)		20.0 g
Glucose		5.0 g
L-Cysteine hydrochloride		1.0 g
Ascorbic acid		0.2 g
Sodium chloride		5.0 g
Potassium phosphate, dibasic		1.0 g
Potassium phosphate, monobasic		0.6 g
Distilled water	q.s.	875 mℓ

Dissolve salts in 600 mℓ distilled water. Add Trypticase, Panmede and glucose, dissolving each before adding the next ingredient. Add and dissolve remaining ingredients. Bring final volume to 875 mℓ with water. Adjust pH to 6.8 with 1 *N* NaOH. Clarify by passage through one layer of Whatman #1 paper. Dispense in 175 mℓ portions to Erlenmeyer flasks. Autoclave. Cool to RT. TP broth can be obtained as a dry mix from commercial sources (Consult Appendix). It has a shelf life of a year or more.

Vitamin mixture NCTC 107[74] — The mixture is composed of five stock solutions. The directions for their preparation, as presented below, differ from those given by Evans et al.,[74] and later by Diamond,[51] and reflect the greater solubility of some of the presently available vitamins.

- *Solution 1a*. Dissolve niacin, 62.5 mg and *p*-aminobenzoic acid, 125 mg in distilled water. Bring final volume to 150 mℓ.
- *Solution lb*. Dissolve niacinamide, 62.5 mg; pyridoxine hydrochloride, 62.5 mg; pyridoxal hydrochloride, 62.5 mg; thiamine hydrochloride, 25 mg; calcium pantothenate, 25 mg; i-inositol, 125 mg; and choline chloride, 1250 mg in distilled water. Bring final volume to 150 mℓ.
- *Solution 1c*. Dissolve riboflavin, 25 mg in 75 mℓ distilled water. Bring final volume to 100 mℓ. Combine solutions 1a, b, and c, bringing final volume to 500 mℓ with distilled water.
- *Solution 2*. Dissolve D-biotin, 30 mg in 200 mℓ distilled water with aid of 1 *N* NaOH added drop by drop. Bring final volume to 300 mℓ.
- *Solution 3*. Dissolve folic acid 30 mg in 200 mℓ distilled water with aid of 1 *N* NaOH. Bring final volume to 300 mℓ.
- *Solution 4a*. Dissolve vitamin D_2 (calciferol), 300 mg in 63 mℓ ethanol (95% v/v). Add vitamin A (retinol), 300 mg and dissolve.
- *Solution 4b*. Dissolve vitamin K (menadione sodium bisulfite), 60 mg in 300 mℓ aqueous solution of Tween 80 (5% v/v). Combine solution 4b with 4a, bringing final volume to 3000 mℓ with distilled water.
- *Solution 5*. Dissolve vitamin E (alpha tocopherol acetate), 25 mg in 250 mℓ distilled water.
- *Working solution*. Mix solution 1, 500 mℓ; solution 2, 250 mℓ; solution 3, 250 mℓ; solution 4, 2,500 mℓ; and solution 5, 250 mℓ. Filter sterilize. Store at −20°C. The mixture should be clear before freezing and after thawing. If turbidity develops, which is usually an indication of an excess use of sodium hydroxide in dissolving some of the vitamins, the mixture should be discarded.

Complete medium — Aseptically add to each flask of autoclaved TP broth 20 mℓ inactivated bovine serum and 5 mℓ vitamin mixture NCTC 107. Dispense to culture tubes, filling to 75 to 80% of capacity. Store at 4°C. Use within 96 hr for isolation of new strains, growth assays and experimental studies, and within 10 days for maintenance of stock cultures. Final pH 6.5. Osmolality, 410-436 mOsm/kg.

b. TYI-S-33 (Trypticase-Yeast Extract-Iron-Serum) Medium (Diamond et al.)[54]

Nutrient broth TYI

Trypticase (BBL)*	20.0 g
Yeast extract (BBL)#*	10.0 g
Glucose	10.0 g
L-Cysteine hydrochloride	1.0 g
Ascorbic acid	0.2 g
Sodium chloride	2.0 g
Potassium phosphate, dibasic	1.0 g
Potassium phosphate, monobasic	0.6 g
Ferric ammonium citrate**	22.8 mg
Distilled water	870 mℓ

#Biosate (BBL), 30 g can be substituted.
*Baker's yeast. Brewer's yeast (Difco) can be substituted.
** Purified brown pearls (Mallinckrodt). Green form cannot be substituted.

Dissolve salt and phosphate buffers in 600 mℓ distilled water. Add and dissolve in order presented Trypticase, yeast extract, glucose, cysteine, ascorbic acid and ferric ammonium citrate. Bring final volume to 870 mℓ. Adjust pH to 6.8 with 1 *N* NaOH. Clarify by passage through one layer of Whatman No. 1 paper. Dispense in 174 mℓ portions to Erlenmeyer flasks. Autoclave, cool to RT. TYI broth is commercially available (Consult Appendix).

Vitamin-Tween 80 mixture — stock solutions:

- Solution A — Vitamin mixture NCTC 107 (see under TP-S-1).
- Solution B — Vitamin B12. Dissolve 40 mg in distilled water, bringing final volume to 100 mℓ.
- Solution C — DL-6, 8 Thioctic acid (reduced). Dissolve 100 mg in absolute ethanol, bringing final volume to 100 mℓ.
- Solution D — Tween 80. Dissolve 50 g in absolute ethanol, bringing final volume to 100 mℓ.
- Working solution — Combine 1000 mℓ of A, 12 mℓ of B, 4 mℓ of C, 4 mℓ of D, and 180 mℓ distilled water. Filter sterilize. Store in dark at 4°C.

Complete medium — Aseptically add to each flask of autoclaved TYI broth 20 mℓ bovine serum, and 6 mℓ of vitamin-Tween 80 mixture. Dispense to culture tubes, filling to 75 to 80% of total capacity. Store at 4°C in the dark because of sensitivity to light. Storage life is similar to TP-S-1. Final pH, 6.6. Osmolality, 370 to 400 mOsm/kg.
Remarks:

It is important to fill each culture tube to within 75 to 80% of its full capacity. Too small a volume of air can be as detrimental to the amebae as too large a volume.

Sera from the following sources have been tested in my laboratory for their ability to support axenic growth of *E. histolytica:* bovine, goat, horse, human, rabbit, sheep, and fetal calf.[75] The best and most consistent growth was obtained with bovine serum, especially serum taken from steers ranging between 18 and 24 months of age. No growth was obtained with fetal calf serum. Growth in human serum was extremely poor and the amebae produced were morphologically abnormal. Serum from young animals, i.e., calf, kid, and lamb, did not support growth to the extent supported by adult serum.

Most of the *E. histolytica* strains I have studied can be grown in media containing 10% serum, a few require serum concentrations of 15 to 20%. This difference in serum requirement must be kept in mind when attempting axenization of new strains.

Biological products used in culture media tend to vary from lot to lot in their ability to support cellular growth. While, this may be of little consequence in cultivating less fastidious organisms than *E. histolytica* it can be a major problem in growing the ameba. Significant variations have been encountered with use of Panmede and Trypticase,* and to a lesser extent with yeast extract and serum. Before accepting a new lot from a distributor I test at least one sample using the following minimal standard of acceptability. The product must be able to support growth of selected strains of *E. histolytica* through six subcultures, the yield of amebae obtained with the last subculture being equal to or better than the yield from the initial culture.

TP-S-1, presently the most widely employed medium for axenic cultivation of *Entamoeba*, is gradually being replaced by TYI-S-33. The latter requires smaller inocula to maintain the cultures, the resulting yields of amebae are greater, growth is more uniform and in the case of *E. histolytica* the generation time has been reduced approximately one-third.

* Trypticase manufactured after August 1981 supports axenic growth of *E. Histolytica* and *G. lamblia* poorly or not at all. BBL Casein Digest #97023 is a satisfactory replacement.

2. Techniques of Cultivation

Occasionally, the first attempt to establish an axenic culture of any *Entamoeba* is successful. More often several attempts are involved. Twenty-two trials were made in the case of one strain of *E. histolytica* before it was axenized in my laboratory. One must not be daunted by a few failures.

Axenized *Entamoeba*, especially *E. histolytica*, are fastidious organisms and highly sensitive to changes in the in vitro environment. Thus, undue care must be exercised in management of the cultures and amebic growth constantly monitored to ensure successful serial culture. Best results are obtained when routine maintenance of the stock cultures is entrusted to one person.

To date, all axenic cultures have been initiated with amebae derived from monxenic cultures.[57] The bacterial associate is eliminated through use of antibiotics. The predatory activity of the amebae plays the major role in clearance of the trypanosomatids. In the case of amebae isolated from homothermic hosts this is aided by the temperature of incubation which is incompatible with growth of the flagellate. In cultures of amebae from poikilothermic hosts, e.g., *E. invadens*, predation is apparently the only mechanism at play. For this reason, only the slower reproducing *T. cruzi* can be used as the associate for these amebae. The more rapidly growing *Crithidia* sp. overgrows the culture and eventually eliminates the ameba. Both the bacterial and trypanosomatid associates are eliminated from the system usually by the third or fourth subculture.

I have initiated axenic cultures in diphasic media, liquid media, and liquid media containing small amounts of agar.[51,54] In an earlier publication I indicated diphasic media was the most reliable of the three.[51] However, the efficiency of the newer TYI-S-33 liquid medium, especially with the addition of the agar, together with improvements in techniques of managing cultures obviate the need for diphasic media.

In initiating axenic cultures I inoculate liquid medium, preferrably TYI-S-33, with and without the addition of agar. Agar although it facilitates axenization can later prove to be a problem when its elimination is attempted. Even an established culture can be lost in this process.

I prefer to use Oxoid's Ionagar #2 (0.01% w/v). Regrettably, it is no longer being manufactured. Difco's Bacto agar (0.025% w/v) or Noble agar (0.01% w/v) are acceptable substitutes.

a. *E. histolytica:* Initiation of Axenic Cultures

1. Prepare two tubes each of TYI-S-33 with and without the addition of agar. If the source of amebae is to be an S-F type culture add 500 Units penicillin G, potassium and 0.5 mg streptomycin sulfate/mℓ of medium.
2. Inoculate one tube of each set with 150,000 amebae and the other with 300,000 amebae, using 72-hr monoxenically cultivated amebae.
3. Incubate cultures in vertical position at 35.5°C, 72 hr.
4. Remove cultures from incubator and gently invert tubes several times to disperse amebae. Replace at an angle of 5° for 15 min to allow settling of amebae. Clumping of amebae will occur if allowed to remain in this position longer. Examine culture with inverted microscope as described earlier. Subculture those showing the greatest growth.
5. Chill cultures 5 min in ice-water bath. Invert tubes several times to dislodge attached amebae. Centrifuge 5 min, 275 g. Remove supernate and discard. Resuspend amebae in 1 mℓ fresh medium. Divide into two equal portions and transfer to two tubes of fresh medium.
6. Incubate 72 hr.
7. Repeat step 4. If multiplication of the amebae is apparent repeat step 5, otherwise,

refeed cultures as follows: Chill 5 min, centrifuge 5 min. Remove and discard supernate. Replace with fresh medium. To prevent excess accumulation of agar in cultures in which it is used, refeed with agar-free medium, but subculture to agar containing medium.

8. Incubate cultures 72 hr. Examine cultures as in step 4, subculture or refeed as required.

Eventually, subculturing can be performed regularly at alternating intervals of 72 and 96 hr, with fewer amebae required as an inoculum.

As a rule, amebae multiply exuberantly for the first few transfers in newly initiated cultures. After the third or fourth transfer multiplication frequently declines sharply and does not return to its former rate until after the seventh or eighth transfer. During this period of decline the cultures are highly vulnerable and are easily lost. For this reason in those cultures in which agar has been included no attempt should be made to eliminate it until after this period of vulnerability has passed and the cultures stabilized. Undiminished multiplication for 10 consecutive subcultures is an indicator of cultural stability. A similar decline in growth may accompany the change from one medium to another, or in shifting from one lot of a major ingredient of the medium to another lot.

Inclining the cultures during incubation favors growth. Usually, after the first 10 subcultures I incline the culture tubes at 10° later at 5°.

b. Established Culture Management

Proper selection of the cultures to be used as sources of inocula for new subcultures is extremely critical. It is based on observation of the amebae *in situ* with the inverted microscope. In performing this examination the amebae which tend to grow in a discontinuous line on the lower side of the inclined tube must be dispersed to a degree permitting microscopic observation. The method of dispersion consisting of inverting the tubes and allowing the amebae to resettle has been described (see Section II.D.2.a).

A culture suitable for transfer is characterized as follows: (1) The medium overlying the amebae is relatively clear and devoid of mucoid strands which indicates overt microbial contamination; (2) The majority of amebae are either attached to the surfaces of the tube, or attached to each other forming clumps of various size; (3) Individual amebae display active pseudopod formation which may or may not be accompanied by locomotion; (4) A few multinucleate giant amebae may be present. Strains characterized by the presence of numerous giant cells in culture are difficult to maintain; (5) Dead amebae denoted by highly granular and shrunken cytoplasm as well as a ragged cell membrane are rare.

Cultures selected for transfer are chilled 5 min and inverted several times. Portions of the suspension are transferred immediately to tubes of fresh medium warmed to RT. The number of amebae transferred is determined by trial and error and as the amebae adapt to the medium frequent adjustments are usually necessary. Some strains can be maintained in serial subculture by transfer of a few hundred organisms per mℓ of medium, while others require a few thousand.

Ideally, inocula in which the number of amebae is known should be used to initiate the subcultures. Since this is impractical for routine maintenance, especially when a large number of strains are involved, the following scheme has been developed. Three subcultures are made from each of two parent cultures. The first tube of medium receives a small amount of amebic suspension, the second receives 1.5 × or 2 ×, and the third tube receives 2 × or 3 × depending on the amount placed in the second tube of the series. Thus, inocula of three different sizes are used. The usual range in the size of the amebic suspension is between a few hundreths and 1 mℓ. If more than 2 mℓ are needed to maintain a culture the inoculum should be prepared from a suspension concentrated by centrifugation.

The cultures which have not been selected for subculture are retained as reserve stock and are held at a temperature of 33°C until the next subculture is performed, then discarded if not needed. At this temperature the rate of amebic multiplication is reduced, and viability of the culture prolonged.

On occasion, for reasons not always apparent, the cultures will not grow out despite the presence of viable amebae. In this event, subculturing is discontinued and refeeding of the culture resorted to until normal multiplication resumes. It is helpful to incubate the cultures vertically during this period of crisis.

c. Mass Cultivation

Glassware — Various shaped containers can be used. The most commonly employed in my laboratory are Micro-Fernbach (Bellco) and Erlenmeyer flasks designated as having a capacity of 125 mℓ. These are filled with medium to within 70 to 75% of their true capacity. The Micro-Fernbach flasks are preferred because their flat bottom facilitates microscopic examination of the cultures. Flask cultivation is not as efficient as tube cultivation in regard to maximum utilization of medium. Larger inocula are required and fewer amebae are produced per unit of medium. On the other hand, they offer a more practical method of obtaining large numbers of amebae and require less handling of the organisms during harvesting of the cultures, resulting in smaller losses of the amebae due primarily to lysis.

Procedures — Inocula used for mass cultures are obtained from pooled tube cultures in log phase of growth (= 72 hr). However, if the density of the amebic population becomes too great in these cultures an adverse effect will be noted on the subsequent growth and physiology of the amebae in the flask cultures. To avoid this, counted inocula are used to seed the tube cultures, and only those cultures yielding less than 200,000 amebae/mℓ after incubation are used to inoculate the flasks. These in turn are inoculated with 5 to 10,000 amebae/mℓ of medium, and are incubated 72 hr. Afterward they are inspected for appearance of the amebae and the presence of gross microbial contamination. To harvest, chill selected flasks in ice water for 15 min. Swirl gently to suspend amebae. Pour contents into suitable sized centrifuge containers, centrifuge 10 min, 275 g. Remove supernate and discard. Resuspend sedimented amebae in wash fluid. The composition of this fluid will depend on the intended use of the amebae. Axenized *E. histolytica* is sensitive to repeated washing and centrifuging and each cycle results in increased lysis. Three washes are maximum, losses of 25% can be expected.

d. *E. histolytica*-Like Amebae: Initiation and Maintenance of Cultures

As far as can be ascertained only two strains of these amebae, Huff and Laredo, have been axenized. The first cultures were obtained in my laboratory and were initiated in a diphasic medium containing chick embryo extract (TTY-S-$CEEC_{25}$).[51] In this medium the amebae could be grown at either 25°C or 35.5°C. However, in the absence of the extract, growth could be obtained only at the lower temperature. The immediate source of the amebae used in these studies were monoxenic A-T cultures.

Later, TP-S-1 containing 0.01% Ionagar #2 was used for initiation of axenic cultures. Inocula consisted of amebae taken from 72-hr A-T or A-C cultures. An incubation temperature of 35.5°C was used for these primary cultures and the first two subcultures, made at 48-hr intervals. The cultures were then maintained at 25°C indefinitely. The succeeding third and fourth subcultures were performed at weekly intervals. All subsequent transfers were made between 10 and 14 days. Lastly, the amebae were adapted to grow in agar-free medium by gradually decreasing the amount of agar used in each succeeding subculture. The higher initial temperature was used as an aid in the elimination of the trypanosomatid associate. Growth of *E. histolytica*-like amebae in TYI-S-33 medium is more luxuriant than in TP-S-1 and is presently the recommended medium.

e. Colonal Growth in Soft-Agar Medium
(Gillin and Diamond)[55,56]

The technique provides a simple, rapid, and efficient method for obtaining cloned colonies of *Entamoeba* without exerting undue selective genetic pressure. Moreover, the sensitivity of the method and the excellent proportionality between the number of viable cells introduced into the medium and the colony forming efficiency make it especially useful for studies in which quantitative assessment of amebic survival is critical, e.g., drug testing, temperature, and oxygen tension effects.

Concentrated agar suspension (5% w/v) — Place dry Bacto agar (Difco) in flask. Add TYI broth. Dissolve agar and sterilize mixture in one operation by autoclaving using standard procedures. Cool and store at 4°C.

Soft-agar medium: method of use — Melt agar concentrate. Maintain in water bath (65 to 70°C). Place tubes of TYI-S-33 (13 mℓ) in second bath (42 to 44°C), equilibrate 30 min. Immediately before inoculating with amebae add required molten agar to TYI-S-33 using a warmed pipet. Mix by inverting tubes five times, taking care to prevent introduction of air bubbles. Inoculate with amebae. Immediately invert 10 times to disperse cells, gel agar rapidly by chilling in ice water, 10 min. The final volume in each tube is 15 mℓ. Incubate. Colony formation is dependent on (1) final concentration of agar employed, (2) filling tubes with medium to maintain reduced oxygen tension, and (3) rapid gelling of agar at 0°C to form a suitable matrix. The optimal agar concentration must be determined not only for each species, but for each strain of a species. This usually lies between 0.3 and 0.8% agar, and for a given *Entamoeba* the range is very narrow. Therefore, it is recommended that increments of 0.05% be used in testing for the optimal concentration. At suboptimal concentrations colonies are diffuse or may fail to form, whereas at concentrations above optimum colonies are smaller and fewer.

Colonies of *E. histolytica* are usually well defined by the fourth or fifth day of incubation, some not until the seventh day. Colonies of *E. invadens* and other reptilian *Entamoeba* are not visible before 7 days. *Entamoeba histolytica*-like amebae fail to form colonies visible to the unaided eye. The number of colonies formed is proportional to the number of amebae inoculated. However, colony forming efficiency (CFE) is related to the age of the culture used as a source of amebae. Best results are obtained with young cultures, i.e., less than 72 hr old. The percent CFE (number of colonies counted/number of cells inoculated X 100) ranges between 10 and 80% depending on the species and/or strain used. Because of difficulties encountered in counting more than 200 colonies/tube, it is recommended that the inoculum be adjusted so no more than this number will form. Clones can be obtained by taking the following precautions. The inoculum must contain only single cells. It should be adjusted so that no more than 10 colonies will form. Microscopic examination of the freshly gelled agar should reveal single well-isolated amebae. After formation of the colonies no amebae wandering between the agar core and walls of the tube should be present. The colonies are picked using a pipet with a finely drawn tip and transferred to tubes of TYI-S-33. Colony picking is performed with the aid of a dissecting microscope. After 3 to 6 days a sufficient number of amebae will have migrated from the agar fragments to establish a culture. The medium is decanted without prior chilling, and replaced with fresh medium. After another week of incubation the clones can be maintained by routine subculturing. Since it is possible, despite all precautions taken, that a colony arose from more than one ameba, it is recommended that the colonies be recloned a second time.

f. Microbial Contaminants: Detection and Elimination

For convenience of discussion contaminants are divided into two groups: primary contaminants, those present in material serving as the immediate source of amebae used in initiating axenic cultures; secondary contaminants, those introduced into an axenic culture accidentally.

The nutritional richness of TP-S-1 and TYI-S-33, and the presence of reducing agents in these media make them ideal for growth of many aerobic and anaerobic bacteria, fungi, and mycoplasmas. The majority of microbial contaminants multiply rapidly and are readily detected by the turbidity produced. On the other hand, cryptic infections associated with slow growing microbes can be detected only through use of sterility test media. For this reason I recommend routine monitoring of cultures using sterility test media designed for detection of both aerobes and anaerobes. The most sensitive medium in my experience and one which I recommend highly is the medium used by Barile and his associates to screen cell cultures for mycoplasmas, bacteria, and fungi.[76] The formula and preparation of this medium as presented below is used by the Media Supply Section, NIH.

Most secondary bacterial contaminants encountered in my laboratory killed the amebae, a few co-existed with them, some enhanced amebic growth. Although I have never detected mycoplasmas in cultures I have axenized, I have received cultures from other laboratories which were infected. Identification studies when performed have shown the mycoplasma contaminants to be associated with human infections, suggesting the person managing the cultures is the primary source. The possibility of introducing mycoplasmas with serum is reduced by heat inactivation.

Secondary fungal contaminants are a rarity in my laboratory. Usually no more than one tube is infected and the problem is dealt with by discarding the culture. Mycostatin, 100 μg/mℓ medium, can be used if necessary.

Bacterial and mycoplasmal contaminants can be eliminated usually by use of combinations of penicillin, streptomycin, gentamycin, kanamycin, and tetracycline. Mycostatin and the broad spectrum antibiotics are amebistatic if not amebicidal and must be used with caution. In some situations it may be necessary to treat the cultures for five or more subcultures before ridding them of the contaminants.

Medium for Detection of Microbial Contaminants

1.	*Basal medium*	(Barile et al.)[76]
	Mycoplasma broth base (BBL)	21.0 g
	Noble agar (Difco)	1.0 g
	Phenol red	0.02 g
	Fresh yeast extract — 25% (MAB)	100 mℓ
	Horse serum (inactivated)	200 mℓ
2.	*Supplements*	
	Eagles vitamin B mixture — 100x	10.0 mℓ
	L-glutamine	1.2 g
	L-arginine hydrochloride	8.4 g
	Dextrose	2.0 g
	Nicotinamide adenine dinucleotide (ICN)	0.15 g
	Distilled water	690 mℓ
3.	*Complete medium*	

Add mycoplasma broth base, Noble agar, and phenol red to 640 mℓ distilled water. Autoclave, cool. Add filter sterilized fresh yeast extract and horse serum. To 50 mℓ distilled water add ingredients listed under supplements, filter sterilize and add to basal medium. Adjust pH to 7.4 with sterile 1 *N* NaOH. Tube in 10 mℓ amounts.

To use the medium introduce 0.4 mℓ of culture to be tested with a 1 mℓ pipet. Start release of fluid at bottom of tube and continue releasing as pipet is withdrawn. Incubate a minimum of 21 days at 24 and 36°C. Since, the medium also supports amebic growth, the culture to be tested is centrifuged 3 min, 275 g, and the relatively ameba-free supernate used as inoculum. All growth in Barile's medium is examined microscopically, preferably using Gram's stain and then subcultured to appropriate diagnostic media. I make it a practice to submit to a specialist trained in diagnostic microbiology all amebic cultures for certification of their freedom from contaminants after I am satisfied that axenic cultivation has been achieved.

E. Unsolved Problems

Four major problems await solution. (1) There is no defined medium for axenic cultivation of *Entamoeba*. A defined medium would facilitate elucidation of the nutritional requirements of these parasites and further such studies as biolipid chemistry. (2) No method exists which permits direct axenization of *Entamoeba* from host material. To date, all axenized strains of *E. histolytica* and *E. histolytica*-like amebae have been derived from monoxenic cultures, and *E. invadens* from monoxenic cultures or cultures in which fresh tissue served as the chief source of nutrients. The interval of time between isolation of the amebae from host material, establishment in monoxenic cultures (usually but not always via xenic cultures), and axenization is inordinately long. The number of generations ensuing is enormous with the attendant opportunity for selection and/or mutation to occur during a period of high vulnerability. This interval must be reduced in order to conserve traits present in the wild population at the time of separation of the parasite from the host. (3) There is no method for inducing encystation of *E. histolytica* under axenic conditions. Encystation of this ameba does not occur spontaneously in axenic cultures as it does in xenic culture systems. Since an understanding of the basic biology of the cyst is prerequisite to developing more rational programs for controlling transmission of the parasite, the highest priority is assigned solution of the problem of encystation in axenic medium. Techniques described for inducing cyst formation in *E. invadens* in an axenic environment have not been applicable to *E. histolytica*.[53] Techniques for axenic cultivation of *E. coli* and *E. hartmanni* await development.

F. Cultures — A Research Tool

The literature on the research accomplished with in vitro cultures of *Entamoeba* is so vast that it cannot be reviewed here. An appreciation of the research conducted on *E. histolytica* since introduction of TP-S-1 medium for axenic cultivation in 1968 may be gained from examination of Table 1.

III. TRICHOMONADS

Trichomonas vaginalis, T. tenax, and *Pentatrichomonas hominis* are parasites of man. The first has been recorded only in this host, and the second from man, chimpanzees, and monkeys. *Pentatrichomonas hominis* infections have been reported from several nonhuman primates, rodents, dogs, cats, and cattle.[6] *Trichomonas gallinae* is primarily a parasite of domestic pigeons. However, other columbids, gallinaceous birds, raptors and gulls harbor the flagellate.[106] *Tritrichomonas foetus* is a cattle parasite. Morphologically similar protozoa have been found in swine, deer, and horses. Their possible identity with *Tritrichomonas foetus* is discussed by Levine.[6] *Trichomonas vaginalis, T. gallinae,* and *Tritrichomonas foetus* are pathogens, *T. tenax* and *P. hominis* are commensals.

A. Classification

The trichomonads are grouped in the family Trichomonadidae, order Trichomonadid, class Zoomastigophora, phylum Sarcomastigophora.

B. Geographic Distribution and Incidence

Trichomonads are distributed world-wide. Reported incidences of *T. vaginalis* infections

Table 1
EXAMPLES OF RESEARCH CONDUCTED WITH AXENIC CULTURES OF *ENTAMOEBA HISTOLYTICA*

Biochemistry and Molecular Biology	
Bioenergetics	Reeves[77,78]
	Weinbach et al.[79]
	Weinbach et al.[80]
Nucleic acids	Booden et al.[81]
	Booden et al.[82]
	Boonlayangoor et al.[83]
	Gelderman et al.[84]
Surface membranes	Pinto da Silva et al.[85]
	Trissl et al.[86]
	Aust-Kettis & Sundquist[87]
	Aley et al.[88]
Virulence	
In vitro, In vivo assay	Diamond et al.[89]
	Mattern et al.[90]
	McCaul et al.[91]
Mechanisms of cytopathogenicity	Lusbaugh et al.[92]
	Bos[93]
	Mattern et al.[94]
	Ravdin et al.[95]
Viruses	Diamond and Mattern[96]
Drug testing	Diamond & Bartgis[97]
Serology and immunology	
Serodiagnosis	Patterson et al.[98]
Epidemiology	Kagan[99]
Cell-mediated immunity	Savant et al.[100]
	Ortiz-Ortiz et al.[101]
	Harris et al.[102]
Complement activation	Huldt et al.[103]
Protective immunity	Krupp et al.[104]
	Meerovitch et al.[105]

range from 1 to 70%. In normal populations the average is probably close to 10%. Females are more frequently infected than males. The incidence of *T. tenax* is high, more than half of a population can be infected. In contrast, the incidence of *P. hominis* infections is low, usually between 1 and 5%. Incidences of *Tritrichomonas foetus* in herds of cattle range from 6 to 30%. *Trichomonas gallinae* is extremely common in columbid birds, more than 50% of a population may be infected. Turkeys and to a lesser extent chickens are commonly infected. The incidence in flocks of turkeys, especially in young birds, may reach levels of 80%.[6]

C. Life Cycles and Disease

Life cycles of trichomonads are simple and direct. There is only one stage, the trophozoite. True cysts are not formed. Reproduction is by longitudinal binary fission. *Trichomonas vaginalis* and *Tritrichomonas foetus* inhabit the genitourinary tract of their hosts and are venerally transmitted. *Trichomonas gallinae* infects the upper digestive tracts of birds. In columbid birds it is passed from parent to offspring through feeding of 'pigeon milk', and by contamination of drinking water. Contaminated water is the source of infection for most other birds.[106] *Trichomonas tenax* is a parasite of the oral cavity and is transmitted through direct contact. *Pentatrichomonas hominis* lives in the colon and/or cecum and is passed from host to host through drinking water and food contaminated with feces.

Infections with *T. vaginalis* are usually latent, more so in the male than the female. Nevertheless, because of the high incidence of infections in some populations morbidity can be high. This manifests itself in females as vaginitis and/or urethritis in varying degrees of severity, and in males as urethritis, prostatitis, and/or epididymitis. In *Tritrichomonas foetus* infections abortion of the first fetus is a common feature in the cow and results in heavy economic losses to cattlemen. *Trichomonas gallinae* infections in turkey and chicken flocks are often associated with high mortality and are a cause of major economic losses for poultry farmers. Honigberg reviews the current knowledge of the important trichomonads of man and domestic animals.[107,108]

D. Cultivation

Each of the trichomonads discussed can be grown under xenic and axenic conditions, and the ease with which the pathogenic species can be cultured axenically has led to increasing use of these cultures for diagnosis of infection as well as source material for research. A situation far different from that pertaining to *E. histolytica*.

In general, media used for cultivation of *Entamoeba* are quite satisfactory for trichomonads, little or no modifications being required. This applies both to xenic and axenic cultures. Nonetheless, several media have been devised specifically for one or more species of trichomonads, the majority of these being developed for *T. vaginalis* and *Tritrichomonas foetus* and to a lesser extent for *T. gallinae, T. tenax*, and *P. hominis*. Undoubtedly, this is a reflection of the importance of each species to the welfare of man and his domestic animals.

1. Historical Review

The review which follows is concerned primarily with development of axenic cultures, particularly those studies using media specifically devised for trichomonads. The literature on xenic cultivation has been adequately reviewed by Taylor and Baker.[1,2] A limited number of monoxenic cultures in which bacteria served as the associate have been established. These have played only a minor role in trichomonad research (consult Taylor and Baker for references.[1,2]).

Trichomonas gallinae was cultivated axenically for the first time by Bos in 1933,[109] *Tritrichomonas foetus* by Witte in 1933,[110] *P. hominis* by Stone and Reynolds in 1939,[111] *T. vaginalis* by Trussel in 1940,[112] and *T. tenax* by Diamond in 1962.[113]

Although it is possible on rare occasion to obtain axenized cultures of trichomonads without resorting to special procedures,[109,110,112] most isolates must be manipulated to rid them of associated microorganisms. Several procedures have been employed. Early investigators taking advantage of trichomonad motility used migration techniques with considerable success. Stone and Reynolds used an intricately designed capillary tube to axenize *P. hominis*.[111] Glaser and Coria employed a V-shaped tube, which they had devised earlier for free-living protozoa, to axenize several isolates of *Tritrichomonas foetus*.[114] Adler unable to free *P. hominis* from all associated bacteria after V-tube migration, treated the partially cleansed flagellates with a sulfonamide and introduced them into the peritoneal cavity of a mouse. Twenty-four hours later he obtained sterile trichomonads which grew out when inoculated into appropriate media.[115] Rees obtained axenic cultures of *Tritrichomonas foetus* by means of microisolation techniques.[116] Use of antibiotics to sterilize trichomonads was introduced by Adler and Pulvertaft, who employed penicillin to axenize *T. vaginalis*.[117] Streptomycin was introduced by Quisno and Foter.[118] Mycostatin has been employed to eliminate yeasts,[119] and kanamycin for mycoplasma.[120]

The earliest workers engaged in the cultivation of trichomonads used media devised for other protozoa, especially modifications of Boeck and Drbohlav's LES medium (consult Taylor and Baker for references.[1]). Later investigators have tended to employ media especially designed for growing one or more trichomonad species.

Probably the most widely used medium for cultivation of *T. vaginalis* is Johnson and Trussell's CPLM.[121] This is a liquid medium consisting essentially of Difco's Bacto peptone, liver infusion, maltose, and serum. Cysteine and a small amount of agar are added to maintain a low redox potential. McEntegart,[122] Feinberg,[123] Feinberg and Whittington,[124] and Lowe[119,125] devised media based on modifications of CPLM. McEntegart eliminated the agar, which interfered with harvesting of the trichomonads. However, in its absence growth could only be obtained by incubating the cultures anaerobically.[122] Feinberg replaced the liver infusion with Panmede, a highly soluble papain digest of ox-liver, and devised a medium for mass cultivation.[123] Later he and Whittington simplified his medium for use in diagnosis and reported success in freeze-drying it.[124] The method used to accomplish this was not presented. Lowe modified Feinberg's original medium, replacing the cysteine with ascorbic acid. By adding agar (0.22% w/v) he produced a semisolid medium suitable for transport and diagnosis.[119,125]

For cultivation of *Tritrichomonas foetus*, Schneider devised a complicated thermostabile diphasic medium.[126] A simpler medium was developed by Plastridge.[127] This medium, especially as modified by Fitzgerald et al.,[128] has been widely employed.

Trichomonas tenax was grown axenically by Diamond in a modified version of TTY-S-$CEEC_{25}$, one of several media he devised for cultivation of *E. histolytica*.[113]

The media discussed above were developed with a specific trichomonad in mind. Diamond devised TYM as a general purpose medium for axenic cultivation of trichomonads and grew the following. All of the species considered in this chapter (with the exception of *T. tenax*), *Trichomonas bactrachorum*-type flagellates from feces of swine and cattle, *Tetratrichomonas gallinarum*, and *Tritrichomonas eberthi* of gallinaceous birds.[48] TP-S-1 developed for *Entamoeba* is an excellent medium for isolation and maintenance of *T. vaginalis*.

Methods for clonal growth of trichomonads have been described. Most of these have been for *T. vaginalis*. Asami et al., grew *T. vaginalis* as colonies in deep agar tubes and on the surface of plates.[129] Ivey obtained colonies of this species with a single agar layer plating technique,[130] and Hollander using a similar technique but a different medium obtained colonies with distinct morphologies.[131] These techniques most probably can be used for clonal growth of other species as the work of Samuels indicates. Using a double layer method of plating he not only produced colonies of *T. vaginalis* but also of the following species: *T. gallinae, Tetratrichomonas gallinarum, Tritrichomonas foetus, Tritrichomonas augusta* from amphibia, and *P. hominis*.[132]

The media mentioned so far contain serum, peptones, yeast extract, or a liver preparation, each one a poorly defined biological product and as such an impediment to elucidation of the nutritional requirements of the parasites in question. Many investigators have devoted considerable effort to development of defined synthetic media and on the whole have fared better than those who have worked on the *Entamoeba*. Several semidefined media in which one or more of these biologicals have been replaced with defined chemicals have been reported for short-term serial subculture. Among them can be listed SL medium of Shorb and Lund,[132] and SLM of Lund and Shorb,[133] for cultivation of *T. gallinae, T. vaginalis*, and *Tritrichomonas foetus*; and media 5E and 7E developed by Lee et al. for culture of trichomonas from poikilothermic hosts.[135] Shorb has reviewed these studies.[136] Recently, Linstead announced a major breakthrough in cultivation of *T. vaginalis*.[137] He developed a semidefined medium, 'modified' CMRL 1066, and two defined media, DL7 and DL8 which support indefinite serial subculture.

2. *Media Preparation and Techniques of Cultivation*

Investigators wishing to cultivate axenically the trichomonads considered in this chapter will find they have available a number of excellent undefined media from which to choose. In general they are quite similar differing only in minor details. They usually contain at least one peptone, a liver or yeast preparation, serum, a mono- or disaccharide, one or more reducing agents, phosphate buffers, and sodium chloride. In addition, some may contain agar. Although devised for a specific trichomonad these media can usually be employed for more than one species. For example, media designed for *T. vaginalis* can be used to grow other species merely by altering the pH upward to neutrality. Four undefined media for isolation and maintenance of trichomonads are presented below. Linstead's DL8 is described for those in need of a defined medium for *T. vaginalis*. Fortified with human serum it can be used for *Tritrichomonas foetus*. Linstead's 'modified' CMRL 1066 is also described because of its potential value in development of defined media for other trichomonads.

a. Media

1. Undefined

a. CPLM (Cysteine-Peptone-Liver-Maltose) Medium
(Johnson and Trussell, Trussell and Johnson)[121,138]

Liver infusion — Place 330 mℓ distilled water in a 600 mℓ beaker. Add 20 g Bacto Liver infusion (Difco) powder and infuse 1 hr, 50°C. Raise temperature to 80°C for 5 min to coagulate protein. Filter through coarse paper in a Buchner funnel.

Ringer's solution — Sodium chloride, 0.6%; sodium bicarbonate, potassium chloride, and calcium chloride, 0.01% each.

Complete medium

Bacto Peptone (Difco)	32.0 g
Bacto Agar (Difco)	1.6 g
Cysteine hydrochloride	2.4 g
Maltose	1.6 g
Liver infusion	320 mℓ
Ringer's solution	960 mℓ

Mix Ringer's solution and liver infusion. Add peptone, maltose, and cysteine hydrochloride, dissolving one by one. Add agar and heat mixture in water bath until agar melts. Filter through coarse paper. Add 0.7 mℓ of 0.5% (w/v) aqueous methylene blue. Adjust pH to 5.8 to 6.0 with 1 *N* NaOH or HCl. Distribute in 8 mℓ amounts to culture tubes. Autoclave. Cool and add aseptically 2 mℓ human serum to each tube. Test incubate 48 hr. Store at RT. Use as long as amber anaerobic zone is present. CPLM medium devised for *T. vaginalis* can be used for cultivation of *T. gallinae*, *Tritrichomonas foetus*, and *P. hominis* by adjustment of its pH to 7.0.

b. TYM (Trypticase-Yeast Extract-Maltose) Medium
(Diamond)[48]

Nutrient broth

Trypticase (BBL)	20.0 g
Yeast extract (BBL)	10.0 g
Maltose	5.0 g
L-Cysteine hydrochloride	1.0 g
L-Ascorbic acid	0.2 g
Potassium phosphate, dibasic*	0.8 g
Potassium phosphate, monobasic*	0.8 g
Agar (Bacto)	0.5 g
Distilled water	900 mℓ

* The buffer salts were inadvertently omitted from the list of ingredients in the published paper, but were included as an addendum with the reprints which were distributed. TYM was devised as a general purpose medium for cultivation of trichomonads.

Dissolve salt and phosphate buffers in 600 mℓ water. Add and dissolve remaining ingredients with exception of agar in order presented. Adjust pH with 1 *N* NaOH to 6.0 to grow *T. vaginalis*, and 6.8 to 7.0 for other trichomonads. Distribute in 180 mℓ amounts to Erlenmeyer flasks and add agar. Autoclave to melt agar and sterilize medium. Cool to 45°C.

Complete medium — Add aseptically to each flask 20 mℓ of sheep or bovine serum. Dispense while still warm in 10 mℓ portions to culture tubes. Store at 4°C. Use within 10 days.

c. Semisolid Diagnostic and Transport Medium for *T. vaginalis* (Lowe)[125]

Basal medium

Panmede, liver digest (P & B)	12.5 g
Sodium chloride	2.5 g
Maltose	0.5 g
Agar	1.25 g
Distilled water q.s.	500 mℓ

Dissolve first three ingredients in water. Adjust pH to 6.2. Add agar. Steam 1.5 hr, 100°C, mixing thoroughly at intervals. Cool to 54°C and add aseptically the following sterile solutions:

Horse serum	50.0 mℓ
Fildes extract (Oxoid)	0.5 mℓ
Ascorbic acid (Aqueous soln, 4% w/v)	10.0 mℓ
Chloramphenicol (Aqueous soln, 1% w/v)*	5.0 mℓ
Gentamycin sulfate (Aqueous soln, 1% w/v)	5.0 mℓ
Amphotericin (Aqueous soln, 0.01% w/v)	5.0 mℓ

Mix thoroughly and distribute aseptically to 0.25 oz Bijoux bottles by filling up to the shoulder. The gel should remain intact when bottle is inverted, but should break up readily on shaking. Readjust the agar concentration, if necessary, to meet these specifications. A standard screw-cap culture tube can be used in place of the bottle and should be filled to the shoulder. Incubate overnight for sterility check.

d. TTYS-CEEC$_{25}$ (Trypose-Trypticase-Yeast Extract-Serum-Chick Embryo Extract, Crude 25%): Isolation and Maintenance Medium for *T. tenax* (Diamond)[113]

Basal broth TTY — Prepare according to directions in Section II.D.2.b, but bring total volume to 1600 mℓ. Distribute in 160 mℓ amounts to Erlenmeyer flasks. Add 0.1 g Bacto agar (Difco). Autoclave. Cool to 45 C.

Buffered saline for embryo extract

Sodium chloride	5.0 g
Potassium phosphate, monobasic	1.6 g
Sodium phosphate, dibasic	1.6 g
Distilled water q.s.	1000 mℓ

Sterilize by autoclaving.

CEEC$_{25}$ — Aseptically harvest 11- to 12-day-old chick embryos. Remove eyes and beak. Weigh and place in sterile Waring®-type blender. Add chilled buffered saline, 2 mℓ/g tissue.

* Honigberg cautions against use of this antibiotic and suggests replacing it with penicillin and streptomycin.[108]

Blend 2 min. Refrigerate 1 hr. Remove liquid beneath froth and place in screw-capped centrifuge tubes. Centrifuge 20 min, 850 g, 4°C. Collect supernate, pool and distribute in 10 mℓ amounts to culture tubes. Quick-freeze in dry-ice alcohol bath. Store at −20°C.

Complete medium — To each flask of TTY broth and liquified agar add aseptically 20 mℓ horse or bovine serum and 10 mℓ vitamin mixture NCTC 107 (consult Section II.D.2.c). Distribute in 9.5 mℓ amounts to culture tubes. Store at 4°C. Use within 10 days. Just before inoculating with trichomonads, warm to RT and add 0.5 mℓ thawed $CEEC_{25}$.

2. Semidefined and Defined

a. "Modified" CMRL 1066 for *T. vaginalis* (Linstead)[137]

CMRL 1066, 10x conc (Gibco)	100 mℓ
HEPES/KOH 1M, pH 7.1	20 mℓ
Glucose 9% (w/v), Ascorbic acid 1% (w/v)	100 mℓ
Ammonium ferric citrate, 3g/ℓ	10 mℓ
Glass distilled water*	770 mℓ

Adjust pH to 7.1 with 1 *M* NaOH. Sterilize by passage through 22 μm filter. Final osmolality, 348 mOsm/kg. Can be stored at 4°C for 2 weeks and up to 3 months at −20°C. To use, distribute in 6 mℓ amounts to 8 × 100 mm screw-capped tubes and add 0.5 mℓ fetal calf serum.

b. DL8 Defined Medium for *T. vaginalis* (Linstead)[137]

Basal medium

Salts	Concentration/ℓ
NaCl	2000 mg
KCl	2000 mg
KH_2PO_4	1000 mg
K_2HPO_4	500 mg
Ferrous gluconate (Serva)	55 mg
$MgSO_4 \cdot 7H_2O$	750 mg
$CaCl_2 \cdot 6H_2O$	350 mg
Trace elements soln. 200x conc.	5 mℓ
$FeSO_4 \cdot 7H_2O$	1000 mg
$ZnSO_4 \cdot 7H_2O$	1000 mg
$CuSO_4 \cdot 7H_2O$	500 mg
$CuSO_4 \cdot 5H_2O$	200 mg
$MnSO_4 \cdot 4H_2O$	100 mg
$Na_2MoO_4 \cdot 2H_2O$	100 mg
KI	6 mg
Amino acids	
Glutamine	440 mg
MEM amino acid soln 50x (Gibco)	30 mℓ
MEM nonessential amino acid soln 100x (Gibco)	15 mℓ

* Use within 24 hr of preparation.

Nucleic acid precursors	
Adenine sulfate	70 mg
Guanosine	30 mg
Uridine	50 mg
Cytidine	50 mg
Thymidine	15 mg
Carbohydrates	
Maltose	5000 mg
DL-a-glycerophosphate diNa·$6H_2O$	100 mg
Ascorbic acid	1000 mg
D(+) glucosamine HCl	100 mg
D(+) galactosamine HCl	50 mg
N-acetyl-D-glucosamine	10 mg
N-acetyl neuraminic acid (Sigma type VI)	10 mg
Vitamins	
Stock soln C (100x conc)	10 mℓ
d-Biotin	20 mg
D-pantothenic acid hemi-calcium salt	100 mg
Nicotinamide	50 mg
Pyridoxamine HCl	50 mg
Riboflavin	10 mg
Thiamine HCl	10 mg
p-Amino benzoic acid	5 mg
Coenzyme Z_{10} (dissolve in ethanol)	1 mg
Choline chloride	10 mg
i-Inositol	100 mg
Reduced glutathione	100 mg
Folic acid (dissolve with NaOH)	20 mg
Vitamin B12	10 mg
Deoxyribose	50 mg
Dithiothreitol	100 mg
Tween 80	100 mg

Adjust final pH to 7.1 with 1 *M* NaOH. Sterilize by passage through 22 μm filter. Final osmolality is 220 mOsm/kg.

Storage — Trace element solution was adjusted to pH 2.0 with 0.1 *M* sulfuric acid and stored at 4°C. Vitamin solution C was adjusted to pH 7.4 and stored frozen and in the dark at −20°C in aliquots of 10 mℓ. Single strength basal medium DL8 was stored in the dark at −20°C in aliquots of 100 mℓ.

Artifical serum — Stock solutions:

a. Cholesterol (chromatographic grade, Sigma). Prepare by adding 20 mg/mℓ to analytical grade absolute ethanol and warming to 37°C.
b. DL-glyceryl-1-palmitate— 2-oleate-3-stearate (GPOS, Calbiochem). Prepare as in stock solution a.
c. Bovine albumin. Prepare 5% (w/v) solution in distilled water. Adjust pH to 7.1 with 0.1 *M* NaOH. Filter sterilize.

To prepare 5 mℓ of artificial serum, place 50 μℓ of stock solutions a and b in sterile 20 mℓ screw-cap bottle. Place 5 mℓ of stock solution c in second sterile bottle. Warm both bottles to 37°C. Using aseptic technique, rapidly add contents of bottle containing stock c to bottle containing stocks a and b, and mix thoroughly. Resulting emulsion can be stored at 4°C for several months. Linstead warns that optimum results depend on following these directions carefully.

Complete medium — Dispense basal medium in 6 mℓ amounts to 8 × 100 mm screw-capped culture tubes. Add 0.5 to 0.6 mℓ artificial serum.

b. Techniques of Cultivation

1. In Undefined Media

a. Collection of Clinical Specimens

Trichomonas vaginalis — Sterile cotton swabs impregnated with charcoal are used frequently to obtain material from the vagina or urethra.[125] Sediments obtained after centrifugation of vaginal washings can be used.

Trichomonas gallinae — Cotton swabs impregnated with cellulose acetate (dissolved in amyl alcohol) have been used to obtain material from the mouth and crop of birds.[48]

Tritrichomonas foetus — Sediments obtained after centrifugation of preputial or vaginal washings can be used for inocula.[128]

Intestinal trichomonads — Fecal or cecal samples are emulsified in 1 or 2 mℓ physiological saline and 0.5 mℓ inoculated per tube of medium.

b. Initiation and Maintenance of Cultures Except for *T. tenax*.

In initiating trichomonad cultures I prefer to inoculate medium with and without the addition of small amounts of agar. As in the case of the axenization of *Entamoeba*, the presence of agar frequently enables establishment of cultures which would otherwise fail. The agar, if desired, can be eliminated after the culture has become established.

To eliminate associated microorganisms from the culture I employ mixtures of the antibiotics listed in Section II.D.2.b. For heavily contaminated material such as feces, I use all those listed. In contrast, for vaginal or urethral secretions I use combinations of penicillin, streptomycin, kanamycin in the concentrations listed, plus mycostatin (100 μg/mℓ) to eliminate *Candida*.

Cultures are incubated in a vertical position. I prefer a temperature of 35.5°C. They should be examined daily for 4 days. It is rare for a trichomonad culture to grow out after this period. Cultures in liquid media can be examined with an inverted microscope in the manner outlined for *Entamoeba*. Those growing in semisolid medium are examined by placing a drop of medium on a slide, covering it with a coverslip, and searching for trichomonads with a conventional microscope.

Trichomonad cultures are relatively easy to maintain. Growth often declines after initial exhuberant multiplication, as in the case of the *Entamoeba*. However, after recovery these cultures are rarely a problem. Subcultures are made by transferring 1 to 3 × 10^3 organisms/mℓ of medium. A culture of *T. vaginalis* can be expected to yield between 2 to 3 × 10^6 cells/mℓ in 72 hr of incubation in TYM and twice the number in TP-S-1 medium. Transfers are performed at alternating intervals of 48 and 72 hr, or 72 and 96 hr.

c. *Trichomonas tenax*

This is the only species discussed which is not readily cultivated axenically. As far as I am aware this has been accomplished only in my laboratory. Moreover, unlike the other trichomonads axenic cultures have been initiated only with organisms grown monoxenically with *Trypanosoma cruzi*. Repeated attempts to axenize *T. tenax* with material taken directly from the host or xenic cultures have failed. In this respect there is a parallelism between this parasite and *E. histolytica*.

The monoxenic culture system used originally in our studies was patterned after that of

Phillips et al. for *E. histolytica*.[36] Later the system described by Diamond for the same ameba was employed (Section II.D.2.b).

To establish axenic cultures, 2-4 × 10^5 trichomonads from 72 hr monoxenic cultures are inoculated into tubes containing either TTY-S-$CEEC_{25}$ or TP-S-1 with 0.025% Bacto agar, plus a supplement of $CEEC_{25}$ (1 pt to 9.5 pts TP-S-1). Cultures are incubated at 35.5°C in a vertical position. Subsequent management of the cultures parallels that described for *E. histolytica* (Section II.D.2.c). After establishment of the cultures the chick embryo extract can be eliminated. *Trichomonas tenax*, as does *T. gallinae*, and *T. vaginalis*, tends to grow attached to the walls of the culture tube as well as in the form of free swimmers. Unlike the *Entamoeba*, the attached cells are not readily detached by chilling, especially *T. tenax*. Thus, in harvesting cultures many flagellates are not recovered. Despite this, established cultures of *T. tenax* inoculated with 2.5 × 10^4 flagellates/mℓ medium yield harvests in 96 hr of 1-3 × 10^6/mℓ in TTY-S-$CEEC_{25}$ and one-third more in TP-S-1.

2. In Semidefined and Defined Media

Linstead initiated cultures in semidefined 'modified' CMRL 1066 supplemented with 8% (v/v) fetal calf serum using 1 × 10^4 cells/mℓ taken from the undefined medium of Kulda et al.[139] Cell densities in established cultures reached 6-8 × 10^5/mℓ in 72 hr. Cultures in defined medium DL8 supplemented with 'artificial' serum reached average cell densities of 1 × 10^6 and could be serially subcultured (50×). Moreover, DL8 supplemented with 8% (v/v) human serum was found suitable for isolation of *T. vaginalis* and *Tritrichomonas foetus* from host material. Linstead's paper should be consulted for further details including results obtained with a more refined serum replacement consisting of defatted bovine albumin, fraction V, and three purified fatty acids.[137]

3. Colonial Growth on Agar Plates
(Samuels)[132]

The double layer technique is the most versatile technique described in respect to the number of species which can be cloned.

Preparation of base layer — TYM medium (pH 7.0) is recommended for cloning of *Tritrichomonas* spp. and CPLM (pH 6.0 to 6.5) for *Trichomonas* spp., and *Pentatrichomonas* spp. Pour 20 mℓ of the appropriate serum-free medium containing 1.6% (w/v) agar from large bore tubes (20 to 25 mm) into sterile petri dishes (100 mm). Just before pouring add 0.2 mℓ of a mixture containing 5000 units penicillin and 5000 μg streptomycin/mℓ, and mix by gentle swirling. Immediately place each plate in an air-tight box with 5 g piece of dry-ice. Retain in box at least 15 min after pouring last plate to harden agar and absorb CO_2.

Preparation of overlay — Melt 10 mℓ of appropriate medium with 0.8% (w/v) agar contained in 20 to 25 mm tube and cool in water-bath to 40°C. Add 0.1 mℓ warmed antibiotic mixture, and 1.0 mℓ sheep serum. Bring medium to RT or 37°C depending on species to be cloned. Inoculate with 0.5 mℓ of trichomonad containing liquid medium and immediately pour over base layer. Set plate in CO_2 box to harden and absorb gas.

Preparation of dessicators (Corning #3118 or #3120) for incubation — Place in bottom of dessicator 100 mℓ 10% NaOH (w/v) and 10 g pyrogallic acid for each liter of volume. Place piece of dry-ice (2 to 3 g) in dessicator, then the plates. Seal lid with Dow high vacuum silicone grease and evacuate to a pressure of 500 mmHg. Incubate 37°C, 3 to 5 days for flagellates from homiothermic hosts, and at 25°C, 5 to 7 days for flagellates from poikilothermic hosts.

Depending on the number of organisms inoculated well-isolated colonies or confluent growth can be obtained. The most typical colonies are found within the overlay. Colonies are picked with a platinum wire. To obtain growth from these colonies when transferred to liquid medium, the trichomonads must be freed of the agar. This is performed by gently mashing the agar matrix containing the organisms.

The technique can be used to isolate and axenize *T. vaginalis* from clinical material in one step. Drug assays also can be performed.

IV. *GIARDIA*

The *Giardia* with which we are concerned are characterized morphologically by a claw-shaped median body, and although the possibility exists that all *Giardia* with this type of body form a single species, i.e., *Giardia duodenalis*, I shall use this name for the flagellates of the cat, chinchilla, guinea pig, and rabbit, and *G. lamblia* for the flagellate of man. For further discussion on nomenclature of the genus, consult Levine's publications.[6,140] To date, only *Giardia* with claw-shaped median bodies have been cultivated.

A. Classification

Members of the genus *Giardia* are grouped in the family Hexamitidae, order Diplomonadida, class Zoomastigophora, phylum Sarcomastigophora.

B. Geographic Distribution and Incidence

Giardia are encountered world-wide. *Giardia lamblia* in man is common. Surveys show a world-wide incidence ranging from 3 to 68%. The parasite appears to be more common in children than adults. In the U.S. it is the most common pathogenic intestinal parasite and has been the cause of major outbreaks of waterborne diarrheas. These are discussed by Craun.[141]

C. Life Cycles and Disease

Giardia life cycles are simple and direct. Two stages occur: the trophozoite and the cyst. The latter is involved in transmission and infection occurs by ingestion of this stage.

Giardia lamblia inhabits the duodenum and upper jejunum. Reproduction of the trophozoite is by binary fission. Encystment takes place during passage of the flagellate from the small to large bowel. Relatively few infected persons display recognizable symptoms. Wolfe describes an acute stage lasting 3 to 4 days which is characterized by explosive watery, foul-smelling diarrhea, abdominal distension, flatulence, and nausea. Low-grade fever may precede or accompany the diarrhea. Subacute and chronic stages are characterized chiefly by recurrent brief episodes of loose foul stools.[142] *Giardia duodenalis* infections in small mammals are apparently asymptomatic.

D. Cultivation

1. Historical Review

Unlike the cultivation of *Entamoeba* and trichomonads, the cultivation of *Giardia* dates back little more than two decades. In 1960, Karapetyan reported growing *G. lamblia* in the presence of *Candida guilliermondi* and chick fibroblasts.[143] Two years later he reported cultivating *G. duodenalis* from the rabbit in monoxenic culture with the yeast *Saccharomyces cerevisiae*, and *G. lamblia* in dixenic culture with two fungal species.[144] Meyer and Pope were next to report cultivation of *Giardia*. They also grew *Giardia* from the rabbit and chinchilla but only after modifying Karapetyan's original technique.[145]

Axenic cultivation was achieved by Meyer who in 1970 reported growing *Giardia*, presumably *G. duodenalis*, from rabbit, chinchilla, and the cat,[146] and in 1976 growing *G. lamblia* from man.[147] *Giardia* established in monoxenic culture were freed of their yeast associate by a migration technique, then maintained separated from the yeast by a dialysis membrane. After a period of weaning the *Giardia* could be grown axenically. Fortess and Meyer axenically cultivated a *Giardia* from the guinea pig.[148]

The next advance in cultivation of *Giardia* was made by Bingham and Meyers who discovered that excystment was pH dependent.[149] Treating mature cysts with hydrochloric acid (pH 2.0) was followed by excystation. Trophozoites obtained in this manner could be cultivated axenically without resorting first to monoxenic cultivation. Prior to these studies all cultures of *Giardia* had been initiated with trophozoites isolated from the small bowel of the host.

Human serum used in Meyer's media (HSP-1 and HSP-2) for axenic cultivation varies from lot to lot in ability to support growth of *Giardia*. Visvesvara in an attempt to circumvent this problem found he could adapt Meyer's Portland strain of *G. lamblia* to grow in filter sterilized TP-S-1 containing 10% bovine serum. Growth was more uniform and luxurious in this medium compared to HPS-1.[150] Gillin and Diamond described clonal growth of *G. lamblia* in a semisolid agarose medium prepared with TP-S-1.[151] Bhatia and Warhurst described a technique for routine initiation of axenic cultures of *G. lamblia* with cysts.[152] They combined Bingham and Meyer's method for inducing excystation with Visvesvara's use of TP-S-1. In place of bovine serum they employed horse serum, used by Diamond in his original description of TP-S-1 for axenic cultivation of *E. histolytica*. For additional references on cultivation of *Giardia* consult Meyer's publication on propagation of these flagellates.[153]

2. *Media Preparation and Techniques of Cultivation*

a. Media

The media to be described serve a dual function, they can be used for both monoxenic and axenic cultivation.

1. M3 Medium
   (Meyer)[146]

The medium consists of an autoclaved component (A), a filter sterilized component (B) and a reducing solution.

Component A	
Hank's balanced salt soln.*	140 mℓ
Yeast extract (Difco)	2.0 g
L-Cysteine hydrochloride	0.2 g
Agar	1.1 g
Autoclave, and cool to 50°C in water-bath.	
Component B	
Medium NCTC-109	10.0 mℓ
Human serum	50.0 mℓ
Penicillin G, potassium (100,000 units/mℓ)	1.0 mℓ
Streptomycin sulfate (0.34 g/mℓ)	0.6 mℓ
Each ingredient is presterilized by filtration.	

Combine A and B in a sterile container. Adjust pH to 6.6 with 0.75% (w/v) sterile $NaHCO_3$ soln. Store at 4°C.

Reducing solution	
Glutathione (reduced)	0.1 g
L-Cysteine hydrochloride	0.1 g
Hank's balanced salt soln.*	10 mℓ

* NaCL, 8.0 g; KCl, 0.4 g; $CaCl_2$, 0.14 g; $MgSO_4{\cdot}7H_2O$, 0.2 g; $NaHPO_4{\cdot}12H_2O$, 0.15 g; KH_2PO_4, 0.06 g; $NaHCO_3$, 0.35 g; glucose, 1.0 g; distilled water to make 1000 mℓ soln. Filter sterilize.

Adjust pH to 9.0 with 1 *N* NaOH. Sterilize by passage through Millipore filter. Store at 4°C in small bottle. Do not use after 4 days.

Complete medium — To 7 mℓ of M3, add 0.3 mℓ of reducing solution just before inoculation. Resultant medium has pH 6.8 and O-R potential of −100 mV.

2. HSP-1
   (Meyer)[147]

Nutrient broth

Phytone peptone (BBL)	10.0 g
Glucose	0.5 g
L-Cysteine hydrochloride	1.0 g
Hank's balanced salt soln.	850 mℓ

Autoclave 10 min, 15 lbs, 121°C. Store at 4°C. Use within 1 month.

Complete medium — Prepare on day of use. To 85 mℓ of broth add 15 mℓ filter sterilized human serum, 50,000 units penicillin G, potassium, and 0.05 g streptomycin sulfate. pH (not adjusted) 6.8 to 7.0.

3. HSP-2
   (Meyer)[147]
   To 100 mℓ HSP-1 medium add following sterile ingredients: 7.5 mℓ tissue culture medium NCTC-135 (Gibco); and 2.5 mℓ glutathione-cysteine reducing solution used with M3 medium.

4. TP-S-1
   For cultivation of *Giardia* all ingredients must be filter sterilized. Horse serum or bovine serum (10% v/v) can be used.

b. Techniques of Cultivation

The methods employed by Meyer to obtain the first axenic cultures of *Giardia* are reminiscent of the methods employed for *Entamoeba*. *Giardia* isolated from the host are induced to grow monoxenically with yeasts naturally associated with the flagellates or from domesticated yeast cultures, i.e., *Saccharomyces cerevisiae*. They are then freed of the yeast by use of migration techniques, and maintained separated from the yeast by a dialysis membrane. After a period of several weeks the *Giardia* can be grown axenically.[146,147]

Fortess and Meyer and later Bhatia and Warhurst described similar techniques for elimination of the intermediate step in which the *Giardia* are maintained apart from the yeast by the dialysis membrane.[148,152] This shortens the period between isolation of the flagellates and their axenic cultivation.[148,152] Both techniques make use of the yeasts naturally associated with the *Giardia* for monoxenic cultivation of the flagellates. Fortess and Meyer used HSP-1 medium supplemented with human serum, and Bhatia and Warhurst used TP-S-1 supplemented with horse serum.

I have chosen to present Bhatia and Warhurst's technique because collection of cysts from feces is a noninvasive procedure and TP-S-1 produces more luxuriant growth of *Giardia*. It should be possible to use this technique for initiating cultures with trophozoites collected from the duodenum, if the necessity arises.

1. Initiation of Axenic Cultures of *G. lamblia* with Cysts

1. Collect feces from patient and store before use at 4°C between 5 and 14 days.
2. Harvest and clean cysts.
3. Add 1 vol of clean cysts to 10 vols of hydrochloric acid (pH 2.0). Incubate at 37°C (5 to 30 min).
4. Centrifuge *Giardia* suspension, 5 min, 700 g.
5. Remove and discard supernate. Resuspend pellet in 10 vols water.
6. Centrifuge. Remove supernate. Resuspend pellet in small amount of TP-S-1 (10% horse serum), transfer to culture tubes of TP-S-1 containing gentamycin (40 μg/mℓ).
7. Incubate cultures at 37°C in slanted position.
8. Examine daily up to 2 weeks. Cysts may hatch in as short a period as 30 min or longer than 1 week. Newly hatched trophozoites attach to the walls of the culture tube and grow slowly.
9. After attachment of newly hatched trophozoites, rotate tubes axially through 90°. This permits the debris and yeasts to fall to the other side of the tube and the trophozoites to continue growth on the tube surface.
10. Refeed cultures every 3 to 4 days after hatching by decanting old medium and replacing with fresh TP-S-1. This procedure gradually rids the cultures of the slower growing yeast. Alternatively, the cultures can be incubated in a vertical position. The free-swimming, yeast-free *Giardia* near the surface are collected and used to initiate axenic cultures.

Method of collecting and cleaning *Giardia* cysts (Bingham et al.)[154] — Dilute feces in tap water. Filter through 820 μm-aperture nylon mesh (Tetko) to remove coarse debris. Place 3 to 5 mℓ filtrate on 3 mℓ chilled sucrose (0.85 *M*) contained in a 15 mℓ conical glass centrifuge tube. Centrifuge at RT, 5 min, 600 g. Remove water-sucrose interface, dilute it with water 1:10 and centrifuge 5 min. Resuspend pellet in 3 mℓ water and repeat sucrose gradient centrifugation. Remove water-sucrose interface, dilute it 1:10 with water and filter through 20 μm-aperture nylon mesh. Centrifuge filtrate 5 min. Use pellet to inoculate medium.

2. Established Cultures: Management

In my laboratory *G. lamblia* is grown in TP-S-1 with 10% (v/v) bovine serum. Tubes are filled to 80 to 85% capacity to minimize the gas volume above the surface of the medium. Cultures are incubated at 35.5°C in a slanted position. Subculturing is performed in a manner similar to the method described for *E. histolytica* (Section II.D.2.c). However, in cultures of *G. lamblia* approximately 60 to 80% of the flagellates during log phase growth are attached to the walls of the tube, the remainder swim free. Attached flagellates are more adherent than *Entamoeba* and must be chilled longer to free them prior to subculturing. Cultures selected for transfer are chilled 10 to 15 min, then gently inverted several times. Portions of the suspension are transferred to tubes of fresh medium warmed to RT. Inocula of 4000 flagellates/mℓ yield 10^6 organisms/mℓ in 72 hr. Subcultures are performed at alternating intervals of 72 and 96 hr.

3. Colonial Growth in Semisolid Agarose Medium (Gillin and Diamond)[151]

Preparation of inoculum — Select cultures from log phase growth (24 to 96 hr). Decant the medium with free-swimming flagellates and collect in a clean second tube. Add fresh medium to first tube which still contains attached organisms. Chill both tubes 10 min in ice-water, then invert them 10 times to suspend flagellates. Using the microscope examine samples from both tubes. Approximately 90% of the formerly attached cells will be found isolated from each other. In contrast, approximately 10 to 20% of the free-swimming cells will be in clumps consisting of two or three individuals. Count cells with a hemocytometer or Coulter electronic counter. Make appropriate dilutions with fresh medium.

Concentrated agarose suspension (2.5% w/v) — Place dry agarose (Sigma Agarose III) in flask. Add TP broth (Section II.D.2.c). Dissolve agarose and sterilize mixture in one operation by autoclaving using standard procedures. Place in water bath (65°C).

Semisolid agarose medium: method of use — Place tubes containing 13 mℓ of TP-S-1 medium (Section II.D.2.c) in waterbath (40 to 42°C) and equilibrate 30 min. Add appropriate volumes of 2.5% agarose concentrate with warmed pipet. Mix by inverting. Add suspension of *Giardia* and disperse by inverting tubes 10 times. Gel agar rapidly by chilling in ice-water 10 min. Colony formation is dependent upon concentration of agarose and rapid gelling as it is for *Entamoeba*. The optimal concentration must be determined for each *Giardia* strain. A range between 0.18 and 0.2% should be tested. Colonies become visible to the unaided eye after 5 to 6 days of incubation, and are counted after the 7th day. Colony forming efficiency ranges between 20 and 40%, and is consistently higher for formerly attached cells than for free-swimming cells. See also comments on colonial growth of *Entamoeba* (Section II.D.2.c).

V. APPENDIX: COMMERCIAL SOURCES OF REAGENTS AND EQUIPMENT

BBL: Cockeysville, MD 21030, USA.
Bellco Glass, Inc., Vineland, NJ 08360, USA.
Biofluids, Inc., Rockville, MD 20850, USA.
Calbiochem-Behring Corp., San Diego, CA 92112, USA.
Difco Laboratories, Detroit, MI 48232, USA.
Gibco, Grand Island, NY 14072, USA. Gibco Bio-Cult, LTD., Paisley Scotland.
ICN Nutritional Biochemicals, Cleveland, OH 44128, USA.
K.C. Biologicals, Inc., Lenexa, KA 66215, USA. On request will provide in bulk quantities TP broth base (powder) and vitamin supplement NCTC 107 (sterile solution) for TP-S-1 medium; and TYI broth base (powder) and modified vitamin supplement NCTC 107 (sterile solution) for TYI-S-33 medium.
Mallinckrodt, Inc., St. Louis, MO 63160, USA.
MA Bioproducts, Walkersville, MD 21793, USA.
Nucleopore Corp., Pleasanton, CA 94566, USA.
Oxoid, LTD., Basingstoke, Hampshire, England, UK.
Paines and Byrne, LTD., Greenford, England, UK.
Selas Flotronics, Spring House, PA 19477, USA.
Serva Fine Biochemicals, Inc., Garden City Park, NJ 11040, USA.
Sigma Chemical Comp., St. Louis, MO 63178, USA.
Tetko, Inc., Elmsford, NY 10523, USA.
United States Biochemicals, Inc., Cleveland, OH 44122, USA.

ACKNOWLEDGMENTS

The author thanks Mr. David B. Keister for critical review of the manuscript and Mrs. Wilma J. Davis for her editorial assistance.

REFERENCES

1. **Taylor, A. E. R. and Baker, J. R.,** *The Cultivation of Parasites In Vitro*, Blackwell Scientific, Oxford, 1968.
2. **Taylor, A. E. R. and Baker, J. R., Eds.,** *Methods of Cultivating Parasites In Vitro*, Academic Press, London, 1978.

3. **Diamond, L. S. and Bartgis, I. L.**, Culture media: protozoa. Part II. Parasitic amoebas, in *Biology Data Book*, 2nd ed., Vol. 1, Altman, P. L., and Dittmer, D. S., Eds., Federation American Societies Experimental Biology, Bethesda, 1972, 432.
4. **Lee, J. J.**, Culture media: protozoa. Part III. Trichomonadidae, in *Biology Data Book*, 2nd ed., Vol. 1, Altman, P. C. and Dittmer, D. S., Eds., Federation American Societies Experimental Biology, Bethesda, 1972, 435.
5. **Dougherty, E. C.**, Introduction to axenic culture of invertebrate metazoa: a goal, *Ann. N.Y. Acad. Sci.*, 77, 27, 1959.
6. **Levine, N. D.**, *Protozoan Parasites of Domestic Animals and of Man*, 2nd ed., Burgess Publ. Co., Minneapolis, 1973.
7. **Goldman, G.**, *Entamoeba histolytica*-like amoebae occurring in man, *Bull. WHO*, 40, 355, 1969.
8. **Levine, N. D., Corliss, J. O., Cox, F. E. G., Deroux, G., Grain, J., Honigberg, B. M., Leedale, G. F., Loeblich, III, A. B., Lom, J., Lynn, D., Merinfeld, E. F., Page, F. C., Poljansky, G., Sprague, V., Vavra, J., and Wallace, F. G.**, A newly revised classification of the Protozoa, *J. Protozool.*, 27, 37, 1980.
9. **Neal, R. A.**, The *in vitro* cultivation of *Entamoeba*, in *Problems of In Vitro Culture*, Taylor, A. E. R., Ed., Blackwell Scientific, Oxford, 1967, 9.
10. **Boeck, W. C. and Drbohlav, J.**, The cultivation of *Endamoeba histolytica*, *Am. J. Hyg.*, 5, 371, 1925.
11. **Dobell, C. and Laidlaw, P. P.**, On the cultivation of *Entamoeba histolytica* and some other entozoic amoebae, *Parasitology*, 18, 283, 1926.
12. **Cleveland, L. R. and Sanders, E. P.**, Encystation, multiple fission without encystment, metastatic development and variation in a pure line and nine strains of *Entamoeba histolytica*, *Arch. Protistenk.*, 70, 223, 1930.
13. **Cleveland, L. R. and Collier, J.**, Various improvements in the cultivation of *Entamoeba histolytica*, *Am. J. Hyg.*, 12, 606, 1930.
14. **Nelson, C. E.**, Alcoholic extract medium for the diagnosis and cultivation of *Entamoeba histolytica*, *Am. J. Trop. Med.*, 27, 545, 1947.
15. **Balamuth, W. and Sandza, J. G.**, Simple standardized culture medium for physiological studies on *Endamoeba histolytica*, *Proc. Soc. Exp. Biol. Med.*, 57, 161, 1944.
16. **Balamuth, W.**, Improved egg yolk infusion for cultivation of *Entamoeba histolytica* and other intestinal protozoa, *Am. J. Clin. Pathol.*, 16, 380, 1946.
17. **Diamond, L. S.**, A new medium for xenic cultivation of *Entamoeba histolytica* and other lumen dwelling protozoa, *J. Parasitol.*, 68, 958, 1982.
18. **Robinson, G. L.**, The laboratory diagnosis of human parasitic amoeba, *Tr. R. Soc. Trop. Med. Hyg.*, 62: 285, 1968.
19. **Kessel, J. F., Lewis, W. P., Molina, Pasquel, C., and Turner, J. A.**, Indirect hemagglutination and complement fixation tests in amebiasis, *Am. J. Trop. Med. Hyg.*, 14, 540, 1965.
20. **Albach, R. A., Shaffer, J. G., and Watson, R. H.**, A comparison of *in vitro* drug testing of strains of *Entamoeba* which grow at 37°C and at room temperature, *Am. J. Trop. Med. Hyg.*, 15, 855, 1966.
21. **Cleveland, L. R. and Sanders, E. P.**, The production of bacteria-free amoebic abscesses in the liver of cats and observations on the amoebae in various media with and without bacteria, *Science*, 77, 149, 1930.
22. **Rees, C. W., Reardon, L. V., Jacobs, L., and Jones, F.**, Problems encountered in the growth of *Endamoeba histolytica* in cultures developed by micro-isolation, *Am. J. Trop. Med.*, 21, 567, 1941.
23. **Chinn, B. D., Jacobs, L., Reardon, L. V., and Rees, C. W.**, The influence of the bacterial flora on the cultivation of *Endamoeba histolytica*, *Am. J. Trop. Med.*, 22, 137, 1942.
24. **Rees, C. W., Bozicevich, J., Reardon, L. V., and Jones, F.**, A preliminary note on the complement fixation test for amoebiasis with antigens prepared from *Endamoeba histolytica* grown with a single species of bacteria, *Am. J. Trop. Med.*, 22, 581, 1942.
25. **Rees, C. W.**, *Problems in Amoebiasis*, Charles C Thomas, Springfield, Ill., 1955, chap. 4,5.
26. **Jacobs, L.**, The elimination of viable bacteria from cultures of *Endamoeba histolytica* and the subsequent maintenance of such cultures, *Am. J. Hyg.*, 46, 172, 1947.
27. **Jacobs, L.**, Further studies on the maintenance of cultures of *Endamoeba histolytica* without viable bacteria, *J. Parasitol.*, 33, 20, 1947.
28. **Shaffer, J. G. and Frye, W. W.**, Studies on growth requirements of *Endamoeba histolytica*. I. Maintenance of a strain of *E. histolytica* through one hundred transplants in the absence of an actively multiplying bacterial flora, *Am. J. Hyg.*, 47, 214, 1948.
29. **Shaffer, J. G., Ryden, F. W., and Frye, W. W.**, Studies on the growth requirements of *Endamoeba histolytica*. III. The growth and multiplication of two strains of *E. histolytica* in a transparent medium without the addition of rice flour or other particulate matter and without demonstrable bacterial growth, *Am. J. Hyg.*, 47, 345, 1948.

30. **Shaffer, J. G., Ryden, F. W., and Frye, W. W.,** Studies on the growth requirements of *Endamoeba histolytica*. IV. Further observations on the cultivation of *E. histolytica* and other intestinal protozoa in a clear medium without demonstrable bacteria multiplication. Some modifications and simplifications of the medium, *Am. J. Hyg.*, 49, 127, 1949.
31. **Shaffer, J. G.,** Studies on the growth requirements of *Endamoeba histolytica*. V. Studies on the nature of some of the factors in the Shaffer-Frye medium that effect the propagation of *E. histolytica, Am. J. Hyg.*, 56, 119, 1952.
32. **Reeves, R. E., Meleney, H. E., and Frye, W. W.,** A modified Shaffer-Frye technique for the cultivation of *Entamoeba histolytica* and some observations on its carbohydrate requirements, *Am. J. Hyg.*, 66, 56, 1957.
33. **McDade, J. J. and Shaffer, J. G.,** Studies on the growth requirements of *Entamoeba histolytica*. VI. A modification of the S-F technic which makes possible the utilization of peptones other than Trypticase, *Am. J. Trop. Med. Hyg.*, 8, 540, 1959.
34. **Reeves, R. E. and Bischoff, J. M.,** Classification of *Entamoeba* species by means of electrophoretic properties of amebal enzymes, *J. Parasitol.*, 54, 594, 1968.
35. **Phillips, B. P.,** Cultivation of *Endamoeba histolytica* with *Trypanosoma cruzi, Science*, 111, 8, 1950.
36. **Phillips, B. P. and Rees, C. W.,** The growth of *Endamoeba histolytica* with live and heat-treated *Trypanosoma cruzi, Am. J. Trop. Med.*, 30, 185, 1950.
37. **Phillips, B. P.,** Further studies with ameba-trypanosome cultures, *Am. J. Trop. Med. Hyg.*, 11, 6, 1962.
38. **Pan, C.-T.,** Studies on the monoxenic cultivation of *Entamoeba histolytica* with hemoflagellates, *J. Infect. Dis.*, 106, 384, 1960.
39. **Diamond, L. S.,** Axenic cultivation of *Entamoeba histolytica, Science*, 134, 336, 1961.
40. **Diamond, L. S.,** Improved method for the monoxenic cultivation of *Entamoeba histolytica* Schaudinn, 1903 and *E. histolytica*-like amebae with trypanosomatids, *J. Parasitol.*, 54, 715, 1968.
41. **McConnachie, E. W.,** Modification and elimination of the bacterial flora in cultures of *Entamoeba invadens* Rodhain, 1934, *Parasitology*, 46, 117, 1956.
42. **Balamuth, W.,** Effects of some environmental factors upon growth and encystation of *Entamoeba invadens, J. Parasitol.*, 48, 101, 1962.
43. **Reardon, L. V. and Bartgis, I. L.,** The cultivation of the small race of *Endamoeba histolytica* with a single species of bacteria, *J. Parasitol.*, 35, 218, 1949.
44. **Stoll, N. R.,** Axenic culture of *Entamoeba invadens* in the absence of tissue, *J. Protozool.*, 4 (Suppl.), 6, 1957.
45. **Stoll, N. R.,** Axenic cultivation of *Entamoeba invadens* in a cell-free medium, *Int. Conf. Zool. London, Sect.*, 8, 639, 1959.
46. **Miller, M. J.,** Bacteria-free *Entamoeba invadens, Nature (London)*, 172, 1192, 1953.
47. **Diamond, L. S.,** The axenic cultivation of two reptilian parasites, *Entamoeba terrapinae*, Sanders and Cleveland, 1930, and *Entamoeba invadens*, Rodhain, 1934, *J. Parasitol.*, 46, 484, 1960.
48. **Diamond, L. S.,** The establishment of various trichomonads of animals and man in axenic cultures, *J. Parasitol.*, 43, 488, 1957.
49. **McConnachie, E. W.,** A medium for axenic cultivation of *Entamoeba invadens, Nature (London)*, 194, 603, 1962.
50. **McConnachie, E. W.,** The growth of an axenic strain of *Entamoeba invadens* in different media, *Parasitology*, 58, 733, 1968.
51. **Diamond, L. S.,** Techniques of axenic cultivation of *Entamoeba histolytica* Schaudinn, 1903 and *E. histolytica*-like amebae, *J. Parasitol.*, 54, 1047, 1968.
52. **Van Vliet, H. H. D. M., OP Den Kamp, J. A. F., and Van Deenen, L. L. M.,** Lipid auxotrophy and the effect on lipid composition of *Entamoeba invadens, J. Protozool.*, 22, 428, 1975.
53. **Rengpien, S. and Bailey, G. B.,** Differentiation of *Entamoeba:* a medium and optimal conditions for axenic encystation of *E. invadens, J. Parasitol.*, 61, 24, 1975.
54. **Diamond, L. S., Harlow, D. R., and Cunnick, C. C.,** A new medium for the axenic cultivation of *Entamoeba histolytica* and other *Entamoeba, Tr. R. Soc. Trop. Med. Hyg.*, 72, 431, 1978.
55. **Gillin, F. G. and Diamond, L. S.,** Clonal growth of *Entamoeba histolytica* and other species of *Entamoeba* in agar, *J. Protozool.*, 25, 539, 1978.
56. **Gillin, F. G. and Diamond, L. S.,** Clonal growth of *Entamoeba* in agar: some applications of this technique to the study of their cell biology, *Arch. Invest. Méd. (Méx)*, 9, 237, 1978.
57. **Diamond, L. S.,** Axenic cultivation of *Entamoeba histolytica:* progress and problems, *Arch. Invest. Méd. (Méx)*, 11 (Supl. 1), 47, 1980.
58. **Reardon, L. V. and Rees, C. W.,** The cultivation of *Endamoeba histolytica* without serum, *Am. J. Parasitol.*, 25 (Supl. 1), 13, 1939.
59. **Von Brand, T., Rees, C. W., Jacobs, L., and Reardon, L. V.,** Studies on reducing substances and gas formation in cultures of *Endamoeba histolytica* and a single species of symbiotic bacteria, *Am. J. Hyg.*, 37, 310, 1943.

60. **Stone, W. S.,** A method of producing encystment in cultures of *Endamoeba histolytica, Am. J. Trop. Med.*, 15, 681, 1935.
61. **Spingarn, C. L. and Edelman, M. H.,** The effect of rice powder on the growth of cultures of *Endamoeba histolytica, Am. J. Trop. Med.*, 30, 629, 1950.
62. **Nelson, E. C. and Jones, M. M.,** Some factors related to *Entamoeba histolytica* growth on rice products in a simple medium, *Am. J. Trop. Med. Hyg.*, 4, 822, 1955.
63. **Griffin, A. M. and McCarten, W. G.,** Some methods for the quantitative study of entozoic amoebae in cultures, *J. Parasitol.*, 35, 193, 1949.
64. **Craig, C. F.,** *Laboratory Diagnosis of Protozoan Diseases*, Lea & Febiger, Philadelphia, 1942, 81.
65. **Spingarn, C. L. and Edelman, M. H.,** Further observations on the use of streptomycin and penicillin in the cultivation of *Endamoeba histolytica* from stools, *Am. J. Trop. Med. Hyg.*, 1, 412, 1952.
66. **Norman, L. and Brooke, M. M.,** The use of penicillin and streptomycin in the routine cultivation of amebae from fecal specimens, *Am. J. Trop. Med. Hyg.*, 4, 472, 1955.
67. **Das, S. R. and Singh, B. N.,** Combined use of acriflavine and gentian violet in culture medium for the cultivation of *Entamoeba histolytica* and *E. coli* from feces of man, monkeys and experimentally infected laboratory animals, *Proc. Indian Natl. Sci. Acad.*, 36, 159, 1970.
68. **Dobell, C.,** Researches on the intestinal protozoa of monkeys and man. IV. An experimental study of the *histolytica*-like species of *Entamoeba* living naturally in macaques, *Parasitology*, 23, 1, 1931.
69. **de Moraes-Ruehsen, M., McNeill, R. E., Frost, J. K., Gupta, P. K., Diamond, L. S., Honigberg, B. M.,** Amebae resembling *Entamoeba gingivalis* in the genital tracts of IUD users, *Acta Cytolog. J.*, 24, 413, 1980.
70. **Neal, R. A.,** Experimental production of pure mixed strains of *Entamoeba histolytica, Parasitology*, 42, 40, 1952.
71. **Dobell, C.,** (Completed by R. A. Neal and C. A. Hoare), Researches on the intestinal protozoa of monkey and man. XII. Bacterial factors influencing the life history of *Entamoeba histolytica* in cultures, *Parasitology*, 42, 16, 1952.
72. **Bos, H. J.,** Monoxenic and axenic cultivation of carrier and patient strains of *Entamoeba histolytica, Z. Parasitenk.*, 47, 119, 1975.
73. **Meerovitch, E.,** *Entamoeba, Giardia* and *Trichomonas*, in *Methods of Cultivating Parasites In Vitro*, Taylor, A. E. R. and Baker, J. R., Eds., Academic Press, London, 1978, chap. 2.
74. **Evans, V. J., Bryant, J. C., Fioramonti, M. C., McQuilkin, W. T., Sanford, K. K., and Earle, W. R.,** Studies of nutrient media for tissue cells *in vitro*. I. A protein-free chemically defined medium for cultivation of strain L cells, *Cancer Res.*, 16, 77, 1956.
75. **Diamond, L. S.,** Serum requirements of axenically cultivated *Entamoeba histolytica, J. Parasitol.*, 56, 79, 1970.
76. **Barile, M. F., Hopps, H. E., Grabowski, M. W., and Riggs, D. B.,** The identification and sources of mycoplasmas isolated from contaminated cell cultures, *Ann. N.Y. Acad. Sci.*, 225, 251, 1973.
77. **Reeves, R. E.,** Carbohydrate metabolism in *Entamoeba histolytica*, in *Comparative Biochemistry of Parasites*, Van den Bossche, H., Ed., Academic Press, New York, 1972.
78. **Reeves, R. E.,** Glycolytic enzymes in *E. histolytica, Arch. Invest. Méd. (Méx)*, 5 (Supl. 2), 411, 1975.
79. **Weinbach, E. C., Harlow, D. R., Takeuchi, T., Diamond, L. S., Claggett, C. E., and Kon, H.,** Aerobic metabolism of *Entamoeba histolytica:* facts and fallacies, in *Proc. Int. Conf. on Amebiasis*, Sepulveda, B. and Diamond, L. S., Eds., Instituto Mexicano del Seguro Social, Mexico, 1976, 190.
80. **Weinbach, E. C., Takeuchi, T., Claggett, C. E., Inohue, F., Kon, H., and Diamond, L. S.,** Role of iron-sulfur proteins in the electron transport system of *Entamoeba histolytica, Arch. Invest. Méd. (Méx)*, 11 (Supl. 1), 75, 1980.
81. **Booden, T., Boonlayangoor, P., and Albach, R. A.,** Incorporation of purines and pyrimidines into axenically grown *Entamoeba histolytica, J. Parasitol.*, 62, 641, 1976.
82. **Booden, T., Albach, R. A., and Boonlayangoor, P.,** Uptake of selected pyrimidines by axenically grown *Entamoeba histolytica, Arch. Invest. Méd. (Méx)*, 9 (Supl. 1), 133, 1978.
83. **Boonlayangoor, P., Albach, R. A., Stern, M. L., and Booden, T.,** *Entamoeba histolytica:* uptake of purine bases and nucleosides during axenic growth, *Exp. Parasitol.*, 45, 225, 1978.
84. **Gelderman, A. H., Bartgis, I. L., Keister, D. B., and Diamond, L. S.,** A comparison of genome sizes and thermal-denaturation-derived base composition of DNAs from several members of *Entamoeba* (Histolytica group), *J. Parasitol.*, 57, 912, 1971.
85. **Pinto de Silva, P., Martínez-Palomo, A., and Gonzalez-Robles, A.,** Membrane structure and surface coat of *Entamoeba histolytica*. Topochemistry and dynamics of the cell surface: cap formation and exudate, *J. Cell Biol.*, 64, 538, 1975.
86. **Trissl, D., Martínez-Palomo, A., Arguello, C., de la Torre, M., and de la Hoz, R.,** Surface properties related to concanavalin A-induced agglutination. A comparative study of several *Entamoeba* strains, *J. Exp. Med.*, 145, 652, 1077.

87. **Aust-Kettis, A. and Sundquist, K. G.,** Dynamics of the interaction between *Entamoeba histolytica* and components of the immune response. I. Capping and endocytosis; Influence of inhibiting and accelerating factors; variation of the expression of surface antigens, *Scand. J. Immunol.*, 7, 35, 1978.
88. **Aley, S. B., Scott, W. A., and Cohn, Z. A.,** Plasma membrane of *Entamoeba histolytica, J. Exp. Med.*, 152, 391, 1980.
89. **Diamond, L. S., Phillips, B. P., and Bartgis, I. L.,** A comparison of the virulence of nine strains of axenically cultivated *E. histolytica* in hamster livers, *Arch. Invest. Méd. (Méx)*, 5 (Supl. 2), 423, 1974.
90. **Mattern, C. F. T., Keister, D. B., and Caspar, P. C.,** Experimental Amebiasis III. A rapid *in vitro* assay for virulence of *Entamoeba histolytica, Am. J. Trop. Med. Hyg.*, 27, 882, 1977.
91. **McCaul, T. F., Poston, R. N., and Bird, R. G.,** *Entamoeba histolytica* and *Entamoeba invadens:* chromium release from labeled human liver cells in culture, *Exp. Parasitol.*, 43, 342, 1977.
92. **Lushbaugh, W. B., Kairalla, A. B., Cantey, J. R., Hofbauer, A. F., and Pittman, F. E.,** Isolation of a cyto-toxin-enterotoxin from *Entamoeba histolytica, J. Infect. Dis.*, 139, 9, 1979.
93. **Bos, H. J.,** *Entamoeba histolytica*: cytopathogenicity of intact amebae and cell-free extracts; isolation and characterization of intracellular toxin, *Exp. Parasitol.*, 47, 369, 1979.
94. **Mattern, C. F. T., Keister, D. B., and Natovitz, C. P.,** *Entamoeba histolytica* "toxin": fetuin neutralizable and lectin-like, *Am. J. Trop. Med. Hyg.*, 29, 26, 1980.
95. **Ravdin, J. I., Croft, B. Y., Guerrant, R. L.,** Cytopathogenic mechanisms of *Entamoeba histolytica, J. Exp. Med.*, 152, 377, 1980.
96. **Diamond, L. S. and Mattern, C. F. T.,** Protozoal viruses, in *Advances in Virus Research*, Academic Press, New York, 1976, 87.
97. **Diamond, L. S. and Bartgis, I. L.,** Axenic cultures for *in vitro* testing of drugs against *Entamoeba histolytica, Arch. Invest. Méd. (Méx)*, 2 (Supl. 1), 339, 1971.
98. **Patterson, M., Healy, G. R., and Shabot, J. M.,** Serologic testing for amebiasis, *Gastroenterol.*, 78, 136, 1980.
99. **Kagan, I. G.,** Seroepidemiology of amebiasis, in *Proc. Int. Conf. Amebiasis*, Sepuluveda, B. and Diamond, L. S., Eds., Instituto Mexicano del Seguro Social, Mexico, 1976, 574.
100. **Savanat, T., Viriyanond, P., and Nimitmongkol, N.,** Blast transformation of lymphocytes in amebiasis, *Am. J. Trop. Med. Hyg.*, 22, 705, 1973.
101. **Ortiz-Ortiz, L., Zamacona, G., Sepulveda, B., and Capin, N. R.,** Cell-mediated immunity in patients with amebic abscess of the liver, *Clin. Immunol. Immunopath.*, 4, 127, 1975.
102. **Harris, W. G. and Bray, R. S.,** Cellular sensitivity in amoebiasis-preliminary results of lymphocytic transformation in response to specific antigen and to mitogen in carrier and disease states, *Tr. R. Soc. Trop. Med. Hyg.*, 70, 340, 1976.
103. **Huldt, G., Davies, P., Allison, A. C., and Schorlemmer, H. U.,** Interactions between *Entamoeba histolytica* and complement, *Nature (London)*, 277, 214, 1979.
104. **Krupp, I. M.,** Protective immunity to amebic infection demonstrated in guinea pigs, *Am. J. Trop. Med. Hyg.*, 23, 355, 1974.
105. **Meerovitch, E., Hartman, D. P., and Ghadirian, E.,** Protective immunity and possible auto-immune regulation in amebiasis, *Arch. Invest. Méd. (Meź)*, 9 (Supl. 1), 247, 1978.
106. **Stabler, R. M.,** *Trichomonas gallinae:* a review, *Exp. Parasitol.*, 3, 368, 1954.
107. **Honigberg, B. M.,** Trichomonads of veterinary importance, in *Parasitic Protozoa*, Vol. 2, Krier, J., Ed., Academic Press, New York, 1978, chap. 3.
108. **Honigberg, B. M.,** Trichomonads of importance in human medicine, in *Parasitic Protozoa*, Vol. 2, Krier, J., Ed., Academic Press, New York, 1978, chap. 4.
109. **Bos, A.,** Ueber Trichomoniasis bei Tauben. III. Mitteilung: Weitere Beobachtungen uber die Kultur und Pathogenitat von *Trichomonas columbae, Zentrabl. Bakteriol. Parasitenk. I Abt Orig.*, 130, 220, 1933.
110. **Witte, J.,** Bakterienfreie Züchtung von Trichomonaden aus dem Uterus des Rindes in einfachen Nährboden, *Zentrabl. Bakteriol. Parasitenk.*, 128, 188, 1933.
111. **Stone, W. S. and Reynolds, F. H. K.,** A practical method of obtaining bacteria-free cultures of *Trichomonas hominis, Science*, 90, 91, 1939.
112. **Trussell, R. E.,** Experimental and clinical *Trichomonas* vaginitis, *J. Iowa State Med. Soc.*, 30, 66, 1940.
113. **Diamond, L. S.,** Axenic cultivation of *Trichomonas tenax*, the oral flagellate of man. I. Establishment of cultures, *J. Protozool.*, 9, 442, 1962.
114. **Glaser, R. W. and Coria, N. A.,** Purification and culture of *Tritrichomanas foetus* (Riedmüller) from cows, *Am. J. Hyg.*, 22, 221, 1935.
115. **Adler, S.,** A note on bacteria-free cultures of *Trichomonas hominis, Tr. R. Soc. Trop. Med. Hyg.*, 35, 219, 1942.
116. **Rees, C. W.,** Obtaining bacteria-free pure lines of *Trichomonas foetus* by means of microisolation, *Am. J. Hyg.*, 26, 283, 1937.
117. **Adler, S. and Pulvertaft, R. J. V.,** The use of penicillin for obtaining bacteria-free cultures of *Trichomonas vaginalis* Donné, 1837, *Ann. Trop. Med. Parasitol.*, 38, 188, 1944.

118. **Quisno, R. A. and Foter, M. J.,** The use of streptomycin in the purification of *Trichomonas vaginalis, J. Bacteriol.*, 51, 404, 1946.
119. **Lowe, G. H.,** A comparison of current laboratory methods and a new semi-solid culture medium for the detection of *Trichomonas vaginalis, J. Clin. Pathol.*, 18, 432, 1965.
120. **Honigberg, B. M., Livingston, M. C., and Frost, J. K.,** Pathogenicity of fresh isolates of *Trichomonas vaginalis*. The mouse assay versus clinical and pathologic findings, *Acta Cytol.*, 10, 353, 1966.
121. **Johnson, G. and Trussell, R. E.,** Experimental basis for the chemotherapy of *Trichomonas vaginalis* infestations. I, *Proc. Soc. Exp. Biol. Med.*, 54, 245, 1943.
122. **McEntegart, M. G.,** Application of a haemagglutination technique to the study of *Trichomonas vaginalis* infections, *J. Clin. Pathol.*, 5, 275, 1952.
123. **Feinberg, J. G.,** A method for the bulk growth of a parasitic protozoan, *Nature (London)*, 171, 1165, 1953.
124. **Feinberg, J. G. and Whittington, M. J.,** A culture medium for *Trichomonas vaginalis* Donné and species of *Candida, J. Clin. Pathol.*, 10, 327, 1957.
125. **Lowe, G. H.,** A comparison of culture media for the isolation of *Trichomonas vaginalis, Med. Lab. Technol.*, 29, 389, 1972.
126. **Schneider, M. D.,** A new thermostabile medium for the prolonged bacteria-free cultivation of *Trichomonas foetus, J. Parasitol.*, 28, 428, 1942.
127. **Plastridge, W.,** Cultivation of a bacteria-free strain of *Trichomonas foetus, J. Bacteriol.*, 45, 196, 1943.
128. **Fitzgerald, P. R., Hammond, D. M., and Shupe, J. LeG.,** The role of cultures in immediate and delayed examinations of preputial samples for *Trichomonas foetus, Vet. Med.*, 49, 409, 1954.
129. **Asami, K., Nodake, Y., and Ueno, T.,** Cultivation of *Trichomonas vaginalis* on solid medium, *Exp. Parasitol.*, 4, 34, 1955.
130. **Ivey, M. H.,** Growth characteristics of clones of *Trichomonas vaginalis* in solid medium, *J. Parasitol.*, 47, 539, 1961.
131. **Hollander, D. H.,** Colonial morphology of *Trichomonas vaginalis* in agar, *J. Parasitol.*, 62, 826, 1976.
132. **Samuels, R.,** Agar techniques for colonizing and cloning trichomonads, *J. Protozool.*, 9, 103, 1962.
133. **Shorb, M. S. and Lund, P. G.,** Requirement of trichomonads for unidentified growth factors, saturated and unsaturated fatty acids, *J. Protozool.*, 6, 122, 1959.
134. **Lund, P. G. and Shorb, M. S.,** Steroid requirements of trichomonads, *J. Protozool.*, 9, 151, 1962.
135. **Lee, J. J., Pierce, S., Hutner, S. H., Smith, B. J., and Gurski, D. R.,** Trichomonads from poikilotherms: nutritional and physiological notes, *J. Protozool.*, 9, 445, 1962.
136. **Shorb, M. S.,** The physiology of trichomonads, in *Biochemistry and Physiology of Protozoa*, Vol. 3, Hutner, S. H. and Lwoff, A., Eds., Academic Press, New York, 1964, 383.
137. **Linstead, D.,** New defined and semi-defined media for cultivation of the flagellate *Trichomonas vaginalis, Parasitology*, 83, 125, 1981.
138. **Trusell, R. E. and Johnson, G.,** *Trichomonas vaginalis* Donné. Recent experimental advances, *Puerto Rico J. Pub. Hlth. Trop. Med.*, 20, 289, 1945.
139. **Kulda, J., Honigberg, B. M., Frost, J. K., and Hollander, D. H.,** Pathogenicity of *Trichomonas vaginalis, Am. J. Obstet. Gynecol.*, 108, 908, 1970.
140. **Levine, N. D.,** *Giardia lamblia:* classification, structure, identification, in *Waterborne Transmission of Giardiasis*, Jakubowski, W. and Hoff, J. C., Eds., U.S. Environmental Protection Agency, Cincinnati, 1979, 2.
141. **Craun, G. F.,** Waterborne outbreaks of giardiasis, in *Waterborne Transmission of Giardiasis*, Jakubowski, W. and Hoff, J. C., Eds., U.S. Environmental Protection Agency, Cincinnati, 1979, 127.
142. **Wolfe, M. S.,** Managing the patient with giardiasis: clinical diagnostic and therapeutic aspects, in *Waterborne Transmission of Giardiasis*, Jakubowski, W. and Hoff, J. C., Eds., U.S. Environmental Protection Agency, Cincinnati, 1979, 39.
143. **Karapetyan, A. E.,** Methods of *Lamblia* cultivation, *Tsitologiia*, 2, 379, 1960, cited in Karapetyan, A., *In vitro* cultivation of *Giardia duodenalis, J. Parasitol.*, 48, 337, 1962.
144. **Karapetyan, A.,** *In vitro* cultivation at *Giardia duodenalis, J. Parasitol.*, 48, 337, 1962.
145. **Meyer, E. A. and Pope, B. L.,** Culture *in vitro* of *Giardia* trophozoites from the rabbit and chinchilla, *Nature (London)*, 207, 1417, 1965.
146. **Meyer, E. A.,** Isolation and axenic cultivation of *Giardia* trophozoites from the rabbit, chinchilla, and cat, *Exp. Parasitol.*, 27, 179, 1970.
147. **Meyer, E. A.,** *Giardia lamblia:* isolation and axenic cultivation, *Exp. Parasitol.*, 39, 101, 1976.
148. **Fortess, E. and Meyer, E. A.,** Isolation and axenic cultivation of *Giardia* trophozoites from the guinea pig, *J. Parasitol.*, 62, 689, 1976.
149. **Bingham, A. K. and Meyer, E. A.,** *Giardia* excystation can be induced *in vitro* in acidic solutions, *Nature (London)*, 277, 301, 1979.
150. **Visvesvara, G. S.,** Axenic growth of *Giardia lamblia* in Diamond's TPS-1 medium, *Tr. R. Soc. Trop. Med. Hyg.*, 74, 213, 1980.

151. **Gillin, F. D. and Diamond, L. S.,** Clonal growth of *Giardia lamblia* trophozoites in a semi-solid agarose medium, *J. Parasitol.*, 66, 350, 1980.
152. **Bhatia, V. N. and Warhurst, D. C.,** Hatching and subsequent cultivation of cysts of *Giardia intestinalis* in Diamond's medium, *J. Trop. Med. Hyg.*, 84, 45, 1981.
153. **Meyer, E. A.,** The propagation of *Giardia* trophozoites *in vitro*, in *Water Transmission of Giardiasis*, Jakobowski, W. and Hoff, J. C., Eds., U.S. Environmental Protection Agency, Cincinnati, 1979, 211.
154. **Bingham, A. K., Jarroll, Jr., E. L., and Meyer, E. A.,** *Giardia* sp: physical factors of excystation *in vitro* and excystation vs. eosin exclusion as determinants of viability, *Exp. Parasitol.*, 47, 284, 1979.

Chapter 3

LEISHMANIA

K.-P. Chang and Wallace R. Fish

TABLE OF CONTENTS

I. INTRODUCTION

The genus *Leishmania* consists of many species of which *L. brasiliensis, L. donovani, L. mexicana, L. tropica,* and their subspecies or strains are pathogenic to man. The morphology of leishmania is very similar among different species. Differentiation of leishmania species, which has relied entirely on clinical symptoms, geographic origin of the parasites and elaborate culture experiments, is now based on more dependable characters, e.g., isoenzyme patterns and DNA buoyant density.[1] The sequelae of infections by different leishmanial species present a spectrum of severity, ranging from innocuous cutaneous lesions caused by *L. tropica* to facial disfigurement by *L. brasiliensis* to fatal consequence by *L. donovani*. The drugs of choice for treating leishmaniasis remain largely the antimonial compounds, which are not altogether satisfactory, but no new antileishmanial agents have been developed in the last several decades. All leishmaniases are vector-born diseases transmitted by the blood-sucking females of phlebotomid sandflies.

That leishmaniases are a world-wide problem is underlined by the fact of their occurrence in all continents except Australia. Most serious of all is the debilitating and often fatal visceral leishmaniasis or kala azar caused by *L. donovani*. This disease, once rampant in central China, has been virtually eradicated in this formerly endemic region in the 1950s by a combination of the following persistent and thorough measures: elimination of dogs (reservoir animals), mass treatment of patients, and eradication of sandflies.[2] No similar eradication program of this magnitude has been undertaken or is planned in any other part of the world. The diseases have either escaped attention, never reached epidemic proportions, or have been somewhat under control by certain measures for other parasitic diseases, e.g., mosquito control. There are now strong indications that all forms of leishmaniases are on the rise and future difficulties in dealing with these diseases are predicted. More detailed information is available in several recent review articles.[3-5]

All leishmania species pathogenic to man have two developmental stages: promastigotes in sandflies and amastigotes in the mammalian hosts. Promastigotes are slender motile cells with anterior flagella (Figures 1 and 2), while amastigotes are small nonmotile intracellular entities in mononuclear phagocytes of the reticuloendothelial system (Figures 3 and 4). Leishmaniae, like other trypanosomatid flagellates, possess organelles such as flagella, flagellar pockets, subpellicular microtubules, and kinetoplasts — the DNA containing regions of mitochondria (Figures 2 and 4). The two different forms of leishmania have basically an identical cellular organization, except for the length of the flagellum which extends beyond the flagellar pocket only in promastigotes (Figure 2). Metabolism of leishmanial organisms has not been thoroughly studied. Only the promastigote stage has been used for such investigations, which show no striking deviations from related trypanosomatids.[6-8] Animal immunity to leishmania infections is believed to be cell-mediated, but the precise mechanism is still under investigation.[9-11] *Leishmania tarentolae*, a parasite of lizards, has proven to be a favorable source of material for the study of kinetoplast DNA.[12]

Two unique features associated with leishmanial infection are the promastigote to amastigote transformation and intracellular parasitism. The mechanisms underlying both phenomena are not clearly understood, but they are apparently interrelated events, which culminate in all the pathogenicity of these diseases. Of particular interest is the fact that leishmanial amastigotes parasitize macrophages, the very cell type responsible for the first line of defense and one of the principal cellular elements in animal immunity. Even more surprising is that leishmanial amastigotes should live in the lysosomal compartment of macrophages — the cell organelles whose very function is to digest invading microorganisms. To investigate these basic and other more practical questions, there is a need to culture both promastigotes and amastigotes in vitro to provide assay systems as well as sufficient materials. In this chapter, we shall review the past attempts and the current methods for the cultivation of both forms of leishmaniae.

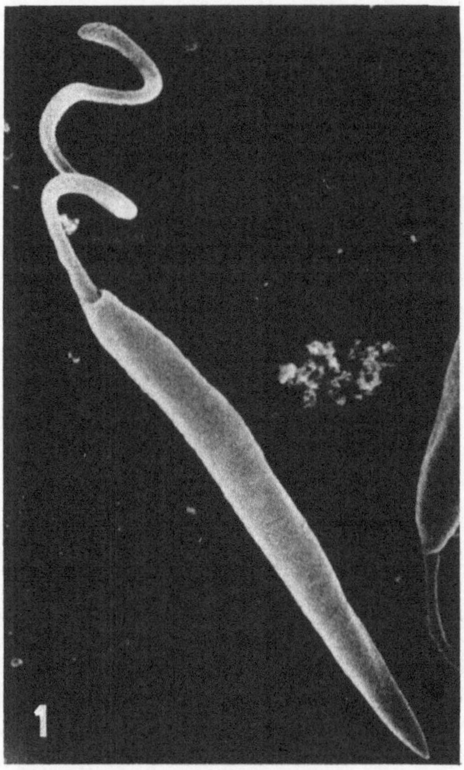

FIGURE 1. Promastigote of *Leishmania donovani* under scanning electron microscope. Note the long curled flagellum originating from the flagellar pocket at the anterior end of the somatic portion (7,500 ×).

From an historical point of view, technology for cultivation of leishmanial promastigotes has generally followed that developed for trypanosomes. In 1904 Rogers[13] first observed the transformation of amastigotes from clinical samples into promastigotes in a simple saline solution. Four years later, Nicolle[14] succeeded in culturing leishmania in a blood agar medium used earlier for growing trypanosomes. Since then culture media have evolved from biphasic, semidefined to defined monophasic forms for other trypanosomatid flagellates, which have been adopted or modified for leishmaniae. Apparently, the resemblance of leishmanial promastigotes to their close relatives in biosynthetic capability renders it unnecessary to design specific medium for their growth in vitro, regardless of the purposes for diagnosis or for laboratory investigations. The development of methods for the cultivation of leishmanial amastigotes has followed an entirely different path and has not been fully explored until recently. In keeping with their property of being intracellular parasites, the cultivation of leishmanial amastigotes in ordinary media has been unsuccessful and necessitates the use of tissue culture cells as substrates. This will be discussed first in the following section.

II. CULTIVATION OF AMASTIGOTES

In comparison with the cultivation of other intracellular protozoa, efforts devoted to leishmanial amastigotes have been relatively moderate and progress has not been made until recently. In general, there are two major reasons for keeping leishmanial amastigotes in vitro: short-term cultivation of these parasites as an assay system and their long-term cultivation as a substitute for passages in laboratory animals as well as for mass production of

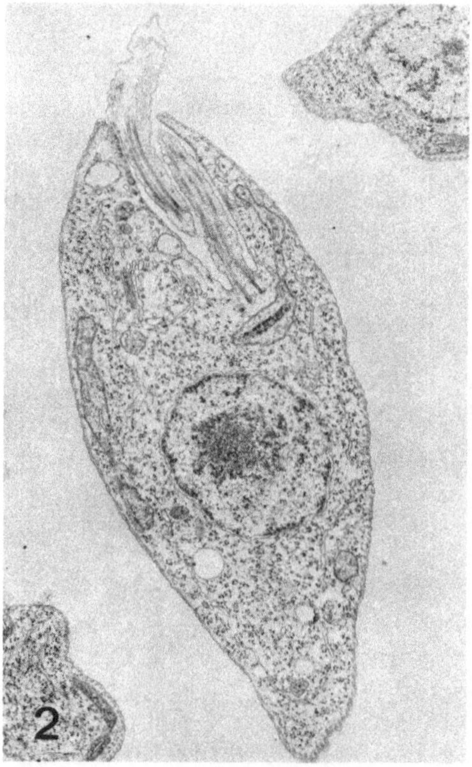

FIGURE 2. Longitudinal section of *Leishmania donovani* promastigote showing its cellular organization. Note the subpellicular microtubules beneath the plasma membrane, organelle-rich cytoplasm (endoplasmic reticulum, ribosomes, vacuoles, mitochondrion, and Golgi apparatus) and centrally located nucleus, above which is the kinetoplast with mitochondrial DNA fibrils. Flagellum arises from kinetosome above the kinetoplast and extends beyond the opening of the flagellar pocket (11,112 ×).

parasites. In this section, we shall summarize the techniques and problems in an attempt to achieve these goals with emphasis on the mammalian tissue cultures which have been more extensively and successfully used than others. There has been no extensive review of this subject, except for accounts in several articles primarily devoted to other intracellular protozoa.[15-17]

A. Tissue or Cell Cultures

1. Tissue Explants

Early tissue culturists employed pieces of tissue or organ for in vitro studies. This technique represents the prototype of today's tissue culture that has been used for studying host-parasite interactions in vitro. Results derived from this type of work are summarized in Table 1. Since hamsters are frequently used as laboratory hosts for experimental leishmaniases, organs or tissues of these animals have been favorable sources of explants.[18-23] Other sources of tissues for such studies include chicken embryos,[19,24-25] guinea pig, and frog.[26] The most often used tissues or organs are skin, liver and spleen which are the sites of infection in vivo. Explanted tissues may have been already infected in vivo[18-23] or infected in vitro[24-26] with promastigotes or amastigotes. Tissues were normally cultured in saline supplemented

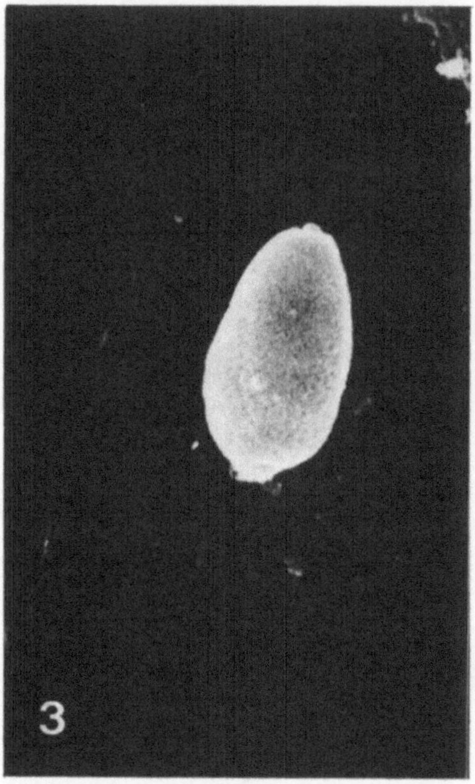

FIGURE 3. Scanning electron microscopy of *Leishmania donovani* amastigotes free from its host cell. Note the barely visible stump of flagellum at the opening of the flagellar pocket (15,000 ×).

with animal sera and/or embryo extracts on plasma clots or hanging drops and then incubated at 27 to 37°C for weeks. At different intervals, cultures were stained to assess the results. In such preparations, most authors observed intracellular entry of promastigotes and their subsequent transformation into amastigotes. It is noteworthy that promastigotes of *L. donovani* may emerge in cultures of infected spleen explants at 37°C.[20] The most successful cultures were obtained by Tchernomoretz[22] who observed intracellular and extracellular multiplication of amastigotes in spleen explants of hamster or calf and succeeded in subculturing these for many passages.

The value of primary tissue explants for the study of host-parasite interactions in leishmaniasis is very limited because of the heterogeneous cellular populations in such preparations and the lack of means for quantitative studies. The only advantage of tissue explants lies in the possibility that they may retain some organ or tissue characteristics which would be lost in cell lines. It should also be recognized that these studies represent the earliest attempts to culture leishmanial amastigotes with the best methods available at the time.

2. Macrophages

As already mentioned, leishmanial amastigotes normally parasitize mononuclear phagocytes or macrophages of their mammalian hosts. It is, therefore, not surprising that leishmania-macrophage in vitro systems have been widely used in several types of investigations, e.g., host-parasite interactions, cellular immunity in leishmaniasis, and assessment of antileishmanial agents. The sources of macrophages have been the peritoneal cavity of host

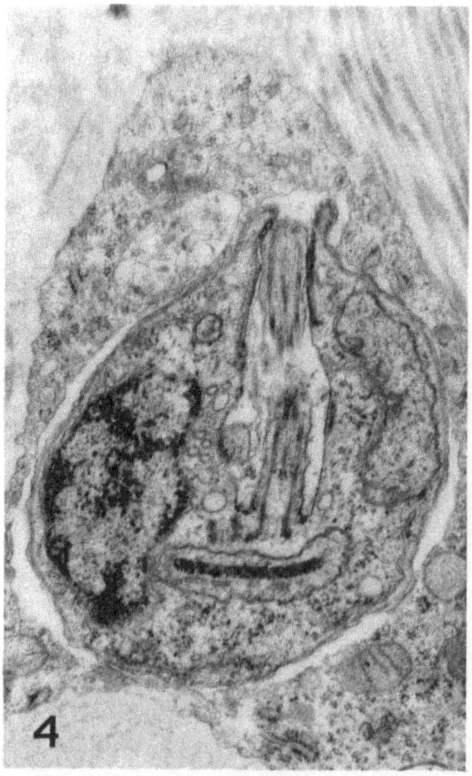

FIGURE 4. Portion of a histiocyte in hamster skin infected with *Leishmania brasiliensis*. The parasite in the form of amastigote is located in a parasitophorous vacuole, and has a prominent nucleus on the left and kinetoplast in the center, above which flagellum originates and extends into flagellar pocket, but not beyond its opening. Note the collagen fibers around the histiocyte (28,450 ×).

animals. Recently, some workers have begun to use human peripheral blood monocytes. These macrophages can be readily obtained in reasonable numbers within a short time. In vitro, these cells retain their phagocytic activity, spread out on the substratum and live for a long period of time without cell division. Thus, they are adequate host cells for the aforementioned experiments, but not the ideal substrate for mass propagation of amastigotes in vitro. For studying leishmaniae, different authors tend to develop their own in vitro systems to fit their own needs for different types of work. Tables 2 to 4 summarize the relevant information of these leishmania-macrophage systems, which are categorized according to the macrophage sources. Experiments involving the use of macrophages from hamsters,[23,27-38] mice,[39-63] and other sources[21,23,39,64-67] are included in Tables 2, 3, and 4, respectively. It should be noted that very little of the work included here was undertaken with the aim of improving leishmanial growth. Because of its potential application in many areas of investigation, some aspects of the general procedures and experimental design are discussed below.

a. The Parasites

Cultured macrophages have been infected with either amastigotes or promastigotes, depending on the type of work and the species concerned. Promastigotes can be cultured in

Table 1
CULTIVATION OF LEISHMANIA AMASTIGOTES IN TISSUE EXPLANTS

Animal	Tissue	Parasite	Culture medium	Temp. & period	Observations	Ref.
Hamster	Cutaneous tis., kidney, liver, spleen	*L. donovani* Pm[a]	HEE[c] HPC[d]	37°C 10 d	Dividing forms	18
	Cutaneous tis., omentum	*L. donovani* Am[b]	CEE[e] in CPC[f]	37°C 35 d	Survival	19
	Spleen	*L. donovani* Am and Pm	20% HaS[g] or RS[h] 5% CEE in CPC	37°C 33 d	Survival; extracellular promastigotes	20
	Spleen & liver	*L. braziliensis* Pm	?	37°C 3 d	Transformation and survival	21
	Spleen	*L. infantum* Am and Pm	10% RS in HPC	37°C 21 d	Multiplication & subcultures	22
	Spleen	*L. donovani*	5% HoS[i], 5% RS 0.1% Yeast ext. in HBBS[j]	37°C 10 d	Dividing amastigotes	23
Chicken embryo	Minced tis., heart, brain	*L. donovani* Pm	Tyrode solution	27-37°C	Promastigotes only at 27°C	24
	Muscle, spleen	*L. donovani* Am	CEE in CPC	37°C 35 d	Survival	19
	Muscle	*L. tropica* Pm major	15% FBS in M-199 40% HuS[k] in HBSS	37°C 4 d	Multiplication of amastigotes; Destruction of host cells	25
Guinea pig	Lung & spleen	*L. tropica* Pm	25% HoS in Tyrode solution Agar slant	33-37°C 10 d	Extracellular division of amastigotes & promastigotes	26
Frog (*Rana pipiens*)	Spleen					

[a] Pm, Promastigotes.
[b] Am, Amastigotes.
[c] HEE, Hamster embryo extract.
[d] HPC, Hamster plasma clot.
[e] CEE, Chicken embryo extract.
[f] CPC, Chicken plasma clot.
[g] HaS, Hamster serum.
[h] RS, Rabbit serum.
[i] HoS, Horse serum.
[j] HBSS, Hank's Balanced Salt Solution.
[k] HuS, Human Serum.

any one of the media described in another portion of this chapter. To be discussed later are several disadvantages associated with the use of these parasitic forms, although they can be obtained readily in substantial numbers. More adequate for growth studies in macrophages are amastigotes, which can be obtained from spleens and skin lesions of laboratory animals with visceral and cutaneous leishmaniases, respectively. Isolation of amastigotes from these materials can be achieved more readily from heavily infected samples taken 1 to 3 months after inoculation. Spleens and skin lesions should be removed aseptically and then separated from tissue debris.

For visceral leishmania, e.g., *L. donovani*, infected spleens are first ground in saline in a Ten Broek type homogenizer and then subjected to differential centrifugation according to a method modified from Simpson.[68] Intact cells and tissue debris are removed by two

Table 2
LEISHMANIAL INFECTION OF HAMSTER PERITONEAL MACROPHAGES IN VITRO

Parasite	Stimulation[a]	Culture medium	Temp. & period	Max.[b] growth	Type of study	Ref.
L. donovani Pm	—[c]	5% HoS, 5% RS, 0.1% yeast extract in HBSS	37°C 2 d	?	Biol.	23
L. donovani Pm (Khartoum)	HBSS	20% HoS in MEM	37°C 2 d	3 × (?)	Biol.	27
L. donovani Pm (Sudan)	Mineral oil	20% FBS in Paker's M-199	Variable 1 wk	?	Biol.	28–30
L. donovani Pm (Sudan 2S)	Starch	McCoy's 5A	35°C 2 d	?	Isotope incorporation	31
L. donovani Am (Sudan 1-3S)	Saline	40% RS or HaS in HBBS	37°C 2 wks	3.4 ×	Biol.	32–33
L. donovani Pm (Sudan 1S)	Culture medium	10% HoS, 10% FBS, 0.5% lactalbumin hydrolysate in HEPES-199	37°C 2 wks[d]	No	Biol.	34–35
L. donovani Am (Sudan 1S)	As above	As above	37°C 1 wk	4 ×	Biol.	36–37
L. donovani Am	As above	As above	37°C 5 wks	9 ×	Cultivation	38

[a] Stimulation, injection of stimulants into peritoneal cavity.
[b] Maximal growth of intracellular amastigotes in folds of increases calculated by the authors.
[c] No stimulation.
[d] Not clearly indicated in the original article.

cycles of centrifugation at 100 g for 5 to 10 min at 4°C; the supernatants obtained are spun at 3500 g for 5 min. The resulting pellets consist largely of amastigotes, provided heavily infected spleens are used. The remaining red blood cells in the pellets can be lysed with the addition of two volumes of 0.87% ammonium chloride in water to one volume of the parasite suspension in divalent cation-free buffer. The amastigotes obtained by this method are essentially free from the contamination by intact host cells, but not by cell organelles. Further purification by ultra-centrifugation in bovine serum albumin gradients is possible,[68] but is not necessary for culture work.

For cutaneous leishmania, nonulcerating closed lesions should be obtained by using proper host-parasite combinations (see below) and by caging infected animals individually. Some strains of *L. tropica* and *L. mexicana* produce a large histiocytoma which, when teased or homogenized in saline, releases large numbers of amastigotes. As described above, relatively clean amastigotes of these species can also be obtained by differential centrifugation and used for infection experiments with macrophages. Other methods for partial purification of cutaneous leishmania have been described.[69-72] Parasites from cultures or from infected animals are washed and then suspended at an appropriate cell density in the medium used for the infection experiments.

Whether amastigotes or promastigotes should be used in in vitro studies with macrophages depends on the type of work and the species of leishmania in question. Since promastigotes can be obtained readily from cultures without the need of purification, it appears that these parasites are the better choice for experimentation. It is known, however, that promastigotes of most species tend to lose infectivity to macrophages after cultivation in vitro. In this regard, there are considerable variations among different species of leishmaniae: *L. donovani* once transformed into promastigotes often shows poor survival in macrophages,[32,35] whereas

Table 3
LEISHMANIAL INFECTION OF MOUSE PERITONEAL MACROPHAGES IN VITRO

Parasite	Mouse strain	Stimulation	Culture medium	Temp. & period	Max. growth	Type of study	Ref.
L. donovani (Sudan) Am	NCS	—	20% FBS in M-199[a] or MEM[b]	37°C 7 d	Survival	Biol.	39
L. donovani (Sudan) Am	Swiss Webster	—	20% HoS or NBCS[c] or 10% each in MEM	37°C 3 d	2 ×	Biol.	40
L. donovani (Sudan) Am	C57BL/6J	—	40% HoS, 2% BEE[d] in NCTC 135	37°C 2 d	2 ×	Immunol.	41
L. donovani (Sudan, Khartoum) Am	Balb/c	Peritoneal infection	20% FBS in MEM	37°C 4 d	?	Immunol.	42
L. donovani *L. tropica* Am & Pm	?	—	2.5% bovine albumin in M-199	37°C 8 d	Yes	Biol. Chemotherapy	43–46
L. tropica (LRC-L137) Pm	Balb/c CBA/H DBA/2 C3H/He	Thioglycollate broth	10% FBS in McCoy's	37°C 26 d	Yes; Destruction of macrophages	Biol. Immunol.	47–49
L. tropica (LRC-L137) Pm *L. Enriettii* (LRC-L144) Pm	C3H/He	Thiglycollate broth	10% FBS in McCoy's	32°C 4 d	8 ×	Biol.	50
L. tropica major Pm *L. enriettii* Am	C57BL	Starch	10% FBS in D-MEM[e]	37°C 2-3 d	Survival 2 ×	Immunol.	51–54
L. tropica Pm	?	—	20% FBS in M-199	37°C 1 d	5—9 ×	Immunol.	55
L. mexicana mexicana Am & Pm	NIH	—	10% CS[f] in M-199	37°C 3 d	Destruction of macrophages	Biol.	56–57
L. mexicana mexicana (LV4) *L. mexicana amazonensis* (LV79) Am & Pm	C57 TFW	Mineral oil or none	20% FBS in HEPES M-199	31—37°C 2—4 d	Yes	Biol.	58–59
L. Tropica Pm *L. donovani* *L. enriettii*	Out bred	2 × HBSS	10% FBS in M-199	37°C 1 d	Survival	Biol.	60–61
L. brasiliensis Pm	?	Starch	10% FBS in M-199	33°C 3 d	Yes (?)[g]	Biol.	62
L. enriettii Am	DBA/2 C3H	—	?	37°C 3½ d	16 × (??)[g]	Biol.	63

[a] M-199, Medium 199.
[b] MEM, Minimal Essential Medium.
[c] NBCS, New born calf serum.
[d] BEE, Bovine embryo extract.
[e] D-MEM, Dulbecco MEM.
[f] CS, Calf serum.
[g] (?) (??), Claims questioned by the present authors.

certain strains of *L. tropica*[47-50] and *L. mexicana*[56-59] appear to retain good infectivity in these cells even after prolonged cultivation as promastigotes. What may account for the variations among different leishmania promastigotes in their ability to survive in macrophages has yet to be studied. In contrast to promastigotes, amastigotes of all species always produce excellent infections in macrophages, provided that the right donor animals are used (see below). It is, therefore, recommended that promastigotes should be used where they remain

Table 4
LEISHMANIAL INFECTION OF MACROPHAGES FROM OTHER SOURCES IN VITRO

Source of macrophage	Parasite	Culture medium	Temp. & period	Max. growth	Type of study	Ref.
Human serous exudate	*L. donovani* Pm	5% HoS, 5% RS, 0.1% yeast extract in HBSS	37°C 2 d	Killed	Biol.	23
Human peripheral blood	*L. donovani* Am	20% FBS in RPMI-1640 HEPES	37°C 14 d	Survival	Biol.	39
Human peripheral blood	*L. donovani*Am	10% FBS in RPMI-1640	37°C 1 wk	3 ×	Biol.	64
Guinea pig peritoneal cavity (starch induced)	*L. enriettii* Am	30% GS[a] in M-199	36°C 2 d	2 ×	Immunol.	65
Guinea pig peritoneal cavity (starch induced)	*L. enriettii* Am	10% GS in D-MEM	37°C 4 d	Survival	Immunol.	66
Guinea pig (male, Hartley outbred) Peritoneal cavity (caseinate or paraffin oil induced)	*L. enriettii* Am	10% FBS in E-MEM[b]	37°C 2 d	Survival	Immunol.	67
Cotton rat Peritoneal cavity (mineral oil induced)	*L. brasiliensis* Pm	?	37°C 3 d	Transformation	Biol.	21

[a] GS, Guinea pig serum.
[b] E-MEM, Earle's MEM.

infectious or where their use is needed for a particular type of work. Otherwise, amastigotes should be employed in spite of the laborious procedure needed to isolate them from animals.

b. The Macrophages

A detailed procedure for collection and maintenance of peritoneal macrophages from small rodents is available.[73] Here, we shall only describe the methodology briefly with emphasis on the steps which are related to the leishmania-macrophage systems. Peritoneal exudate cells are collected by lavage of the cavity with HBSS (Harks' Balanced Salt Solution) or culture medium supplemented with 10 units per mℓ of heparin. While several million cells can usually be obtained from a mouse peritoneal cavity, other rodents provide much fewer cells. Yields can be tremendously increased by injecting stimulants several days before harvest. Solutions used for stimulation include saline, culture medium, starch or glycogen suspensions, thioglycollate broth, or endotoxin. These elicited macrophages differ from resident macrophages and from each other in cellular activity according to the means of stimulation. Peritoneal exudate cells from unstimulated animals largely consist of macrophages with small numbers of lymphocytes. In addition to these cells, polymorphonuclear phagocytes and other types of cells are also present after stimulation. Macrophage populations can be selected out by their adherent property after cultivation. Nonadherent cells, e.g., dead polymorphs and lymphocytes, are washed away, leaving behind a homogeneous population of macrophages. Regular mammalian tissue culture media containing 10 to 20% heat-inactivated fetal bovine serum (HIFBS) will support mouse peritoneal macrophages during long-term cultivation. Cells from other rodents may have additional requirements. For example, hamster peritoneal macrophages survive much longer on polylysine-coated glass surfaces in HEPES-buffered Medium 199 supplemented with 10% horse serum, 10% HIFBS, and 0.5% lactalbumin hydrolysate.[34] Macrophage monolayers for leishmanial infection can be prepared on glass coverslips in Leighton tubes or petri dishes, or directly cultured in tissue culture flasks or plates. Incubation is normally done at 37°C with 5% CO_2 in air, favorable conditions for mammalian cells in general. Temperatures slightly lower than 37°C should be used for certain cutaneous leishmaniae.[74]

For infection experiments with leishmaniae, macrophages have been collected from hamsters,[23-38] mice,[39-63] guinea pigs,[65-67] cotton rats,[21] and human peripheral blood.[39,64] The selection of macrophage donor animals for these studies is largely based on their susceptibility to particular species of leishmaniae and the type of work. On the basis of susceptibility, the most frequently used combinations are *L. donovani*-hamster (Table 2), *L. mexicana*-mouse,[56-59] *L. tropica*-mouse,[47-55] and *L. enriettii*-guinea pig.[65-67] Unnatural or less well matched combinations, e.g., *L. enriettii*-mouse,[50-54] or *L. donovani*-mouse[39-46] are usually chosen for particular types of study. Although the *L. donovani*-human monocyte combination is adequate, it has been explored only recently.[39,64] One report using the *L. brasiliensis*-cotton rat combination involves only short-term observations of host-parasite interactions.[21] The importance of the right host-parasite combination in leishmania-macrophage in vitro systems is suggested by several observations. For example, amastigotes of *L. donovani* grow in hamster macrophages,[33,36-38] but not in those of mice.[39] Promastigotes of *L. tropica* (LRC-L137) disappear from cultures of macrophages from C57BL/6 or N2B/BL mice, yet they multiply in those from other strains.[47] The differences among various macrophages from different animals responsible for their ability to support the growth of intracellular amastigotes are unknown.

Some immunological properties of macrophages should also be taken into consideration in their use for experiments with leishmaniae. Ideally, peritoneal macrophages from inbred animals should be used without stimulation of the peritoneal cavity. Pooling of peritoneal exudate cells from outbred animals may provoke a mixed lymphocyte reaction and stimulation of the peritoneal cavity is known to heighten the "state" of macrophages with regard to H_2O_2 production and lysosomal enzyme levels. Both of these may be deleterious to the parasites. Although this is not true in the *L. donovani*-hamster macrophage system,[36-38] such effects may be manifested in other host-parasite combinations. Whether it is appropriate to use peritoneal macrophages for studying leishmania in vitro may be questioned. Their use is based on convenience rather than adequacy, as leishmaniae normally parasitize histiocytes, Küpffer cells of the liver, spleen and bone marrow macrophages.

c. Quantitation of Infection and Growth

The numbers of macrophages and parasites before use are counted in a hemacytometer. A bacterial counting chamber is more adequate for smaller amastigotes, e.g., those of *L. donovani*. There is an alternative method for counting amastigotes: known numbers of red blood cells are mixed with parasites and the number of the latter can then be calculated on the basis of their ratio to the red cells determined in Giemsa stained smears. For infection, known numbers of macrophages and parasites are brought together, in most cases, by the addition of a parasite suspension to adherent macrophages already in culture for some time, or they are mixed before plating. More accurate quantitation can be achieved with the aid of micropipettes fitted with sterile tips. Parasite to macrophage ratios should not exceed 20 to 1 to avoid overparasitization which may hamper accurate counting of parasites by microscopy. In our laboratory, we regularly mix parasites with macrophages as follows: 10^6 to 10^5 leishmania in 0.1 $m\ell$ medium per 15 mm^2 glass coverslip placed in a petri dish. The mixture is allowed to stay on the coverslip without overflow and incubated about 4 hr at 34 to 37°C with 5% CO_2 in air to allow phagocytosis of leishmanial parasites, which takes place without the need of opsonization. At the end of the incubation period, prewarmed medium is carefully added to the culture at 1 $m\ell$ per coverslip. For *L. donovani*, we regularly obtain a level of 70 to 80% infected cells, each with about 5 to 10 amastigotes after a 24-

hr period. The necessity of washing infected cultures (to remove extracellular uningested parasites) and the initial incubation period for phagocytosis depend on the species of parasite, kind of macrophage, and the type of work. The medium is changed every 2 to 3 days when necessary. The glass coverslips used for the above experiments are cleaned with acid followed by thorough rinsing, or with a nontoxic detergent, e.g., 7 ×. Penicillin G and streptomycin at 50 to 200 units or μg per mℓ have been used to prevent bacterial contamination without evident adverse effects to the intracellular leishmania. Amphotericin B, mycostatin, or other antifungal drugs should be avoided, as they are leishmanicidal at dosages as low as 1 to 5 μg per mℓ.[44]

At different time points after infection, samples (preferably in triplicate) are withdrawn and processed for quantitation of the parasite growth. In nearly all studies, this depends on the counting of intracellular parasites in macrophages. Confluent monolayers, or samples with high cell density on coverslips or culture vessels, are fixed when wet in Bouin's, Carnoy's, or Zenker's fixatives, stained and mounted on slides after dehydration. Preparations with low cell density can be dried quickly (with the aid of a hot air blower or by vacuum aspiration), stained and air-dried for mounting. The standard stain for leishmania, as for other related protozoa, is a diluted solution of Giemsa or its variations at neutral pH. Stain concentration and staining period vary with the number of cells and the leishmania species. The percent of infected cells and the average number of amastigotes per cell are determined by examining 10 to 20 microscope fields or 100 to 200 macrophages. Alternatively, quantitation can be done by counting the total parasite numbers in 100 cells or categorization of amastigote numbers per cell into histogram bars. The level of parasite growth is indicated by the increase of parasite counts in the macrophages with time of incubation.

There are several inherent difficulties associated with this method of quantitation. First, preparation of samples and microscopic examination are time consuming, laborious, and cumbersome. Second, the method lacks sensitivity, as microscopic examination cannot distinguish moribund parasites from living ones. Third, direct counting of a representative area by microscopy has an inherent error of about 20%. Two methods other than microscopic counting for leishmania in macrophages have been described. In the first method, infected macrophages were treated briefly with a low concentration of sodium dodecyl sulfate to release amastigotes, which were then allowed to transform into promastigotes in culture medium before counting.[75] The second method used (^{3}H)-thymidine which is incorporated by leishmaniae, but not by macrophages, as they do not synthesize DNA.[48] Both methods have not been widely used and need to be critically evaluated for general application to all leishmania-macrophage systems.

In spite of the shortcomings of the current quantitative method for intracellular leishmaniae, it is generally adequate for most types of investigations. Here, we shall limit our discussion to the leishmania-macrophage systems which support the intracellular multiplication of leishmanial amastigotes. It should be emphasized again that careful quantitation of leishmanial infection in macrophages for a prolonged period of time has been conducted only in very few cases. Herman[33] compared the intracellular growth of *L. donovani* amastigotes in hamster peritoneal macrophages maintained in Hank's balanced salt solution (HBSS) supplemented with sera from different animals. Rabbit and cotton rat sera supported the growth of this parasite, whereas chicken serum did not. In the same host-parasite system, more consistent growth of amastigotes was made possible by improving the culture conditions for the hamster macrophages.[34] The intracellular parasites gained several-fold-increase in numbers during the first week, after which time they gradually disappeared from the culture.[36-37] However, when the amastigotes were released at the end of the first week and used to infect freshly isolated hamster macrophages, a new cycle of growth results and this could be repeated at least five times.[38] This finding suggests that, for unknown reasons, hamster macrophages, after prolonged in vitro culture, become inadequate host cells for the growth of amastigotes. In mouse peritoneal macrophages, *L. donovani* amastigotes have been reported to increase in number several fold in 2 to 3 days[40-41] or to grow so rapidly that host cells are lysed and free amastigotes are released into the culture medium.[43-46] However, other workers could not confirm these observations and found only very limited, if any, multiplication of the parasite.[39-42] Recently, Berman et al.[64] also found evidence of multiplication of *L. donovani* amastigotes in human peripheral blood monocytes which had been matured in culture for 6 days before infection. Earlier, one of the present authors (K-.P. Chang) observed no growth, but survival of this parasite in cells from the same source.[39] In this work, infection was initiated by plating a monocyte-amastigote mixture in RPMI 1640 with 20% HIFBS on glass coverslips and was allowed to continue for 1 week.

From the above studies, it is evident that multiplication of *L. donovani* in cultured macrophages does not match their more rapid growth in hamsters. *Leishmania mexicana* and *L. tropica* (certain strains) appear to be much less fastidious and proliferate with a high growth rate in mouse peritoneal macrophages.[50-54,57,59] In these in vitro systems, there is apparently an increase with incubation time in the percent of infectivity, suggesting rupture of the infected cells and infection of previously uninfected ones by the amastigotes released. However, the lack of direct evidence for this and other uncertainties in the experimental conditions described do not exclude the possibility that the increased rate of infection is due to the preexisting extracellular parasites.

The optimal conditions for intracellular growth of amastigotes in macrophages have not been achieved for any leishmania species. There is a general tendency to accept culture conditions for macrophages on the assumption that whatever is optimal for the host cells should be also good for the parasites. Clearly, this is not the case and additional requirements are necessary for the parasites to reach a growth rate in vitro comparable to that in vivo. In considering the nutrition of leishmanial amastigotes, one should note the fact that they are parasites of macrophage lysosomes,[36-37,57,58] cell organelles accessible to extracellular macromolecules sequestered by macrophages via pinocytosis and phagocytosis. In addition to the metabolites of macrophages, these molecules or their breakdown products by the action of host lysosomal enzymes may be utilized by the parasites as nutrients. For better growth of leishmaniae in macrophages, further considerations may be given to the use of different media, gas phases and the sources of host cells from organs or tissues normally parasitized by these parasites in vivo.

3. *Primary Cell Lines*

To prepare suspensions of individual cells like peritoneal macrophages for cultivation, organs or tissues must be disrupted by various means, e.g., trypsinization. These cells can then adhere to culture vessels and form monolayers. In addition, monolayers can also be prepared from the outgrowth of tissue explants. Cells of these primary cultures or lines normally remain diploid and either do not divide, like macrophages, or have finite cycles of cell division. These primary cell cultures have been used as substrates for growing amastigotes of *L. donovani*,[76-80] *L. brasiliensis*[77,81-82] and *L. tropica*[83] (Table 5). Where applicable, methodology for cultivation and quantitation is generally the same as for macrophages. In most cases, promastigotes were used for inoculation. Invasion of the cultured cells by these promastigotes and their subsequent transformation into amastigotes were observed. Multiplication of amastigotes was often stated on the basis of morphological grounds, e.g., the presence of dividing forms in the cells. Gleiberman and Belova[83] found selective destruction of parasitized cells, which curiously were not heavily infected. Lelijveld and Atanasiu[87] reported the release of amastigotes from burst cells as a result of infection. However, quantitative results are presented only in two separate studies with promastigotes of *L. donovani* and *L. brasiliensis* in human skin fibroblasts[82] and amniotic cells.[77] In human

Table 5
LEISHMANIAL INFECTION OF PRIMARY CELL LINES IN VITRO[a]

Parasite	Cell type & source	Culture medium	Temp. & period	Observations growth	Ref.
L. donovani Pm	Human amniotic	0.5% lactalbumin hydrolysate, 0.1% yeast extract in HBBS	37°C 4 d	Invasion & transformation	76
L. donovani Pm (strains P & S)	Human amniotic	20% FBS in M-199	33°C 2—3 mon	Heavy, but with promastigotes in medium	77
L. donovani Pm	Hamster or monkey kidney	10% CS, 2.5% 10 X amino acid for Puck medium in IC_{65}	37°C 3 d	Invasion & transformation	78
L. donovani Am (Khartoum)	Hamster liver infected in vivo	10% CS in M-199	37°C 2 wks	Survival	79
L. donovani Am (Sudan 1-3S)	Chicken embryonic fibroblasts	M-199	37°C 2 d	Limited	80
L. brasiliensis Pm (Venezuelan CV)	Human amniotic	20% FBS in M-199	28°C 4 mon	Good, elongated amastigotes	77
L. brasiliensis Pm pifanoi (Bahia)	Baby hamster kidney	10% RS in E-MEM	33°C 2 d	Heavy (?)	81
L. brasiliensis Pm	Human skin fibroblasts	20% FBS in F-12	33—37°C 3 wks	Invasion & transformation	82
L. donovani Pm *L. tropica major*	Chicken skin & muscle, Newborn dog, cat, rabbit, guinea pig, kidney	FBS, Lactalbumin hydrolysate in M-199	37°C 3 d	Cytopathic effects due to amastigote growth	83

[a] See footnotes of Tables 1-4 for abbreviations.

fibroblasts, *L. donovani* was found not to invade the cells, whereas *L. brasiliensis* achieved infection, but failed to survive subculture. In human amniotic cells, both species not only gained intracellular entry, but also increased in numbers upon subculture. However, intracellular infection persists only at low temperatures with a simultaneous extracellular multiplication of promastigotes in the cultures with *L. donovani*. This raises some doubts as to whether there is a true intracellular growth of *L. donovani* amastigotes in these cells. In this same study, horse serum was found to have a deleterious effect on the intracellular parasites.

The general impression from these studies is that promastigotes of *L. brasiliensis* infect cells readily but grow only at lower temperatures, e.g., 33°C, whereas those of *L. donovani* may have difficulty in entering cells and, when they do gain entry, show limited multiplication.

4. *Permanent Cell Lines*

Cells that can be cultured indefinitely in vitro become established as permanent lines. Unlike cells of the primary lines, these are often polyploid and may lose some of their original characteristics. Table 6 provides a list of some permanent lines that have been used for infection experiments with promastigotes and/or amastigotes of *L. donovani*,[83-94] *L. mexicana*,[93-95] *L. brasiliensis*,[94,96] *L. tropica*,[25,83,93] and other species.[83] Not included in the table are numerous cell lines of mice (L, CCLI, NTCT, KB), human (Chang's liver, CCL2, Girardis heart, Detroit series of marrow cells), and others, whose usefulness are either ambiguous or marginal for leishmanial infection.[46,78,83,97] Some of the cell lines in Table 6 apparently supported the intracellular growth of amastigotes during the first several days to

Table 6
LEISHMANIAL INFECTION OF PERMANENT CELL LINES IN VITRO[a]

Parasite	Cell line	Culture medium	Temp. &period	Maximum growth	Ref.
L. donovani Am (Sudan)	Monkey kidney LLC-MK2	2% CS in E-MEM	37°C 5 d	2 ×	84
L. donovani Pm (Tunisia)	Mosquito *Aedes albopictus*	Mitsuhashi & Maramorosch Medium	24°C 3 d	10 × (?)	85
L. donovani Pm	HeLa J-111 human leukocytic	?	37°C ?	Growth in J-111	86
L. donovani Am (Sudan 1S)	Mouse peritoneal macrophage P388D1	10% FBS in E-MEM	37°C 2 wks	5 ×	87
L. donovani Am (Sudan 1S)	Mouse peritoneal macrophage J774G8	20% FBS in E-MEM, M-199, RPMI 1640, or McCoy's	37°C 2 wks	4 ×	88–89
L. donovani Am Pm	Sticker Dog Sarcoma 503	20% ChS[b] in M-199	37°C 5 mon	Continuous passages	46, 90–92
L. donovani Am HV3 (LV9) *L. mexicana mexicana*(M379) (LV4) Pm *L. tropica* Pm (P) (LV39)	Sticker Dog Sarcoma 503 & Hamster peritoneal HP	10% FBS in M-199 or Puck's	32—35°C 1—2 wks	Yes	93
L. donovani Pm (Strains S & G) *L. mexicana* Pm *L. brasiliensis pifanoi* Am Pm	Sticker Dog Sarcoma 503	10% FBS in M-199	33—36°C 1 yr	Continuous passages	94
L. mexicana Am *amazonesis* Pm	Mouse peritoneal macrophage J774G8	20% FBS in RPMI 1640	35°C 6 mon	Continuous passage	95
L. brasiliensis panamensis	Vero (African green monkey kidney)	E-MEM	32°C 6 d	?	96
L. tropica Pm *major*	HeLa, Monkey heart muscle	15% FBS in M-199 or 40% HuS in HBSS	37°C 4 d	Cytopathic effect	25
L. donovani Pm *L. canis*, *L. tropica*, *L. enriettii*	Hep-2, Monkey kidney, Monkey liver, Monkey heart muscle	FBS, Lactalbumin hydrolysate in M-199	37°C 3 d	Cytopathic effect due to infection	83

[a] See footnotes of Tables 1—4 for abbreviations.
[b] ChS, chicken serum.

2 weeks, after which time subcultures were either not attempted[25,83,85-87,96] or carried out, but parasites disappeared.[84,88,89]

There are reports of using cells from a cold-blooded vertebrate (turtle heart cells) and from insects (mosquito) for *L. mexicana* and *L. donovani*, respectively. The former was mentioned only in passing in a paper otherwise devoted to a different subject.[58] Of interest is the latter in that promastigotes of Tunisian *L. donovani* transform into amastigotes and grow in the insect cells at 24°C.[85] The unusual culture conditions, however, cast doubts on this extraordinary finding.

Most encouraging results have been obtained by using cell lines of mononuclear phagocytes. This is consistent with the fact that they are the natural host cells for leishmania in vivo. Lamy et al.[90] grew *L. donovani* amastigotes in Sticker Dog Sarcoma cells of histiocyte origin. Amastigotes persist in these cells after repeated subcultures for at least 5 months, indicating that the host cells and the parasites have entered into a stable relationship. This constitutes the first instance of continuous cultivation of leishmania amastigotes in tissue culture cells. Confirming Lamy's observation, Frothingham and Lehtimaki[94] extended the culture period to more than a year and the use of these sarcoma cells to other species of leishmaniae. In this work, the serum requirement and temperature in relation to the intracellular growth of amastigotes were carefully examined. The optimal growth temperatures were determined to be 36 and 33°C for visceral and South American cutaneous leishmaniae, respectively. Heat-inactivated fetal bovine serum was found superior to hamster or rabbit sera and 10 to 20 rather than 5% HIFBS favored the growth of *L. mexicana*. The cutaneous species grow much better than the visceral species under the respective optimal conditions. At 36°C, both *L. brasiliensis* and *L. mexicana* disappeared from the cultures. The evidence presented for the multiplication of amastigotes in these cells indicates that the last two species of leishmaniae in this culture should be a good system for producing amastigotes in vitro. Other macrophage cell lines have been subsequently employed for leishmanial infection. Amastigotes of *L. donovani* were reported to grow in J111, a human leukocytic monocyte series, but no details were given.[86] This same parasite apparently grows in and eventually destroys monolayers of P388D1, a chemically transformed mouse peritoneal macrophage line.[87] In another mouse macrophage line, J774G8, a similar result was obtained, but no destruction of cell cultures resulted from the infection.[88]

In our laboratory, one of us (K.-P. Chang) has investigated further the potential of this macrophage cell line for possible large-scale production of amastigotes.[89,95] The J774G8 cells are routinely maintained in MEM with 10% HIFBS as a suspension in spinner bottles or in RPMI 1640 with 10% HIFBS as a monolayer in tissue culture flasks. In the latter, cells are loosely adhered to the culture vessels and can be readily washed off by vigorous aspiration. We have infected these cells with some 12 different strains or isolates of leishmania spp., of which only *L. mexicana amazonensis* gives very promising results. Here, we shall summarize some of our results with *L. donovani* and *L. mexicana*. Amastigotes of *L. donovani* from infected hamsters or promastigotes of *L. mexicana* from cultures were mixed with J744G8 cells at a ratio of 10 or 20 to 1 and then cultured in tissue culture flasks or in petri dishes on flying coverslips. For *L. donovani*, we have examined the infections of the floating cells in the flasks and of the adherent cells on the glass coverslips. Media used included MEM, McCoy's, Medium 199 and RPMI 1640, each with a supplement of 20% HIFBS.[89] Under all culture conditions used, amastigotes gained a several-fold-increase in numbers during the first week and then were gradually reduced within one month. Repeated experiments suggest to us that the amastigotes of *L. donovani* are unable to keep pace with the growth of host cells and thus become diluted with time of incubation or subcultures.

Cultivation of *L. mexicana amazonensis* in J774G8 has been much more successful, and this parasite has now been kept as amastigotes in continuous culture for about 6 months. Infected cells were incubated in tightly capped tissue culture flasks at 35°C in RPMI 1640 plus 20% HIFBS. Every 3 days, cells were removed from the culture flask by vigorous aspiration for medium renewal and for quantitation of parasites. For medium renewal, cultured cells were spun into a pellet, to which fresh medium was added; the resuspended cells were then returned to the original flask. For quantitation of parasites, the total number of amastigotes per culture was calculated as follows: the total number of cells × % infected cells × average numbers of amastigotes per cell. The total number of cells was determined from the total volume of the culture and the cell number per mℓ as counted in a hemacytometer. The percentage of infected cells and the average number of amastigotes per cell was determined by examining 100 cells in thin wet mount preparations of fresh material by phase-contrast microscopy under an oil immersion lense. Since the amastigotes of this leishmanial species are large in size and produce very striking vacuoles in the macrophages, accurate counting of these intracellular parasites is possible by this simple method. In moderately infected cell cultures, the best generation time of the amastigotes was calculated to be 72 hr. The best culture gives 10^8 amastigotes from 10^7 promastigotes per 10^6 cells in approximately 2 weeks. To ensure a uniform infection, amastigotes are normally released from heavily infected cultures by needle passages or vigorous vortexing in phosphate buffered saline with 2 m*M* EDTA and then used to initiate a new culture. Large-scale cultures have now been initiated in large tissue culture flasks and in roller bottles, from which we hope to collect heavily infected cells for isolation of amastigotes.

B. Cultivation in Cell-Free Conditions

The appearance of "amastigotes" in the absence of mammalian cells has long been noted as casual observations during unrelated experiments. It is common knowledge that a small number of amastigote-like forms are often present in promastigote cultures. In 1922, Row[98] observed a spontaneous reversal of promastigotes to "amastigotes" in his 3N blood agar culture with a concomitant recovery of infectivity to laboratory animals. Adler[99] in 1926 also reported that animal-derived *L. donovani* and *L. tropica* grew in blood agar medium as clumps of "amastigotes" in the presence of antisera. In our laboratory, we have observed on several occasions that species/strains of South American origin transformed into small aflagellated entities without any specific treatment. While these cursory observations raise the possibility that promastigotes might, under certain conditions, transform into "amastigotes" in the absence of mammalian cells, deliberate efforts to produce such transformation and to grow the latter form have been few.

The first effort was made by Trager[100] in 1953 who incubated promastigotes or amastigotes of *L. donovani* (Khartoum strain) in red cell extract made in saline and supplemented with animal serum at 37°C with 5% CO_2 in air. He noted a 4.5-fold increase of leishmanial forms intermediate in morphology between amastigotes and promastigotes in the first 4 days. Thereafter, the culture degenerated. The best growth was obtained by using human red cell extract in combination with human or hamster serum. Sera from rabbit, duck, guinea pig, or infected hamster were unsatisfactory. The author concluded that the failure to cultivate the true intracellular amastigotes suggests stringent environmental requirements of these parasites for growth. Subsequently, Lemma and Schiller[101] in 1964 claimed to have succeeded in extracellular cultivation of amastigotes by thermal acclimatization of promastigotes to 34°C in blood agar biphasic medium with HBSS as an overlay. Leishmania species successfully adapted by these authors to grow as "amastigotes" include *L. donovani* (Khartoum strain), *L. tropica*, and *L. brasiliensis*. While the morphological evidence provided appears convincing, these cultured "amastigotes" behave rather like promastigotes than true amastigotes in infectivity to laboratory animals. Moreover, a complete acclimatization takes about 6 months, which is much too long to be considered as natural. Further, their finding could not be confirmed by others with Sudan and Khartoum strains of *L. donovani* under identical conditions, which produced thermally adapted promastigotes instead of amastigotes.[97,102] It appears certain then, that promastigotes of visceral leishmania cannot be enticed to transform into and/or grow as amastigotes by thermal acclimatization in cell-free cultures.

There is evidence, however, that leishmania species of South American origin may respond more favorably to a simple temperature increase without a prolonged period of acclimatization in complex medium. Greenblatt and Glaser[103] found that *L. enriettii* became amastigote-like and completed one cycle of cell division before degeneration when the incubation temperature was increased from 26 to 35 or 37°C. Molyneux et al.[104] observed a similar phenomenon with *L. mexicana mexicana* by changing the temperature from 20 to 32°C, except that there was no apparent growth. From the descriptions provided, it appears that a great majority of promastigotes remain intact after transformation into what looks like amastigotes at least in the first 24 hr. A more detailed study has been made recently by Hendricks et al.[105] Amastigote-like forms were obtained from promastigotes of *L. brasiliensis panamanensis* in

Schneider's medium or Medium 199 with 30% HIFBS by a change of ambient temperature from 28.5 ± 1.5 to 34°C. In 2 to 4 days, depending on the media used, parasites gained a 2 to 4-fold increase in numbers with 80 to 90% of them converted to "amastigote" forms. In subsequent work[106] promastigotes of the same species were cultured initially in Schneider's medium for insect cells at 24°C and then shifted to Medium 199 for mammalian cells at 34°C. In 3 days at 34°C, more than 95% of the promastigotes appears to resemble amastigotes. In both cases, only the "amastigote" forms derived from cultures with Medium 199 at 34°C were said to retain antigenicity for fluorescent antibody tests. It is contended that the above in vitro manipulation may mimic the change from insect to mammalian environments encountered in nature for such transformation. The authors further leave the impression that most, if not all, species of leishmania respond similarly to this simple treatment. In similar experiments on some nine different strains or isolates of *L. donovani* and *L. tropica*, a temperature change from 27 to 35°C produced only degenerative parasites (personal observations). An exception to this generality is *L. mexicana amazonensis*, which becomes amastigote-like in 2 to 3 days (as observed by others) when cultured at 35°C in Steiger's RE III medium with 20% HIFBS at pH 5.5 to 6.0. The best culture gave 1 cycle of cell division during the first 3 days, after which time the parasites degenerated progressively in the next 10 days despite frequent renewal of culture media.

From these preliminary results, the authors tend to agree with the observations that some species of leishmania from South America can respond to a temperature increase (32 to 35°C) and remain as viable cells which closely resemble amastigotes. However, their identity with the true intracellular amastigotes must await confirmation by immunological or biochemical criteria. It should be stressed that production of amastigote-like forms by raising the temperature is effective for only certain species and does not result in continuous cultivation of these parasites.

III. CULTIVATION OF PROMASTIGOTES

In vitro cultivation of *Leishmania* promastigotes was first accomplished by Nicolle[14,107,108] who isolated *L. donovani, L. infantum,* and *L. tropica* in modified versions of the biphasic medium designed by Novy and McNeal for trypanosomes.[109] This medium is frequently referred to as NNN or 3N after the names of the original authors. Since then, an unceasing succession of modifications of these media have been concocted for isolation and maintenance of different species. In addition, many monophasic liquid media have also been devised mainly for the cultivation of leishmaniae in quantities sufficient for experimental work. This section will be presented in two major parts, i.e., biphasic and monophasic media with further subdivisions of the latter into undefined, semidefined, and defined liquid media. The voluminous literature accumulated during the last 70 years and the ambiguity of many authors make any attempt to prepare a comprehensive review in this area an almost impossible task, especially for the biphasic media. In this section, we include as many references as possible, but the coverage is by no means complete.

A. Biphasic Media

These media consist of organic nutrients, e.g., tissue extracts or infusions and blood, blended into agar slants, normally in test tubes. This medium is biphasic because a liquid phase is formed upon cooling as a water condensate or by the addition of a liquid overlay. This represents the basic design of a biphasic medium, a great variety of which have been described in the literature, in many cases differing only slightly in composition. Table 7 lists some of these media, their basic components and leishmania the species cultured.[107,108,110-138]

General methods for preparing components of biphasic media are provided as footnotes

Table 7
BIPHASIC MEDIA FOR CULTIVATION OF *LEISHMANIAE*

Authors	Solid phase[a]	Liquid phase	Organisms[b]	Ref.[c]
Nicolle (NNN-#1)	Beef infusion;[d] 0.75% Peptone; 25.00% Rabbit blood	Water condensate	Li	107
Nicolle (NNN-#2)	25.00% Rabbit blood	Water condensate	Ld,Lt	14,108
Pedrosa	15.00% Rabbit blood	Water condensate	Lt	110
Salle and Schmidt	Beef infusion; 0.8% Proteose peptone; 0.4% Glucose; 18.00% Rabbit RBC lysate (1/2 in dd H_2O)	Water condensate	Lt	111
Ray	6.00% Beef infusion; 1.70% Peptone; 1.70% Glucose; 0.5% Glysine; 0.85% Asparagine; 16.00% Rabbit blood	Water condensate	Lt,Lb,Ld,Li	112
Yang	25% Soy bean infusion; 1.25% Peptone; 0.5% Glucose; 50% Rabbit blood	Water condensate	Ld	113
Senekjie	4.25% Beef infusion; 1.7% Neopeptone; 15% Rabbit blood	Water condensate Water condensate	Lt,Lg,Ld	114 (115,116,117)
Chang (all values approximate)	0.8% Bactotryptose; 0.66% Liver extract; 0.17% Glucose; 17% horse serum; 0.017% horse hemoglobin ($RBC's/H_2O$;1/3)	Same as solid phase minus agar except: 27% horse serum; 0.034% horse hemoglobin	Ld,Lb,Lt	118
Tobie et al.	0.225% Beef infusion; 0.375% Bacto peptone; 25% Rabbit blood	Locke's solution[e]	Ld,Lm	119 (120)

Table 7 (continued)
BIPHASIC MEDIA FOR CULTIVATION OF *LEISHMANIAE*

Authors	Solid phase[a]	Liquid phase	Organisms[b]	Ref.[c]
Sorouri	4.5% Beef infusion; 0.45% Neopeptone; 0.25% Glucose; 10% Rabbit or horse blood	Locke's solution	Lt	121 (122)
Citri and Grossowicz (% approximate)	1.7% Neopeptone; 0.57% Beef infusion; 0.17% Glucose; 4.7% Human hemoglobin (RBC/H_2O; 1:3)	8.9% human plasma 1.8% Tomato juice in a solution of the same composition as the solid phase	Lt	123
Grant, et al.	Senekjie's with citrated bovine blood	Locke's solution	Lt,Lb,Ld	124
Neal and Miles	Nutrient agar; 10% Rabbit blood; 1.0% Glucose	0.85% NaCl	Ld,Lb,Len,Lt	125
Rioux, et al.	4.7% Brain-heart infusion; Agar (Difco); 10% Rabbit blood	Water condensate	Unclear (possibly Ld)	126
Rezai, et al.	2.8% Brain heart infusion; 2.5% Rabbit blood	Saline (0.87% NaCl)	Ld,Lt,Len	127
Kuwahara, et al.	BHI agar (Difco); 1.8% Glucose; 10% Rabbit blood	Locke's solution	Ld	128
Schnur and Zuckerman	0.9% Peptone; 10% Rabbit blood; 0.27% Bovril (?)	0.9% NaCl	Ld, Lt, Lspp. (variety of isolate)	129
Kilgour, et al.	Nutrient agar (Oxoid); 15% Human blood	Locke's solution	Lm, Ld, Lt, Lspp.	130
Chance, et al.	Nutrient agar (Oxoid); 13% Human blood	Locke's solution	La, Ld, Lt, Li, Lbp, Lp, Lh, Lbb, Ltmi, Lm, Lmm, Ldc, Len, Lm, Lbg. Over 140 strains and species	1, 131

Pampiglione, et al.	?% Trypticase Soy Agar (BBL); 15% Rabbit blood	Brain Heart Infusion Broth and Hank's balanced salts (1:1)	Ld	132
Walton, et al.	3.4% Blood Agar Base (Difco); 15% Rabbit blood	Water condensate	Lbb; Lm	133 (134)
Evans (4N)	3.32% Blood agar base (Oxoid); 17% Rabbit blood	Locke's solution	Lh, Ld, Lm, Lb	135 (136)
Mayrink, et al.	NNN	"LIT" (see Table 8)	Dermal Leishmania	137
Rassam, et al.	Brain heart infusion agar; 13% Rabbit blood; 1% Glucose	Locke's solution	Iraqi dermal leishmaniae and strains of Ld, Lt ma, Lt	138

[a] See text under Biphasic media and footnote d for the general preparative method to assemble this part of the medium. The final concentrations of ingredients given in % are calculated from the original formulae. Common ingredients excluded in the table are agar (1.5—2.5%), NaCl (0.2—0.6%) and other salts (Ray's, Chang's and Citri and Grossowizs'). Components without % were not given by the original author(s) and were presumably prepared according to the manufacturer's suggestions. Blood used is defibrinated by shaking with glass beads for 30 min or longer.

[b] La, *L. aethiopica;* Lb, *L. braziliensis*; Lbb, *L. braziliensis braziliensis;* Lbg, *L. braziliensis guyanensis;* Lbp, *L. braziliensis panamensis;* Ld, *L. donovani*; Ldc, *L. donovani chagasi;* Le, *L. enriettii;* Lh, *L. hertigi;* Li, *L. infantum;* Lm, *L. mexicana;* Lma, *L. mexicana amazonensis;* Lmm, *L. mexicana mexicana;* Lp, *L. pifanoi;* Lt, *L. tropica;* Ltma, *L. tropica major;* Ltmi, *L. tropica minor;* Lsp and Lspp, unidentified *Leishmania* species.

[c] References in parentheses are for leishmaniae grown in the original medium by other workers.

[d] Concentrations of beef infusion given represent the equivalent amounts of Bacto-Beef calculated from: 500 g lean beef = 100 g Bacto-Beef. To prepare beef infusion, an appropriate amount of Bacto-Beef is stirred in dd H_2O at 50°C for 1 hr and then at 80°C for 5 min. The solution is then cleared by paper filtration.

[e] NaCl, 8.0; KCl, 0.2; $CaCl_2$, 0.2; KH_2PO_4, 0.3; glucose, 2.5 in g/ℓ.

in the table. In general, Bacto-beef (Difco) or an equivalent product is infused into aqueous solution to which other protein products (e.g., peptone), salts, and agar are added and dissolved by heating. This beef infusion is filtered, autoclaved and cooled to 45 to 50°C. An appropriate amount of defibrinated whole blood is then added to this solution, which is solidified in a slanted position. Complete medium may be incubated at 37°C for checking sterility and then stored at 4°C until use. Clinical samples or single drops of fluid from old cultures are inoculated into the fluid phase of the media which are incubated at about 25°C for leishmania growth. When used for diagnosis, media should include antibiotics, e.g., penicillin and streptomycin, each at no more than 1,000 U and μg per mℓ, respectively.[139] Since biphasic media are used mostly for diagnosis or laboratory maintenance of leishmaniae, yields of parasites have rarely been given in quantitative terms. Figures provided by some authors show wide differences, ranging from 10^6-10^8 cells per mℓ, depending on the media, organisms and number of passages.

Beef infusion and peptone in the original formulae of biphasic media have been successfully substituted with other nutrient bases. It is rather difficult to assess the advantages of such modifications, as they were frequently made on an empirical basis without comparison to the originals. Salle and Schmidt[111] substituted casein or gelatin for the peptone in their medium for *L. tropica*. Cleveland and Collier[140] obtained satisfactory results for *L. tropica* and *L. infantum* with a medium containing horse erythrocyte lysate, proteose peptone, and glucose as nutritive components. They also found liver infusion to have growth-stimulating activity in suboptimal media. Other substances used as nutrients in the solid phase of biphasic media include brain heart infusion,[126-128,131,138] soybean extract or digests,[113,132] goat liver infusion,[141] and spleen extract.[141]

More attention has been paid to the blood components of the biphasic media with respect to animal sources, processing methods, and concentrations for the growth of leishmaniae. The best source of blood for these media remains controversial, however, rabbit blood has been most widely used. This source was found more reliable than that of human, dog, or hen in a medium for *L. donovani*.[142,143] In the same medium, blood from the monkey supports reasonable growth of these parasites, but that of dog or hen does not.[143] In a medium for four species of leishmaniae, Ray[112] found human to be a better source of blood than rabbit, guinea pig, or horse, with sheep being the worst. Rioux et al.[126] confirmed the inferior quality of sheep blood as a component of the media used, but this was contradicted by the earlier work of Heyneman and Mansour.[139] However, such comparisons may not be valid, as different modifications of biphasic media have been used by these workers. In related studies, Sorouri[121] found no difference between rabbit and horse blood, and Grant et al.[124] found bovine blood satisfactory. Perhaps the most careful study on blood sources for biphasic media of leishmaniae was that of Heyneman and Mansour.[139] Defibrinated blood of guinea pig, donkey, and sheep was found adequate in place of rabbit blood, particularly at a 20 to 30% concentration. At an equal concentration, donkey blood consistently outperformed sheep blood due presumably to the significantly higher hematocrit of the former (40% vs. 25%) and hence its higher nutritive value. It appears from the above findings that, although rabbit blood is used by the majority of workers, other blood sources are equally satisfactory as a constituent of biphasic media for leishmaniae.

There has been no concensus of opinion as to the concentration of blood needed for the best growth of leishmaniae in biphasic media. Most workers used 25% rabbit blood, according to the original formula of Nicolle.[14] In media with this blood source, a concentration greater than 15% was found to have deleterious effects upon growth,[111] which was not noted by Pedrosa[110] until it reached 30%. On the other hand, Yang[113] and Mohan et al.[141] cultured *L. donovani* and *L. tropica*, respectively, in media with 50% rabbit blood. Heyneman and Mansour[139] found that 30% rather than 10% guinea pig blood gave much better growth of leishmaniae, whereas the difference was marginal between media with 10% and 30% donkey blood. The inability of some leishmaniae to grow in biphasic media with a high concentration of rabbit blood may be due to the presence of natural antibody.[144] The fact that the whole blood, but not washed erythrocytes even at 35 to 40%, sometimes inhibits leishmania growth in biphasic media supports this suggestion.[111]

Pretreatment of the whole blood to avoid clotting has little effect on the growth of leishmania promastigotes in biphasic media. Defibrination of blood by shaking with glass beads is the method of choice of most workers. Defibrination gave somewhat better results over heparinization applied to donkey blood in a medium for growing six strains of *L. donovani*.[139] Citrated blood has been successfully used in a medium to culture more than 140 isolates of 15 mammalian and some reptilian leishmaniae.[1,131]

Attempts to substitute whole blood in biphasic media with erythrocytes or sera have resulted in inconsistent and inconclusive findings. Rabbit serum was used at 10% in a semisolid medium for cultivation of *L. brasiliensis, L. infantum,* and *L. tropica*,[145-147] and as the sole nitrogenous nutrient for 60 strains of three mammalian and three reptilian leishmania species.[148,149] Better growth of leishmaniae was reported in a medium with 35 to 40% washed rabbit erythrocytes than in that with an optimal amount of whole blood.[111] However, Ray[112] found fair to no growth at all of four pathogenic leishmania species in a medium containing only serum or washed red blood cells. Donkey and sheep erythrocytes alone compare unfavorably with the whole blood or its lysates at various concentrations in a medium tested for survival, serial subcultures of leishmaniae and frequency of their successful cultivation.[139]

Whole blood can generally be replaced with its hemolysate in biphasic media without much effect on the growth of leishmaniae. To prepare hemolysates, whole blood or washed erythrocytes are lysed by repeated freeze-thawing or by osmotic shock in water and then cleared by centrifugation. Salle and Schmidt[111] obtained better growth of *L. tropica* in medium with 18% rabbit blood lysate than in that with an equivalent amount of whole blood. Ray[112] observed that hemolysate was equivalent to or slightly worse than whole blood as a component of a biphasic medium, depending on the species of leishmaniae used. These observations were generally confirmed by Heynemen and Mansour.[139] In general, 10 to 25% defibrinated whole blood or its lysate has been used in biphasic media for cultivation of leishmaniae.

Introduction of an overlay to increase the volume of the liquid phase in biphasic media for leishmaniae, apparently first used by Mathis,[150] is now a common practice. The composition of some liquid overlays are given in Table 7. The most commonly used are Locke's solution and saline (0.8 to 0.9% NaCl), which have been substituted with more complex nutrient solutions. Chang[118] obtained excellent growth of leishmaniae by using an overlay of essentially the same composition as that of the solid phase. Pampiglione et al.[132] covered their trypticase soy agar-blood base with a mixture containing equal volumes of brain heart infusion broth and Hank's Balanced Salt Solution for cultivation of *L. donovani*. A cutaneous leishmania was successfully isolated in 3N medium with LIT (see below) overlay, but not in that with a simple saline solution.[137] Evidently, Locke's solution or saline as an overlay for biphasic media is adequate for routine maintenance of established cultures, but isolation of more fastidious strains may be facilitated by the use of rich nutrient solutions.

In addition to the variability associated with the media for growing leishmaniae, the ease of cultivation may also depend on individual isolates from different sources or geographic origins — a problem related to the application of biphasic media for diagnosis. Barbosa et al.[151] experienced considerable difficulty in growing a Brazilian cutaneous leishmania during the initial period, after which the organism grew profusely in LIT. Different cutaneous strains from Libya showed wide variation in their ability to grow in vitro, yet they were clearly the same strain of leishmania, as determined by contemporary taxonomic indicators, e.g.,

DNA buoyant density, isoenzyme patterns, and serology of excretory factors.[152] Analogous phenomena have been noted for leishmaniae from different sources.[133,134,137,152] In contrast to this is *L. hertigi* (a parasite of porcupines): four strains are readily cultivable in 4N medium, but show strain-related differences in amastigote size, isoenzyme patterns, serotypes of excretory factors, and other immunological criteria.[136] Child et al.[117] inoculated three different biphasic media with hamster-derived amastigotes of five different species/strains of New World leishmaniae. One of these media appears to be better than the rest, but they are all inferior to semidefined liquid media (see next section).

The continuing use of biphasic media has never been proved necessary and, in fact, has become a way of practice, no more than a custom for cultivation of leishmaniae. As will be made apparent in the next sections, there are many advantages in the use of liquid media as a substitute.

B. Monophasic Media

1. Undefined Liquid Media

As early as 1910, Laveran and Petit[154] used a peptone blood liquid medium for growing *L. donovani*. In 1912, Row[155] succeeded in cultivating several species of leishmaniae in a simple hemoglobinized saline solution (lysed rabbit blood in 0.86% NaCl) — a medium evaluated to be adequate for diagnostic purposes.[156] A liquid medium prepared from the basal nutritive components of the generally used biphasic media should undoubtedly support the growth of most, if not all, species. Indeed, Salle and Schmidt[111] in 1928 did this by using their medium without agar. Table 8 lists some of these undefined liquid media and provides the general preparative methods in the footnotes.[157-170]

In addition to the components of the biphasic medium, these media are often enriched with other nutrient bases for the cultivation of bacteria, e.g., tryptose,[135,157-162] Bacto-peptone,[154-166] Neopeptone,[166-168] brain heart infusion,[1,169] liver fractions,[135,158-160,164-165,170] and yeast extract.[161,162,166,167] Whole blood lysates or a lysate of washed red blood cells and/or serum are added instead of whole blood to provide additional nutrients and also as a source of heme — an essential growth factor for all trypanosomatid flagellates. The resulting rich broth has been used as a general culture medium or for mass propagation of leishmania promastigotes with high yields for laboratory experiments.

Of the media listed in Table 8, only Mansour's blood broth[163] was specifically developed to replace biphasic media for diagnostic purposes. Freeze-dried preparations of this liquid medium can be shipped to and reconstituted in distant places without losing growth-supporting activity, thereby increasing shelf-life and ease of handling. In addition, some of these liquid media apparently are nutritionally more rich than biphasic media and thus may be a better diagnostic tool. Panmede[129-131,152,171] and BHI-BL[1] may be considered as such media which, indeed, support the growth of numerous isolates of different leishmania species, although it is not clear whether they have ever been submitted to extensive field trials. The application of these liquid media to diagnosis or primary isolations of parasites from clinical samples should follow the general conditions already described for biphasic media, e.g., the incubation temperature and the inclusion of antibiotics.

Liquid media are clearly better suited than biphasic media for mass culture of leishmania promastigotes. Exuberant growth of parasites in these media often requires adequate aeration which can be provided best by agitation in rotary shakers (W. R. Fish, unpublished), fermentation chambers[172] or chemostatic culture vessels,[161] or, alternatively, by providing a large surface area for growth with a thin layer of medium in culture vessels such as Erlenmeyer flasks, Fernbach flasks, Roux bottles, or large tissue culture flasks. Many species of leishmaniae often form a pellicular surface growth seen as a whitish floating sheet. Since provisions are often made to prepare these liquid media relatively free of large particulate matter, cultured organisms can be readily harvested by simple centrifugation, e.g., 1,500

Table 8
UNDEFINED LIQUID MEDIA FOR CULTIVATION OF *LEISHMANIAE*

		Ingredients[a]					
		Blood Component					
Medium	Nutrient base[b]	Lysate[c]	Serum[d]	Others[e]	Organism[f]	Max. growth	Ref.
GLSH	0.3% Lactalbumin; Hydrolysate; 0.1% Tryptose; 0.45% Bacto-beef blood	—	10% CS	0.1% Glucose 20 μg/mℓ Biotin	Ld (Pm), Lb, Lm, Lt, Le	1.5 g/ℓ	157
LIT	0.5% Ox liver; Infusion; 0.5% Tryptose	0.5% BRBCL SRBCL RRBCL	10% CS	0.02% Glucose	Lb, Ld (PM)	?	158 (121, 159)
EBLB	1.5% Tryptose; 1% Casein hydrolysate; 1% Liver digest	HRBCL	—	0.15% Proline 0.04% EDTA	Lb (Pm) Ld, Lm, Lt, Le	?	135, 160
Schaefer, et al.	1.5% Bacto-tryptose, 1 % Yeast extract	0.2% RRBCL	13% FBS or RS	0.2% Glucose	Lb (Pm) Ld, Lm, Lt	12 × 10^7 per mℓ	161, 162
Mansour, et al.	5 % Bacto-beef; 0.1% Oxoid Liver; 0.5% Bacto-peptone	5—10% RBL	—	0.25% Glucose	Lb (Am) Ld, Lt	16 × 10^6 per mℓ	163
Fulton & Joyner	0.17% Bacto-peptone; 4.1% Ox Liver extract	1.1% RBL	8.3% RS	—	Ld (Pm)	17 × 10^6 per mℓ	164
Dwyer	2.5% Bacto-beef; 0.5% Bacto-peptone; 0.1% Oxoid Liver	12.5% RBL	10% FBS	0.18% Glucose	Ld (Pm)	10^8 per mℓ	165
NEO-YE	1 % Neopeptone; 0.05% Yeast extract	—	—	0.5% Glucose 10 μg/mℓ Hemin	Lta (Pm)	10^8 per mℓ	166, 169
Senekjie	5 % Bacto-beef	—	—	—	Lt (Pm)	?	168
BHI	3.75% brain heart infusion	—	—	25 μg/mℓ Hemin	Lta (Pm)	10 × 10^8 per mℓ	169
BHI-BL	3.75% brain heart infusion	10% HBL	—	—	Numerous isolates (Am, Pm) of Lb, Ld, Lm, Lt & reptilian species	5 g/ℓ	1
Panmede	0.18% Proteolysed liver (Panmede Powder)	—	7.3% RS	0.18% Glucose	Numerous (Am, Pm) isolates of various species	35 × 10^6 per mℓ	170 (129, 131, 152)

[a] The ingredients are prepared as separate solutions, which are then sterilized and assembled as below. The basal nutrients are dissolved or infused generally according to manufacturer's suggestions in double distilled H_2O (also see footnotes for Table 7), to which glucose, salts, hemin solutions (1—2 mg/mℓ in *N* NaOH), etc., are added. This portion of the medium is adjusted to pH 7—7.8 with *N* NaOH and sterilized by autoclaving at 120°C, 15 lb/in² for 30 min. To this, appropriate portion of filter-sterilized blood components are finally added to make up the final concentration of each component shown. Concentrations are in v/v for BL, RBCL and serum, and in w/v for others.

[b] The sources of these ingredients are: lactalbumin hydrolysate, Nutritional Biochemicals Co., Cleveland, Ohio; brain heart infusion, Bacto-peptone, Bactobeef, Neopeptone, yeast extract, liver extract and tryptose, Difco

Table 8 (continued)
UNDEFINED LIQUID MEDIA FOR CULTIVATION OF *LEISHMANIAE*

Laboratory, USA or Oxoid Ltd., England; and Panmede powder, Paines & Byrne Ltd., Pabyrne Laboratory, Greenford, Middlesex, England.

[c] B-, H-, R-, and S-RBCL, Bovine, human, Rabbit and sheep red blood cell extract; H- and R- BL, Human and rabbit whole blood lysate. For BL, citrated or defibrinated whole blood is diluted with appropriate volumes of dd H_2O or freeze-thawed to lyse red blood cells and then used directly or, sometimes, after centrifugation at high speed, e.g., 40,000 g, to remove particulate matter. For RBCL, red blood cell pellets are subjected to the same treatment as for BL. Sterilization is done by using a 0.45 μm filter disc or by observing aseptic procedures during preparation.

[d] Sera are purchased from commercial sources, except for RS which, in many laboratories, is derived from defibrinated whole blood obtained by intracardiac puncture from laboratory rabbits. It is recommended to heat-inactivate sera at 56°C for 30 min before use. Sterilization is done by filtration.

[e] Various concentrations of NaCl, KCl, $CaCl_2$, phosphate buffer salts, etc., are also included in addition to those shown. We are of the opinion that a phosphate buffered saline or 0.8% NaCl with 25 m*M* HEPES buffer should be sufficient in view of the grossly undefined composition of these media.

[f] Lb, *L. braziliensis;* Ld, *L. donovani;* Lm, *L. mexicana;* Lt, *L. tropica;* Lta, *L. tarentolae;* (Am) and (Pm) indicate whether amastigotes and/or promastigotes are used as inocula in these media.

to 3,500 g for 5 to 10 min. This presents a distinct advantage of liquid broth over biphasic media, from which promastigotes are often contaminated with red blood cells or other particles derived from the solid phase.

2. *Semidefined Liquid Media*

The first partially defined liquid medium described for the cultivation of leishmaniae was probably that originally designed for *Trypanosoma cruzi* by Citri and Grossowicz.[173] Initially, they grew *L. tropica* in an undefined medium consisting of casein hydrolysate as a nutrient base and tomato juice as a source of vitamins.[173] Subsequently, the latter component of the undefined medium was substituted with a mixture of vitamins, RNA, a nucleotide, and a nucleoside.[123] This semidefined medium, however, supports only the growth of *L. tropica* and reptilian leishmania species.

Recently, leishmania promastigotes of most species have been found to grow well in many commercially available media originally designed for the cultivation of mammalian or insect cells.[105,106,174-179] Compositions of some of these media are listed in Table 9. A single supplement of these media with 5 to 30% HIFBS allows continuous and good growth of most species.[105,106,117,134,174,177,178] Multiple supplements, e.g., blood lysate and HIFBS[175] or hemin, lactalbumin hydrolysate, yeast powder and liver extract,[176] produce excellent growth of some species, e.g., *L. donovani*. It is noteworthy that the latter supplemented medium requires no HIFBS. Also, it should be mentioned that, of the three insect tissue culture media used for leishmania, only Grace's for moth cells is completely defined.[180] Schneider's medium for *Drosophila* cell lines contains yeastolate,[181] and Mitsuhashi and Maramoroschs' medium has yeastolate and lactalbumin hydrolysate.[182] These two media, especially the latter, are thus not defined as are those for mammalian cells, e.g., Medium 199.[183] Defined media specifically designed for leishmania can serve the same purpose when supplemented with 2 to 10% HIFBS or Bacto-peptone.[178,179] The general culture conditions for leishmania in these media are as described for biphasic or undefined liquid media in the previous sections.

The use of commercially available media for growing leishmaniae is attractive in terms of simplicity and convenience of medium preparation, but most of these media require a

Table 9
SEMIDEFINED LIQUID MEDIA FOR CULTIVATION OF *LEISHMANIAE*[a]

	Undefined portion				
Defined portion[b]	Serum[c]	Others	Organisms	Maximal growth (X10^{-6}/mℓ)	Ref.
95% NCTC-135 (or M-199, MEM, RPMI 1640)	5% FBS	—	Lb (Pm)	?	174
M-199	5% RS	5% RBL	Ld (Pm)	80	175
90% M-199 (HEPES-buffered)	—	0.25% Liver Ext.; 0.5% Yeast Powder; 1% Lacto. Hydro.; 10 μg/mℓ Hemin	Ld (Pm)	60—100	176
85% Schneider's[d]	15% FBS	—	Lb (Am)	200	134
80% Schneider's, Grace's, Mitsuhashi-Maramoroschs'[d]	15% FBS	—	Lb (Am) Lma, Lm	124	117
70% Schneider's	30% FBS	—	Ld (Am, Pm) Lb, Lp, Lm, Lt, Lc, Le (121 isolates)	70	105–106
90% HOSMEM	10% FBS	—	Ld (Pm) Lta	40	177
RE III	2—10% FBS	—	Lb (Pm)	>10	178
RE III	—	0.02—0.1% Defatted BP	Lb (Pm) Ld, Le, Lm, Lt	22—55	179

[a] See footnotes for Table 8 for abbreviations not provided.

[b] These media, except the last two, are available from commercial sources in ready-to-use liquid form and some (M-199, NCTC-135, RPMI 1640) are also supplied as pre-mixed powder in amounts for preparing 1 ℓ or more. The average shelf-life of liquid media is 1—2 months. It is much less expensive to use powder mix in laboratories where a large quantity of these media are routinely used. See footnote in Table 10 for formulations of RE III and HOSMEM.

[c] FBS should be screened for growth-supporting activity before use.

[d] These two media are not defined: Schneider's medium contains 0.2% yeastolate, and Mitsuhashi and Maramoroschs' medium has 0.65% lactalbumin hydrolysate and 0.5% yeastolate.

substantial supplement of HIFBS, an increasingly expensive item. Insect tissue culture media plus HIFBS, however, prove to be superior to biphasic media, in many instances, for the primary isolation of South American leishmanial species from animal lesions or human biopsy samples.[105,106,117,134,184] The possibility of replacing HIFBS with less expensive nutritional supplements, e.g., other types of sera or lactalbumin hydrolysate, Bacto-peptone, liver extract, and the like should be examined to reduce the cost of using these media for diagnostic purposes. For laboratory experiments, most species of leishmaniae can probably be gradually adapted to grow in these media with much lower concentrations of HIFBS than those currently in use, provided that additional hemin is included. At least *L. donovani* is known to grow after adaptation in HEPES-buffered Medium 199 supplemented with only 0.5% HIFBS plus hemin.[185]

Based on some of the information presented in Table 3 there have been speculations that media designed for the cells of insects and mammals are apt to support leishmanial growth because of their roles in nature as vectors and hosts for these parasites.[105,106,117] One should accept this assumption with caution in considering the microenvironments of leishmaniae: the lysosomal compartment of macrophages for amastigotes[36,57,59] and the sandfly gut for promastigotes, both of which bear no obvious resemblance to the aforementioned media in chemical composition and physical properties. The ability of leishmania promastigotes to grow in a wide variety of media only indicates their flexibility or adaptability to different nutritional and culture environments.

3. Defined Liquid Media

Chemically defined media are normally composed of known nutrients, e.g., amino acids, vitamins, sugars, nucleotides, nucleosides or their bases and trace elements in a buffered salt solution. Three defined media have been developed for growing leishmania promastigotes, i.e., Medium C of Trager,[186] RE III of Steiger and Steiger,[187,188] and HOSMEM II of Berens and Marr.[189] The composition and the preparative methods for these media are presented in Table 10.

Amastigotes isolated from infected laboratory animals or promastigotes cultured in undefined or semidefined media have been used as inocula. For conclusive evidence of their cultivability in defined media, organisms must be subcultured for at least 5 and preferably 10 passages with persistent and reasonably good growth. These passages ensure the dilution of nutrients carried with the inocula or the organisms to an insignificant amount which does not contribute to their growth in the defined media. Defined media for leishmaniae often remain stable for only 1 to 2 months at 4°C. The shelf-life of these media can be prolonged to 1 year or more when stored at −20°C as component stock solutions.

Cultivation of leishmaniae in defined media provides a powerful tool to probe their biosynthetic capabilities. This area has been reviewed in several articles devoted primarily to other trypanosomatid flagellates.[190-194] Here, we shall limit our discussion to the three defined media for the cultivation of leishmaniae. At 27°C, Medium C supports good growth of a reptilian leishmania, *L. tarentolae* (6-7 $\times$ 10^7 cells/mℓ in a week)[186] and other lower trypanosomatids,[195] but not leishmania species pathogenic to man.[68,186] On the other hand, both RE III and HOSMEM II can be used for the cultivation of visceral and cutaneous leishmaniae.[187-189] The maximal cell density is about 4 $\times$ 10^7 per mℓ in 6 days at 26°C for *L. donovani* in HOSMEM II and 5 $\times$ 10^7 per mℓ in 5 days at 27°C for *L. donovani* and *L. brasiliensis* in RE III. Subsequent study shows that all leishmania species used can be adapted to grow in RE III without supplements and yield adequate numbers for biochemical analysis.[179] HOSMEM II is prepared basically by mixing a commercially available medium (MEM) and its component solutions for mammalian cells, whereas Medium C and RE III are constructed from individual nutritional components. Cultivation of leishmaniae in these defined media provides valuable information concerning their nutritional requirements, which are discussed below according to individual categories of nutrients.

a. Amino Acids

As already pointed out by Steiger and Steiger,[188] the failure of Medium C to support the growth of higher leishmaniae may be accounted for by the omission of glutamine and possibly cystine or cysteine, which are included in both RE III and HOSMEM II. Nutritional experiments showed, indeed, that these amino acids are indispensible for *L. donovani* and *L. brasiliensis*.[188] The concentration of amino acids may be reduced by at least half in RE III and Medium C, as indicated by the much lower level of these compounds in HOSMEM II and by the analytical data of their utilization in spent medium.[196] Alanine, asparagine, glycine, serine, aspartic acid, and glutamic acid can apparently be deleted from HOSMEM II and Medium C, as none of these amino acids is included in RE III.

b. Vitamins

The composition of most water soluble vitamins is very similar between RE III and HOSMEM II, except for biotin, which is 10 times less in the latter medium and this amount

Table 10
COMPOSITION OF DEFINED MEDIA FOR *LEISHMANIA*

Component	Concentration (mg/ℓ)		
	Medium C[a]	RE III[b]	HOSMEM II[c]
Na_2HPO_4	1250	60 ($2H_2O$)	0
$NaH_2PO_4 \cdot H_2O$	0	0	1400
KH_2PO_4	500	60	0
$NaHCO_3$	0	1000	1000
HEPES[d] (60 m*M*)	0	14250	0
MOPS[e] (30 m*M*)	0	0	6280
Phenol red	0	5[f]	10
NaCl	2000	8000	6800
KCl	0	400	400
$CaCl_2$	26	70	0
$MgSO_4$	1000 ($7H_2O$)	200 ($7H_2O$)	98
$CoSO_4 \cdot 7H_2O$	0.44	0	0
$CuSO_4$	0.5	0	0
$FeSO_4 \cdot 7H_2O$	10	0	0
H_3BO_3	0.11	0	0
$MnSO_4 \cdot H_2O$	3	0	0
$ZnSO_4 \cdot 7H_2O$	22	0	0
L-Arginine HCl	300	200	189
L-Cystine	0	0	12 + 31(2HCl)
L-Cysteine HCl	0	50	0
L-Glutamine	0	300	438
L-Histidine	150 (HCl)	100	63 ($HCl \cdot 2H_2O$)
L-Isoleucine	(DL-)600	100	78
L-Leucine	1500	300	78
L-Lysine HCl	1250	250	109
L-Methionine	(DL-)300	50	23
L-Phenylalanine	(DL-)400	100	49
L-Threonine	(DL-)500	400	72
L-Tryptophan	(DL-)200	50	15
L-Valine	(DL-)500	100	69
L-Alanine	(DL-)700	0	9
L-Asparagine·H_2O	0	0	15
L-Aspartic acid	(DL-)1200	0	13
L-Glutamic acid	1900	0	15
Glycine	100	0	8
L-Proline	500	300	12
L-Serine	(DL-)400	0	11
L-Tyrosine	400	50	18 52 (Na_2 salt)
p-Aminobenzoic Acid	3	0	1
Biotin	0.2	1	0.1
Choline-Cl	3	1[f]	1
Inositol	3	2	2
Nicotinamide	5	1	1
Pyridoxine-HCl	2	0	0
Pyridoxal-HCl	2	1	1
Pyridoxamine-2 HCl	2	0	0
Ca-pantothenate	8	1	1
Riboflavin	2	1	1
Thiamine-HCl	2	1	1
Folic acid	1.6	10	12.5
Ascorbic acid	0	0	0.2

Table 10 (continued)
COMPOSITION OF DEFINED MEDIA FOR *LEISHMANIA*

	Concentration (mg/ℓ)		
Component	Medium C[a]	RE III[b]	HOSMEM II[c]
Cyanocobalamin	0	0	0.2
Lipoic acid	0	0.4[f]	0.4
Retinol Acetate	0	0	0.4
Menadione	0	0	0.4
Glucose	5000	2000	3000
Na-pyruvate	0	0	110
Hemin	25	5	6.25
Adenine	1.7	0	0
Guanine	1.7	0	0
Xanthine	1.7	0	0
Hypoxanthine	0	0	15
Uracil	1.7	0	0
Cytidylic acid	0.5	0	0
Adenosine	0	20	0
Bovine serum albumin Serum albumin fraction V (Defatted)	0	15[f]	12.5
pH	7.8—8.0	7.3—7.4	7.2
Osmolarity (mOsm.)	?	?	420

[a] Medium of Trager for *L. tarentolae.*[186] This medium is assembled from the following component solutions: (1) 100 × solution of $MgSO_4$; (2) 100 × solution of $CaCl_2$; (3) 100 × solution of trace elements; (4) 30 × solution of folic acid at pH 6.7; (5) 10 × solution of glucose, vitamins and nucleic acid bases; biotin and bases are first dissolved by heating in doubly distilled water (dd water) at slightly acidic (pH 6.8) and alkaline pHs, respectively. (6) 2 × solution of amino acids, hemin and buffer salts; hemin at 0.5% (w/v) is dissolved in 0.05 *N* NaOH before adding to the amino acid solution, which is heated to dissolve and then adjusted to pH 7.8—8.0. Solutions 1, 2, 4 and 6 are autoclaved and solutions 3 and 5 are filter sterilized.

[b] Medium of Steiger and Steiger for leishmaniae.[187,188] This medium is made up of the following parts: (1) 2 × solution of salts; (2) 5 × solution of amino acids; (3) 100 × solution of water soluble vitamins; (4) 400 × solution of hemin in N NaOH; (5) 100 × solution of folic acid in *N* Na; (6) Lipoic acid in ethanol (8 mg/ℓ). The remaining ingredients are dissolved in dd water and mixed with proper proportions of the above stock solutions. The final medium is sterilized by filtration through 0.22—0.45 μm membranes.

[c] Medium of Berens and Marr for leishmaniae.[189] This medium is prepared from a powder mix and component solutions of Eagle's Minimal Essential Medium (MEM) from a commercial source with a variety of additives as below: (1) 100 × solution of hypoxanthine, ascorbic acid, vitamin B12 (cyanocobalamin), bovine serum albumin, menadione, lipoic acid (thioctic acid) and vitamin A (retinol acetate); the last three are first dissolved in 95% ethanol; (2) 400 × solutions of hemin and folic acid in 0.05 *N* NaOH (final pH = 8); (3) the remaining components are dissolved in 800 mℓ dd water by the addition of the following materials: 10.58 g of stock # F-14 powder of Eagle's MEM for suspension culture (Grand Island Biological Co., GibCO), 10 mℓ of 50 × MEM amino acid solution (GibCO), 10 mℓ of 100 × solution of nonessential amino acids (GibCO), 11 mℓ of 1% (w/v) sodium pyruvate, 2 g glucose, and $NaHCO_3$, MOPS, biotin, *p*-aminobenzoic acid at concentrations as indicated in the table. The above stock solutions in proper proportions are mixed, brought up to 1 ℓ and filter-sterilized with 0.22 μm membranes. A gas phase of 5% CO_2 in air is used for this medium to maintain a proper pH by the original authors.

[d] *N*-2-hydroxyethyl piperazine-*N*-2-ethane sulfonic acid.

[e] 3-(*N*-morpholino) propanesulfonic acid.

[f] These components can be deleted without effect on the growth of *L. donovani* and *L. mexicana*.

is probably adequate. *p*-aminobenzoic acid, ascorbic acid, cyanocobalamin, menadione, and fat-soluble vitamins in HOSMEM II are apparently not essential. In this laboratory, we have deleted lipoic acid from RE III and observed no reduction in the growth of leishmaniae.

c. Purines

All trypanosomatids are incapable of synthesizing purines de novo and thus require an exogeneous supply of these bases.[173,174] This requirement can be satisfied by a supplement of individual nucleosides or one of their bases, e.g., adenosine in RE III and hypoxanthine in HOSMEM II.

d. Other Essential Growth Factors

One nutritional peculiarity of virtually all trypanosomatids is their stringent requirement for heme,[193,194] as they have an incomplete biosynthetic pathway for this compound.[197] Thus, hemin or hematin must be included in the defined culture media for these organisms, including leishmaniae. Some of the hemin in the three defined media probably also serve as an iron carrier, as suggested by the fact that the hemin concentration can be greatly reduced in the presence of chelated iron in a defined medium for *Crithidia fasciculata*.[198] The small amount of bovine serum albumin, whose beneficial function is ill-defined in RE III and HOSMEM II, can be omitted from the former medium according to our experience.

e. Calcium and Trace Elements

Calcium is a peculiar omission from HOSMEM II. This divalent cation is either supplied by chemical contamination or is not required at a high level by leishmaniae. Calcium is amply supplied in both Medium C and RE III. No trace metals are included in RE III and HOSMEM II; presumably these are present as contaminants in other components of the media.

f. Buffer System

For this, HEPES and MOPS are used in RE III and HOSMEM II, respectively. These synthetic buffer systems have many advantages over conventional ones.[199] Since no gassing of culture flasks is needed for both RE III[188] and HOSMEM II,[179] the buffer capacity of HEPES and MOPS used in these media apparently renders the pH independent of the bicarbonate system.

Clearly, RE III and its simplified version (see Table 10 footnote f) are suitable media for in vitro studies of leishmania promastigotes on account of their rational design, simple composition and successful rates of cultivation. This is apparently the medium of choice for further study to determine the minimal nutritional requirements of these parasites. A critical evaluation of the nutritional requirements of different species in this medium should prove valuable in defining the basic biosynthetic capability of these organisms. The minimal defined medium obtained would facilitate experiments involving labeled precursors and chemical and antigenic analyses of structural or excreted macromolecules without the interference of extraneous factors.

IV. CURRENT PROBLEMS AND PROSPECTS

A. Cultivation of Amastigotes

Problems associated with the in vitro cultivation of leishmanial amastigotes can be discussed in two different contexts based on the experimental aims. A short-term cultivation of leishmanial amastigotes in macrophages has been used to serve as an assay system. This aim has been accomplished with varying degrees of success by different workers. The intrinsic difficulties of this system have been discussed in the relevant section, from which it is apparent that a more sensitive method is needed for more efficient and accurate quantitation of infection and intracellular growth of amastigotes. Attention should also be paid to the culture conditions which are optimal for both macrophages and parasites. The methods currently in use are those developed for the former, and conditions favorable to the macrophages may not be sufficient to encourage the growth of parasites. Continuous cultivation for mass production of leishmanial amastigotes and their maintenance is the second apparent aim of in vitro culture experiments. This has unfortunately not received sufficient attention in the past. The reported transformation of certain species from promastigotes to amastigotes in a cell-free system may satisfy the goal of producing amastigotes in vitro and thus deserves attention. However, the identity of the amastigote-like forms needs confirmation by comparison to the genuine intracellular amastigotes. Cultivation of these in tissue culture cells is a feasible alternative to their procurement from experimental animals.

From several reports since the 1960s, it has been obvious that certain leishmania species of South American origin are more adaptable to cultivation in some cell lines.[78,94] The potential of these results for larger-scale operation has not been explored until recently in mouse peritoneal macrophage lines. Recent findings with *L. mexicana* in J774G8 cells are particularly encouraging, although considerable efforts are needed to make this into an efficient system on a routine basis. Especially, less costly and laborious means must be developed to justify the continuous maintenance of leishmanial amastigotes in tissue cultures instead of laboratory animals. Additional efforts should be continued to find a similar system for other leishmania species, especially *L. donovani*. Although it was the first species grown in prolonged culture in Dog Sticker Sarcoma cells,[90] the overall yield in this system has been found unsatisfactory.[94] Beyond morphological aspects, we know very little about the amastigote stage of all leishmania species. The stumbling block has been the inability to obtain sufficient materials for investigations. It can be stated with confidence that this obstacle will be soon overcome at least for certain species through in vitro culture in macrophage lines.

B. Cultivation of Promastigotes

The cultivation of leishmanial promastigotes, with few exceptions, presents no particular difficulty, as attested to by their growth in many media, most of which are not designed for this group of organisms. Existing problems, thus, lie much less in the cultivability of promastigotes than in the efficient application of available media for different purposes; e.g., diagnosis, laboratory maintenance, and production of infective parasites for various investigations.

Efficient culture techniques for diagnosis of leishmaniasis require an in vitro environment for rapid conversion of small numbers of amastigotes into a population of motile promastigotes demonstrable by microscopy. An ideal diagnostic medium is, therefore, one which can effect an accelerated amastigote to promastigote transformation followed by rapid growth of the latter. Further considerations should be given to the cost of medium ingredients and preparation and the ease of subsequent handling. To substitute for the long standing choice of biphasic media, likely candidates to meet the above criteria would be a rich undefined broth prepared from inexpensive, autoclavable nutrient bases, preferably without blood or serum. Examples close to this formulation can be found in Table 8, and, indeed, some of the media listed support the growth of many isolates from different geographic origins.[1,129,131,152] These media can be easily made from inexpensive ingredients and have all the advantages of the semidefined media,[105,106,117] recently developed for diagnostic use in place of biphasic media. In testing the relative efficiency of different media for possible diagnostic value, it is imperative to standardize medium components, especially those used for the biphasic media.

For routine maintenance of leishmania species, biphasic media are often considered to be better than others. Supposedly, the organisms live longer in these media and thus less attention is needed for subculturing, due presumably to a gradual release of nutrients from the solid phase. While this virtue may justify the continued use of biphasic media, readers should be informed that leishmanial promastigotes can be readily stored in and retrieved from cryopreservation in liquid nitrogen. For quantitative experiments, biphasic media have limited value because of the inexactness in preparative methods and variability in sources of medium components, compounded by the problems of random modifications and misquotations prevailing in the literature.

By cultivation in liquid media, a reasonable quantity of clean promastigotes can be obtained from most species of leishmaniae. Through the use of this material, significant progress has been made in some aspects of leishmania studies, e.g., molecular and biochemical taxonomy for the differentiation of leishmania species.[1,200] Of equal importance is the cultivation of leishmaniae for studies aiming at the solutions of problems more directly related to leishmaniases. In this regard, whether the cultured promastigotes are adequate material for investigation is, in many cases, a question worthy of attention. Investigations for relevant information should be directed at infective promastigotes — presumably the equivalent of the proboscis forms,[201] which the sandfly vectors introduce into human hosts, leading to the sequelae of the diseases. These infective parasites may be inferred to exist in cultured populations because of their infectivity to laboratory animals, but no direct evidence has established this. Moreover, the ability of many leishmania species to infect laboratory animals is lost progressively with the period of cultivation,[202-203] suggesting an in vitro selection of noninfective vs. infective forms. While the mechanism underlying this phenomenon is not clear, it indicates that the relevance of any study to leishmaniasis using long-term cultured promastigotes may be considered dubious without proof of animal infectivity by these parasites. Unqualified extrapolation of information derived from cultured promastigotes to amastigotes must be also considered highly inappropriate, as significant differences in metabolism are expected between these two leishmanial forms in view of their completely different natural habitats. Thus, production of infective leishmaniae for investigations still presents a considerable problem, which is often overlooked due to the ease of their cultivation. The lack of easy solutions contributes to the reluctance of many workers to consider this problem in their studies. There are several potential means to circumvent this difficulty: e.g., (1) use of leishmanial species which have a lesser tendency of losing infectivity in vitro; (2) completion of experiments before infectivity is lost; (3) periodic recycling of cultured promastigotes through mammalian hosts or reestablishment of promastigote cultures from amastigotes; (4) periodic recycling of promastigotes through in vitro cultured macrophages[46] or insect tissue culture cells; (5) cloning of stable infective promastigotes;[204] (6) cell-free culture of promastigotes in media favorable to the retention or generation of infective forms. Although some of the above practices have been in use, none can be considered as a well-established method and much remains to be studied. Of obvious importance is the need to determine precisely the number of generations of parasites in culture until the loss of animal infectivity occurs for each species.

C. Leishmania Morphogenesis In Vitro

In vitro techniques have been applied to study the morphological changes of leishmaniae known to occur during the natural transmission cycle, as already described in several places in this chapter. Amastigote to promastigote transformation that occurs naturally in the sandfly gut has been observed in vitro under different conditions. Pertinent points can be summarized as follows: (1) the process of transformation is accompanied by the appearance of new antigens, an increased utilization of oxygen[68] and an expansion, but not regeneration, of the mitochondrial apparatus with the cell volume;[205] (2) the organisms require a source of amino acids and carbohydrates;[68] (3) its occurrence is inhibited by cyclic AMP[206] and delayed by mediators released from lymphocyte cultures.[207] Common difficulties that have yet to be overcome for more detailed study of this transformation are the lack of synchronization in the process and of simple methods for purification of amastigotes. Promastigote to amastigote transformations that occur in the mammalian host have been studied in cell-free media or tissue culture cells, as already discussed in relevant sections. It is generally true that an intracellular environment and/or a high temperature (34 to 37°C) favor the occurrence of this event, in keeping with the fact that they are intracellular parasites of mammalian cells.

Biochemical and molecular events underlying the mechanism of leishmania morphogenesis are unknown, pending the development of adequate in vitro systems for detailed studies. Besides academic interests, such studies may provide valuable information with practical applications. When amastigote to promastigote transformation becomes more fully understood and then standardized, it can be used for determining amastigote viability needed in different assay systems. The mechanism of promastigote to amastigote transformation has significant implications with regard to chemo- and immuno-prophylaxes of leishmaniases. Blockage of this event, directly or indirectly, through the intervention of drugs or host immune mechanisms is a strategy, perhaps, more logical than attempts to attack the highly refractory intracellular amastigotes.

V. ADDENDUM

Since the submission of this manuscript, there has been a number of papers published concerning the cultivation of amastigotes and promastigotes.

As pointed out in the main body of the text, leishmania-macrophage in vitro models for adequate intracellular growth of amastigotes must be developed to meet the requirement for various types of investigations. It is satisfying to find that this area has received due attention during the last 2 years. Some recent findings pertinent to this area are summarized below. In the *L. enriettii*-guinea pig model, peritoneal macrophages from different animal strains and of different sedimentation velocities were found to differ in their ability to support the intracellular growth of amastigotes.[208] However, the strain-specific difference is obvious only during the first 1 to 2 days. In a *L. mexicana*-mouse model, amastigotes were found to grow in peritoneal macrophages, but were killed in those from bone marrow.[209] Since the latter macrophage population undergoes replication in vitro, it is thought that certain cell cycle dependent changes of macrophage physiology may be detrimental to the parasites.[209] Further investigations are needed to determine the nongenetic mechanism(s) underlying the heterogeneity of macrophage subpopulations in relation to the intracellular multiplication of amastigotes. In a *L. tropica*-mouse model, peritoneal macrophages were found to support better growth of amastigotes when made less adherent by cultivation in polypropylene tubes.[210] This in vitro system proved adequate for studying lymphokine fractions, which mediate leishmanistatic and leishmanicidal activities of macrophages.[211] In *L. donovani*-mouse models, three independent studies showed that amastigotes survived, but failed to grow to any significant extent in peritoneal macrophages.[212,213,214] These parasite survival systems were used for assaying leishmanicidal lymphokines released by concanavalin A-[213] and antigen-[214]stimulated splenic lymphocytes. It would be of interest to determine if *L. donovani* amastigotes can be encouraged to grow in peritoneal macrophages cultured under different physical conditions, as found with *L. tropica*.[210] Berman and his co-workers have used the leishmania-human macrophage system developed by him earlier for screening potential antileishmanial drugs,[215] for studying the expression of parasite antigens in the surface of infected cells,[216] and for determining the temperature sensitivity for the growth of *L. tropica*.[217] In the same in vitro system, freshly transformed promastigotes of *L. donovani* (Sudan strain) were found to gain entry into macrophages where they differentiated into and multiplied as amastigotes.[218] This finding, if confirmed, would be a significant contribution, as it is well known that the Sudan strains of *L. donovani* survive poorly in macrophages when fully differentiated promastigotes are used for inoculation. Whether the culture conditions, the source of macrophages and/or the degree of parasite differentiation may contribute to the reported success[218] remains to be determined. A complete cycle of leishmanial differentiation can now be accomplished with ease in an in vitro system, consisting of *L. mexicana amazonensis* and a macrophage line of BALB/c mouse origin, J774G8.[219] In this system, intracellular differentiation of leishmanias from promastigotes to amastigotes takes at least 7 days and extracellular differentiation in the reverse direction only 2 days; tubulin biosynthesis is changed, consistent with the morphological alteration of flagella and cytoskeleton known to accompany the transformation.[220] The simplicity of this system makes possible the maintenance of both promastigotes and amastigotes in vitro. Infected macrophages are available at all times for experimental use, e.g., screening for leishmanicidal drugs.[221] Finally, it deserves mentioning that centrifugation in Percoll gradients has been successfully used for the purification of intracellular amastigotes of *L. mexicana* from animal lesions[222] as well as from infected macrophages grown in vitro.[219]

For the cultivation of leishmanial promastigotes, there have been several reports concerning the application and improvement of available media, but no new design of medium formulation. Schneider's medium plus heat-inactivated fetal bovine serum has been successfully used for the diagnosis of visceral leishmaniasis[223] and for mass cultivation of *L. mexicana amazonensis*.[224] However, the universal application of this medium for diagnostic use was questioned by Shaw and Lainson[225] who found that it was inferior to a modified NNN medium for growing *L. brasiliensis*. Of interest is the report that discrete colonies of promastigotes can be produced on blood agar plates upon which impression smears are made using the spleen of a hamster infected with *L. donovani*.[226] This suggests the possibility of cloning leishmanias directly from animal biopsies and clinical samples. The RE series of Steiger's defined medium has been further simplified: RE IX contains neither bovine albumin nor lipoic acid with a reduced amount of both folic acid and hemin; with RE X, however, a further deletion of glucose is made.[227] *Leishmania donovani* promastigotes can be grown continuously in both media.

ACKNOWLEDGMENTS

We are grateful to Prof. W. Trager for his suggestions and to James Stanorski and Francis Schopick for their skillful preparation of this manuscript. KPC is a Hirschl Career Scientist Awardee supported by U.S. PHS, NIAID Grant #AI-15183 and UNDP/World Bank/WHO Special Program for Research and Training in Tropical Diseases, Contract #780040.

REFERENCES

1. **Chance, M. L., Shchory, L., and Peters, W.,** Biochemical taxanomy of leishmania. I. Observations on DNA, *Ann. Trop. Med. Parasitol.*, 68, 307, 1974.
2. **Chung, H.-L.,** A resume of Kala azar work in China, *Chin. Med. J.*, 71, 421, 1953.
3. **Zuckerman, A. and Lainson, R.,** Leishmania, in *Parasitic Protozoa*, Vol. I, Kreier, J. P., Ed., Academic Press, New York, 1977, 58.
4. **Hommel, M.,** The genus leishmania: biology of the parasites and clinical aspects, *Bull. Inst. Pasteur*, 75, 5, 1978.
5. **Williams, P. and Coelho, M. D. V.,** Taxonomy and transmission of leishmania, *Adv. Parasitol.*, 16, 1, 1978.
6. **Gutteridge, W. E. and Coombs, G. H.,** *Biochemistry of Parasitic Protozoa*, MacMillan, London, 1977.
7. **Mukkada, A. J.,** Tricarboxylic acid and glyoxylate cycle in the leishmania, *Acta Trop.*, 34, 167, 1977.

8. **Marr, J. J.**, Carbohydrate metabolism in leishmania, in *Biochemistry and Physiology of Protozoa*, Vol. III, Levandowsky, M. and Hutner, S. H., Eds., Academic Press, New York, 1980, 313.
9. **Mauel, J. and Behin, R.**, Cell mediated and humoral immunity to protozoan infections, *Transplant. Rev.*, 19, 121, 1974.
10. **Zuckerman, A.**, Current status of the immunology of blood and tissue protozoa. I. Leishmania, *Exp. Parasitol.*, 38, 370, 1975.
11. **Mauel, J. and Behin, R.**, Immunology of leishmaniasis, in *Biochemistry and Physiology of Protozoa*, Vol. III, Levandowsky, M. and Hutner, S. H., Eds., Academic Press, New York, 1980.
12. **Simpson, L.**, Structure and function of kinetoplast DNA, *J. Protozool.*, 20, 2, 1973.
13. **Rogers, L.**, Preliminary note on the development of *Trypanosoma* in cultures of the Cunningham-Leishman-Donovani bodies of cachexial fever and Kalaazar, *Lancet*, 2, 215, 1904.
14. **Nicolle, C.**, Culture du parasite du bouton d'Orient, *C. R. Acad. Sci. Paris*, 146, 842, 1908.
15. **Pipkin, A. C.**, Avian embryos and tissue culture in the study of parasitic protozoa. II. Protozoa other than *Plasmodium*, *Exp. Parasitol.*, 9, 167, 1960.
16. **Zuckerman, A.**, Propagation of parasitic protozoa in tissue culture and avian embryos, *Ann. N.Y. Acad. Sci.*, 139, 24, 1966.
17. **Lamy, L. H.**, Protozoaires intracellulaires en culture cellulaire: Interest, possibilities, limits, *Ann. Biol.*, 11, 145, 1972.
18. **Gavrilov, W. and Laurencin, S.**, Application d'une méthode de culture de tissus dans l'étude des protozoaires, *Ann. Soc. Belge Méd. Trop.*, 18, 41, 1938.
19. **Pai, H. C. and Hu, C. H.**, Attempts to grow *Leishmania donovani* in tissue cultures, *Proc. Soc. Exp. Biol. Med.*, 46, 606, 1941.
20. **Hawking, F.**, Growth of protozoa in tissue culture. V. *Leishmania donovani*, *Trans. R. Soc. Trop. Med. Hyg.*, 41, 545, 1948.
21. **Pipkin, A. C. and Coles, W. C.**, Growth of *Leishmania brasiliensis* in tissue cultures of rodent cells, *J. Protozool.*, 7, 24, 1960.
22. **Tchernomoretz, I.**, The tissue culture of L.D. bodies of *Leishmania donovani* and *Leishmania infantum* from flagellates, *Harefuah*, 30, 87, 1946.
23. **Pulvertaft, R. J. V. and Hoyle, G. F.**, Stages in the life cycle of *Leishmania donovani*, *Trans. R. Soc. Trop. Med. Hyg.*, 54, 191, 1960.
24. **Yen, A. C. H. and Chung, H. L.**, Cultivation of *Leishmania donovani* in media of embryonic chick tissues, *Proc. Soc. Exp. Biol. Med.*, 31, 1258, 1934.
25. **Degtiavera, S. M. and Zasukhin, D. M.**, Cultivation of a causative agent of cutaneous leishmaniasis of the desert type in tissue culture, *Med. Parasitol. Moskva*, 28, 706, 1959.
26. **Weinman, D.**, Factors affecting the morphology of *Leishmania tropica* — the production of *Leishmania* forms in culture, *Parasitology*, 31, 185, 1939.
27. **Miller, H. C. and Twohy, D. W.**, Infection of macrophages in culture by leptomonads of *Leishmania donovani*, *J. Protozool.*, 14, 781, 1967.
28. **Akiyama, H. J. and Taylor, J. C.**, Effect of macrophage engulfment and temperature on the transformation process of *Leishmania donovani*, *Am. J. Trop. Med. Hyg.*, 19, 747, 1970.
29. **Akiyama, H. J. and Haight, R. D.**, Interaction of *Leishmania donovani* and hamster peritoneal macrophages. A phase-contrast microscopical study, *Am. J. Trop. Med. Hyg.*, 20, 539, 1971.
30. **Akiyama, H. J. and McQuillen, N. K.**, Interaction and transformation of *Leishmania donovani in vitro* cultured cells. An electron microscopical study, *Am. J. Trop. Med. Hyg.*, 21, 873, 1972.
31. **Bhattacharya, A. and Janovy, J., Jr.**, *Leishmania donovani*: autoradiographic evidence for molecular exchanges between parasites and host cells, *Exp. Parasitol.*, 37, 353, 1975.
32. **Herman, R.**, Fluorescent antibody studies on the intracellular form of *Leishmania donovani* grown in cell culture, *Exp. Parasitol.*, 17, 218, 1965.
33. **Herman, R.**, Studies of the numbers and morphology of the intracellular form of *Leishmania donovani* grown in cell culture, *J. Protozool.*, 13, 408, 1966.
34. **Chang, K.-P.**, Hamster peritoneal macrophages *in vitro:* substratum adhesion, spreading phagocytosis and phagolysosome formation, *In Vitro*, 14, 663, 1978.
35. **Chang, K.-P.**, *Leishmania donovani:* Promastigote-macrophage surface interactions *in vitro*, *Exp. Parasitol.*, 48, 175, 1979.
36. **Chang, K.-P. and Dwyer, D. M.**, Multiplication of a human parasite *(Leishmania donovani)* in phagolysosomes of hamster macrophages *in vitro*, *Science*, 193, 678, 1976.
37. **Chang, K.-P. and Dwyer, D. M.**, *Leishmania donovani* — hamster macrophage interactions *in vitro:* cell entry, intracellular survival and multiplication, *J. Exp. Med.*, 147, 515, 1978.
38. **Chang, K.-P.**, Intracellular multiplication of *Leishmania donovani* during repeated passages in primary cultures of hamster peritoneal macrophages, *J. Parasitol.*, 64, 931, 1978.
39. **Chang, K.-P.**, Survival and growth of *Leishmania donovani* in human and murine mononuclear phagocytes, *J. Parasitol.*, 63, 43, 1977.

40. **Miller, H. C. and Twohy, D. W.,** Cellular immunity to *Leishmania donovani* in macrophages in culture, *J. Parasitol.*, 55, 200, 1969.
41. **Skov, C. B. and Twohy, D. W.,** Cellular immunity to *Leishmania donovani*. II. Evidence for synergy between thymocytes and lymph node cells in reconstitutions of acquired resistance to *L. donovani* in mice, *J. Immunol.*, 113, 2012, 1974.
42. **Herman, R. and Farrell, J. F.,** Effects of cyclophosphamide on visceral leishmaniasis in the mouse, *J. Protozool.*, 24, 429, 1977.
43. **Lamy, L., Wonde, T., and Lamy, H.,** Comportement de *Leishmania donovani* en présence de macrophages de souris entretenus *in vitro, C. R. Acad. Sci. Ser. D.*, 263, 671, 1966.
44. **Lamy, L., Wonde, T., and Lamy, H.,** Activité de l'amphotéricine B sur *Leishmania donovani* en multiplication dans des macrophages de souris entretenus *in vitro, Bull. Soc. Pathol. Exot.*, 59, 964, 1966.
45. **Lamy, L., Wonde, T., and Lamy-Roux, L.,** Influence de la température sur la formation réciproque des formes flagellées et aflagellées de Leishmania en présence de macrophages péritoneaux de souris *in vitro, C. R. Acad. Sci. Sec. D.*, 264, 1889, 1967.
46. **Lamy, L.,** Cycle biologique complet de Leishmania *in vitro* et étude de divers comportements, *Protistologica*, 3, 359, 1967.
47. **Handman, E., Ceredig, R., and Mitchell, G. F.,** Murine cutaneous leishmaniasis: disease patterns in intact and nude mice of various genotypes and examination of some differences between normal and infected macrophages, *Aust. J. Exp. Biol. Med. Sci.*, 57, 9, 1979.
48. **Handman, E. and Spira, D. T.,** Growth of *Leishmania* amastigotes in macrophages from normal and immune mice, *Z. Parasitenk.*, 53, 75, 1977.
49. **Handman, E. and Burgess, A. W.,** Stimulation by granulocyte-macrophage colony-stimulating factor of *Leishmania tropica* killing by macrophages, *J. Immunol.*, 122, 1134, 1979.
50. **Handman, E. and Greenblatt, C. L.,** Promotion of leishmania infections in non-permissive host macrophages by conditioned medium, *Z. Parasitenk.*, 53, 143, 1977.
51. **Behin, R., Mauel, J., Biroum-Noerjasin, and Rowe, D. S.,** Mechanisms of protective immunity in experimental cutaneous leishmaniasis of the guinea pig. II. Selective destruction of different leishmania species in activated guinea-pig and mouse macrophages, *Clin. Exp. Immunol.*, 20, 351, 1975.
52. **Mauel, J., Buchmüller, Y., and Behin, R.,** Studies on the mechanism of macrophage activation. I. Destruction of intracellular *Leishmania enriettii* in macrophages activated by cocultivation with stimulated lymphocytes, *J. Exp. Med.*, 148, 393, 1978.
53. **Buchmüller, Y. and Mauel, J.,** Studies on the mechanism of macrophage activation. II. Parasite destruction in macrophages activated by supernates from concanavalin A-stimulated lymphocytes, *J. Exp. Med.*, 150, 359, 1979.
54. **Buchmüller, Y. and Mauel, J.,** Inhibitor of serine esterases enhance lymphokine-induced microbicidal activity in macrophages, *J. Reticuloendothel. Soc.*, 27, 89, 1980.
55. **Farah, F. S., Samra, S. A., and Nuwayri-Salti, N.,** The role of the macrophage in cutaneous Leishmaniasis, *Immunology*, 29, 755, 1975.
56. **Alexander, J.,** Effect of the antiphagocytic agent cytochalasin B on macrophage invasion by *Leishmania mexicana* promastigotes and *Trypanosoma cruzi* epimastigotes, *J. Protozool.*, 22, 237, 1975.
57. **Alexander, J. and Vickerman, K.,** Fusion of host cell secondary lysosomes with the parasitophorous vacuoles of *Leishmania mexicana* infected macrophages, *J. Protozool.*, 22, 502, 1975.
58. **Lewis, D. H.,** Infection of tissue culture cells of low phagocytic ability by *Leishmania mexicana, Ann. Trop. Med. Parasitol.*, 68, 327, 1974.
59. **Lewis, D. H. and Peters, W.,** The resistance of intracellular *Leishmania* parasites to digestion by lysosomal enzymes, *Am. Trop. Med. Parasitol.*, 71, 295, 1977.
60. **Ardehali, S. M. and Khoubyar, K.,** Uptake of different *Leishmania* by mouse peritoneal exudate cells, *Trans. R. Soc. Trop. Med. Hyg.*, 72, 645, 1978.
61. **Ardehali, S. M., Khoubyar, K., and Rezai, H.,** Studies on the effect of the anti-phagocytic agent cytochalasin B on leishmania-macrophage interaction, *Acta Trop.*, 36, 15, 1979.
62. **Merino, F., Ajjam, E., Hernandez, A., Dawidowicz, K., and Merino, E. J.,** *In vitro* infection of murine macrophage by *Leishmania braziliensis* mechanism of penetration, *Int. Arch. Allergy Appl. Immunol.*, 55, 487, 1977.
63. **Rahman, A. A. and Sethi, K. K.,** Intracellular behavior of *Leishmania enriettii* within murine macrophages, *Experientia*, 34, 598, 1978.
64. **Berman, J. D., Dwyer, D. M., and Wyler, D. J.,** Multiplication of *Leishmania* in human macrophages *in vitro, Infect. Immunol.*, 26, 375, 1979.
65. **Bryceson, A. D. M., Bray, R. S., Woltencroft, R. A., and Dumonde, D. C.,** Immunity in cutaneous leishmaniasis of the guinea-pig, *Clin. Exp. Immunol.*, 7, 301, 1970.
66. **Mauel, J., Behin, R., Noerjasin, B., and Rowe, D. S.,** Mechanisms of protective immunity in experimental cutaneous Leishmaniasis of the guinea-pig. I. Lack of effects of immune lymphocytes and of activated macrophages, *Clin. Exp. Immunol.*, 20, 339, 1975.

67. **Belehu, A., Poulter, H. W., and Turk, J. L.,** Modification of cutaneous leishmaniasis in the guinea-pig by cyclophosphamide, *Clin. Exp. J.*, 24, 125, 1976.
68. **Simpson, L.,** The leishmania leptomonad transformation of *Leishmania donovani* nutritional requirements, respiration changes and antigenic changes, *J. Protozool.*, 15, 201, 1968.
69. **Childs, G. E., McRoberts, M. J., and Foster, K. A.,** Partial purification of amastigotes from cutaneous lesion of American leishmaniasis, *J. Parasitol.*, 62, 676, 1976.
70. **Show, J. J. and Lainson, R.,** A simply prepared amastigote leishmanial antigen for use in the indirect fluorescent antibody test for leishmaniasis, *J. Parasitol.*, 63, 384, 1977.
71. **Brazil, R. P.,** Isolation of the intracellular stages of *Leishmania mexicana amazonensis* using cellulose column, *Am. Trop. Med. Parasitol.*, 72, 579, 1978.
72. **Infante, R. B., Hernandez, A. G., Riggione, F., and Dawidowicz, K.,** A new method for partial purification of leishmania amastigotes from cutaneous lesions, *Parasitology*, 80, 105, 1980.
73. **Stuart, A. E., Habeshaw, J. A., and Davison, A. E.,** Phagocytes, in *Handbook of Experimental Immunology*, Vol. 2, 3rd ed., Blackwell Scientific, Oxford, 1978, chap. 31.
74. **De Castro, M. P. and Pinto, S. C.,** Influence of the temperature on growth of *Leishmania enriettii* in tissue culture, *J. Protozool.*, 7, 7, 1960.
75. **Mauel, J., Behin, R., and Biroum-Noerjasin, X.,** Quantitative release of live microorganisms from infected macrophages by sodium dodecyl sulfate *Nature (London) New Biol.*, 244, 93, 1973.
76. **Belle, E. A.,** Cultivation of *Leishmania donovani* in human amnion epithelial cell tissue cultures. A preliminary report, *Can. Med. Assoc. J.*, 79, 726, 1958.
77. **Frothingham, T. G. and Lehtimaki, E.,** Leishmania in primary cultures of human amniotic cells, *Am. J. Trop. Med. Hyg.*, 16, 658, 1967.
78. **Lupasco, G., Bossie, A., Dincoulesco, M., Epurean, E., and Profeta, A.,** Contributions a l'etude de la culture et de l'action cytopathogene de *L. donovani* sur cultures de tissus, *Arch. Roum. Pathol. Exp. Microbiol.*, 27, 641, 1968.
79. **Pipkin, A. C.,** Avian embryos and tissue culture in the study of parasitic protozoa. II. Protozoa other than plasmodium, *Exp. Parasitol.*, 9, 167, 1960.
80. **Holbrook, T. and Palczuk, N.,** *Leishmania* in the chick embryo. IV. The effects of embryo age and hatching and behavior of *L. donovani* in culture of chick fibroblasts, *Exp. Parasitol.*, 37, 398, 1975.
81. **Lelijveld, J. and Atamasiu, P.,** Multiplication de *Leishmania braziliensis* sur culture cellulaire de rein de hamster, *Ann. Inst. Pasteur*, 100, 788, 1966.
82. **Chang, K.-P.,** Leishmania infection of human skin fibroblasts *in vitro:* absence of phagolysosomal fusion after induced phagocytosis of promastigotes and their intracellular transformation, *Am. J. Trop. Med. Hyg.*, 27, 1084, 1978.
83. **Gleiberman, S. G. and Belova, E. M.,** General morphology of the cytopathogenic effect of Leishmania in tissue cultures, *Med. Parazit. (Moskva)*, 33, 650, 1964.
84. **Herman, R.,** Acriflavin-induced dyskinetoplastic *Leishmania donovani* grown in monkey kidney cell culture, *J. Protozool.*, 15, 35, 1968.
85. **Dedet, J.-P. and Gaudin, O.-G.,** *Leishmania donovani* multiplication in a cell line of *Aedes albopictus*, *Trans. R. Soc. Trop. Med. Hyg.*, 70, 535, 1976.
86. **Miller, H. C. and Twohy, D. W.,** Invasion of cultured cells by leptomonads of *Leishmania donovani*, *J. Protozool.*, 13, 19, 1966.
87. **Berens, R. L. and Marr, J. J.,** Growth of *Leishmania donovani* amastigotes in a continuous macrophage-like cell culture, *J. Protozool.*, 26, 453, 1979.
88. **Chang, K.-P.,** Interactions of Leishmanias with mammalian cells *in vitro*, in *In Vitro Cultivation of Major Tropical Disease Pathogens*, Tropical Diseases Research Series No. 2, Schwabe & Co., Basel, 1980.
89. **Chang, K.-P. and Michl, J.,** *Leishmania donovani* amastigotes in a macrophage cell line *in vitro:* ultrastructural study of ingestion and multiplication phases, in press.
90. **Lamy, L., Samso, A., and Lamy, H.,** Installation, multiplication et entretien d'une souche de *Leishmania donovani* en culture cellulaire, *Bull. Soc. Pathol. Exot.*, 59, 16, 1964.
91. **Lamy, L. H.,** La transformation réciproque des formes mastigotes et amastigotes de *Leishmania* et son déterminisme en présence de cellules vivantes *in vitro*, *Am. Inst. Pasteur, Paris*, 117, 545, 1969.
92. **Lamy, L. H., Fromentin, H., and Lamy, H.,** Comparison, perte et récupération du pouvoir infectieux par des *Leishmania* en l'absence et en présence de cellules vivantes, *Protistologica*, 7, 435, 1971.
93. **Mattock, N. M. and Peters, W.,** The experimental chemotherapy of *Leishmaniasis*. I. Techniques for the study of drug action in tissue culture, *Ann. Trop. Med. Parasitol.*, 69, 349, 1975.
94. **Frothingham, T. E. and Lektimaki, E.,** Prolonged growth of *Leishmania* species in cell culture, *J. Parasitol.*, 55, 196, 1969.
95. **Chang, K.-P.,** Human cutaneous leishmania in a mouse macrophage line: propagation and isolation of intracellular parasites, submitted.
96. **Walton, B. C., Brooks, W. H., and Arjona, I.,** Serodiagnosis of American leishmaniasis by indirect fluorescent antibody test, *Am. J. Trop. Med. Hyg.*, 21, 296, 1972.

97. **Wonde, T. and Honigberg, B. M.**, Morphology and infectivity of *Leishmania donovani* cultivated in non-living media at elevated temperatures, *Am. J. Trop. Med. Hyg.*, 20, 828, 1971.
98. **Row, R.**, On the reversion of the flagellate form of *Leishmania donovani* and *Leishmania tropica, Ind. J. Med. Res.*, 10, 476, 1922.
99. **Adler, S. and Theodor, O.**, The identity of *Leishmania tropica*, Wright, 1903 and *Herpetomona papatasii*, Adler, 1925, *Am. Trop. Med. Parasitol.*, 20, 355, 1926.
100. **Trager, W.**, The development of *Leishmania donovani in vitro* at 37°C. Effects of the kind of serum, *J. Exp. Med.*, 97, 177, 1953.
101. **Lemma, A. and Schiller, E. L.**, Extracellular cultivation of the leishmanial bodies of species belonging to the protozoan genus *Leishmania, Exp. Parasitol.*, 15, 503, 1964.
102. **Gillig, C. J.**, III, Adaptation of *Leishmania* to *in vitro* cultivation at 37°C, *J. Protozool.*, 24, 406, 1977.
103. **Greenblatt, C. L. and Glaser, P.**, Temperature effect on *Leishmania enriettii in vitro, Exp. Parasitol.*, 16, 36, 1965.
104. **Molyneux, D. H., Lewis, D. H., and Killick-Kendrick, R.**, Aspects of the microecology of Leishmania, in *Ecologie des Leishmanioses*, Colloques Internationaux du C.N.R.S., No. 239, 1978, 31.
105. **Hendricks, L. D., Wood, D. E., and Hajduk, M. E.**, Hemoflagellates: commercially available liquid media for rapid cultivation, *Parasitology*, 76, 309, 1978.
106. **Hendricks, L. D. and Childs, G. E.**, Present knowledge of the *in vitro* cultivation of Leishmania, in *In Vitro Cultivation of Major Tropical Disease Pathogens*, Tropical Diseases Research Series No. 2, Schwabe & Co., Basel, 1980.
107. **Nicolle, C.**, Culture des corps de Leishman isolés de la rate dans trois cas d'anémie splénique infantile, *Bull. Soc. Path. Exot.*, 1, 121, 1908.
108. **Nicolle, C.**, Nouvelles acquisitions sur le Kala-azar: cultures, inoculation au chien, étiologie, *C. R. Acad. Sci. Paris*, 146, 498, 1908.
109. **Novy, F. G. and MacNeal, W. J.**, On the trypanosomes of birds, *J. Infect. Dis.*, 2, 256, 1905.
110. **Pedrosa, A. M.**, Notes on the biology of *Leishmania tropica, Am. J. Trop. Med.*, 3, 47, 1923.
111. **Salle, A. J. and Schmidt, C. L. A.**, The metabolism of *Leishmania tropica, J. Infect. Dis.*, 43, 378, 1928.
112. **Ray, J. C.**, Cultivation of various *Leishmania* parasites on solid medium, *Ind. J. Med. Res.*, 20, 355, 1932.
113. **Yang, F.-H.**, Ueber Züchtung der *Leishmania donovani* auf festem Nährboden, *Contrib. Inst. Physiol. Natl. Acad. Sci. Peking*, 3, 37, 1933.
114. **Senekjie, H. A.**, Studies on the culture of *Leishmania tropica, Trans. R. Soc. Trop. Med. Hyg.*, 33, 267, 1939.
115. **von Brand, T. and Johnson, E. M.**, A comparative study of the effect of cyanide on the respiration of some trypanosomatidae, *J. Cell. Comp. Physiol.*, 29, 33, 1947.
116. **Korn, E. D. and Greenblatt, C. L.**, Synthesis of α-linolenic acid by *Leishmania enriettii, Science*, 142, 1301, 1963.
117. **Childs, G. E., Foster, K. A., and McRoberts, M. J.**, Insect tissue culture media for cultivation of New World *Leishmania, Int. J. Parasitol.*, 8, 255, 1978.
118. **Chang, S. L.**, Studies on haemoflagellates. I. A semi-solid medium and a fluid medium with a solid base for growing various species of *Leishmania* and *Trypanosoma cruzi, J. Infect. Dis.*, 80, 164, 1947.
119. **Tobie, E. J., von Brand, T., and Mehlman, B.**, Cultural and physiological observations on *Trypanosoma rhodesiense* and *Trypanosoma gambiense, J. Parasitol.*, 36, 48, 1950.
120. **Janovy, J. Jr. and Poorman, A. E.**, Temperature and metabolism in *Leishmania*. I. Respiration in *L. donovani, L. mexicana*, and *L. tarentolae, Exp. Parasitol.*, 25, 276, 1969.
121. **Sorouri, P.**, The nuclear cytology of *Leishmania tropica, J. Morphol.*, 97, 393, 1955.
122. **Simon, M. W., Martin, E., and Mukkada, A. J.**, Evidence for a functional glyoxylate cycle in the leishmaniae, *J. Bacteriol.*, 135, 895, 1978.
123. **Citri, N. and Grossowicz, N.**, A partially defined culture medium for *Trypanosoma cruzi* and some other haemoflagellates, *J. Gen. Microbiol.*, 13, 273, 1955.
124. **Grant, P. T., Sargent, J. R., and Ryley, J. F.**, Respiratory systems in the Trypanosomatidae, *Biochem. J.*, 81, 200, 1961.
125. **Neal, R. A. and Miles, R. A.**, Heated blood agar medium for the growth of *Trypanosoma cruzi* and some species of *Leishmania, Nature (London)*, 198, 210, 1963.
126. **Rioux, J.-A., Lanotte, G., Dedet, J.-P., and Martini-Dumas, A.**, Utilisation du milieu "coeur-cerveau-sang de mouton" pour la culture en masse des formes promastigotes de Leishmanes, *Ann. Parasitol. Paris*, 45, 381, 1970.
127. **Rezai, H. R., Sher, S., and Gettner, S.**, *Leishmania tropica, L. donovani* and *L. enriettii:* immune rabbit serum inhibitory *in vitro, Exp. Parasitol.*, 26, 257, 1969.
128. **Kuwahara, S. S., Pinto, L. J., and Kazan, B. H.**, Occurrence of a 5,7 diene sterol in *Leishmania donovani, J. Gen. Microbiol.*, 66, 375, 1971.

129. **Schnur, L. F. and Zuckerman, A.,** Leishmania excreted factor (EF) serotypes in Sudan, Kenya and Ethiopia, *Ann. Trop. Med. Parasitol.*, 71, 273, 1977.
130. **Kilgour, V., Gardener, P. J., Godfrey, D. G., and Peters, W.,** Demonstration of electrophoretic variation of two aminotransferases in *Leishmania*, *Ann. Trop. Med. Parasitol.*, 68, 245, 1974.
131. **Chance, M. L., Schnur, L. F., Thomas, S. C., and Peters, W.,** The biochemical taxonomy of *Leishmania* from the Aethiopian zoogeographical region of Africa, *Ann. Trop. Med. Parasitol.*, 72, 533, 1978.
132. **Pampiglione, S., Manson-Bahr, P. E. C., LaPlaca, M., Borgatti, M. A., and Musumeci, S.,** Studies in Mediterranean leishmaniasis. III. The leishmanin skin test in kala-azar, *Trans. R. Soc. Trop. Med. Hyg.*, 69, 60, 1975.
133. **Walton, B. C., Shaw, J. J., and Lainson, R.,** Observations on the *in vitro* cultivation of *Leishmania braziliensis*, *J. Parasitol.*, 63, 1118, 1977.
134. **Walton, B. C., Intermill, R. W., and Hajduk, M. E.,** Differences in biological characteristics of three *Leishmania* isolates from patients with Espundia, *Am. J. Trop. Med. Hyg.*, 26, 850, 1977.
135. **Evans, D. A.,** Kinetoplastida, in *Methods of Cultivating Parasites In Vitro* A. E. R. Taylor and D. A. Evans, Eds., Academic Press, New York, 1978, 55.
136. **Croft, S. L., Schnur, L. F., and Chance, M. L.,** The morphological, biochemical and serological characterization of strains of *Leishmania hertigi* from Panama and Brazil and their differentiation, *Ann. Trop. Med. Parasitol.*, 72, 93, 1978.
137. **Mayrink, W., Williams, P., Coelho, M. V., Dias, M., Martins, A. V., Mogalhaes, P. A., DaCosta, C. A., Falcao, A. R., Melo, M. N., and Falcao, A. L.,** Epidemiology of dermal leishmaniasis in the Rio Doce Valley, State of Minas Gerais, Brazil, *Ann. Trop. Med. Parasitol.*, 73, 123, 1979.
138. **Rassam, M. B., Al-Mudhaffar, S. A., and Chance, M. L.,** Isoenzyme characterization of *Leishmania* species from Iraq, *Ann. Trop. Med. Parasitol.*, 73, 527, 1979.
139. **Heyneman, D. and Mansour, N. S.,** Leishmaniasis in the Sudan Republic. VII. Testing of various animal bloods and concentrations in cultures of *Leishmania* spp. with notes on leptomonad survival in chick embryo cultures, *J. Egypt. Pub. Hlth. Assoc.*, 37, 187, 1962.
140. **Cleveland, L. R. and Collier, J.,** The cultivation and differentiation of haemoflagellates in autoclaved media, *Am. J. Hyg.*, 12, 614, 1930.
141. **Mohan, K., Suri, J. C., and Madan, M.,** A comparative study for cultivation of *Leishmania tropica*, *Ind. J. Med. Res.*, 63, 1280, 1975.
142. **Cornwall, J. W. and LaFrenais, H. M.,** A contribution to the study of Kala-azar (I), *Ind. J. Med. Res.*, 3, 698, 1916.
143. **Cornwall, J. W. and Menon, T. K.,** A contribution to the study of Kala-azar (III), *Ind. J. Med. Res.*, 4, 672, 1917.
144. **Schmunis, G. A. and Herman, R.,** Characteristics of so-called natural antibodies in various normal sera against culture forms of *Leishmania*, *J. Parasitol.*, 56, 889, 1970.
145. **Noguchi, H. and Lindenberg, A.,** The isolation and maintenance of *Leishmania* on the medium employed for the cultivation of organisms of the leptospira group of spirochetes, *Am. J. Trop. Med.*, 5, 63, 1925.
146. **Noguchi, N. and Tilden, E. B.,** Comparative studies of herpetomonads and leishmania, *J. Exp. Med.*, 44, 307, 1926.
147. **Kligler, I. J.,** The cultural and serological relationships of *Leishmania*, *Trans. R. Soc. Trop. Med. Hyg.*, 19, 330, 1925.
148. **Adler, S. and Theodor, O.,** The identity of *Leishmania tropica*, Wright, 1903, and *Herpetomonas papatasii* 1925, *Ann. Trop. Med. Parasitol.*, 20, 355, 1926.
149. **Adler, S.,** Culture of *Leishmania* and other trypanosomatidae in haemoglobin-free media, *Trans. R. Soc. Trop. Med. Hyg.*, 28, 201, 1934.
150. **Mathis, C.,** Cultures de *Leishmania infantum* et *L. tropica*, sur milieux au sang chauffes, *C. R. Soc. Biol.*, 71, 538, 1911.
151. **Barbosa, W., Moreira de Souza, M. do C., Razzi, D. M., Gerais, B. B., and Oliviero, R. L.,** Note on the classification of the *Leishmania* sp. responsible for cutaneous leishmaniasis in the East Central region of Brazil, *Ann. Trop. Med. Parasitol.*, 70, 389, 1976.
152. **Ashford, R. W., Chance, M. L., Ebert, F., Schnur, L. F., Bushwereb, A. K., and Drebi, S. H.,** Cutaneous leishmaniasis in Libyan Arab Republic: distribution of the disease and identity of the parasite, *Ann. Trop. Med. Parasitol.*, 70, 401, 1970.
153. **Bray, R. S. and Munford, F.,** On the maintenance of strains of *Leishmania* from the Guianas, *J. Trop. Med. Hyg.*, 70, 23, 1967.
154. **Laveran, A. and Petit, A.,** Culture de la *Leishmania donovani* en milieu liquide, *C. R. Soc. Biol.*, 68, 114, 1910.
155. **Row, R.,** A simple haemoglobinized saline culture medium for the growth of *Leishmania* and allied protozoa, *Br. Med. J.*, 1119, 1912.
156. **Paktar, N. A.,** Value of Row's medium for culture of leishmania in kala-azar, *Ind. Physicians*, 8, 261, 1949.

157. **Jadin, J. B. and Le Ray, D.,** Les milieux de cultures des Leishmanies, in *Ecologie des Leishmaniases,* Colloques Internationaux du C.N.R.S., No. 239, Editions du C.N.R.S., Paris, 1977, 21.
158. **Castellani, O., Ribeiro, L. V., and Fernandes, J. F.,** Differentiation of *Trypanosoma cruzi* in culture, *J. Protozool.,* 14, 447, 1967.
159. **Camargo, E. P., Coelho, J. A., Moraes, G., and Figueiredo, E. N.,** *Trypanosoma* spp. *Leishmania* spp. and *Leptomonas* spp.: enzymes of ornithine-arginine metabolism, *Exp. Parasitol.,* 46, 141, 1978.
160. **Al-Taqi, M. and Evans, D. A.,** Characterization of *Leishmania* spp. from Kuwait by isoenzyme electrophoresis, *Trans. R. Soc. Trop. Med. Hyg.,* 72, 56, 1978.
161. **Schaefer, F. W., III, Bell, E. J., and Etges, F. J.,** *Leishmania tropica:* chemostatic cultivation, *Exp. Parasitol.,* 28, 465, 1970.
162. **Martin, E., Simon, M. W., Schaefer, F. W., III, and Mukkada, A. J.,** Enzymes of carbohydrate metabolism in four human species of *Leishmania:* a comparative survey, *J. Protozool.,* 23, 600, 1976.
163. **Mansour, N. S., Hady, J., and McConnell, E.,** A modified liquid medium for *Leishmania, J. Parasitol.,* 59, 1088, 1973.
164. **Fulton, J. D. and Joyner, L. D.,** Studies on protozoa. I. The metabolism of Leishman-Donovan bodies and flagellates of *Leishmania donovani, Trans. R. Soc. Trop. Med. Hyg.,* 43, 273, 1949.
165. **Dwyer, D. M.,** A monophasic medium for cultivating *Leishmania donovani* in large numbers, *J. Parasitol.,* 58, 847, 1972.
166. **Simpson, L. and Berliner, J.,** Isolation of the kinetoplast DNA in the form of a network, *J. Protozool.,* 21, 382, 1974.
167. **Leon, W., Brun, R., and Krassner, S. M.,** Effect of Berenil on growth mitochondrial DNA and respiration of *Leishmania tarentolae* promastigotes, *J. Protozool.,* 24, 444, 1977.
168. **Senekjie, H. A.,** Biochemical reactions of *Leishmania tropica, Am. J. Trop. Med.,* 19, 601, 1939.
169. **Gaughan, P. L. Z. and Krassner, S. M.,** Hemin deprivation in culture stages of the hemoflagellate, *Leishmania tarentolae, Comp. Biochem. Physiol.,* 39B, 5, 1971.
170. **Schnur, L. F., Zuckerman, A., and Greenblatt, C. L.,** Leishmanial serotypes, as distinguished by the gel diffusion of factors excreted *in vitro* and *in vivo, Israel J. Med. Sci.,* 8, 932, 1972.
171. **Enders, B., Brauns, F., and Zwisler, O.,** Biochemical and technical considerations regarding the mass production of certain parasitic protozoa, *Bull. W. H. O.,* 55, 393, 1977.
172. **Citri, N. and Grossowicz, N.,** Growth requirements of *Leishmania tropica* and other leishmanias, *Trans. R. Soc. Trop. Med. Hyg.,* 49, 603, 1955.
173. **Citri, N. and Grossowicz, N.,** A liquid medium for the cultivation of *Trypanosoma cruzi, Nature (London),* 173, 1100, 1954.
174. **Merino, F. and Gabaldon, I.,** *In vitro* culture of *Leishmania brasiliensis* in a semi-defined medium, *Acta Cient. Venezolana,* 29, 181, 1978.
175. **Semprevivo, L. H.,** Exometabolites of *Leishmania donovani* promastigotes. I. Isolation and initial characterization, *Proc. Soc. Exp. Biol. Med.,* 159, 105, 1978.
176. **Chang, K.-P., Steiger, R. F., Dave, C., and Cheng, Y. C.,** Effects of methylglyoxal *bis* (guanylhydrazone) on trypanosomatid flagellates: inhibition of growth and nucleoside incorporation in *Trypanosoma brucei, J. Protozool.,* 25, 145, 1978.
177. **Berens, R. L., Brun, R., and Krassner, S. M.,** A simple, monophasic medium for axenic cultivation of hemoflagellates, *J. Parasitol.,* 62, 360, 1976.
178. **Fish, W. R., Holz, G. G., Jr., and Beach, D. H.,** Cultivation of trypanosomatids, *J. Parasitol.,* 64, 546, 1978.
179. **Beach, D. H., Holz, G. G., Jr., and Anekwe, G. E.,** Lipids of *Leishmania* promastigotes, *J. Parasitol.,* 65, 203, 1979.
180. **Grace, T. D. C.,** Establishment of four strains of cells from insect tissues grown *in vitro, Nature (London),* 195, 788, 1962.
181. **Schneider, I.,** Cell lines derived from late embryonic stages of *Drosophila melanogaster, J. Embryl. Exp. Morph.,* 27, 353, 1972.
182. **Mitsuhashi, J. and Maramorosch, K.,** Leafhopper tissue culture: embryonic, nymphal and imaginal tissues from aseptic insects, *Contrib. Boyce Thompson Inst.,* 22, 435, 1964.
183. **Morgan, J. F., Mortan, H., and Parker, R. C.,** Nutrition of animal cells in tissue culture. I. Initial studies on a synthetic medium, *Proc. Soc. Exp. Biol. Med.,* 73, 1, 1950.
184. **Hendricks, L. and Wright, N.,** Diagnosis of cutaneous leishmaniasis by *in vitro* cultivation of saline aspirates in Schneider's drosophila medium, *Am. J. Trop. Med. Hyg.,* 28, 962, 1979.
185. **Dwyer, D. M.,** Personal communication, 1979.
186. **Trager, W.,** Nutrition of a hemoflagellate *(Leishmania tarentolae)* having an interchangeable requirement for choline or pyridoxal, *J. Protozool.,* 4, 269, 1957.
187. **Steiger, R. F. and Steiger, E.,** A defined medium for cultivating *Leishmania donovani* and *L. braziliensis, J. Protozool.,* 62, 1010, 1976.

188. **Steiger, R. F. and Steiger, E.,** Cultivation of *Leishmania donovani* and *L. braziliensis* in defined medium: nutritional requirement, *J. Parasitol.*, 64, 437, 1976.
189. **Berens, R. L. and Marr, J. J.,** An easily prepared defined medium for cultivation of *Leishmania donovani* promastigotes, *J. Parasitol.*, 64, 160, 1978.
190. **Lwoff, M.,** The nutrition of parasitic flagellates (Trypanosomatidae, Trichomonadinae), in *Biochemistry and Physiology of Protozoa*, Lwoff, A., Ed., Academic Press, New York, 1951.
191. **Zeledon, R.,** Cultivation and transformation of hemoflagellates. A review, *Rev. Biol. Trop.*, 19, 197, 1971.
192. **Trager, W.,** Cultivation and nutritional requirements, in *Infectious Blood Diseases of Man and Animals*, Vol. 1, Weinman, D. and Ristic, M., Eds., Academic Press, New York, 1968, chap. 8.
193. **Trager, W.,** Nutrition and biosynthetic capabilities of flagellates: problems of *in vitro* cultivation and differentiation, in *Trypanosomiasis and Leishmaniasis with Special Reference to Chagas' Disease*, CIBA Foundation Symposium 20 (New Series), Elsevier-Excerpta-North Holland, Amsterdam, 1974, 225.
194. **Hutner, S. H., Bacchi, C. J., and Baker, H.,** Nutrition of the kinetoplastida, in *Biology of the Kinetoplastida*, Vol. 2, Lumsden, W. H. R. and Evans, D. A., Eds., 1979.
195. **Chang, K.-P. and Trager, W.,** Nutritional significance of symbiotic bacteria in two species of hemoflagellates, *Science*, 183, 531, 1974.
196. **Steiger, R. F. and Meshnick, S. R.,** Amino-acid and glucose utilization of *Leishmania donovani* and *L. braziliensis*, *Trans. R. Soc. Trop. Med. Hyg.*, 71, 441, 1977.
197. **Chang, K.-P., Chang, C. S., and Sassa, S.,** Heme biosynthesis in bacterium-protozoon symbioses: enzymic defects in host hemoflagellates and complemental role of their intracellular symbiotes, *Proc. Natl. Acad. Sci. USA*, 72, 2979, 1975.
198. **Shaprio, A., Hutner, S. H., Katz, L., Bacchi, C. J., Tamburro, K. O., and Baker, H.,** Dense *Crithidia* growth and heme sparing: relation to Fe, Cu, Mo chelation, *J. Protozool.*, 25, 530, 1978.
199. **Good, N. E., Winget, G. D., Winter, W., Connolly, T. N., Izawa, S., and Singh, R. M. M.,** Hydrogen ion buffers for biological research, *Biochemistry*, 5, 467, 1966.
200. **Chance, M. L., Gardener, P. J., and Peters, W.,** Biochemical taxonomy of Leishmania as an ecological tool, in *Ecologie des Leishmanioses*, Colloques Internationaux du C.N.R.S., No. 239, 1978, 53.
201. **Killick-Kendrick, R.,** Biology of Leishmania in phlebotomine sandflies, in *Biology of the Kinetoplastida*, Lumsden, W. H. R. and Evans, D. A., Eds., Academic Press, London, 1979, chap. 8.
202. **Giannini, M. S.,** Effects of promastigote growth phase, frequency of subculture and host age on promastigote initiated infections with *Leishmania donovani* in the golden hamster, *J. Protozool.*, 41, 521, 1974.
203. **Keithly, J.,** Infectivity of *Leishmania donovani* amastigotes and promastigotes for golden hamsters, *J. Protozool.*, 23, 244, 1976.
204. **Dwyer, D. M.,** *Leishmania donovani:* surface membrane carbohydrates of promastigotes, *Exp. Parasitol.*, 41, 341, 1977.
205. **Brun, R. and Krassner, S. M.,** Quantitative ultrastructural investigations of mitochondrial development in *Leishmania donovani* during transformation, *J. Protozool.*, 23, 493, 1976.
206. **Walter, R. D. and Ebert, F.,** Effect of cyclic AMP on transformation and proliferation of leishmania cells, *Tropenmed. Parasitol.*, 29, 439, 1978.
207. **Brun, R., Berens, R. L., and Krassner, S. M.,** Inhibition of *Leishmania donovani* transformation by hamster spleen homogenates and active human lymphocytes, *Nature (London)*, 262, 689, 1976.
208. **Gorczynski, R. M., MacRae, S., Kuba, R., and Price, G. B.,** Macrophage heterogeneity and Ir-gene control as factors involved in the immune response of guinea pigs to infection with *Leishmania enrietti*, *Cell. Immunol.*, 60, 367, 1981.
209. **Alexander, J.,** Interaction of *Leishmania mexicana mexicana* with mouse macrophages *in vitro*, in *Heterogeneity of Mononuclear Phagocytes*, Forster, O. and Landy, M., Eds., Academic Press, New York, 1981, 447.
210. **Nacy, C. A. and Diggs, C. L.,** Intracellular replication of *Leishmania tropica* in mouse peritoneal macrophages: comparison of amastigote replication in adherent and non-adherent macrophages, *Infect. Immun.*, 34, 310, 1981.
211. **Nacy, C. A., Meltzer, M. S., Leonard, E. J., and Wyler, D. J.,** Intracellular replication and lymphokine-induced destruction of *Leishmania tropica* in C3H/HeN mouse macrophages, *J. Immunol.*, 127, 2381, 1981.
212. **Herman, R.,** Cytophilic and opsonic antibodies in visceral leishmaniasis in mice, *Infect. Immun.*, 28, 585, 1980.
213. **Haidaris, C. G. and Bonventre, P. F.,** Elimination of *Leishmania donovani* amastigotes by activated macrophages, *Infect. Immun.*, 33, 918, 1981.
214. **Chang, K.-P., and Chiao, J. W.,** Cellular immunity of mice to *Leishmania donovani in vitro:* lymphokine-mediated killing of intracellular parasites in macrophages, *Proc. Natl. Acad. Sci. USA*, 78, 7083, 1981.
215. **Berman, J. D. and Wyler, D. J.,** An *in vitro* model for investigation of chemotherapeutic agents in leishmaniasis, *J. Infect. Dis.*, 142, 83, 1980.

216. **Berman, J. D. and Dwyer, D. M.,** Expression of Leishmania antigen on the surface membrane of infected human macrophages *in vitro*, *Clin. Exp. Immunol.*, 44, 342, 1981.
217. **Berman, J. D. and Neva, F. A.,** Effect of temperature on multiplication of Leishmania amastigotes within human monocyte-derived macrophages *in vitro*, *Am. J. Trop. Med. Hyg.*, 30, 318, 1981.
218. **Pearson, R. D., Romito, R., Symes, P. H., and Harcus, J. R.,** Interaction of *Leishmania donovani* promastigotes with human monocyte-derived macrophages: Parasite entry, intracellular survival and multiplication, *Infect. Immun.*, 32, 1249, 1981.
219. **Chang, K.-P.,** Human cutaneous leishmania in a mouse macrophage line: propagation and isolation of intracellular parasites, *Science*, 209, 1240, 1980.
220. **Fong, D. and Chang, K.-P.,** Tubulin biosynthesis in the developmental cycle of a parasitic protozoan, *Leishmania mexicana:* changes during differentiation of motile and non-motile stages, *Proc. Natl. Acad. Sci. USA*, 78, 7624, 1981.
221. **Carson, D. A. and Chang, K.-P.,** Phosphorylation and anti-leishmanial activity of formycin B, *Biochem. Biophys. Res. Comm.*, 100, 1377, 1981.
222. **Hart, D. T., Vickerman, K., and Coombs, G. H.,** A quick, simple method for purifying *Leishmania mexicana* amastigotes in large numbers, *Parasitology*, 82, 345, 1981.
223. **Hockmeyer, W. T., Kager, P. A., Rees, P. H., and Hendricks, L. D.,** The culture of *Leishmania donovani* in Schneider's insect medium: its value in the diagnosis and management of patients with visceral leishmaniasis, *Trans. R. Soc. Trop. Med. Hyg.*, 75, 861, 1981.
224. **Childs, G. E., McRoberts, M. J., and Moussa, M. A.,** Systems for the *in vitro* large-scale propagation of New World Leishmania, *Ann. Trop. Med. Parasitol.*, 73, 395, 1979.
225. **Shaw, J. J. and Lainson, R.,** The *in vitro* cultivation of members of the *Leishmania braziliensis* complex, *Trans. R. Soc. Trop. Med. Hyg.*, 75, 127, 1981.
226. **Keppel, A. D. and Janovy, J., Jr.,** Morphology of *Leishmania donovani* colonies grown on blood agar plates, *J. Parasitol.*, 66, 849, 1981.
227. **Steiger, R. F. and Black, C. D. V.,** Simplified defined media for cultivating *Leishmania donovani* promastigotes, *Acta Trop.*, 37, 195, 1980.

Chapter 4

PLASMODIUM

James B. Jensen

TABLE OF CONTENTS

I. INTRODUCTION

Malaria organisms are protozoan parasites of the subphylum Apicomplexa belonging to the genus *Plasmodium*, of the suborder Haemospororina, which is further differentiated from the Eimeriorina (coccidia) and the Adeleorina (haemogregarines) by the fact that the microgamont produces a moderate number (usually eight) of microgametes. *Plasmodium* share the suborder Haemospororina with two other genera, *Haemoproteus* and *Leucocytozoon*, neither of which will be discussed in this chapter. The plasmodia are rather closely related to parasites of the order Piroplasmorida of which parasites from the genera *Babesia* and *Theileria* will be discussed in other chapters of this book. All species of *Plasmodium* are heteroxenous having asexual reproduction (schizogony) in vertebrate hosts and sexual (sporogony) in the invertebrate mosquito hosts. Gametogony, the production of gametes, actually bridges the two hosts, beginning in the vertebrate and ending in the mosquito midgut. This chapter will concern itself only with those species of *Plasmodium* that have played a role in the studies of in vitro cultivation or that have, to date, been successfully cultivated. Malaria is not a disease of significant importance in animal husbandry except where poultry may be exposed to endemic infections in wild birds. Malaria in man, however, is of great medical importance and thus our discussion will cover principally the human species.

Malaria, until recent times, has been the greatest killer of the human race, claiming half the total mortality of mankind.[1] The tertiary malignant form, falciparum malaria, continues to rank chief among the infectious diseases with an estimated world incidence of over 300 million cases per annum. Africa alone accounts for nearly half of these cases with an annual mortality of 1 million, principally among children under 14 years of age.[2] Of all the parasitic diseases, malaria is not only the greatest killer but is also the most widely spread. Approximately 2,015 million people live in areas where malaria has either been eradicated or where control measures are being attempted. Some 343 million people live in holoendemic areas where no specific programs of protection are being developed. The majority of these people live in Africa. Alarming as these figures are, the most disturbing aspect is the major resurgence of malaria occurring in areas of the world where control of the disease was nearly successful, namely in Sri Lanka, India, Pakistan, Central and South America. This resurgence of malaria, and the failure to consolidate the gains made against malaria in control programs is due, in part, to the emergence of strains of anophelene mosquito vectors that are increasingly resistant to insecticides, chiefly the chlorinated hydrocarbons, but also the newly developed organophosphates. Another disturbing development is the spread of strains of *Plasmodium falciparum* that are resistant to antimalarial drugs, principally the 4-aminoquinolines — chloroquine and amodiaquine.[3] In view of this disheartening epidemiological situation, new measures for the control of malaria are obviously needed. Chief among the new measures is the renewed effort for the development of an antimalarial vaccine. For decades, the hope for a malaria vaccine was frustrated by the failure to grow malaria parasites in culture. This obstacle was finally overcome by the work of Trager and Jensen[4] who succeeded in growing a Southeast Asian strain of *P. falciparum* in continuous culture.

Human malaria is believed to have arisen thousands of years ago in Africa as a zoonosis derived from the higher primates, the gorillas and chimpanzees, which still harbor parasites very similar to those of man today. The four species infecting man, *P. ovale, P. vivax, P. malariae,* and *P. falciparum,* vary in morphology, distribution, and pathology. The first three listed above produce clinical disease of greater or lesser severity, and, although quite debilitating, infections are only occasionally fatal. The latter species, however, ranks as one of man's greatest killers. Its importance in human history can scarcely be underestimated. Human diseases are at times measured by two general criteria, morbidity and mortality — the former measures the incidence of a disease, and the latter the index of death produced by the malady. A disease may have high morbidity but low mortality, such as the common cold, or may have low morbidity but high mortality, like human rabies, which is generally accepted to be 100%. In the case of falciparum malaria the high morbidity coupled with the high mortality (estimated to be 5 to 10% in untreated cases) gives this disease its grimly high rank among infectious diseases.

All four human species share essentially the same general life cycle. Natural infections in man are initiated by the bite of an infected female anophelene mosquito. As the insect begins to feed she injects an anticoagulant salivary secretion, containing the thread-like sporozoite into the wound. The next steps are unknown, but within 40 min the sporozoites appear in the hepatic parenchymal cells where they undergo the pre-erythrocytic asexual multiplication known as tissue-stage schizogony. In avian plasmodia tissue schizogony occurs in reticuloendothelial cells and gives rise to merozoites, some of which are infectious to erythrocytes, while others continue the exoerythrocytic cycle. Thus in birds this cycle is capable of indefinite propagation and accordingly lends itself to methods of continuous in vitro cultivation. In *P. falciparum* and *P. malariae,* the agents of nonrelapsing malaria, there is only one episode of pre-erythrocytic schizogony resulting in red cell-infective merozoites that initiate the clinical blood infection of malaria. In the relapsing species, *P. vivax* and *P. ovale,* there appears to be two types of sporozoites — one which initiates the first clinical episode of malaria by developing through the pre-erythrocytic schizogony within 8 to 9 days following the introduction of the sporozoites, and a second type, termed a hypnozoite,[5] becomes dormant and upon reactivation initiates another clinical episode or "relapse" of the disease.[6] The blood-stage cycle of all four human malaria parasites is essentially the same, but there are some differences that may influence the cultivability of the different species. *Plasmodium vivax* and *P. ovale* have a marked predilection for reticulocytes and young erythrocytes, whereas *P. malariae* invades chiefly mature cells. Only *P. falciparum* shows no preference for cell type, invading even the nucleated normoblasts often seen in severe anemia produced by the disease. The number of merozoites per erythrocytic generation also varies with the species — *P. malariae* and *P. ovale* average eight merozoites (range 4-16) per schizont while *P. vivax* and *P. falciparum* average 16 merozoites (range 8-32) per schizont. Generally then, the erythrocytic cycle begins with the attachment of the merozoite to the erythrocyte plasmalemma with subsequent endocytic events leading to complete internalization of the parasite into the erythrocyte. The merozoite undergoes organelle dedifferentiation to produce the ring stage trophozoite which begins feeding on the hemoglobin of its host-cell and grows until it nearly fills the erythrocyte. Schizogony begins with nuclear division and progresses through the development of organelles particular to the merozoite, ending with segmentation of the parasite cytoplasm into a cluster of new merozoites. In *P. falciparum, P. vivax,* and *P. ovale* the merozoite to merozoite cycle takes 48 hr, whereas *P. malariae* takes 72 hr to complete one erythrocytic cycle. Although the actual inductive mechanism for gametogony is poorly understood, some ring-stage trophozoites evolve into the sexual precursor stages, the male microgametocyte and the female macrogametocyte. These stages do not develop further in the blood of man or other vertebrates. However, when they are ingested by the proper mosquito species the macrogametocytes and microgametocytes are released in the insect midgut from their erythrocytic envelopes and mature into the macro- and microgametes, respectively. Union of the two gametes produces a motile zygote, the oökinete, initiating the sporogonic cycle. The oökinete penetrates through the epithelial cell layer of the mosquito gut to rest finally between the basement membrane of the gut wall and the epithelial cells. A cyst wall forms around the next stage, the oocyst, where a meiotic division, followed by sporogonic mitotic divisions, finally result in the production of numerous sporozoites. Rupture of the oocyst releases the sporozoites into the haemocoele of the mosquito. The sporozoites migrate to all parts of the body, but some enter the acinal cells of the salivary glands. When the infected mosquito begins to feed, the sporozoites enter the salivary duct and are subsequently introduced into the vertebrate host blood stream. Specific details of schizogony, gametogony and sporogony that are relevant to the in vitro cultivation of the respective stages, will be discussed in the appropriate sections below.

II. HISTORICAL REVIEW OF CULTIVATION

A. Cultures of Erythrocytic Stages

In 1912 Bass and Johns[7] were the first to report in vitro development of malaria parasites. Defibrinated blood, obtained from patients infected with *P. falciparum* and *P. vivax*, was cultured at 37°C in glass vials to which a small amount of glucose had been added. They observed newly developed rings after one generation and occasionally after one or two additional cycles. Following a period of optimism among malariologists the techniques of Bass and Johns were largely abandoned due to the fact that little or no reinvasion of erythrocytes occurred after the maturation of the parasites derived from the infected patient. It should be noted, however, that even this brief period of in vitro development was useful for studies on chemotherapeutic agents[8] and is presently used as the basis for the in vitro test for drug resistance in malaria.[9,10]

After an intervening period of 20 to 25 years attempts at cultivation were renewed with several species of avian and primate malarias. The best results were short-term cultures using *P. lophurae* in chicken or duck erythrocytes[11,12] and *P. knowlesi* in rhesus monkey erythrocytes.[13,14] The medium initially used in these short-term cultures of *P. knowlesi*, known as the Harvard growth medium,[13,15] was based on a chemical analysis of monkey plasma. Subsequently, improved growth and multiplication were obtained with various modifications of the Harvard medium,[16-19] but not withstanding such improvements, this medium and its various modifications would not support more than a few erythrocytic cycles in vitro. In these early experiments, blood from an infected animal was mixed with uninfected erythrocytes of the appropriate host (duckling blood for *P. lophurea*, rhesus blood for *P. knowlesi*) to give an initial parasitemia of about 1%. The washed cells, suspended in the Harvard medium, or one of the modifications of it, was placed in a culture vessel having considerable surface area (such as 50-mℓ Erlenmeyer flask) and gently agitated on a rocking platform, usually at 16 cycles per minute, to maintain the cells in suspension. These rocker-dilution techniques, as they were called, give consistent results — some would say consistently discouraging results, but the consistency was sufficient to stimulate further studies.Usually, the parasites derived from the experimental animal infections would mature through the first cycle and often would reinvade into new erythrocytes, with little or no increase in parasite numbers. At best, the parasites would develop for a few days but with decreasing numbers, after an initial increase, until the cultures no longer contained viable parasites.[20,21]

These earliest in vitro experiments, with both *P. lophurae* and *P. knowlesi*, began to reveal specific nutritional requirements such as the requirement by *P. lophurae* for calcium-pantothenate[22] and by *P. knowlesi* for para-aminobenzoic acid(PABA).[23] This finding of PABA requirement for malaria parasites was predicted by the studies of Maier and Riley[24] where the antimalarial effect of sulfonamides against *P. lophurae* in ducklings was antagonized by PABA. This work has had far-reaching ramifications in the use of long-acting sulfa drugs as antimalarials and in the present understanding of the folate metabolism of malaria parasites.[25] Other specific requirements demonstrated by short-term cultures include biotin, methionine, and purines (see Trager, 1977).[26] Finally, short-term cultures of *P. knowlesi* and *P. lophurae*, as well as similar cultures of *P. gallinaceum* of chickens, *P. coatneyi* of rhesus monkeys, and even *P. falciparum* of man, have been used to study certain aspects of metabolism of the parasite[27] and the mode of action of antimalarial drugs.[28-30]

Numerous attempts to prolong cultivation times were made, chiefly based on the presumption that cultured erythrocytes experienced metabolic lesions and rapidly became unsuitable as a host-cell for the parasites. Thus *P. lophurae* was maintained in culture for 8 days by the frequent addition of fresh erythrocytes with initial increases of 20-fold. However, the rate of parasite multiplication could not keep ahead of the rate of erythrocyte dilution and the culture died out.[22] Similar results were obtained by Geiman and his colleagues[14] with *P. knowlesi*, having been kept for 6 continuous days. Trigg and Shakespeare[31] later demonstrated the progressively less multiplication with each successive cycle of development in vitro. It would be redundant to mention all of the numerous experiments that were ultimately unsuccessful. It should be sufficient to say that from the early 1940s to the early 1970s little progress was made toward continuous cultivation.

Several events can now be seen clearly as milestones along the way to the successful continuous cultivation of *P. falciparum*. The first was the demonstration by Young et al.[32] for *P. vivax*, and Geiman and Meagher[33] for *P. falciparum*, that the new world monkey, *Aotus trivirgatus* could serve as an experimental host for human malaria parasites. Although these monkeys have never been in plentiful supply, they were, nonetheless, far better than splenectimized chimpanzees as laboratory sources for malaria parasites. This discovery made possible experimentation on human plasmodia with the exploitation of newly developed culture media and techniques. Since it seemed evident that the rocker-dilution methods, long in vogue, had failed to lead to continuous cultures, a new approach involving perfusion techniques should be tried. Several early perfusion methods proved to be quite cumbersome, without giving significantly better results (see Bertogna et al.).[34] However, Trager[35] reasoned that, at least for such species of *P. coatneyi* and *P. falciparum* which spend much of their 48-hr developmental cycle sequestered in deep tissues, the constant agitation of the rocker-dilution methods might be deleterious to the developing parasites, and hence a settled layer of infected erythrocytes, constantly fed by a slow flow of medium, might more closely approximate the in vivo environment. Early experiments with the so-called flow vial were only mildly encouraging. However, when a new medium, RPMI 1640, originally developed by George Moore and his associates for culture of human leukocytes,[36] was found to be significantly superior to the Harvard medium in rocker-dilution cultures,[37] the stage was set for the combination of factors that brought about the first continuous cultures of any malaria parasite.[4]

Plasmodium falciparum, strain FVO, originally isolated from a patient from Vietnam to Oak Noll, California, provided by the kindness of Dr. W. A. Siddiqui, was isolated from an experimental infection in an *Aotus* monkey, washed free of monkey plasma and leukocytes, and mixed with washed human type AB + erythrocytes collected in ACD blood preservative. The original monkey blood, with a parasitemia of 15%, was diluted to 0.32% with washed human erythrocytes. The mixture of infected monkey and uninfected human cells, further diluted with medium RPMI 1640 containing 15% human AB + serum, was placed in the bottom of a flow-vial, previously developed by Dr. Trager[35] (Figure 1). Fresh culture medium was introduced by peristaltic pump as a continuous flow of 2 mℓ/hr. A humidified gas mixture of 7% CO_2, 5% O_2 and 88% N_2 delivered through silicone rubber tubing was passed over the cell layer and the entire apparatus was set in a 37°C incubator. Every fourth day 2 to 3 mℓ of freshly washed human AB + erythrocytes, suspended at 45% in RPMI 1640

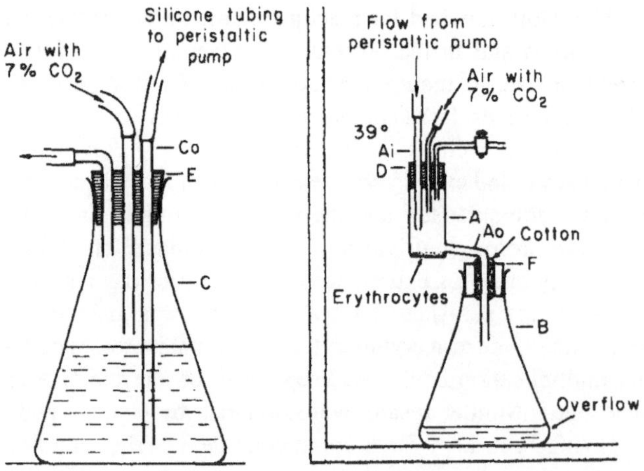

FIGURE 1. Diagram of flow-vial used for the first continuous culture of *P. falciparum*. The apparatus consists of a glass vial (A) filled with a silicone-rubber stopper (D) containing the medium delivery tube (Ai) and a gas inflow tube. Located 2 to 3 mm from the bottom is the medium outflow tube (Ao) which passes through a cotton bounded hole in silicone stopper (F) fitted into a 125 mℓ Erlenmeyer overflow flask (B). During operation a small settled layer of infected erythrocytes in the bottom of the vial (A) is continuous supplied with culture medium delivered by peristaltic pump from the medium reservoir (C) through the medium delivery tube (Ao). Excess medium flows out of flow vial into overflow flask B. A specific gas mixture, analyzed to be 2% CO_2, 10% O_2 and 88% N_2, flows through the flow vial and out the cotton plug in stopper F. (From Trager, W., *J. Protozool.*, 18, 239, 1971. With permission.)

and 15% human AB+ serum, were added to the contents of the flow vial, mixed well, and then the same volume of cells removed. This dilution method was effectively a subculture every 4 days. Results of the first 48 days of this procedure are found in Table 1. Initially the rate of parasite growth over 4 days was just sufficient to keep ahead of the rate of dilution so that by day 24 in culture the parasite count was 0.59% (as compared to 0.32% on the first day). However, when the dilution rate was calculated this figure actually represented a net increase of nearly 2,500× the starting material. The addition of fresh erythrocytes to the cultures every 4 days provided 1 to 2 mℓ of infected blood for further experiments. This material was initially diluted to a 12% cell suspension in the RPMI 1640-HS 15% medium, pipetted into 35 mm plastic petri dishes and placed in a CO_2 incubator with 5% CO_2 and air. Under these conditions the parasites did well for 1 to 2 cycles but eventually they died out. Because one of us (JBJ) had previously cultured coccidia *(Eimeria magna)* in petri dishes placed in a candle jar[38] this method was used with the *P. falciparum* cultures. Results were immediately satisfying with parasite growth rates equaling six to sevenfold increases per cycle.[39] The real advantage of this petri dish-candle jar method was its simplicity and expandability, thus many experimental parameters of parasite growth could be readily examined. Experiments using the candle jar method soon led to improvements in the flow vial so that eventually the rate of parasite multiplication in both systems was comparable. These initial *P. falciparum* cultures produced parasites that were normal in every aspect (Figure 2) except that they did not show the synchronous development of natural malaria infections.

B. Extracellular Culture of Erythrocytic Stages

The avian malaria parasite *P. lophurae* is the only obligate intracellular protozoan that has been kept alive and developing extracellularly in vitro.[40-42] This parasite was originally

Table 1
DATA ON THE INITIAL DEVELOPMENT OF *PLASMODIUM FALCIPARUM* DURING THE FIRST 53 DAYS OF CULTURE

Days in culture	Rate of dilution	Parasites per 10,000 erythrocytes			
		R	T	S	Total
		Line A			
0	1	32	0	0	32
2		6	23	28	53
4		18	5	7	30
4	3	1	1	4	6
6		27	7	4	38
11	36	5	2	2	9
13		19	13	6	38
14		40	9	1	50
22	1,728	9	2	1	12
24		30	20	9	59
25		29	32	21	82
28	18,000	11	3	0	14
31		43	25	7	66
46	10×10^6	3	1	0	4
48		22	5	4	31
		Line B			
28	18,000	11	3	0	14
31		54	14	9	77
35	580,000	6	3	0	9
38		56	9	4	69
38	2×10^6	5	1	0	6
41		29	9	4	42
46	32×10^6	8	1	2	11
50		170	48	74	292
50	128×10^6	7	5	5	17
53		144	76	22	242

Note: Line A was maintained in culture continuously to day 48. Line B was derived from line A on day 28 and, in addition, was grown through several cycles in petri dishes. Rates of dilution refers to the reduction of parasites by the addition of freshly washed human erythrocytes. Abbreviations: R, rings without pigment; T, uninucleated forms with pigment; S, multinucleated schizonts.

isolated by Dr. Lowell Coggeshall from the Borneo fire-backed pheasant, *Lophura igniti* kept in the New York Zoological Park in 1938.[43] No other isolations of *P. lophurae* have been made or natural infections discovered, but the original strain has been extensively studied in several abnormal hosts, particularly in ducks. In addition to the numerous studies of erythrocytic stage cultivation and cell-free short-term cultivation, the exoerythrocytic cycle has been cultured continuously (see exoerythrocytic cultures, below). The extracellular cultivation of *P. lophurae* was a major achievement that allowed the investigator, for the first time, to dissect the host-cell metabolism from that of the parasite and to thus examine, as could not otherwise be possible, the nutritional requirements of an obligate intracellular parasite. The details of techniques and interesting findings will be discussed below.

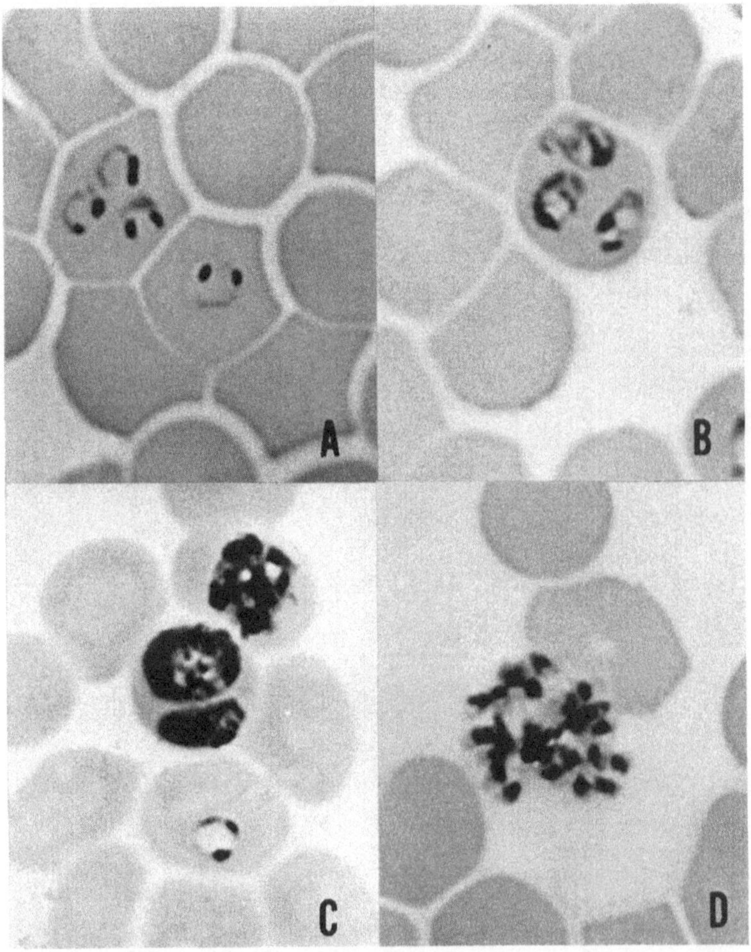

FIGURE 2. Photomicrograph of malarial parasites from the first continuous cultures of *P. falciparum*. Parasites grown in culture are indistinguishable from those seen from natural infections. (A) Ring stage parasites showing one cell containing three parasites. Multiply infected cells are a common feature of settled erythrocyte cultures. These ring stages last about 18 to 24 hr after which the pigmented nonvacuolated trophozoite stages are seen (B). Nuclear division marks the beginning of schizogony (C) which leads to segmentation of the schizonts into the infective merozoite stages, (D). Compare these with parasites seen in Figure 6. (From Jensen, J. B. and Trager, W., *Isr. J. Med. Sci.*, 14, 563, 1978. With permission.)

C. Culture of Avian Exoerythrocytic Stages

The exoerythrocytic stages of Plasmodia, as the term implies, are those developmental stages occurring in vertebrate host-cells other than erythrocytes or their immediate precursors. In avian species of *Plasmodium* these stages may be pre-erythrocytic, developing from sporozoites with cryptozoic (primary) and metacryptozoic (secondary) schizogony producing cryptozoites and metacryptozoites, respectively; or they may be initiated from erythrocytic forms that invade tissue cells to become phanerozoites. These terms, cryptozoites, metacryptozoites, and phanerozoites are used to denote exoerythrocytic zoites according to their origin, but otherwise these stages are essentially indistinguishable. In avian species the exoerythrocytic stages occur in cells of the reticuloendothelial system, whereas in mammalian species they develop in hepatic parenchymal cells. It is generally accepted that mammalian exoerythrocytic schizogony has one, or at best, a few cycles. Accordingly, unlike avian species, mammalian exoerythrocytic cultivation would predictably be of short duration, and to date this has been the case.

The exoerythrocytic stages of avian Plasmodia have been grown routinely in tissue culture for over 30 years. Detailed reviews of early work have been published by Porter,[44] Hawking,[45] Pipkin and Jensen,[46] and Huff,[47] thus only a brief historical account will be given here. The first attempts at exoerythrocytic parasite cultivation by Huff and Bloom[48] consisted of cultures of bone marrow cells from canaries infected with *P. elongatum*. Although these investigators saw no parasites in their cultures, the material remained infective when injected into susceptible canaries after 48 hr in vitro. Hawking[49,50] demonstrated that cultures started from a rich source of exoerythrocytic parasites and maintained in a susceptible cell line, could be grown abundantly in vitro for long periods of time. His starting material originated from chicks infected with *P. gallinaceum* sporozoites and consisted chiefly of splenic or peripheral blood macrophages. His cultures were infective for young chicks for up to 89 days in vitro. This work was confirmed by Zuckerman,[51] who in addition, made the significant observation that chick embryonic spleen was a far richer source of exoerythrocytic stages than young chicks. The first long-term cultures were those of Meyer and Musacchio[52] who grew the exoerythrocytic stages of *P. gallinaceum* for four continuous years in explants of chick embryonic myocardial cells using alternately the plasma-clot and hanging-drop techniques. Concurrent with these long-term cultures in tissue explants was the development of serial passages of avian parasites in embryonic duck, chicken, and turkey eggs. The most successful techniques were those of supra-chorioallantoic implantation of infected tissue from other infected embryos. This method is presently used to maintain several species of avian exoerythrocytic stages. Tissues from infected embryos were an excellent source of seed material for initiating cultures in monolayers of avian cells and led to the development by Huff and his colleagues of the techniques used today for long-term cultures.[49,53]

Cultures of avian exoerythrocytic parasites have generally been started from tissues of infected birds or embryonated eggs, but may also be initiated from mosquito-derived sporozoites, and perhaps also from erythrocytic stages. Meyer[54] demonstrated exoerythrocytic stages in embryonic chick brain cells after the addition of erythrocytic stages to them. Starting cultures from sporozoites is, however, a difficult procedure owing to the difficulty of obtaining these stages free of bacterial and fungal contaminants encountered when dissecting the parasites from mosquito salivary glands.

D. Culture of Mammalian Exoerythrocytic Stages

The exoerythrocytic stages of mammalian Plasmodia occur in hepatic parenchymal cells, and unlike those of avian species, develop principally from sporozoites. They are then pre-erythrocytes, having no phanerozoites, and it is becoming generally accepted that most mammalian species have one, or at most, only a few exoerythrocytic schizogonous cycles.[55] Early attempts at culturing exoerythrocytic forms in various tissue explants from monkeys infected with *P. cynomolgi*,[56] and in human bone marrow explants from patients infected by sporozoites of *P. vivax* and *P. falciparum*[57,58] were uniformly unsuccessful. Recently, however, Foley and his co-workers[59] were successful at culturing rat liver cells infected with exoerythrocytic forms of *P. berghei* for up to 44 hr and found the cultures to be infective when inoculated into recipient rodents. They were able to rule out the presence of erythrocytic stages or sporozoites in their cultures by showing that the blood from the donor rats was not infective when injected into susceptible rodents and by immunizing the recipient rats against *P. berghei* sporozoites. They concluded therefore that the primary liver explants contained viable exoerythrocytic schizonts that had developed for 44 hr in vitro. This report is reminiscent of that by Huff and Bloom[48] who cultured bone marrow explants of canaries infected with *P. elongatum* for 48 hr and found them to be still infective to susceptible birds.

Attempts to culture mammalian exoerythrocytic stages by introducing sporozoites directly to cultured liver cells were initially frustrated by the difficulty in culturing the hepatic parenchymal cells. Dubin and colleagues[60] introduced sporozoites of *P. vivax* to liver tissue culture but obtained no parasite development. Examination of the cultures revealed only fibroblasts and Kupffer cells and thus they explained their failure to the lack of parenchymatous liver cells. Recently, however, using newly developed techniques for liver tissue culture, Doby and Barker[61] repeated the earlier experiments with *P. vivax*. They reported intracellular inclusions that they suspected were exoerythrocytic parasites. Unfortunately their material was lost soon after their original observations thus no further examination was possible.

Beaudoin et al.[62] recently demonstrated that the presumed host-cell specificity of exoerythrocytic-staged parasites was a myth by their successful cultivation of avian plasmodia, *P. fallax* and *P. lophurae* in embryonic mouse liver parenchymal cells. The parasites continued to develop for several cycles in these mouse liver cells until the cultures became over parasitized. These studies illustrated that cultured mouse parenchymal cells could support exoerythrocytic parasites, and that the presumed host specificity does not necessarily hold true for in vitro systems. These studies additionally laid the ground work for the first in vitro cultures of mammalian exoerythrocytic stages. Strome et al.[63] reported the first successful cultivation of mammalian exoerythrocytic stages when *P. berghei* was grown in monolayers of embryonic turkey brain cells and embryonic mouse liver parenchymal cells. Their work was subsequently confirmed by Sinden and Smith[64] who successfully infected embryonic rat brain cells, as well as human liver (Chang), human lung (L-132), and embryonic rat liver cells. Somewhat surprisingly, these latter cultures were less suitable than the embryonic rat brain cells, where the parasites are not known to develop in vivo. No secondary schizonts were reported by either of these two laboratories.

E. Culture of Sporogonic Stages

Unlike the other cycles discussed so far, the exoerythrocytic cycle (chiefly in avian Plasmodia) and the erythrocytic cycle, the sporogonic cycle, as it is erroneously termed, is not actually a cycle at all, but a progression of stages beginning with gametocytogenesis in the vertebrate, host to the mosquito midgut for gametogenesis and fertilization, to oocyst development and sporozoite formation in the insect haemocoele, and finally to sporozoite maturation in the salivary glands. Not only is sporogony not cyclic, but it involves radical shifts in environment with apparently equally radical shifts in metabolism for the parasite. For these reasons and numerous unknown others, the cultivation of sporogonic stages is still in its infancy — but not for lack of effort, which has been considerable. An excellent review of the early work and a discussion of the prevailing problems has been recently published by Vanderberg et al.[65] Since this review, however, a few advances have been made. Initially, Jensen[66] reported that although gametocytes of *P. falciparum* develop readily in cultures of the erythrocytic stages of this parasite, these gametocytes did not appear completely mature and, despite the early optimistic report by Carter and Beach,[67] these cultured gametocytes were not infective to *Anopheles stephansi* mosquitoes via membrane feeding experiments. Recently, however, Dr. Meuwissen of the University of Njimegen, The Netherlands reported consistent in vitro exflagellation of cultured *P. falciparum* microgametocytes and one successful experimental infection of *A. gambiae* mosquitoes with cultured gametocytes. In these experiments the erythrocytic cultures were maintained continuously at 37 ± 1°C for 10 days or more until the gametocytes were mature and then fed via membranes to the mosquitoes. Although no published reports exist at this writing, progress is being made.

It should be mentioned here that Speer et al.[68,69] reported successful short-term cultures of the erythrocytic stages of *P. berghei* in cells other than erythrocytes — cells that theoretically could be propagated in vitro. However, after considerable effort by researchers at Parke-Davis, Detroit, these findings could not be substantiated. Perhaps subtle, unknown factors were missing in the subsequent studies.

Table 2
COMPOSITION OF MOORE'S RPMI 1640 (DATA DERIVED FROM GIBCO CATALOGUE)

Inorganic salts	mℓ/ℓ
$Ca(NO_3)_2 \cdot 4H_2O$	100.0
KCl	400.0
$MgSO_4$	48.84
NaCl	6000.0
$NaHCO_3$	2000.0[a]
Na_2HPO_4	800.00
Other components	
Glucose	2000.0
Glutathione (reduced)	1.0
Phenol Red	5.0
Amino acids	
L-Arginine (free base)	200.00
L-Asparagine	50.0
L-Aspartic Acid	20.0
L-Cystine	65.0 (2 HCl)
L-Glutamic Acid	20.0
L-Glutamine	300.0
Glycine	10.0
L-Histidine (free base)	15.0
L-Hydroxyproline	20.0
L-Isoleucine (allo free)	50.0
L-Leucine (Methionine free)	50.0
L-Lysine HCl	40.0
L-Methionine	15.0
L-Phenylalanine	15.0
L-Proline (Hydroxy L-Proline free)	20.0
L-Serine	30.0
L-Threonine (allo free)	20.0
L-Tryptophane	5.0
L-Tyrosine	28.94 (Na Salt)
L-Valine	20.0
Vitamins	
Biotin	0.20
D-Ca Pentothenate	0.25
Choline Cl	3.00
Folic Acid	1.00
i-Inositol	35.00
Nicotinamide	1.00
Para-Aminobenzoic Acid	1.00
Pyridoxine HCl	1.00
Riboflavin	0.20
Thiamine HCl	1.00
Vitamin B_{12}	0.005

[a] Added before use.

III. TECHNIQUES FOR CULTIVATION OF ERYTHROCYTIC STAGES

A. The Medium

The medium is prepared from powdered RPMI 1640 (Grand Island Biologicals) containing L-glutamine, but without bicarbonate (Table 2). For 1 ℓ, 10.4 g of the powdered pre-mix is added to 900 mℓ of glass redistilled water, 5.94 g of HEPES (N-2-hydroxyethylpiperazine-N′-2-ethane-sulfonic acid) is dissolved and the volume is adjusted to 960 mℓ. This medium should then be sterilized through a 0.45 μM Millipore filter by either positive or negative

Table 3
COMPARISON OF THE GROWTH OF *P. FALCIPARUM* CULTURE SUPPLEMENTED WITH FRESH HUMAN TYPE A$^+$ SERUM AND COMMERCIALLY AVAILABLE ANIMAL AND HUMAN SERA

Sera	Percent growth in comparison to control	Comments
Control serum[a]	100%	The $\bar{m}$ parasitemia @ 96 hr for controls in all experiments was 6.8%, representing an average increase of 60-70 ×.
Freshly collected fetal bovine	35%[b]	The first few cycles were encouraging, but growth declined with each subculture.
Adult bovine	7.2%	
Newborn bovine	1.0%	
Horse	19.0%	
Sheep	2.3%	
Swine	14.0%	After 48 hr cultures showed some agglutination of rbc's; at 96 hr considerable amount of hemolysis.
Human frozen[c]	27.0%	
Human lyophilized (Miles)	4.4%	Hemolysis began after 72 hr in vitro; became extensive at 96 hr.
Human lyophilized[d] (Difco)	17.6%	Hemolysis began after 48 hr in vitro; became extensive at 96 hr.
Human lyophilized in laboratory	100%	Stored at 23°C for 6 wks before reconstituting by addition of distilled water.

Note: All sera, except controls, were heat inactivated 30 min at 56°C. Parasites were grown in petri dishes in a candle jar.

[a] Fresh human serum obtained from freshly clotted blood supplied by New York Blood Center.
[b] The average of all samples tested for the first 96 hr in vitro.
[c] Commercially pooled "fresh" frozen human serum from Miles Laboratories, Inc.
[d] Reconstituted and added to culture medium at equivalent of 20% serum.

filtration, depending upon the apparatus used. The sterile medium is then aseptically distributed into 100 mℓ portions in tightly stoppered glass medium bottles and can be stored for 4 to 5 weeks at 4°C. The medium is completed by the addition of 4.2 mℓ of sterile 5% w/v $NaHCO_3$. After addition of the $NaHCO_3$ to the yellow medium, it should slowly turn to orange over a 30 to 60 sec period. If the color changes immediately, or changes beyond orange to pink, this indicates that the $NaHCO_3$ solution is no good, having lost most or all of its CO_2. Pink medium, while no good for cultivation, can be used to wash erythrocytes.

No useful modifications of the medium has yet been found. Medium 199, and Ham's F-12 Nutrient Mixture might be useful and have actually been used,[70,71] but there is no advantage in using these media. In fact, the M-199 used by Chen et al.[71] was supplemented with adenosine-5-triphosphate, coenzyme A, adenosine, creatinine, inositol, glutathione, glyclyglycine, glucose, and vitamine C, but growth rates were no better than other laboratories achieve in RPMI 1640.

B. Serum

Initially we used 15% v/v human serum. However, experience taught us that human serum varies widely in its nutrient status (Table 3) but even the poorest sera were adequate when used at 10%.[72] Recently, we have begun to pool the human serum until at least two different sera have been collected. This pooled serum is used at 5%. Blood for serum is drawn in a 500 mℓ collection bag devoid of anticoagulants and incubated for 1 to 2 hr at 37°C, then stored for 24 hr at 4°C to allow the clot to form and then to shrink maximally. The serum is drawn off aseptically using a plasma transfer set, the cells removed by centrifugation, and stored at −20°C until used. Old serum, or serum from outdated blood, is not suitable. However, plasma drawn in citrated blood preservatives (ACD or CPD) can be used if it is fresh and the clotting factors are first removed by the addition of sterile $CaCl_2$ (10% w/v solution added at 1 mℓ to 100 mℓ plasma). If gently stirred on a magnetic-stirring plate for 30 to 60 min at ambient temperatures the clot will form around the magnet. The ABO blood type of serum used depends upon the type of erythrocytes to be used and the experimental design. When initiating new cultures where the blood type of the patient is unknown, or when *Aotus* infections are the source of the parasites, AB+ serum should be used. This type is compatible with any type of human erythrocytes and is only slightly incompatible with *Aotus* cells. For routine maintenance of *P. falciparum* cultures, and for most experiments, type A+ serum is recommended because both type A and O erythrocytes can be used. These two cell types predominate (over 80%) in caucasians and thus type A+ cells and sera are readily available in USA and Europe. Most blood banks prefer not to provide type O+ blood for experimental purposes because this blood is the universal donor type and can be transfused to most patients.

New Zealand white rabbit serum is the only animal serum successfully used to date.[73] The rabbit serum must be prepared and stored the same way as has been outlined above for the human serum. Commercially supplied animal sera (and human serum, other than blood bank supplied) have not proven suitable for continuous cultures.[72] Ifediba and Vanderberg[74] reported that commercial calf serum supplemented with either Neopeptone (Difco) or Protease-Peptone #3 (Difco) at 1.0% of a 15% w/v solution could be used in place of 10% human serum. Their initial cultures were adapted from human to supplemented calf serum over a 9-month period. However, subsequent lines were adapted to calf serum over a 1-month period by gradually moving from 10% human; 0% calf to 10% calf; 0% human serum. Chen et al.[71] claim to be successful in using freshly collected newborn calf serum in their highly supplement M-199 medium (see media, above) with good results. They do not mention, however, whether their cultures were short-term or continuous. Freshly collected fetal bovine serum and pooled goat serum gave excellent results for short-term cultures (Jensen, unpublished) but would not sustain continuous cultures.

The complete medium with serum is best prepared fresh by aseptic addition of sterile serum to sterile medium. If necessary, however, contaminated serum can be clarified by centrifugation (1500 × G for 30 min) before filtering through a 0.45 m*M* filter or by mixing with RPMI 1640 medium and then filtering the mixture through an appropriate filter. The complete medium with serum can be stored at 4°C for 1 week.

In malarious countries it will be important to avoid two possible sources of unsuitable serum: (1) serum from individuals with a high degree of immunity to falciparum malaria; (2) serum from individuals who have recently taken any drug with an antimalarial effect. Unpublished results from several laboratories already show that both above kinds of serum have inhibitory effects.

C. Erythrocytes

Since *P. falciparum* in nature is exclusively a parasite of man, since human erythrocytes have a longer in vivo life time than those of *Aotus trivirgatus*, the only suitable experimental host, since more is known about human erythrocytes than about any other cell and since in any case human red cells are the easiest to obtain in quantity, they are the host cell of choice for in vitro cultivation of *P. falciparum*.

Blood can usually be obtained from local blood banks in either ACD (acid-citrate-dextrose; per liter: 8 g citric acid, 22 g sodium citrate, 4.5 g glucose; for each 100 mℓ of whole blood add 15 mℓ ACD) or CPD (citrate-phosphate-dextrose; per liter: 3.27 g citric acid, 26.3 g sodium citrate, 25.5 g glucose, 2.22 g sodium monobasic phosphate; for each 100 mℓ whole blood add 14 mℓ CPD). Cells can be conveniently distributed into 50 to 100 mℓ portions and stored at 4°C for up to 35 days. Accordingly, as has been reported,[39] erythrocytes outdated by blood banking standards can be used for culturing *P. falciparum*. These cells should, however, be stored as whole blood and not as packed cells, the latter of which become less suitable after 21 days of storage. Erythrocytes from some individuals support cultures better than do others and storage for 7 to 10 days appears to improve the cells for parasite cultures.

The ABO blood type has no effect on suitability of the erythrocytes for *P. falciparum* cultures. The type used will depend somewhat on the experiment and upon the serum available. Type A+ cells are routinely used with A+ serum; however, O+ cells are also compatible with A+ serum.

Erythrocytes are prepared for culture by pipetting a selected volume into a sterile centrifuge tube and spinning at 350 G for 10 min. Using a sterile pipette the citrated plasma and buffy coat are discarded. Erythrocytes stored for 7 to 10 days will have no viable leukocytes or platelets, however a buffy coat of small micro-clots can often be seen and these should be removed after each centrifugation. After discarding the plasma-buffy coat the cells are resuspended in 5 volumes of RPMI 1640 *sans* serum and recentrifuged as before. This washing procedure is repeated for the second wash after which the cells are suspended at 50% in complete medium with serum. To this 50% washed cell suspension is added an appropriate volume of *P. falciparum* infected cells to give a pre-culture parasitemia of approximately 0.1 to 0.2%.

D. Culture Techniques

In general, cultures of *P. falciparum* grow well when the suspensions of infected erythrocytes are diluted to an 8% hematocrit with complete medium with appropriate serum and placed in a flat-bottomed vessel at a total depth of 3 to 4 mm. The gas phase should contain an O_2 concentration of 17% or less, and of a CO_2 concentration of 2 to 3% (the balance being N_2). The culture medium can be supplied by continuous flow or replaced at intervals of 24 hr or less, depending upon the level of parasitemia. Several techniques that fulfill these parameters have been devised,[39,75-79] some of which will be described in detail below.

1. Petri Dish-Candle Jar Method[39]

Generally, infected blood, reduced to a starting parasitemia of 0.1 to 0.2% by the addition of freshly washed erythrocytes (see above) is further diluted to 8% hematocrit with complete medium containing serum and distributed at 1.5 mℓ per 35 mm culture dishes or 4 mℓ per 60 mm dishes. The cultures are placed into a glass desiccator, equipped with a stopcock on the lid, containing a lighted paraffin candle. The lid is sealed with silicone grease with the stopcock open to vent the heated air from within the desiccator. When the candle goes out the candle jar is set at 37°C in an incubator. If, after 24 hr, the culture dishes have considerable condensation on their lids this indicates a temperature gradient within the incubator. This condition can sometimes be corrected by using only one candle jar per incubator. If the problem persists a new incubator should be used. Experience has indicated that recirculating forced-air or water jacketed incubators provide the most uniform temperature and thus are superior to convection-type incubators.

The culture medium should be renewed every 24 hr by gently tipping the dish and aspirating off the medium with a cotton-stuffed sterile pasteur pipette. The fresh medium should be warmed to 37°C before being added to the dishes. Once the medium has been added to the cultures they are gently swirled to resuspend the cells, allowing them to resettle to an even layer when placed back into the incubator. While the medium is being renewed the candle jar should be kept warm in the incubator. The candle wicks should be kept trimmed to

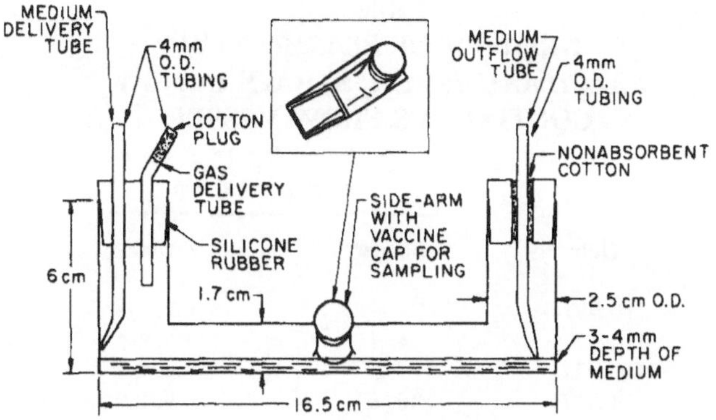

FIGURE 3. New continuous flow vessel for cultivation of *P. falciparum*. This vessel is an improved version of the original flow-vial pictured in Figure 1. Medium is delivered through the Medium Delivery Tube and leaves the vessel through the Medium Outflow Tube. The level of the medium in the vessel is controlled by the height of the Medium Outflow Tube. Parasites are added and removed through the Side Arm which is capped with a serum stopper. Vessel is custom made. (From Trager, W., *J. Protozool.*, 26, 125, 1979. With permission.)

prevent smoking after they go out. Blood films can be made from the cultures if care is taken to sample only concentrated cells after the old medium has been removed. Too much medium taken with the cells produces a film that will be difficult to examine or count. Generally, a lower hematocrit results in a thinner settled cell layer and in turn will result in a higher parasitemia, but no overall increase in parasites number. Thus, higher parasitemias (>20%) can be achieved with 4% hematocrits that are changed every 12 hr. Fresh erythrocytes should be added every 4 to 5 days, but should be added more often if parasitemias are kept between 5 to 12%. For experiments where substrates, antimalarial compounds, etc. are to be tested, more linear growth rates can be obtained by starting cultures near 0.1% parasitemia. If culture conditions are correct, most strains of *P. falciparum* will increase from 0.1% to 6 to 8% over a 96-hr (two cycles) period, an average increase of 8 times per cycle.

One variation on the petri dish method is to use multi-well micro-titer plates in place of the petri dishes. If 96 well plates are used the parasites should be diluted to 4% hematocrit and added to the wells at 200 μliters per well. The medium can be readily removed with a 22-gauge needle. The plates are placed in the candle jars as are the petri dishes.

2. *Continuous Flow Methods*

The flow vial used for the first continuous cultivation of *P. falciparum*[4] has been superseded by a model that is larger and more convenient to use (Figure 3).[76] This new flow vessel is constructed out of 20 × 30 mm rectangular glass tubing, has a chimney at each end, with a vaccine-stoppered side arm for the removal of infected and addition of uninfected erythrocytes. The chimneys are fitted with #4 silicone rubber stoppers containing medium delivery and outflow glass tubes, respectively, as well as cotton-filled gas delivery and exit vents. The culture medium flows into the vessel via a peristaltic pump set to deliver at a rate of ~2 mℓ per hour and is removed using a separate channel on the pump employing tubing having a larger diameter than the delivery tubing. This arrangement provides for a continuous flow of medium through the vessel, the depth of which is maintained by the position of the outflow tube. The fresh medium, held in a 500 mℓ Erlenmeyer flask, is stored in a small refrigerator and is delivered to the peristaltic pump and from the pump to the flow vessel in Teflon®-lined tubing (Teflon TFE, ID 3/16 inches). Silicone rubber or surgical tygon

Table 4
GROWTH OF *PLASMODIUM FALCIPARUM*, FCR_3 STRAIN IN NEW CONTINUOUS FLOW VESSEL

Date	Percent parasitemia	
	Before[a]	After
11/25/77		0.2, 0.3
11/28/77	4, 4	1.5, 1.4
11/30/77	7, 5	2.4, 2.1
12/02/77	10, 8	0.9, 0.9
12/05/77	9, 9	0.8, 1.1
12/07/77	6, 6	0.8, 0.8
12/09/77	8, 11	1.0, 0.7
12/12/77	11, 9	0.7, 0.6
12/14/77	3.5, 3.0	1.7, 1.0
12/16/77	11, 7	1.5, 0.7
12/19/77	20, 10	1.1, 0.5
12/21/77	12, 9	0.6, 0.5
12/23/77	8, 6	0.8, 1.0
12/26/77	6, 5	2.6, 2.2
12/28/77	7, 5	1.0, 1.1
12/30/77	6, 6	0.6, 0.9
01/02/78	7, 7	0.9, 1.5
01/04/78	6, 10	2.0, 2.6
01/06/78	10, 11	0.5, 0.8
01/09/78	13, 10	0.5, 0.8
01/11/78	5, 7	5.0, 0.7
01/13/78	11, 8	1.4, 0.8
01/16/78	14, 10	1.1, 1.2
01/18/78	6, 8	0.5, 0.9
01/20/78	5, 8	0.7, 1.1
01/23/78	18, 12	1.0, 0.9
01/25/78	12, 6	1.0, 0.9

Note: Each vessel contained 12 mℓ of an 8% cell suspension yielding approximately 1.0 mℓ packed-cell-volume of infected erythrocytes.

[a] The figures give the percent parasitemia just before and after addition 3 × per week, usually Monday, Wednesday, and Friday.

From Trager, W., *J. Protozool.*, 26, 125, 1979. With permission.

tubing is used for the portion of the delivery tubing that passes through the peristaltic pump. The Teflon®-lined tubing, unlike silicone rubber, is impermeable to CO_2 and O_2, thus it maintains the proper soluble concentrations of these gases in the culture medium during delivery. The used culture medium is pumped into a separate Erlenmeyer flask. Humidified gas (3% CO_2, 1% O_2, 96% N_2) is delivered through either the Teflon®-lined or surgical

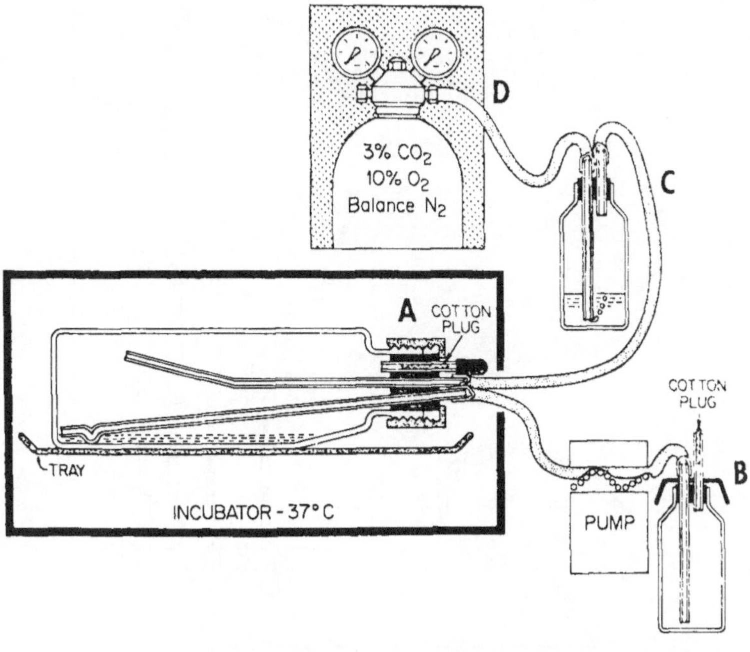

FIGURE 4. Chin's continuous flow system using commercially available tissue-culture flasks with larger capacity than Trager's custom-built vessels in Figure 3. (From Chin, W., *Trans. R. Soc. Trop. Med. Hyg.*, 73, 334, 1980. With permission.)

tygon tubing by passing it first through a large carboy of distilled water and finally through a 50 mℓ Erlenmeyer flask at a rate of 1 bubble/second.

The vessels usually contain 1 mℓ of packed erythrocytes, but this volume varies with the level of parasitemia. Generally, fresh erythrocytes are added to the vessels every 2 or 3 days, with parasitemias maintained between 0.7 to 2.0% to 7 to 14% ($\overline{m}$ = 1.2% and 8.5%) for an average increase of 7.6 times per cycle (Table 4). The addition of fresh erythrocytes (which constitutes an effective subculture) is done usually 3 times per week (Monday, Wednesday, and Friday). The procedure is as follows. The medium delivery pumps are shut down. The parasitemia of the flow-vessel is determined by sampling through the serum-stoppered side port, making a thin film and staining with Giemsa (1:40; pH 7.0). Freshly washed erythrocytes are mixed with complete medium containing serum to an 8% hematocrit. The contents of the flow vessel are removed, except for sufficient seed material, to give a postsubculture parasitemia near 1.0%. An appropriate volume of the 8% uninfected erythrocyte suspension is added through the sideport, mixed with the infected seed material and the flow-vessel left alone for 1 hr to allow the cells to settle, after which the peristaltic pumps are turned on again. Flow vessels are routinely kept "on line" for about 2 months, after which time they are replaced with newly sterilized ones. A variation on this method using commercially available culture flasks with considerably larger capacity has been developed by Chin[77] (Figure 4). The only serious constraint of this culture system is the cost of obtaining high quality peristaltic pumps, such as the Harvard model #1203, of which a minimum of two is needed if one is earnest in developing a continuous semiautomated system of parasite cultivation. The petri dish-candle jar method, described elsewhere in this chapter, is much better suited for the casual investigator or where daily manual changing of culture medium provides no serious difficulties.

3. Large-Scale Laboratory Cultivation Method

Any of the above methods can be scaled up to increase the laboratory production of *P.*

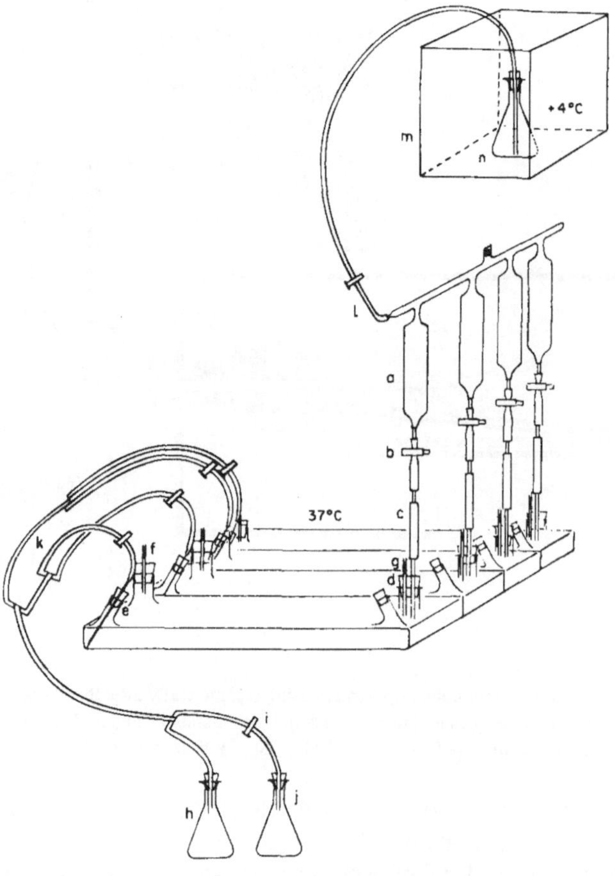

FIGURE 5. Culture vessels for large-scale laboratory cultivation of *P. falciparum*. This apparatus allows for large-scale production of parasites for immunological or biochemical purposes. Although it requires daily manual change of culture medium, this operation takes only about 15 to 20 min and does not require peristaltic pumps and can be accomplished without exposing the cultures to possible microbial contamination. The culture vessels, made from rectangular glass tubing (22 × 75 × 400 mm), with a cylindrical chimney (d) and serum stoppered-side port (e) at each end are connected by glass and silicone tubing (c) to a 4-place manifold whose arms (a) contain the premeasured and prewarmed culture medium. To renew the culture medium it is allowed to flow (by siphon) via the hypodermic needle inserted through the serum stoppered side port (c) into a flask (h) below. All the medium can be thus removed with few, if any, erythrocytes being lost. Fresh medium is delivered to the vessels from the manifold (a) through tubing (c) by opening clamp (b). After the fresh medium is added, the erythrocytes, are evenly suspended by gently rocking the vessels. Usually 2 hr before the medium is changed, the manifold is filled (by siphon) with chilled culture medium through tubing (1) from a reservoir (m) held in a small refrigerator above the incubator. To harvest the parasitized erythrocytes, the medium is first removed, as above, then the clamp (i) is shifted from flask (j) to flask (h) and the vessels are tilted to a 45°C angle which allows the cells to flow into flask (j). Subcultures can be made by leaving sufficient infected erythrocytes in the vessels to seed new erythrocytes added through the serum stoppered-side port located at the manifold end of the vessel. A gas mixture of 2% CO_2, 3% O_2, 95% N_2 flows continuously through the vessels from cotton plugged tube (f) to (g).

Table 5
COMPOSITION OF ERYTHROCYTE-EXTRACT-ENRICHED MEDIUM FOR THE EXTRACELLULAR CULTIVATION OF *PLASMODIUM LOPHURAE*

	Stock solutions		For 100 mℓ of final medium	
No.	Ingredients	Concentration (mg/100 mℓ)	Volume stock solution (mℓ)	Final concentration (mg/100 mℓ)
1	NaCl	6,600	5.0	330.0
	KCl	8,800		440.0
2	$NaH_2PO_4 \cdot H_2O$	550	2.5	13.8
	K_2HPO_4	6,270		156.8
3	$NaHCO_3$	5,000	1.8	90.0
4	$CaCl_2$	260	1.5	3.9
5	$MnSO_4 \cdot H_2O$	170	2.0	3.4
6	Na-acetate	1,500	1.0	15.0
	Glycerol	2,500		25.0
7	Dextrose	5,000	4.3	215.0
8	Hexose diphosphate (Mg salt)	5,000	1.4	70.0
9	Glutathione (reduced)	2,000	5.0	100.0
	L-ascorbic acid	20		1.0
	Nicotinamide	8,000		400.0
115	Lactalbumin hydrolysate	1,500	52.0	750
	Bovine plasma fraction V	1,200		600
12-2	Gelatin (Wilson PF)	36,000	20.0	7,200
13 h	Riboflavin	5	1.0	0.05
	Thioctic acid	1		0.01
	Thiamin·HCl	5		0.05
	Pyridoxin·HCl	5		0.05
	Pyridoxamine·2HCl	2		0.02
	Pyridoxal·HCl	2		0.02
	Vitamin B_{12}	1		0.01
	Inositol	10		0.10
	Nicotinamide	50		0.50
	Choline-Cl	50		0.50
13 b	Calcium pantothenate	200	0.5	1.0
13 d	*p*-Aminobenzoic acid	100	0.1	0.1
14 B	Biotin	5	0.01	0.0005
14 F	Folic acid	5	0.02	0.001
15 c	Adenosine	100	1.0	1.0
	Orotic acid	40		0.4

Note: All solutions were prepared in water redistilled in a Pyrex glass still. Stock solutions 1, 2, 4, 5, 7, 13 b, 13 d, 14 B, and 14 F were placed in tubes and sterilized by autoclaving. Solution 12-2 was placed in tubes and sterilized by steaming 20 min on each of 3 successive days. Solutions 3, 6, 8, 13 h, and 15 c were sterilized by filtration (Porcelain Selas 03 filter) and aseptically placed into tubes. Solution 115 was similarly sterilized and placed, in 52-mℓ amounts, in 250-mℓ Erlenmeyer flasks. The final mixture was prepared by addition aseptically to such a flask of the designated amount of each solution in the order 1, 2, 3, 7, 8, 12-2, 4, 5, 6, 13 h, 13 b, 13 d, 14 B, 14F, and 15 c. Finally freshly prepared and sterilized (through a glass ultrafine filter) solution 9 was added, followed by 15 mℓ of duck serum (from defibrinated blood of ducks not over 3 months old). The stock solutions were stored up to 3 months in a refrigerator, except for 13 h, which was kept frozen at −20°C, and solution 9, which was always freshly prepared. The final mixture was used on the day of preparation.

falciparum parasites. The most readily expandable method is that of the petri dish-candle jar, however, if a very large quantity of parasites are needed manual labor quickly becomes a constraining factor. For large-scale laboratory cultivation of *P. falciparum* the flat-bottomed vessels seen in Figure 5 have been developed. They contain 5 mℓ of packed erythrocytes per vessel and two sets of four vessels each are routinely handled per incubator. Parasitemias of 20% are readily obtained. No peristaltic pumps are needed and, except for the culture vessels, which are easily made in any glass-blowing shop, all components are available from scientific supply companies.

IV. TECHNIQUES FOR EXTRACELLULAR CULTIVATION OF *P. LOPHURAE*

A. The Medium

The medium is prepared from an extract of duck erythrocytes prepared in a complex nutrient mixture (Table 5). Erythrocytes from ducks less than 3 months old are washed free of serum and leukocytes, and quickly frozen as 10 mℓ packed cell volumes. Upon thawing in ice water (usually after holding at −70°C overnight) the hemolysed cells are diluted by the addition of 13.3 mℓ of the nutrient mixture and centrifuged for 50 min of 1800 × G at 20°C. This temperature is essential because the medium contains gelatin. Each tube should yield 8 to 9 mℓ of richly colored red-cell extract containing about 14% hemoglobin (about 1/2 the concentration of the erythrocyte). The nucleated erythrocyte stromata are discarded. The pH of the extract is adjusted to 6.9 by addition of 0.1 *M* HCl. The extract-medium then receives, at a rate of 0.2 mℓ/10 mℓ, a solution of L-malic acid, yeast adenylic acid (2′ and 3′ mixed isomers) and nicotinamide-adenine-dinucleotide, to give final concentrations of 6.0 m*M*, 1.4 m*M* and 0.15 m*M*, respectively. To complete the medium, ATP, pyruvate, coenzyme A, and folinic acid are added as small volumes of concentrated solutions to give a final concentration of 2 m*M*, 5 m*M*, 0.05 m*M*, and 0.01 m*M*, respectively.

B. Free Parasite Preparation

Infected erythrocytes are removed from young Pekin ducklings in 30 mg heparin per 100 mℓ of 0.85% NaCl at 0.1 mℓ/mℓ blood, 14 days postinfection, when the essentially synchronous parasitemia is 90 to 100% uninucleated trophozoites (pigmented parasites). The blood is centrifuged, buffy coat and plasma discarded, and resuspended to 20% hematocrit in the red-cell extract-medium. To 6.3 mℓ of this suspension is added 0.13 mℓ sterile guinea pig serum (as a complement source) and 0.7 mℓ of rabbit antiduck antiserum. The mixture is gently agitated 30 min at 40°C with vigorous shaking at 15 min, transferred to a conical centrifuge tube and spun for 1 min at 200 × G to remove the agglutinated erythrocyte stromata. The resulting supernatant fluid will contain the free parasites (and some erythrocyte nuclei) which are then ready to be placed into culture.

C. Preparations for Cultivation

Each 50 mℓ sterile, silicone stoppered Erlenmeyer culture flask receives 0.25 sterile duck plasma, obtained by carefully bleeding a duckling without anticoagulant and centrifuging the blood. To this is added 0.2 mℓ of chick embryo extract and the flask slowly rotated so that the plasma clot that forms covers the bottom of the flask as a thin layer. Three milliliters of red-cell extract medium is overlaid on the plasma-clot with the appropriate supplements (ATP, pyruvate, CoA, and folinic acid) and finally the inoculum of free parasites (50 to 150 × 10^6 per flask) are added. The culture flasks, placed on a rocker platform and agitated at 16 cycles per minute, are continuously gassed through cotton-stoppered tubes inserted through the silicone stoppers with humidified 5% CO_2 and air. About 1/2 of the parasites will form a thin "scum" layer on the plasma-clot which can be examined by removing a sample with a pasteur pipette. The initial starting material, essentially 90 to 95% uninucleated forms, will mature to 50% multinucleated forms within 24 hr in vitro.

Table 6
COMPOSITION AND SUPPLIERS FOR COMPONENTS OF THE DIPLOID CULTURE MEDIUM USED TO CULTURE THE AVIAN EXOERYTHROCYTIC STAGES OF *PLASMODIUM FALLAX* AND *P. LOPHURAE*

Components	Amounts	Stock	Supplier
Medium 199, with Earle's salts	9.87 g	Powdered base	GIBCO[a]
BME[d] with Earle's salts	9.00 g	Powdered base	GIBCO
BME vitamin mixture	10.0 mℓ	Commercial stock × 100	MBA[b]
Folinic acid; calcium salt	2.0 mℓ	6 mg/100 mℓ H_2O	GIBCO
MEM nonessential amino acids	10.0 mℓ	Commercial stock × 100	MBA
L-glutamine	0.292 g		Sigma[c]
$NaHCO_3$	2.45 g		Sigma
Penicillin-Streptomycin	20.0 mℓ	Stock: 5000 units + 5000 μg/mℓ	
Fetal Bovine Serum (heat inactivated)	10.0 mℓ		GIBCO
Glass redistilled water	Sufficient to total 2000 mℓ		

Note: Adjust final pH to 7.4 with 1.0 *N* NaOH. Filter final solution through 0.45 μ*M* millipore apparatus.

[a] Grand Island Biological Company; Grand Island, New York.
[b] Microbiological Associates; Bethesda, Maryland.
[c] Sigma Chemical Company; St. Louis, Missouri.
[d] Basal Medium (Eagle).

To prolong the cultures the medium is changed after 20 hr and then at 12-hr intervals thereafter. The changing of medium eliminates the nonadherent parasites, which appear essentially normal after 48 hr but begin to degenerate after 72 hr. If the parasites remain within the original erythrocyte plasmalemma they appear normal after 96 hr in vitro and remain infective to ducklings after 5 days.[40,80]

V. TECHNIQUES FOR CULTIVATION OF AVIAN EXOERYTHROCYTIC STAGES

The culture medium is composed of a 1:1 mixture of modified Medium 199 and Diploid Growth Medium whose composition is found in Table 6. Details for cultivation procedures of avian exoerythrocytic stages were given by Davis et al.[81] and updated by Beaudoin.[82] Cultures are initiated from infected embryonic turkey brain tissue infected with *P. fallax* or *P. lophurae* maintained by weekly serial passage from embryo to embryo, or from cryopreserved stabilates of infected embryonic turkey brain cells. A third species, *P. gallinaceum,* has been grown in cultured chicken brain cells with only limited success. The infected turkey embryos are a rich source of exoerythrocytic stages and are maintained by chorioallantoic grafting when the embryos are 7 to 13 days old. Parasitized brain cells can be harvested 6 days later. After locating the embryo by candling the egg, the multilobed brain is exposed by teasing away the skull tissue. Because of its fragility the brain is best removed by sucking it up into a 2.5 mℓ syringe. The brain is broken up and the cells dispersed by repeated passage through a 23 gauge needle into a T-30 tissue culture flask containing 5 mℓ culture medium. After dispersing, the brain cells are divided equally into two T-30 tissue culture flasks with an additional 2.5 mℓ of medium added to each. After 24 hr at 37°C in a CO_2 incubator (5% CO_2 in air), the culture medium should be renewed. If the original medium contains large pieces of brain tissue (which is often the case) these may be recovered by centrifugation, resuspended in fresh medium and readily dispersed by expelling a few times through a hypodermic needle, as before. This procedure usually gives four flasks per embryo with nearly confluent monolayers of infected brain cells within 5 to 7 days.

The infected brain cells are usually subcultured weekly, but must be closely monitored to prevent them from becoming over parasitized. To subculture, replace the culture medium with Earle's balanced salt solution containing 0.05% trypsin and incubate 5 to 10 min until the monolayer begins to crack and loosen. Replace the trypsin solution with 10 mℓ of culture medium and wash cells from surface of the culture vessel by repeated flushings using a serological pipette. Once the cells are uniformly resuspended they are divided between two new flasks.

After the second or third subculture, the cells will become overparasitized with exoerythrocyte forms because the parasites grow faster than the brain cells. When this happens the infected cultures must be diluted with uninfected embryonic turkey brain cells. Therefore, parallel cultures of uninfected turkey brain cells, initiated as described above for infected brain cells, must be maintained along with the infected cultures. Overparasitized cultures are subcultured by mixing the contents from one infected culture flask with an approximately equal cell volume from an uninfected flask. This mixture is then dispersed into four new flasks with the appropriate amount of culture medium.

Because uninfected embryonic turkey brain cells cannot be cultured indefinitely, the overparasitized cultures must be periodically mixed with uninfected cells obtained from primary cultures of turkey brain cells. Thus in order to maintain avian exoerythrocytic parasites in long-term cultures it is necessary to grow parallel cultures of uninfected turkey brain cells which must be reestablished from turkey embryos at regular intervals of several weeks. Both infected and uninfected embryonic turkey brain cells can be readily recovered from frozen storage, thus facilitating the renewal of primary cultures and eliminating the necessity of maintaining serial passage of the parasites in embryonated eggs. In addition to culturing *P. fallax* and *P. lophurae* in turkey brain cells Beaudoin and his co-workers have cultured these species in embryonic mouse hepatic parenchymal cells.[62] These investigators were able to grow the avian parasite in the mammalian cells through several cycles until the cultures became overparasitized. These findings illustrate that cell cultures other than embryonic turkey brain cells can support cultures of exoerythrocytic parasites, and that the presumed host specificity does not necessarily hold true for in vitro systems.

VI. TECHNIQUES FOR CULTIVATION OF MAMMALIAN EXOERYTHROCYTIC STAGES

Unlike the avian exoerythrocytic culture procedures, which are initiated from embryonic turkey brain tissue infections, the mammalian cultures are started from mosquito-derived sporozoites.[63] Accordingly, *P. berghei* (the only species cultured to date) infections maintained in mice, are used to infect *Anopheles stephansi* mosquitoes. Twenty-one days after feeding on infected mice, the mosquitoes are anesthetized by cooling to 4°C for 5 min, individually disinfected by immersing in 70% ETOH for 3 sec and washed in 20 mℓ of sterile Medium 199. Under aseptic conditions the disinfected mosquitoes are dissected and the salivary glands removed and transferred to 1 mℓ of Mouse Liver Medium (Table 7) containing 1% Fungizone. The salivary glands are subsequently centrifuged onto the surface of the recipient host-cells at 1600 × G for 30 min and incubated at 37.5°C in 5% CO_2 95% air.

In the original report on cultivation of *P. berghei* exoerythrocytic stages, Strome et al.[63]

Table 7
CONSTITUENTS OF MOUSE-LIVER-MEDIUM USED FOR CULTIVATION OF EXOERYTHROCYTIC STAGES OF *P. BERGHEI*

Diploid Growth Medium + Medium 199 (Gibco)	100.0 mℓ
Basal Medium Eagle vitamins 100× (Gibco)	1.0 mℓ
Nonessential amino acids 100× (Gibco)	2.0 mℓ
Minimum Essential Medium amino acids 50× (Gibco)	2.0 mℓ
Sodium pyruvate 100 m*M* solution (Micro Associates)	1.0 mℓ
Insulin, protamine zinc, 40 units/mℓ stock (Lilly)	0.25 mℓ
Penicillin/streptomycin solution 5000 units of each (Micro Associates)	1.0 mℓ
Sodium bicarbonate 7.5% wt/vol aqueous solution (Gibco)	3.0 mℓ
	110.25 mℓ

reported successful cultivation in embryonic rat liver and brain as well as embryonic turkey brain (as used in *P. fallax* and *P. lophurae* cultures). In addition to embryonic rat liver and brain cells, Sinden and Smith[64] also grew *P. berghei* exoerythrocytic stages in human liver (Chang) and human lung (L-132) cells. They reported that of the four cell types tried embryonic rat brain supported the parasite cultures best. The embryonic rat brain and liver tissue cultures are initiated by dissecting from 18-day-old albino rat embryos the appropriate organs. After suspending in sterile Medium 199 the organs are cut into 1-mm^2 pieces and trypsinized for 20 min at 37°C. The resulting cell suspensions are recovered, washed and resuspended in 10 mℓ of MLM, inoculated into plastic tissue culture flasks and incubated at 37.5°C in an atmosphere of 5% CO_2 in air. The initial culture medium and resulting cell debris are discarded after 24 hr. Fresh culture medium is renewed every 48 to 72 hr, as needed. The embryonic turkey brain cells are prepared as described above for avian exoerythrocytic stage cultures. The rates of infection at best are generally low and the sporozoites, after invasion, develop through the initial stages to normal appearing schizonts. Whether mature merozoites are produced and whether these are infective to the cultured cells or erythrocytes has not yet been determined.

VII. ACCESSORY PROCEDURES USEFUL IN CULTIVATION EXPERIMENTS

A. Cryopreservation of Erythrocytic Cultures

This procedure is a modification of that reported by Rowe et al.[83] The cryoprotectant is prepared by adding 70 mℓ of glycerol to 180 mℓ of 4.2% sorbitol in 0.85% NaCl and sterilizing by passing through a 0.45 μM filter. The infected erythrocytes are centrifuged and the culture medium discarded. A quantity of cryoprotectant, equal to the packed cell volume is added and allowed to equilibrate for 5 to 10 min at room temperature. One-half milliliter of the suspension is added to 1/2 dram sterile screw cap vials and frozen quickly by agitation in an alcohol-dry ice bath or, less preferably, liquid N_2. Frozen vials are stored in a liquid N_2 refrigerator. To reclaim, vials are quickly thawed in a 37°C water bath, (care should be taken to first vent any liquid N_2 by momentarily loosening the cap) and transferred to a sterile, capped centrifuge tube. One-half milliliter of 3.5% w/v NaCl is added to the red cell-glycerol suspension and gently centrifuged at 500 X G/10 min. The saline-glycerol supernatant is discarded and 1/4 mℓ of 3.5% NaCl added. After recentrifugation the saline is discarded and replaced by 1/2 mℓ RPMI 1640 medium containing human serum, the suspension recentrifuged as before and again the supernatant fluid discarded. One final wash is performed in 5 mℓ of RPMI 1640 and serum to remove all the saline and glycerol. The remaining pellet is mixed with an equal volume of freshly washed human erythrocytes, diluted to an 8% hematocrit by the addition of 6.0 mℓ of complete RPMI 1640 medium, distributed into four 35 mm petri dishes and placed in a candle jar, as outlined elsewhere in this chapter.

B. Concentration of Trophozoites and Schizonts

As described by Jensen,[84] mature stages of *P. falciparum* can be separated from the ring-stage infected and most uninfected erythrocytes by suspending cultured erythrocytes in RPMI 1640 containing gelatin. Recent observations have demonstrated that only cultures containing a preponderance of parasites that cause alterations in the erythrocyte membranes respond to this technique.[85] The gelatin (Sigma swine-skin 300 bloom or 175 bloom) is made up to 2× concentration by dissolving 1 g (300 bloom) or 1.5 g (175 bloom) into 96 mℓ of RPMI 1640 containing 25 m*M* HEPES, but without $NaHCO_3$, by continuous stirring on a magnetic stir-plate warmed to 60°C. When completely dissolved, but before cooling, the gelatin solution is filter-sterilized through a 0.45 μ*M* filter then cooled to room temperature and completed by addition of 4 mℓ of sterile 5% $NaHCO_3$. The 2× gelatin is stored frozen until used. Cultures of *P. falciparum* suspended in complete RPMI 1640 medium to a 20% hematocrit are further diluted with an equal volume of prewarmed 2× gelatin solution and incubated vertically in a 37°C water bath for 30 min. The resulting 10% hematocrit contains a final gelatin concentration of 0.5% w/v for type 300 bloom and 0.75% for type 175 bloom gelatin. After 30 min, or when the top 5 mm of the solution is cleared of cells, the uninfected and ring-stage infected erythrocytes will have settled in the bottom of the tube. Apparently, the gelatin greatly enhances rouleaux formation among the appropriate erythrocytes. The resulting supernatant will be rich in trophozoite- and schizont-infected cells. These stages are recovered by pipetting the supernatant into a sterile centrifuge tube and spinning at room temperature at 350 to 500 × g for 10 min. Both fractions contain fully viable parasites and either or both can be placed back in culture or used as desired.

C. Synchronization of *P. falciparum* Cultures

As described by Lambros and Vanderberg,[86] the normally unsynchronous cultures of *P. falciparum* can be partially synchronized by washing the infected erythrocytes in 5% aqueous sorbitol. Apparently, the uninfected erythrocytes are impermeable to sorbitol, but changes in erythrocyte permeability, induced by the developing parasites, allow this sugar to enter erythrocytes containing the late ring-stage and older parasites, resulting in lysis. The resulting cell suspensions have only ring-stage parasites estimated to be 18 hr or less old. By combining gelatin separation with the sorbitol synchronization, more uniform development can be achieved. To do this, the mature schizont-stages are separated by the gelatin separation procedure and then added to newly washed erythrocytes which are placed into a candle jar. After 4 hr, when many of the merozoites released from the schizonts will have to invade the fresh cells, the cultures are washed with 5% aqueous sorbitol. The resulting erythrocytes should contain only the newly invaded ring-stages having a 4-hr or less development period.

D. Merozoite Harvesting Procedure

Merozoites of *P. falciparum* that fail to penetrate erythrocytes have a very short extra-cellular life span. This makes isolation of this stage difficult, to say the least. However the following procedure helps to preserve the extracellular merozoites. Cultures are first synchronized by the sorbitol method described above and then monitored until the majority of the parasites are multinucleated schizonts, which then can be concentrated to a 80 to 90% parasitemia by the gelatin method and finally, after washing to remove erythrocyte and moribund parasite debris, resuspended in RPMI 1640 containing 25% v/v human red-cell extract and incubated at 37°C with 30-min periodic agitation. After 6 hr the erythrocytes are settled by gentle centrifugation, leaving the supernatant fluid rich in noninvasive, but morphologically intact, merozoites. These can be further cleared from the contaminating erythrocyte stromata by affinity-chromatography on a concanavalin A column or a antihuman erythrocyte affinity column.

VIII. IN VITRO CULTURES AS A RESEARCH TOOL

A. In Vitro Cultures of *P. falciparum*

1. Advances in Erythrocyte-Stage Cultivation

Soon after the original report of continuous cultivation of *P. falciparum* by Trager and Jensen[4] another group, working independently, reported a 21-day culture of *P. falciparum* that was lost due to an unfortunate laboratory accident.[70] These investigators were subsequently unable to repeat their earlier success and have developed a method of cultivation based on the techniques of Trager and Jensen. Since the original report, several strains of *P. falciparum* have been successfully established in culture,[87-89] but as discussed by Chen and Collins,[89] strains differ greatly in the ease of adaption to in vitro conditions. Indeed, some strains apparently do not become established in culture. Only one other human malaria species has been cultured in vitro, and even that one is a mixed culture of *P. falciparum* and *P. malariae*.[90] Using the petri dish-candle jar technique this mixed culture was maintained for 240 days through 42 subcultures by the time of publication. Multiplication rates for *P. falciparum* were lower than those reported by other investigators, 2-7× per 96 hr vs. 40-60× per 96 hr. To date neither *P. vivax* nor *P. ovale*, the other two human species, have been cultured.

There have been three reports of prolonged cultivation of *P. knowlesi*.[71,91,92] Using the petri dish-candle jar method, Butcher cultured *P. knowlesi* for 4 weeks with erratic growth and low parasitemias using rhesus monkey erythrocytes and serum. Chen, et al.[71] were more successful, growing *P. knowlesi* for up to 183 continuous days with a total parasite development equivalent to a 5.4×10^{82} dilution factor. They used the petri dish-candle jar method maintaining parasitemias between 0.92% and 3.6%. Recently, using the semiautomated tipping culture apparatus developed by Jensen et al.,[78] Wickham et al.[92] reported prolonged cultivation of *P. knowlesi*. They reported that by using human type A or B sera at 14% v/v with an increased glucose concentration they were able to achieve growth rates of up to 5-6× per 24 hr cycle. They contained their experiments up to 12 weeks, during which time they noted that although the parasites continued to grow well, they were nonetheless morphologically abnormal. The cultures were maintained in either the erythrocytes of the rhesus monkey *(Macaca mulatta)* or the natural host, the kra monkey *(M. fascicularis)*, and after 31 days in vitro had undergone no significant antigenic variation as determined by the schizont-infected cell agglutination test,[93] although other variants were detected to a small degree.

Another simian species, *P. fragile* which resembles the human species *P. falciparum*, has been cultured in vitro by Chin and co-workers.[94] At the time of publication this species had been in culture for over 7 months in vitro using rhesus monkey cells in RPMI 1640 medium supplemented at 10% v/v with a 1:1 mixture of rhesus and human type AB serum. The growth rate averaged 5× every 3 to 4 days.

2. Studies with Antimalarial Compounds

The culture methods lend themselves to studies of drug sensitivities and two new in vitro methods for in field determination of drug resistance have been developed. The original in vitro drug resistant assay of Reickmann and Lopez-Antuñano,[9] known subsequently as the macrotechnique, was replaced with a newly developed microtechnique based on the petri dish-candle jar method.[10] In this method small multi-well tissue culture plates were substituted for the petri dishes. This method had many advantages over the older method, not the least being that sufficient blood to run the test could be obtained by a finger-prick, as opposed to a larger sample drawn by venepuncture for the macrotechnique. Another advantage was increased repeatability, due in part, to the fact that unlike the older method, the initial parasitemia had little influence on the outcome of the test.

Another method, developed by Dinh and Trager[88] that extends the test from 24 hr in vitro to 48 hr appears to hold promise. The procedure is quite simple, and theoretically, the additional 24-hr period (when compared to the Reickmann microtechnique) may have the advantage of increased sensitivity because all of the parasite developmental stages are exposed to the action of the drug being examined.

The cultures of *P. falciparum* have been used to develop culture lines of the parasites that are resistant to chloroquine[95] and to pyrimethamine[96] from strains originally sensitive to these drugs. Studies of this type may soon elucidate mechanisms of the emergence of drug resistant strains. In these studies the parasites were grown in increasing concentrations of the respective drugs. In a study of drug actions on cultured *P. falciparum*, Richards and Maples[97] pointed out the following advantages of in vitro drug studies: cultures offer a means of studying the intrinsic drug activity quickly and cheaply; higher drug concentrations can be assessed in vitro than would be possible in vivo, allowing observation are less active compounds than would be possible in animal trials; complications of host metabolism are eliminated; only small quantities of difficult to synthesize compounds are required; mechanism of drug action can be investigated; *P. falciparum*, as opposed to simian species, can be tested directly, conserving valuable experimental animals, and specific parasite metabolic cycles can be more readily examined with regard to drug action as in the search of new antimalarial compounds. Finally, Trager et al. has demonstrated the suitability of the culture methods in screening new potential antimalarial compounds.[98]

3. *Studies of Parasite Ultrastructure and Development*

The first definitive study of ultrastructural development of the asexual erythrocytic stages of *P. falciparum* was made possible by the in vitro cultures.[99] Many of the fine-structural details were predictable from studies of other *Plasmodium* species. However, there were several interesting observations made in this study (Figure 6). The unique erythrocyte-membrane alterations, evidently containing proteins synthesized by the parasite,[100] first appear as plaques beneath the erythrocyte plasmalemma in cells harboring the early trophozoite stage (Figure 6C). These knob-like alterations gradually coalesce together and appear to be sloughed into the surrounding milieu. In a different study, Langreth et al.[85] showed that after prolonged in vitro cultivation the cultures began to be populated with parasites that do not produce knobs (Figure 6B). In each developing merozoite anlagen observed in the E.M. study there was an active Golgi apparatus near the anterior end apparently functioning in the production of coated vesicles thought to be the precursors of the polar organelles, the rhoptries and micronemes. In yet another electron microscopic study the invasion of an erythrocyte by a merozoite was observed in the presence of the antimalarial drug chloroquine at concentrations sufficient to inhibit the development of the feeding stages.[101]

Studies on gametocytogenesis[66] and gametogenesis[67] revealed that, unlike the asexual stages, which develop in 48 hr or slightly less, the macro- and microgametocytes require 10 or more days to mature and that gametocyte induction can occur in vitro and may be modulated by cyclic-AMP (L. Miller, personal communication) and other, yet undescribed factors.[102] Apparently, maturation of the gametocytes and their subsequent production of gametes can occur in vitro but all of the elements responsible for consistent reproduction of this phenomenon are not yet known.[67]

Perhaps one of the most interesting of all studies made possible by the in vitro cultures of *P. falciparum* are those dealing with the relationship that exists between this parasite and certain human hemoglobin anomalies. One only need look at a map of the distribution of certain red-cell characteristics — sickle hemoglobin, thalassemia, glucose-6-phosphate dehydrogenase deficiency — to see how remarkably well coordinated these traits are with holoendemic regions of *P. falciparum*. As suggested by Allison,[103] this parasite evidently applies considerable selection pressure and that because of apparent reduced susceptibility of individuals with these hemoglobin anomalies they prevail in the human population. Just

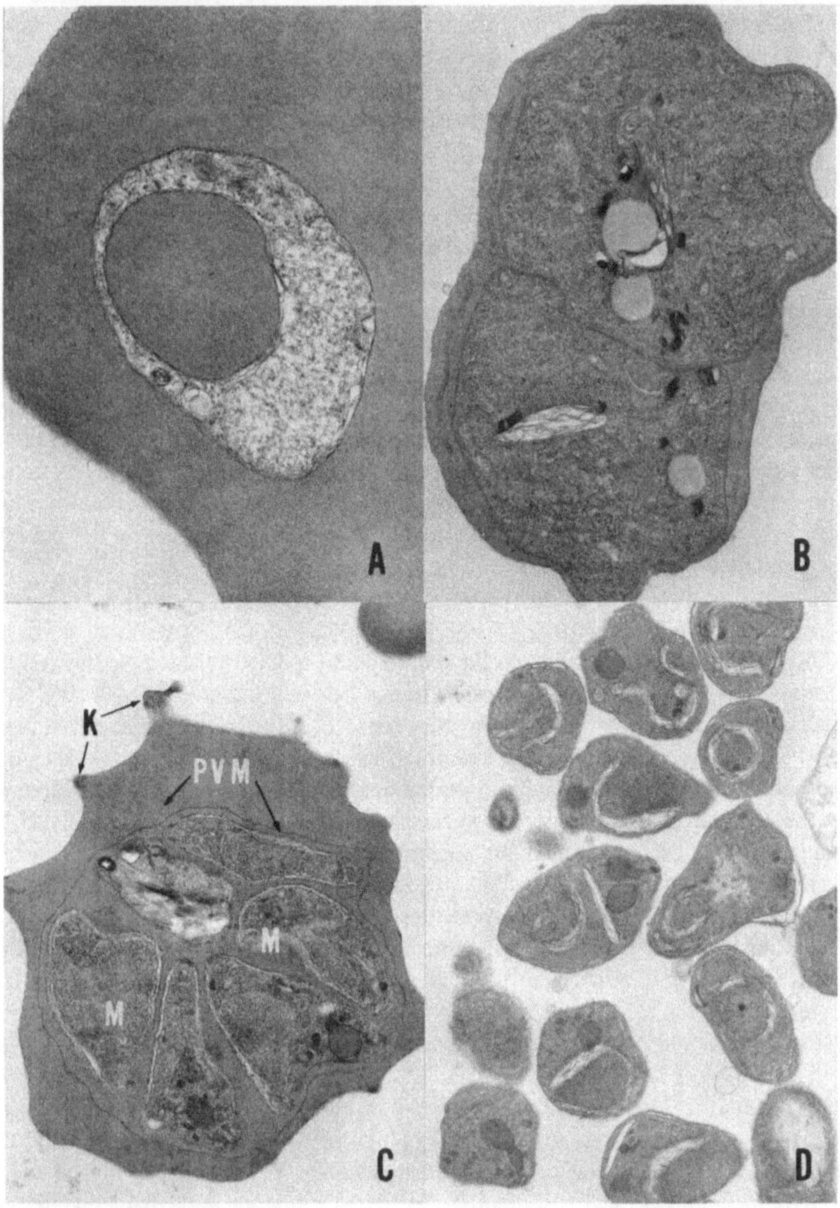

FIGURE 6. Electron micrographs of cultured *P. falciparum*. (A) Young ring stage parasite without pigment granules — indicating that it has not yet begun to metabolize the host-cell hemoglobin. (B) Two trophozoites located within a single erythrocyte. Note that there are no apparent structural changes in host-cell membrane. Generally parasites isolated from natural infections produce marked alterations in the erythrocyte membrane (K, in micrograph C) but after prolonged cultivation these "knobby" infected erythrocytes become fewer in number. (C) A mature segmented schizont showing the expanded parasitophorous vacuole membrane (PVM). (D) The schizonts (C) release the merozoites (D) which are the infective stages of the parasite and, since they are the extracellular stages in the blood stream and are the stages most susceptible to immunological attack they are of great interest as a source of vaccine antigens. (Micrographs provided by Dr. Susan Langreth of Uniformed Services University Medical School, Bethesda, Maryland.)

how persons with sickle cell hemoglobin are protected from *P. falciparum* was the subject of the studies of Friedman,[104] who cultured *P. falciparum* in sickle cell (SS), sickle cell carrier (AS), and normal cells (AA). In the candle jar with 17% oxygen all three cell types supported parasite development to the same degree. However, when the oxygen concentration was reduced to 5% or less considerable sickling of the AS and SS cells occurred and, apparently because of the precipitation of the paracrystalline arrays in the abnormal hemoglobin, caused the death of parasites. Accordingly, in a person infected with *P. falciparum* some of the parasitized erythrocytes would pass through areas of the vascular system where the oxygen concentration was below 5% and the resulting sickling would destroy the parasites. Thus a person with AS or SS hemoglobin would have the advantage over the normal AA in a region where *P. falciparum* was a serious pathogen, as it is in West Africa where sickle-cell is also common. In a subsequent study Friedman and his co-workers showed that infected erythrocytes had a lower intracellular pH which greatly enhanced the sickling rate of the AS or SS cells. In the sickled cells a rapid loss of potassium was the apparent cause of parasite death because parasites in sickled cells grown in medium high in potassium were protected against the fatal effects of sickling. In yet another study, erythrocytes containing hemoglobins C, S and N_{BaLt} had significantly reduced capacity to support the development of the parasite in vitro,[105] giving evidence for the role of *P. falciparum* in the preservation of these hemoglobin anomalies in certain populations.

4. Studies in Parasite Biochemistry

Carefully controlled experiments on the combination of CO_2, O_2, and N_2 showed that the oxygen requirements for in vitro cultivation range between 0.5 to 17% with 3% O_2 being optimum when CO_2 concentrations were between 1 to 2%. Oxygen concentrations lower than 0.5% and higher than 18% were deleterious, suggesting that *P. falciparum* is an obligate microaerophile.[106] These studies and others demonstrating the lack of an active Kreb's cycle suggest that the parasite may depend upon reactions involving oxygenases and hydroxylases in which oxygen is a chemical reactant. Such oxygenases are often metalloproteins which are usually sensitive to chelating agents. Diethyldithiocarbomate, a chelating agent with proven inhibitory activity against many of these enzymes, and especially its diethyl derivative, disulfiram, are not very toxic to mammals. Thus disulfiram was tested in vitro cultures for possible inhibitory activity against *P. falciparum.* It proved to be effective at 0.1 μg/mℓ suggesting that the role of oxygen in *P. falciparum* metabolism may be to interact with certain oxygenases and metalloproteins such as cytochrome oxidase.[107]

5. Studies in Malaria Immunity and Serology

Perhaps the greatest role yet to be played by the in vitro cultivation of *P. falciparum* is the development of a malaria vaccine. A crude parasite vaccine tested by Reese et al.[108] in *Aotus trivirgatus* monkeys was sufficiently effective to kindle the hope that a malaria vaccine might someday be possible. Subsequently, many studies involving antigenic analysis of cultured *P. falciparum* have been published and many more will be forthcoming.[109-112]

Antigens of *P. falciparum,* derived from continuous cultures, are being used for serodiagnosis of malaria in indirect fluorescent antibody,[113] enzyme-linked immunosorbent assay (ELISA),[114] and a solid phase radioimmunoassay techniques.[115]

B. Extracellular Cultures of *P. lophurae*

The real advantage of extracellular cultivation of the obligate parasite, *P. lophurae,* was that this technique allows one to separate the metabolism of the parasite from that of the host-cell. Ultrastructural studies of *P. lophurae* developing extracellularly illustrated the necessity of the host-cell derived parasitophorous vacuole membrane to the developing parasite and how this membrane actually grew as the parasite developed[80] (Figure 7). Studies

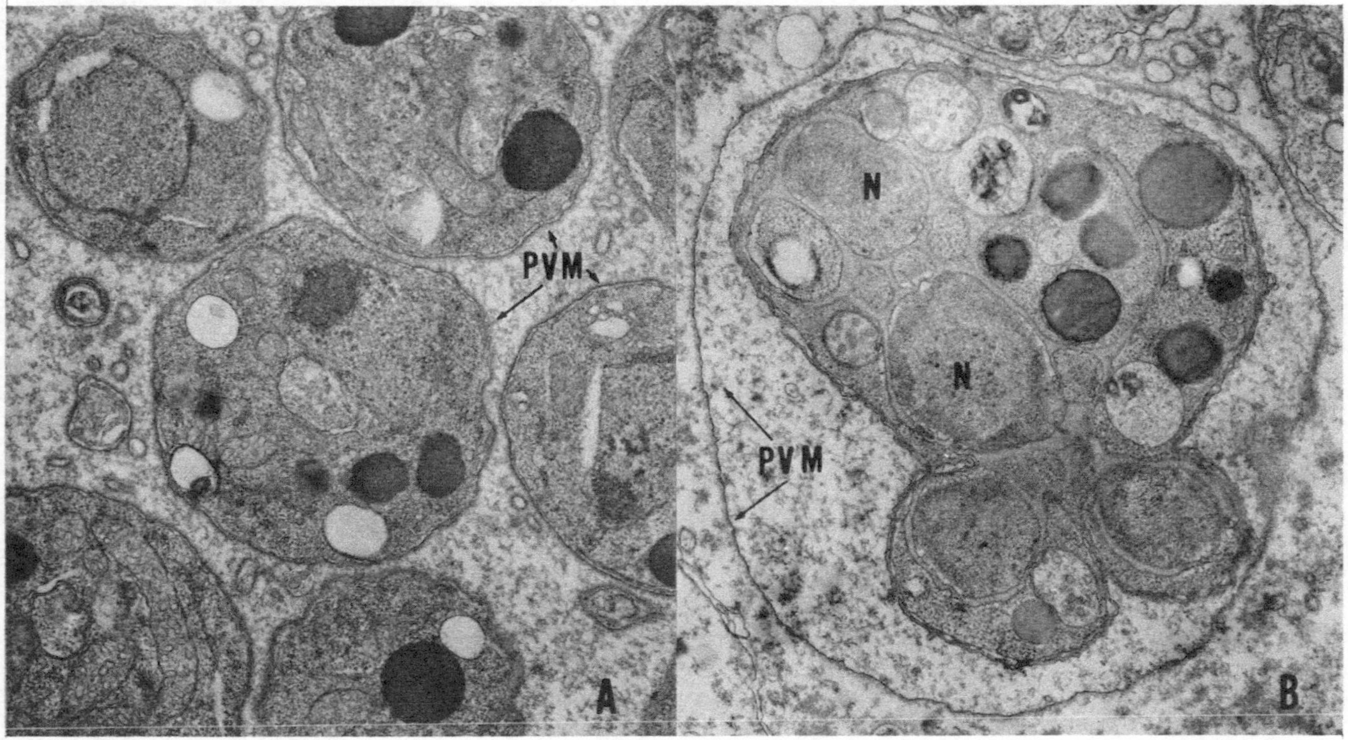

FIGURE 7. Electron micrographs of *P. lophurae* developing extracellularly. (A) Young uninuclear trophozoites with the erythrocyte membrane derived parasitophorous vacuole membrane (PVM) in close apposition to the parasite plasmalemma. (B) After several hours of extracellular-development the parasites become multinucleated schizonts. Note the greatly expanded parasitophorous vacuole whose limiting membrane (PVM) has greatly increased in size and no longer lies adjacent to the parasite. (From Langreth, S. G. and Trager, W., *J. Protozool.*, 20, 606, 1973. With permission.)

Table 8
EFFECT OF CoA IN THE EXTRACELLULAR DEVELOPMENT OF *P. LOPHURAE;* ITS PARTIAL REPLACEMENT BY PHOSPHOPANTETHEINE (PPS) BUT NOT BY PHOSPHOPANTOTHENOYLCYSTEINE (PPC) OR PHOSPHOPANTOTHENIC ACID (PPA)

				Av. % Parasites after 1 day[a]		
Exp.	Flasks	Supplement	Cons. m*M*	>1 nucleus	Degenerate	Incorporation (cpm/100 × 10^6 parasites per hr of exposure)
1	1-3	0		16	9	398
	4-6	PPA	0.06	16	7	414
	7-9	CoA	0.05	21	4	665
2	1-3	0		12	16	207
	4-6	PPA	0.06	16	22	266
	7-9	CoA	0.05	16	7	356
3	1-3	CoA	0.05	30	11	646
	4-6	PPC	0.05	16	16	473
	7-9	PPS	0.05	21	13	602
	10-12	0		15	16	347

Note: The medium was one-third strength erythrocyte extract with all supplements except CoA as indicated. At 16 hr each flask received, in Exps. 1, 2 2.5 μCi of L-{methyl-^{14}C} methionine, in Exp. 3 2.5 μCi of L-{U-^{14}C} isoleucine. Samples were taken 3-4 hr later.

[a] At 0 time % with >1 nucleus was 6, 4, and 9 for Exp. 1-3, respectively.

using radiolabeled substrates showed that the parasites have a requirement for exogeneous sources of ATP and coenzyme A.[42,116] The requirement for coenzyme A (Table 8) seems to rest on a major biosynthetic defect of the parasite because, although the enzymes of the pathway from pantothenic acid to CoA are found in the duck erythrocyte, they could not be demonstrated in *P. lophurae*.[117,118] The exogenous ATP requirement may be related to transport of materials across the parasitophorous vacuole membrane,[119] which contains ATPase on its inner surface.[120] It is clear that only a small beginning has been made toward dissection of the intimate physiological relationships between an intracellular parasite and its host-cell.

C. Exoerythrocytic Cultivation

Because at this writing the exoerythrocytic cultivation of mammalian stages of malarial parasites is still in its infancy only studies dealing with avian exoerythrocytic stage cultures will be discussed. Because these stages are found in vivo in cells of the reticuloendothelial system and are thus not readily observable, the first obvious advantages of cultures of exoerythrocytic stages was the opportunity for detailed studies on morphology and development of these stages (Figure 8, a comparison of avian and mammalian exoerythrocytic culture forms). Hegner and Wolfson[121] were the first to report direct observations of these forms in vitro when they described stages of *P. cathemerium* from canary lung in a hanging drop culture. A thorough study of the morphology and behavior of living exoerythrocytic stages of *P. gallinaceum* and *P. fallax* was reported by Huff et al.[53] using phase-contrast microscopy and time-lapse photography. Although the first ultrastructural studies on exoerythrocytic stages of Meyer and Oliveira-Musacchio[122,123] were lacking in precise detail, an excellent complete study was reported later by Hepler and colleagues.[124]

As has been discussed for the erythrocytic stages of *P. falciparum*, the exoerythrocytic

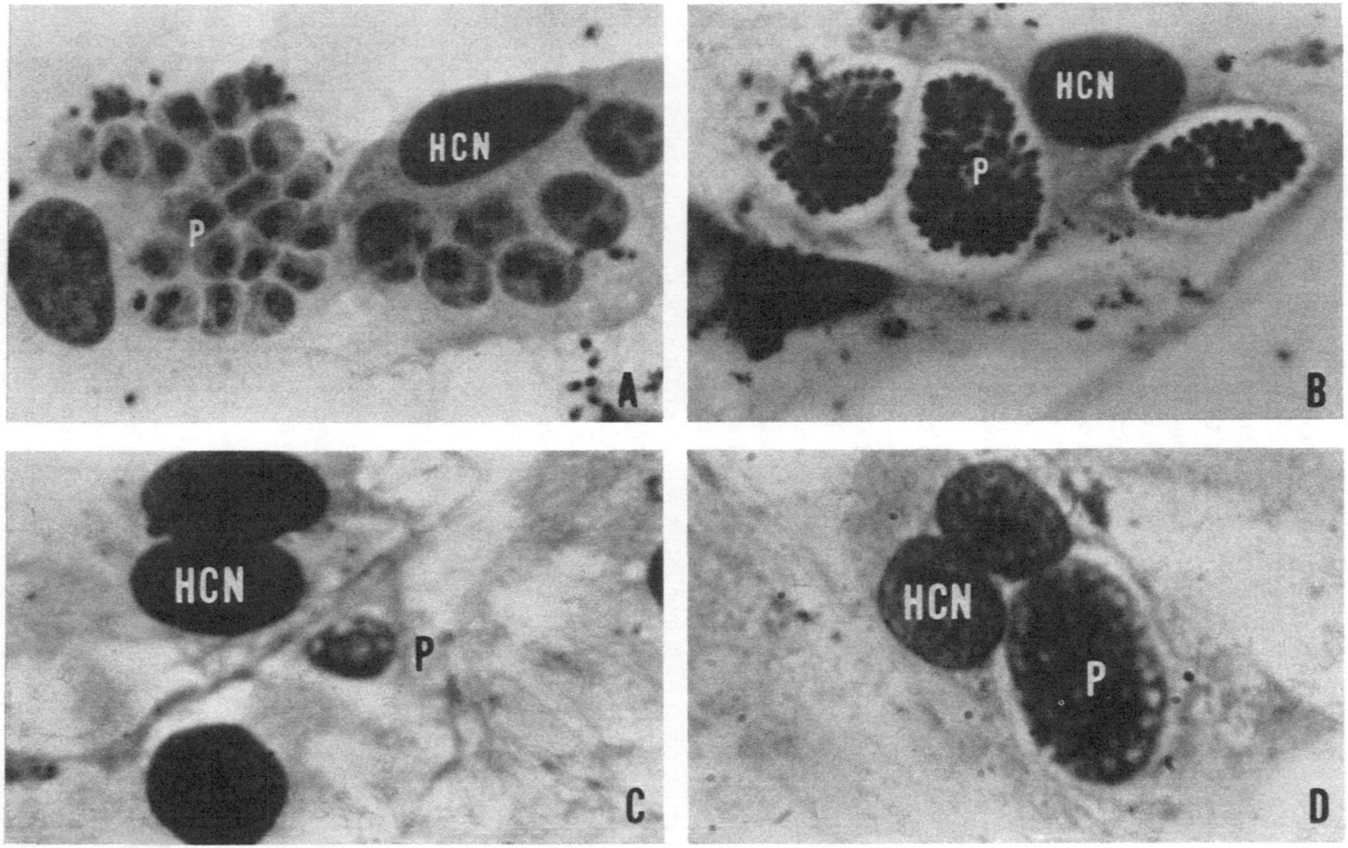

FIGURE 8. Photomicrography of avian *(P. fallax)* and mammalian *(P. berghei)* exoerythrocytic stages (EE) grown in culture. (A) Young EE trophozoites of *P. fallax* growing in turkey brain cells in vitro. (B) Mature segmenting schizonts of *P. fallax* in turkey brain cells. (C) An uninucleated trophozoite of *P. berghei* in embryonic mouse liver cells. (D) A large maturing schizont of *P. berghei*. To date no secondary schizogony has been reported in the mammalian plasmodia grown in vitro. (Photomicrographs provided by Drs. Richard Beaudoin and Michael Hollingdale of the National Naval Medical Research Institute, Bethesda, Maryland.)

cultures provided an excellent tool for studying chemotherapeutic agents on these stages. The first attempts to study the direct actions of drugs on exoerythrocytic stages were reported by Tonkin[125] when she reported the effects of different drugs at various concentrations using primary cultures of chick tissues infected with *P. gallinaceum*. Antiparasitic effects were observed with several compounds, quinine sulfate being only slightly effective and then only at concentrations that were toxic to the cells. This observation explained how quinine, although effective against the blood stages of the parasites, is ineffective in preventing relapses. The 8-aminoquinoline, primaquine that is effective in controlling relapses, was studied by Beaudoin and Aikawa[126] using refined culture techniques and electron microscopy to observe the effects of the drug on *P. fallax* cultured in turkey brain cells. Their study revealed that the site of drug action was apparently the parasite mitochrondria, which, unlike those of the host-cell, were greatly swollen in cultures containing the drug. In another study, an experimental antimalarial showed toxic effects in conjunction with swollen endoplasmic reticulum (Figure 9). In yet another study Beaudoin and his co-workers showed that a strain of *P. gallinaceum* could retain its ability to transmit resistance to pyrimethamine through several subcultures of exoerythrocytic stages in the absence of an erythrocytic cycle.[127] With this demonstration they completed the last link in the life cycle of the parasite through which drug resistance had been transmitted.

The immunology of malaria infections is another area of study that has been greatly expanded by the cultures of exoerythrocytic cultures. Evidence of immunity to exoerythrocytic stages was demonstrated by Graham et al.[128] where turkey poults, inoculated with sera from birds recovering from *P. fallax* infections, were later challenged with exoerythrocytic stages produced in culture. The birds that received serum showed 90% fewer exoerythrocytic forms than the controls. Later, Holbrook and colleagues[129] extended the study by immunizing birds with exoerythrocytic merozoites harvested from cultured turkey fibroblasts. Strong protective immunity was demonstrated in the vaccinated birds; 90% of which survived challenge versus 30% in the controls. Immunization with exoerythrocytic merozoites failed to protect against challenge with erythrocytic forms. Other experiments by Holbrook and colleagues[130] showed that merozoites harvested from cell cultures of *P. fallax*, when used to immunize mice, offered some protection when challenged with sporozoites of *P. berghei*, a mammalian *Plasmodium*. In their conclusions the authors suggested that the immunity may have been nonspecific, but did not rule out the possibility that some cross immunization had occurred.

IX. THE UNSOLVED PROBLEMS

Ideally one would hope that malarial parasites could one day be cultured in vitro through each of its three major "cycles," sporogony, tissue and erythrocytic schizogony, and gametogony. However, as discussed above, this has not been possible and indeed may not be possible for a long time to come. To date, in vitro erythrocytic schizogony has been achieved only with the primate species: *P. falciparum, P. malariae, P. knowlesi,* and *P. fragile*. Accordingly, the next steps to achieve would be routine gametocytogenesis and gametogenesis of one of these species. This may soon be a reality according to unpublished reports that cyclic AMP may play a stimulatory role in gametocyte induction in vitro (L. H. Miller, personal communication). The final maturation of the gametes (called exflagellation in the male gamete) occurs routinely in vitro and fertilization to form the öokinete (zygote) appears to be a readily obtainable procedure, at least in *P. berghei*, a rodent species.[131] Oogenesis, the beginning of sporogony, has proven to be a difficult process to achieve in culture. In truth, getting from the zygote to infective sporozoites has been, to date, a series of frustrating failures (see Vanderberg et al.[65] for a review). However, if infective sporozoites could be produced the next steps look hopeful, thanks to the pioneering work of Beaudoin and his

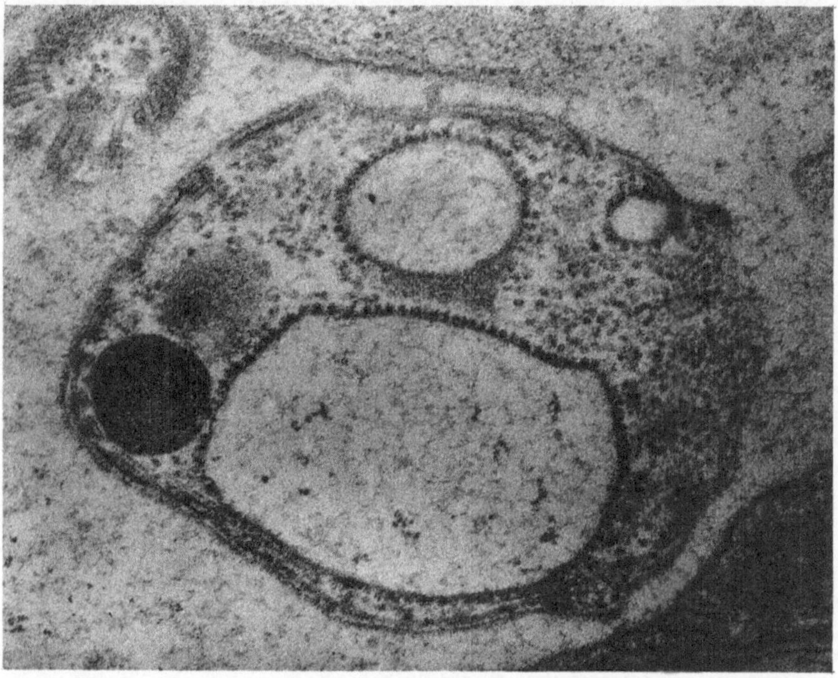

FIGURE 9. Electron micrograph of *P. fallax* showing ultrastructural swelling of the endoplasmic reticulum when the parasites were grown in vitro in the presence of the experimental antimalarial drug, WR 49808, a naphthoquinone under investigation by the Walter Reed Army Institute. (Micrograph provided by Dr. Richard Beaudoin, National Naval Medical Research Institute, Bethesda, Maryland.)

colleagues.[63] Finally, if infective sporozoites were available, and if they could be cultured in the appropriate cell lines, and if these exoerythrocytic cultures produced infective erythrocytic merozoites, we might arrive back at the beginning of our saga, the erythrocytic cycle. Obviously much remains to be done.

X. FUTURE PROSPECTS FOR PLASMODIUM CULTURES

In vitro cultures have always been tremendously useful tools for studying the biochemistry, biology, serology and immunology, and development of the organism in question and such will continue to be so for malarial parasites. However, the greatest hope for the cultures of *P. falciparum* lies with the renewed possibility for a vaccine against one of man's deadliest scourges. The production of such a vaccine may be a difficult process due, in part, to the fact that malarial antigens are poorly immunogenic, may be immunosuppressive, and probably immunopathogenic. The identification of the protective antigens and their purification from the pathogenic ones is the difficult task ahead of us. This job may be aided considerably by developing hybridomas that produce protective monoclonal antibodies. Such monoclonal antibodies, if they prove to be protective, may greatly facilitate the identification and purification of the protective parasite antigens. Further down the road, but still feasible, is the production of protective malarial antigens from bacterial cultures implanted with the appropriate plasmodial genes by recombinant DNA methodologies. However, all of the above achievements will be reached only by careful analysis of cultured *P. falciparum* antigens.

From a practical standpoint, the replacement of the human (or rabbit) serum requirements for in vitro cultures of *P. falciparum* would be helpful for both economic and health reasons. Not only is human serum a valuable commodity, but the pooling of large lots of human serum greatly increases the risk of viral contamination, especially with non A-non B hepatitis viruses. Accordingly, development of a human vaccine using human serum would be a risky enterprise. Current studies on serum replacement using freshly collected, pooled bovine sera supplemented with peptones appears to be promising.

REFERENCES

1. **Garnham, P. C. C.,** *Progress in Parasitology,* Athlone Press, University of London, London, 1971, 67.
2. World Health Organization Technical Report Series, No. 537, Malaria control in countries where time-limited eradication is impractical at present, W. H. O., Geneva, 1974, 9.
3. World Health Organization Technical Report Series, No. 529, Chemotherapy of malaria and resistance to antimalarials, W. H. O., Geneva, 1973.
4. **Trager, W. and Jensen, J. B.,** Human malaria parasites in continuous culture, *Science,* 193, 673, 1976.
5. **Markus, M. B.,** Terminology for invasive stages of protozoa of the subphylum Apicomplex (Sporozoa), *So. African J. Sci.,* 74, 105, 1978.
6. **Krotoski, W. A., Krotoski, D. M., Garnham, P. C. C., Bray, R. S., Killick-Kendrick, R., Draper, C. C., Targett, G. A. T., and Guy, M. W.,** Relapses in primate malaria: discovery of two populations of exoerythrocytic stages, *Br. Med. J.,* 1, 153, 1980.
7. **Bass, C. C. and Johns, F. M.,** The cultivation of malarial plasmodia *(Plasmodium vivax* and *Plasmodium falciparum) in vitro, J. Exp. Med.,* 16, 567, 1912.
8. **Black, R. H.,** The effect of antimalarial drugs on *Plasmodium falciparum* (New Guinea strains) developing *in vitro, Trans. R. Soc. Trop. Med. Hyg.,* 40, 163, 1946.
9. **Rieckmann, K. H. and Lopez Antuanano, F. J.,** Cloroquine resistance of *Plasmodium falciparum* in Brazil detected by a simple *in vitro* method, *Bull. W. H. O.,* 45, 157, 1971.
10. **Rieckmann, K. H., Sax, L. J., Cambell, G. H., and Mrema, J. E.,** Drug sensitivity of *Plasmodium falciparum.* An *in vitro* microtechnique, *Lancet,* 22, 1978.
11. **Trager, W.,** The development of the malaria parasite *Plasmodium lophurae* in red blood cell suspensions *in vitro, J. Parasitol.,* 33, 345, 1947.
12. **McGhee, R.B. and Trager, W.,** The cultivation of *Plasmodium lophurae in vitro* in chicken erythrocyte suspensions and the effects of some constituents of the culture medium upon its growth and multiplication, *J. Parasitol.,* 36, 123, 1950.
13. **Ball, E. G., Anfinsen, C. B., Geiman, Q. M., McKee, R. W., and Ormsbee, R. A.,** *In vitro* growth and multiplication of the malaria parasite, *Plasmodium knowlesi, Science,* 101, 542, 1945.
14. **Geiman, Q. M., Anfinsen, C. B., McKee, R. W., Ormsbee, R. A., and Ball, E. G.,** Studies on malarial parasites. VII. Methods and techniques for cultivation, *J. Exp. Med.,* 84, 583, 1946.
15. **Anfinsen, C. B., Geiman, Q. M., McKee, R. W., Ormsbee, R. A., and Ball, E. G.,** Studies on malarial parasites. VIII. Factors affecting the growth of *Plasmodium knowlesi in vitro, J. Exp. Med.,* 84, 607, 1946.
16. **Geiman, Q. M., Siddiqui, W. A., and Schnell, J. V.,** *In vitro* studies on erythrocytic stages on plasmodia: medium improvement and results with seven species of malarial parasites, *Milit. Med.,* 131, 1015, 1966.
17. **Trigg, P.I. and Gutteridge, W.E.,** A minimal medium for the growth of *Plasmodium knowlesi* in dilution cultures, *Parasitology,* 62, 113, 1971.
18. **Trager, W.,** Coenzyme A and the antimalarial action *in vitro* of antipantothenates against *Plasmodium lophurae, P. coatneyi,* and *P. falciparum, Trans. N.Y. Acad. Sci. Ser. II,* 28, 1094, 1966.
19. **Siddiqui, W. A. and Schnell, J. V.,** *In vitro* and *in vivo* studies with *Plasmodium falciparum* and *Plasmodium knowlesi, Proc. Helm. Soc. Wash.,* 39 (special issue), 204, 1972.
20. **Siddiqui, W. A., Schnell, J. V., and Geiman, Q. M.,** Nutritional requirements for *in vitro* cultivation of a simian malarial parasite, *Plasmodium knowlesi, Milit. Med.,* 134 (special issue), 929, 1969.
21. **Trigg, P. I.,** A new continuous perfusion technique for the cultivation of malarial parasites *in vitro, Trans. R. Soc. Trop. Med. Hyg.,* 62, 371, 1968.
22. **Trager, W.,** Further studies on the survival and development *in vitro* of a malarial parasite, *J. Exp. Med.,* 77, 411, 1943.
23. **Ball, E. G.,** Chemical and nutritional observations on malarial parasites grown *in vitro, Fed. Proc. Fed. Am. Soc. Exp. Biol.,* 5, 397, 1946.
24. **Maier, J. and Riley, E.,** Inhibition of antimalarial action of sulfonamides by p-aminobenzoic acid, *Proc. Soc. Exp. Biol. Med.,* 50, 152, 1942.
25. **Ferone, R.,** Folate metabolism in malaria, *Bull. W. H. O.,* 55, 291, 1977.

26. **Trager, W.**, Cofactors and vitamins in the metabolism of malarial parasites, *Bull. W. H. O.*, 55, 285, 1977.
27. **Polet, H. and Conrad, M. E.**, The influence of three analogs of isoleucine on *in vitro* growth and protein synthesis of erythrocytic forms of *Plasmodium knowlesi*, *Proc. Soc. Exp. Biol. Med.*, 130, 581, 1969.
28. **Ryley, J. F.**, The mode of action of proguanil and related antimalarial drugs, *Br. J. Pharm.*, 8, 424, 1953.
29. **McCormick, G. J., Canfield, C. J., and Willet, G. P.**, *Plasmodium knowlesi: in vitro* evaluation of antimalarial activity of folic acid inhibitors, *Exp. Parasitol.*, 30, 88, 1971.
30. **Siddiqui, W. A., Schnell, J. V., and Geiman, Q. M.**, A model *in vitro* system to test the susceptibility of human malarial parasites to antimalarial drugs, *Am. J. Trop. Med. Hyg.*, 21, 392, 1972.
31. **Trigg, P. I. and Shakespeare, P. G.**, Factors affecting the long-term cultivation of the erythrocytic stages of *Plasmodium knowlesi in vitro*, in, *Biochemistry of Parasites and Host-Parasite Relationships*, van den Bossche, H., Ed., North-Holland, Amsterdam, 1976, 435.
32. **Young, M. D., Porter, J. A., and Johnson, C. M.**, *Plasmodium vivax* transmitted from man to monkey to man, *Science*, 153, 1006, 1966.
33. **Geiman, C. M. and Meagher, M. J.**, Susceptibility of a new world monkey to *Plasmodium falciparum* from man, *Nature (London)*, 215, 437, 1967.
34. **Bertagna, P., Cohen, S., Geiman, Q. M., Haworth, J., Königk, E., Richards, W. H. G., and Trigg, P. I.**, Cultivation of malaria parasites, *Bull. W. H. O.*, 47, 357, 1972.
35. **Trager, W.**, A new method for intraerythrocytic cultivation of malaria parasites *(Plasmodium coatneyi* and *P. falciparum)*, *J. Protozool.*, 18, 239, 1971.
36. **Moore, G. E., Gerner, R. E., and Franklin, H. A.**, Culture of normal human leukocytes, *JAMA*, 199, 519, 1967.
37. **Trager, W.**, Prolonged cultivation of malaria parasites *(Plasmodium coatneyi* and *P. falciparum)*, in *Biochemistry of Parasites and Host-Parasite Relationships*, van den Bossche, H., Ed., North-Holland, Amsterdam, 1976, 427.
38. **Jensen, J. B. and Hammond, D. M.**, Ultrastructure of the invasion of *Eimeria magna* sporozoites into cultured cells, *J. Protozool.*, 22, 411, 1975.
39. **Jensen, J. B. and Trager, W.**, *Plasmodium falciparum* in culture: use of outdated erythrocytes and description of the candle jar method, *J. Parasitol.*, 63, 883, 1977.
40. **Trager, W.**, The nutrition of an intracellular parasite (avian malaria), *Acta Tropica*, 14, 289, 1957.
41. **Trager, W.**, Folinic acid and non-dialyzable materials in the nutrition of malaria parasites, *J. Exp. Med.*, 108, 753, 1958.
42. **Trager, W.**, Malaria parasites *(Plasmodium lophurae)* developing extracellularly *in vitro:* incorporation of labeled precursors, *J. Protozool.*, 18, 392, 1971.
43. **Garnham, P. C. C.**, *Malaria Parasites and Other Haemosporidia*, Blackwell Scientific, Oxford, 1966, 661.
44. **Porter, R. J.**, Studies in tissue culture of exoerythrocytic schizogony in avian malarial parasites. Symposium on exoerythrocytic forms of malarial parasites, *J. Parasitol.*, 34, 300, 1948.
45. **Hawking, F.**, Tissue culture of plasmodia, *Br. Med. Bull.*, 8, 16, 1951.
46. **Pipkin, A. C. and Jensen, D. V.**, Avian embryos and tissue culture in the study of parasitic protozoa. I. Malarial parasites, *Exp. Parasitol.*, 7, 491, 1956.
47. **Huff, C. G.**, Cultivation of the exoerythrocytic stages of malarial parasites, *Am. J. Trop. Med. Hyg.*, 13, 171, 1964.
48. **Huff, C. G. and Bloom, W.**, A malarial parasite infecting all blood and blood-forming cells of birds, *J. Infect. Dis.*, 57, 315, 1935.
49. **Hawking, F.**, Tissue culture of malaria parasites *(Plasmodium gallinaceum)*, *The Lancet*, 246, 693, 1944.
50. **Hawking, F.**, Growth of protozoa in tissue culture. I. *Plasmodium gallinaceum*, exoerythrocytic forms, *Trans. R. Soc. Trop. Med. Hyg.*, 39, 245, 1945.
51. **Zuckerman, A.**, Infections with *Plasmodium gallinaceum* in chick embryos induced by exoerythrocytic and blood stage, *J. Infect. Dis.*, 79, 1, 1946.
52. **Meyer, H. and Musacchio, M. O.**, *Plasmodium gallinaceum* in tissue cultures: results obtained during 4 years of uninterrupted cultivation of the parasite *in vitro*, *Proc. 6th Int. Cong. Trop. Med. Mal.*, (Lisbon, September 9, 1958), 10, 1959.
53. **Huff, C. G., Pipkin, A. C., Weathersby, A. B., and Jensen, D. V.**, The morphology and behavior of living exoerythrocytic stages of *Plasmodium gallinaceum* and *P. fallax* and their host cells, *J. Biophys. Biochem. Cytol.*, 7, 93, 1960.
54. **Meyer, H.**, Cultivation of the exoerythrocytic form of *Plasmodium gallinaceum* in tissue cultures of embryonic chicken brain, *Nature (London)*, 160, 155, 1947.
55. **Garnham, P. C. C.**, The continuing mystery of relapses in malaria, Commonwealth Institute of Helminthology, *Protozool. Abstr.*, 1, 1, 1977.
56. **Hawking, F., Perry, W. L. M., and Thurston, J. P.**, Tissue forms of a malaria parasite, *Plasmodium cynomolgi*, *The Lancet*, 1, 783, 1948.

57. **Dubin, I. N.,** Bodies suggesting exoerythrocytic forms of *Plasmodium vivax* in tissue culture, *Proc. Soc. Exp. Biol. Med.*, 65, 154, 1947.
58. **Dubin, I. N.,** A search for exoerythrocytic forms in human malaria by means of tissue cultures of bone marrow, *J. Natl. Mal. Soc.*, 7, 330, 1948.
59. **Foley, D. A., Kennard, J., and Vanderberg, J. P.,** *Plasmodium berghei:* Infective exoerythrocytic schizonts in primary monolayer cultures of rat liver cells, *Exp. Parasitol.*, 47, 166, 1978.
60. **Dubin, I. N., Laird, R. L., and Drinnon, V. P.,** Further observations on the development of sporozoites of *Plasmodium gallinaceum* into cryptozoites in tissue culture, *J. Natl. Mal. Soc.*, 9, 119, 1950.
61. **Doby, J.M. and Barker, R.,** Essais d'obtention *in vitro* des formes preerythrocytaires de *Plasmodium vivax* en cultures de cellules hepatiques humaines enoculees par sporozoites, *Soc. Biol. Rennes., C. R. Seances*, 170, 661, 1976.
62. **Beaudoin, R. L., Strome, C. P. A., and Clutter, W. G.,** Cultivation of avian malaria parasites in mammalian liver cells, *Exp. Parasitol.*, 36, 355, 1974.
63. **Strome, C. P. A., Desantis, P., and Beaudoin, R. L.,** Cultivation of exoerythrocytic stages of *Plasmodium berghei* from sporozoite to merozoite, *In Vitro*, 15, 531, 1979.
64. **Sinden, R. E. and Smith, J.,** Culture of the liver stages (exoerythrocytic schizonts) of the rodent malaria parasites from sporozoites *in vitro*, *Trans. R. Soc. Trop. Med. Hyg.*, 74, 134, 1980.
65. **Vanderberg, J. P., Weiss, M. M., and Mack, S. C.,** *In vitro* cultivation of the sporogonic stages of *Plasmodium:* a review, *Bull. W. H. O.*, 55, 377, 1977.
66. **Jensen, J. B.,** Observations on gametocytogenesis in *Plasmodium falciparum* from continuous culture, *J. Protozool.*, 26, 129, 1979.
67. **Carter, R. and Beach, R. F.,** Gametogenesis in culture by gametocytes of *Plasmodium falciparum*, *Nature (London)*, 270, 240, 1977.
68. **Speer, C. A., Silverman, P. H., and Schiewe, S. G.,** Cultivation of erythrocytic stages of *Plasmodium berghei* in Leydig tumor cell cultures, *J. Parasitenkd.*, 50, 237, 1976.
69. **Speer, C. A., Silverman, P. H., and Schiewe, S. G.,** Cultivation of the erythrocytic stages of *Plasmodium berghei* in primary bone marrow cells, *J. Parasitol.*, 62, 657, 1976.
70. **Haynes, J. D., Diggs, C. L., Hines, F. A., and Desjardins, R. E.,** Culture of human malaria parasites *(Plasmodium falciparum)*, *Nature (London)*, 263, 767, 1976.
71. **Chen, Z., Gao, M., Li, Y., Han, S., and Zhang, N.,** Studies on the cultivation of erythrocytic stage of *Plasmodium in vitro*, *Chin. Med. J.*, 93, 31, 1980.
72. **Jensen, J. B.,** Some aspects of serum requirements for continuous cultivation of *Plasmodium falciparum*, *Bull. W. H. O.*, 57 (Suppl.), 27, 1979.
73. **Sax, L. J. and Rieckmann, K. H.,** Use of rabbit serum in the cultivation of *Plasmodium falciparum*, *J. Parasitol.*, 66, 1980.
74. **Ifediba, T. and Vanderberg, J. P.,** Peptones and calf serum as a replacement for human serum in the cultivation of *Plasmodium falciparum*, *J. Parasitol.*, 66, 236, 1980.
75. **Jensen, J. B. and Trager, W.,** Some recent advances in the cultivation of *Plasmodium falciparum*, *Israel J. Med. Sci.*, 14, 563, 1978.
76. **Trager, W.,** *Plasmodium falciparum* in culture: an improved continuous flow method, *J. Protozool.*, 26, 125, 1979.
77. **Chin, W.,** A method for large-volume cultivation of malaria parasites based on the principle of the Trager-Jensen culture method, *Trans. R. Soc. Trop. Med. Hyg.*, 73, 334, 1979.
78. **Jensen, J. B., Trager, W., and Doherty, J.,** A semi-automated apparatus for the continuous cultivation of *Plasmodium falciparum*, *Exp. Parasitol.*, 48, 36, 1979.
79. **Siddiqui, W.A.,** Continuous *in vitro* cultivation of *Plasmodium falciparum* in human erythrocytes: description of a single technique to obtain high yields of parasites, in, *Practical Tissue Culture Applications*, Maramorosch, K. and Hirumi, H., Eds., Academic Press, New York, 1979, 267.
80. **Langreth, S. G. and Trager, W.,** Fine structure of the malaria parasite *Plasmodium lophurae* developing extracellularly *in vitro*, *J. Protozool.*, 20, 606, 1973.
81. **Davis, A. G., Huff, C. G., and Palmer, T. T.,** Procedures for maximum production of exoerythrocytic stages of *Plasmodium fallax* in tissue culture, *Exp. Parasitol.*, 19, 1, 1966.
82. **Beaudoin, R. L.,** Should cultivated exoerythrocytic parasites be considered as a source of antigen for a malaria vaccine? *Bull. W. H. O.*, 55, 373, 1977.
83. **Rowe, A. W., Eyster, E., and Kellner, A.,** Liquid nitrogen preservation of red blood cells for transfusion, *Cryobiology*, 5, 119, 1968.
84. **Jensen, J. B.,** Concentration from continuous culture of erythrocytes infected with trophozoites and schizonts of *Plasmodium falciparum*, *Am. J. Trop. Med. Hyg.*, 27, 1274, 1978.
85. **Langreth, S. G., Reese, R. T., Motyl, M. R., and Trager, W.,** *Plasmodium falciparum:* Loss of knobs on the infected erythrocyte surface after long-term cultivation, *Exp. Parasitol.*, 48, 213, 1979.

86. **Lambros, C. and Vanderberg, J. P.,** Synchronization of *Plasmodium falciparum* erythrocytic stages in cultures, *J. Parasitol.*, 65, 418, 1979.
87. **Jensen, J. B. and Trager, W.,** *Plasmodium falciparum* in culture: establishment of additional strains, *Am. J. Trop. Med. Hyg.*, 27, 743, 1978.
88. **Nguyen-Dinh, P. and Trager, W.,** *Plasmodium falciparum* in vitro: determination of chloroquine sensitivity of three new strains by a modified 48 hour test, *Am. J. Trop. Med. Hyg.*, 29, 339, 1980.
89. **Chin, W. and Collins, W. E.,** Comparative studies of three strains of *Plasmodium falciparum* isolated by the culture method of Trager and Jensen, *Am. J. Trop. Med. Hyg.*, 29, 1980 (in press).
90. **Rai-Chowdhuri, A. N., Chowdhury, D. S., and Regis, M. L.,** Simultaneous propagation of *P. malariae* and *P. falciparum* in a continuous culture, *Indian J. Med. Res.*, 70 (Suppl.), 72, 1979.
91. **Butcher, G. A.,** Factors affecting the *in vitro* culture of *Plasmodium falciparum* and *Plasmodium knowlesi*, *Bull. W. H. O.*, 57 (Suppl.), 17, 1979.
92. **Wickham, J.M., Dennis, E. D., and Mitchell, G. H.,** Long term cultivation of a simian malaria parasite (*Plasmodium knowlesi*) in a semi-automated apparatus, *Trans. R. Soc. Trop. Med. Hyg.*, 1980.
93. **Brown, K. N. and Brown, I. N.,** Immunity to malaria: antigenic variation in chronic infections of *Plasmodium knowlesi*, *Nature (London)*, 200, 1286, 1965.
94. **Chin, W., Moss, D., and Collins, W. E.,** The continuous cultivation of *Plasmodium fragile* by the method of Trager-Jensen, *Am. J. Trop. Med. Hyg.*, 28, 591, 1979.
95. **Nguyen-Dinh, P. and Trager, W.,** Chloroquine resistance produced *in vitro* in an African strain of human malaria, *Science*, 200, 1397, 1978.
96. **Golenser, J., Casuto, D., and Pollack, Y.,** *In vitro* induction of resistance to aminopterin in *Plasmodium falciparum*, in press.
97. **Richards, W. H. G. and Maples, B. K.,** Studies of *Plasmodium falciparum* in continuous cultivation. I. The effect of chloroquine and pyrimethamine on parasite growth and viability, *Am. Trop. Med. Parasitol.*, 73, 99, 1979.
98. **Trager, W., Robert-Gero, M., and Lederer, E.,** Antimalarial activity of s-isobutyl adenosine against *Plasmodium falciparum* in culture, *FEBS Lett.*, 85, 264, 1978.
99. **Langreth, S. G., Jensen, J. B., Reese, R. T., and Trager, W.,** Fine structure of human malaria *in vitro*, *J. Protozool.*, 25, 443, 1978.
100. **Kilejian, A.,** Characterization of a protein correlated with the production of knob-like protrusions on membranes of erythrocytes infected with *Plasmodium falciparum*, *Proc. Natl. Acad. Sci. USA*, 76, 4650, 1979.
101. **Langreth, S.G., Nguyen-Dinh, P., and Trager, W.,** *Plasmodium falciparum:* merozoite invasion *in vitro* in the presence of chloroquine, *Exp. Parasitol.*, 46, 235, 1978.
102. **Carter, R. and Miller, L. H.,** Evidence for environmental modulation of gametocytogenesis in *Plasmodium falciparum* in continuous culture, *Bull. W. H. O.*, 57 (Suppl.), 37, 1979.
103. **Allison, A. C.,** Abnormal hemoglobin and erythrocyte enzyme deficiency traits, in *Human Variation and Natural Selection*, Roberts, D. T., Ed., John Wiley & Sons, New York, 1975, 306.
104. **Friedman, M. J.,** Erythrocytic mechanism of sickle cell resistance to malaria, *Proc. Natl. Acad. Sci. USA*, 1994,1978.
105. **Friedman, M. J., Roth, E. F., Nagel, R. L., and Trager, W.,** The role of hemoglobins C, S, and N_{BaLt} in the inhibition of malaria parasite development in vitro, *Am. J. Trop. Med. Hyg.*, 28, 777, 1979.
106. **Scheibel, L. W., Ashton, S. H., and Trager, W.,** *Plasmodium falciparum:* microaerophilic requirements in human red blood cells, *Exp. Parasitol.*, 47, 410, 1979.
107. **Scheibel, L. W., Adler, A., and Trager, W.,** Tetraethylthiuram disulfide (antabuse) inhibits the human malaria parasite *Plasmodium falciparum*, *Proc. Natl. Acad. Sci. USA*, 76, 5303, 1979.
108. **Reese, R. T., Trager, W., Jensen, J.B., Miller, D. A., and Tantravahi, R.,** Immunization against malaria with antigen from *Plasmodium falciparum* cultivated *in vitro*, *Proc. Natl. Acad. Sci. USA*, 75, 5665, 1978.
109. **Campbell, G. H., Mrema, J. E. K., O'Leary, T. R., Jost, R. C., and Rieckmann, K. H.,** *In vitro* inhibition of the growth of *Plasmodium falciparum* by *Aotus* serum, *Bull. W. H. O.*, 57 (Suppl.), 219, 1979.
110. **Miller, L. H., Johnson, J. G., Schmidt-Ullrich, R., Haynes, J. D., Wallach, X., and Carter, R.,** Determinants on surface proteins of *Plasmodium knowlesi* merozoites common to *Plasmodium falciparum* schizonts, *J. Exp. Med.*, 151, 790, 1980.
111. **Reese, R. T. and Motyl, M. R.,** *In vitro* inhibition of *Plasmodium falciparum* grown with immune sera from monkeys and man, *Fed. Proc. Fed. Am. Soc. Exp. Biol.*, 38, 1275, 1979.
112. **Kilejian, A.,** Stage-specific proteins and glycoproteins of *Plasmodium falciparum:* identification of antigens unique to schozints and merozites, *Proc. Natl. Acad. Sci. USA*, 77, 3695, 1980.
113. **Hall, C. L., Haynes, J. D., Chulay, J. D., and Diggs, C. L.,** Cultured *Plasmodium falciparum* used as antigen in a malaria indirect fluorescent antibody test, *Am. J. Trop. Med. Hyg.*, 27, 849, 1978.

114. **Spencer, H.C., Collins, W. E., Chin, W., and Skinner, J. C.,** The enzyme-linked immunosorbent assay (ELISA) for malaria. I. The use of in vitro-cultured *Plasmodium falciparum* as antigen, *Am. J. Trop. Med. Hyg.*, 28, 927, 1979.
115. **Avraham, H., Golenser, J., Spira, D. T., Sutilzeanu, D.,** *Plasmodium falciparum:* assay of antigens and antibodies by means of a solid phase radioimmunoassay with radioiodinated staphylococcal protein A, *Exp. Parasitol.*, 1980.
116. **Trager, W.,** Studies on the extracellular cultivation of an intracellular parasite (avian malaria). I. Development of the organism in erythrocyte extracts and the favoring effect of adenosine triphosphate, *J. Exp. Med.*, 92, 349, 1950.
117. **Trager, W. and Brohn, F. H.,** Coenzyme A requirement of malaria parasites: effect of coenzyme A precursors on extracellular development *in vitro* of *Plasmodium lophurae*, *Proc. Natl. Acad. Sci. USA*, 72, 1834, 1975.
118. **Brohn, F. H. and Trager, W.,** Coenzyme A requirement of malaria parasites: Enzymes of coenzyme A biosynthesis in normal duck erythrocytes and erythrocytes infected with *Plasmodium lophurae*, *Proc. Natl. Acad. Sci. USA*, 72, 2456, 1975.
119. **Trager, W.,** Bongkrekic acid and the adenosinetriphosphate requirement of malaria parasites, *Exp. Parasitol.*, 34, 412, 1973.
120. **Langreth, S. G.,** Electron microscope cytochemistry of host-parasite membrane interactions in malaria, *Bull. W. H. O.*, 55, 171, 1977.
121. **Hegner, R. and Wolfson, F.,** Tissue culture studies of parasites in reticuloendothelial cells in birds infected with *Plasmodium*, *Am. J. Hyg.*, 29, 83, 1939.
122. **Meyer, H. and Oliveira Musacchio, M. De.,** Estudo electron microscopio da forma exo-eritrocitaria do *Plasmodium gallinaceum* en culturas de tecido, *An. Acad. Bras. Cienc.*, 32, 91, 1960.
123. **Meyer, H. and Oliveira Musacchio, M. De.,** An electron microscopic study of the final and initial forms of *Plasmodium gallinaceum* in thin sections of infected tissue cultures, *J. Protozool.*, 12, 193, 1965.
124. **Hepler, P. K., Huff, C. G., and Sprinz, H.,** The fine structure of the exoerythrocytic stages of *Plasmodium fallax*, *J. Cell Biol.*, 30, 333, 1966.
125. **Tonkin, I. M.,** The testing of drugs against exoerythrocytic forms of *Plasmodium gallinaceum* in tissue culture, *Br. J. Pharmacol. Chemother.*, 1, 163, 1946.
126. **Beaudoin, R. L. and Aikawa, M.,** Primaquine-induced changes in morphology of exoerythrocytic stages of malaria, *Science*, 160, 1233, 1968.
127. **Beaudoin, R. L., Strome, C. P. A., and Huff, C. G.,** Persistance of pyrimethamine resistance in exoerythrocytic stages of *Plasmodium gallinaceum*, *Exp. Parasitol.*, 20, 156, 1967.
128. **Graham, H. A., Palczuk, N. C., and Stauber, L. A.,** Immunity to exoerythrocytic forms of malaria. II. Passive transfer of immunity to exoerythrocytic form, *Exp. Parasitol.*, 34, 372, 1973.
129. **Holbrook, T. W., Palczuk, N. C., and Stauber, L. A.,** Immunity to exoerythrocytic forms of malaria. III. Stage specific immunization of turkeys against exoerythrocytic forms of *Plasmodium fallax*, *J. Parasitol.*, 60, 348, 1974.
130. **Holbrook, T. W., Spitalny, G. L., and Palczuk, N. C.,** Stimulation of resistance in mice to sporozoite-induced *Plasmodium berghei* malaria by infections of avian exoerythrocytic forms, *J. Parasitol.*, 62, 670, 1976.
131. **Weiss, M. M. and Vanderberg, J. P.,** Studies on *Plasmodium* öokinetes. II. In vitro formation of *Plasmodium berghei* öokinetes, *J. Parasitol.*, 63, 932, 1977.

Chapter 5

AFRICAN TRYPANOSOMES

George C. Hill and Hiroyuki Hirumi

TABLE OF CONTENTS

I. INTRODUCTION

The African trypanosomiases, a group of diseases caused by infection with certain species of African trypanosomes, have greatly hindered social and economic progress on the African Continent. Human trypanosomiases (sleeping sickness) killed hundreds of thousands of people in various epidemic areas during the early period of this century.[1] A number of new cases still occur each year constituting a serious risk to the health of at least 35 million people with the fear of a classic epidemic outbreak on the continent.[2] Animal trypanosomiases (nagana, souma, surra, and dourine) have also affected millions of livestock every year and greatly restricted the utilization of vast areas of potential rangeland for livestock production. Thus, there is a great need for the development of effective control measures and simple, yet reliable, diagnostic techniques against these important diseases in both man and his domesticated animals.

Causative agents of the diseases are protozoan blood parasites which belong to the genus *Trypanosoma,* Gruby, 1943, in the family Trypanosomatidae, Doflein, 1901. The genus *Trypanosoma* can be further divided into two sections: Stercoraria and Salivaria. Major species of the causative agents belong to the section Salivarian which contains four subgenera: *Duttonella* Chalmers, 1918, *Nannomonas* Hoare, 1964, *Trypanosozoon* Luhe, 1906 and *Pycnomonas* Hoare, 1964.[3] Among the salivarian trypanosomes, species of medical and veterinary importance are *Trypanosoma (Duttonella) vivax* (hosts: bovine, sheep, goat, equine, dog, and antelope, *T. (D) uniforme* (hosts: bovine, sheep, goat, and antelope), *Trypanosoma (Nannomonas) congolense* (hosts: bovine, sheep, goat, equine, pig, dog, and antelope), *T. (N.) simiae* (hosts: pig, warthog; possibly bovine, equine, and camel), *Trypanosoma (Pycnomonas) suis* (hosts: pig, warthog, and bushpig), *Trypanosoma (Trypanozoan) brucei brucei* (hosts: all domestic mammals and antelope), *T. (T.) b. rhodesiense* (hosts: man, bovine, and antelope), *T. (T.) b. gambiense* (hosts: man), *T. (T.) evansi* (hosts: bovine, equine, camel, dog, etc.) and *T. (T.) equiperdum* (hosts: equine). Most species listed are transmitted by tsetse fly vectors *(Glossina* ssp), except *T. evansi* which is transmitted by *tabanid* flies and *T. equiperdum* by direct contact.[4]

In this chapter emphasis will be mainly focused on five major tsetse fly-borne species (three bovine pathogens: *T. vivax, T. congolense,* and *T. brucei brucei,* and two human pathogens: *T. b. rhodesiense* and *T. b. gambiense).* These trypanosomes undergo a complex life cycle, involving mammalian hosts and the insect vectors, and are extracellular parasites throughout their life cycle. Hematozoic trypomastigotes (bloodstream forms), which have subterminal kinetoplasts and are covered with glycoprotein layers (surface coats), are infective for mammals and multiply in the bloodstream and/or tissues of host animals. The bloodstream forms transform into procyclic forms (insect trypomastigotes) and become noninfective for mammals, losing their surface coat when they are ingested with blood meals into the insect vectors. The procyclic trypomastigotes further develop to epimastogotes in which the kinetoplast occupies a postnuclear position. Subsequently, the parasites transform into metatrypanosomes (metacyclic trypomastigotes). At this stage the parasites regain the surface coats and reacquire their infectivity for mammals. The metacyclic trypomastigotes rapidly change into hematozoic trypomastigotes when they are injected into the mammalian hosts by infective fly bites.

The sites of developmental changes in insect vectors vary among the salivarian species. The transformation of all insect stages of *T. vivax* take place in the proboscis of vectors, whereas, bloodstream forms of *T. congolense, T. b. brucei, T. b. rhodesiense,* and *T. b. gambiense* first change to procyclic trypomastigotes in the insect midgut. The transformed procyclic trypomastigotes of *T. congolense* then move into the proboscis where further development takes place. In contrast, the procyclic trypomastigotes of the Brucei-group *(T. b. brucei, T. b. rhodesiense,* and *T. b. gambiense)* move into the salivary glands of vectors where they further develop to epimastigotes and subsequently to the metacyclic trypomastigotes. Hematozoic trypomastigotes of pleomorphic strains of the salivarian species also change their morphology in their mammalian hosts. The metacyclic trypomastigotes injected into the blood circulation of the hosts rapidly transform into the long, slender bloodstream trypomastigotes, then to the intermediate forms and subsequently to the short, stumpy trypomastigotes after a peak of parasitemia.

The behavior of the bloodstream trypomastigotes in mammalian hosts also differs among the salivarian trypanosomes. The hematozoic trypomastigotes of the Brucei-group not only proliferate in the circulatory system but also infiltrate into the connective tissues of various organs, while those of *T. vivax* and *T. congolense* are known to remain within the blood vascular system.* Furthermore, *T. congolense* bloodstream trypomastigotes often bind to the surface of endothelial cells of the capillary walls with their anterior ends.[5] On the other hand, those of the other species freely move within the bloodstream.

In order to better understand the African trypanosomiases, it is important to establish in vitro techniques which support the propagation of various developmental stages of the salivarian trypanosomes, briefly described above, under strictly controlled laboratory conditions. A number of such attempts have been made over the past 70 years. The earlier results indicated that the salivarian trypanosomes, particularly the animal-infective bloodstream forms, might be the most difficult members of the order Kinetoplastida to cultivate in vitro.[6,7] However, major technical problems have recently been overcome and well standardized techniques for the in vitro cultivation of various forms of the salivarian trypanosomes, especially of the Brucei-group, have become available (reviewed by Evans,[7] Hirumi,[8] and Hirumi, et al.[9]). This review is devoted to a discussion of the recently developed in vitro techniques and their application to the fundamental studies of African trypanosomes in vitro.

II. EARLIER INVESTIGATIONS

Since the first attempt to maintain *T. b. brucei* in vitro by Novy and MacNeal, a number of investigations on the cultivation of African trypanosomes have been carried out by many workers during the past 70 years, aimed at the application of in vitro systems to the study of African trypanosomes.[10,11] The earlier investigations have been repeatedly reviewed by a number of authors (Baker,[12] Brooks,[13] Evans,[7] Guttman and Wallace,[14] Hirumi,[8] Hirumi et al.,[9] Hoare,[3] Lwoff,[15] Schneider,[16] Tayler and Baker,[17] Tobie,[18] and Trager[19]). Thus highlights of the previous work are briefly discussed below.

Various forms of salivarian trypanosomes which have been successfully cultivated in vitro may be grouped into (A) insect forms, including (a) procyclic trypomastigotes (insect trypomastigotes) of *T. b. brucei, T. b. gambiense, T. b. rhodesiense, T. congolense,* and *T. vivax,* and (b) epimastigotes of *T. b. brucei, T. congolense,* and *T. vivax,* (B) bloodstream forms of *T. b. brucei, T. b. gambiense,* and *T. b. rhodesiense,* and (C) complete cyclic development of *T. b. brucei.* Methods employed to cultivate these forms to date have been categorized into four systems: (1) host cell-free medium system; (2) co-cultivation with tsetse fly organs, tissues and/or cells; (3) mammalian feeder layer cell system; (4) a combination of above (2) and (3); (5) infective host tissue explant cultures.

A. Insect Forms

1. Host Cell-Free Medium System

In most earlier work, cultures were initiated by inoculating them with bloodstream trypomastigotes which were isolated from infected mammalian hosts and maintained at 25 to

* Recently the appearance of *T. vivax* bloodstream forms in lymph nodes and lymph stream has been demonstrated (Emery, D. L., Barry, J. D. and Moloo, S. K., personal communication).

28°C in the absence of host cells. The bloodstream trypomastigotes transferred to the host cell-free system changed to the procyclic trypomastigotes, morphologically resembling those grown in tsetse fly vectors. The in vitro transformed procyclic trypomastigotes lost their surface coat and became noninfective to mammals.

A number of monophasic, as well as diphasic, media have been used. Among those, modified Tobie's medium (Evans and Brown[20]), HX25 medium (Cross and Manning[21]), HX25M medium (Cross et al.[22]), SDM-77 medium (Brun and Jenni[23]), and EBLB medium (Evans[7]) successfully supported the growth of procyclic forms in vitro.

The history of trypanosome stocks has been thought to be one of the most important factors for obtaining successful cultivation of this form. It has also been pointed out that the organisms, which are newly isolated from host animals infected in the field, are generally the most successful starting materials, whereas, old stocks which have been maintained in laboratory animals by syringe passage are very difficult or impossible to cultivate in vitro.[7] Although bloodstream trypomastigotes cannot be continuously cultivated in the host cell-free system, the procyclic trypomastigotes of *T. b. brucei, T. b. gambiense, T. b. rhodesiense,* and *T. congolense* can now be routinely propagated with a high success rate using this system, and cultures with $>10^7$ trypanosomes $m\ell^{-1}$ are usually obtained.[7]

2. Co-Cultivation with Tsetse Fly Organs, Tissues, and/or Cells

A number of attempts were made in order to achieve the cyclic development of salivarian trypanosomes in vitro through their various developmental stages using tsetse fly organ, tissue, and/or cell cultures (reviewed by Brooks,[13] Schneider,[16] and Trager[19]).

In the pioneering work of this approach, Trager[24] first cultivated various tissues of a tsetse fly vector, *Glossina palpalis,* evaluating the merit of different modifications of Wyatt's insect medium[25] in supporting the growth of tsetse fly tissues. The primary tsetse fly tissue cultures were then inoculated with bloodstream trypomastigotes of *T. b. brucei, T. congolense,* or *T. vivax,* which were isolated from infected sheep and incubated at 30 to 32°C. The bloodstream trypomastigotes of all three species transformed to procyclic trypomastigotes which readily multiplied in these cultures. The appearance of epimastigotes and, subsequently, metacyclic trypomastigotes was observed. These cultures were successfully maintained in vitro for a period of 3 months by subcultivating at 6- to 8-day intervals. The cultures of *T. vivax* were incubated at 38°C for 19 hr at various stages of cultivation and thereafter used to inoculate sheep. Two out of six sheep inoculated were successfully infected by trypanosomes derived from 19- and 39-day-old cultures. Cultured *T. brucei* and *T. congolense* were not infective for sheep, even though the cultures contained organisms which morphologically appeared to be metacyclic forms.

Some 15 years later Trager again attempted to cultivate *T. vivax* in the presence of various organs of *G. morsitans.*[26] Although the initial attempts were made as nearly as possible in the same way as 15 years earlier, the trypanosomes derived from two different isolates (DRV-1 and 36/15) did not give rise to continuous multiplication in cultures. A new approach was, therefore, made using a modified Eagle's MEM-F15 supplemented with 20% sheep serum and an extract of a *G. morsitans* pupa. The cultures containing salivary glands and gut tissues of *G. morsitans* were inoculated with bloodstream forms obtained from the blood of a heavily infected sheep and incubated at 27 to 29°C. The bloodstream forms multiplied first as elongated spiral trypomastigotes. In some successful cultures they further transformed to epimastigotes which multiplied extensively to form dense layers attached to pieces of peritrophic membrane. Many transitional forms, including those approaching the morphology of metacyclic forms were also seen. Although none of the cultures survived much longer than 2 weeks, the results were in some ways considered to be more promising than those of the earlier work. Especially striking was the attachment of the epimastigotes to membranes, somewhat resembling the behavior of the organisms in the proboscis of infected fly vectors.

Unfortunately, the successful cultures could not be tested for their infectivity to mammals due to the short time of Dr. Trager's stay in West Africa.

Trager's pioneering work has been followed by several investigators using tsetse fly organ and/ or tissue cultures. Nicoli and Vattier attempted to culture *T. b. rhodesiense* (EATRO strain 389) in the presence of tissue explants derived from 6- to 13-day-old *G. fuscipes guanzensis* pupae using one of Trager's media.[27] Trypanosomes maintained in vitro were consistently noninfective for mammals. Amrein et al., also co-cultivated trypanosomes of the *Brucei*-group with various organs of *G. morsitans, G. pallidipes,* and *G. brevipalpis* in TC Medium 199 but with little success.[28] Cunningham successfully propagated *T. b. brucei* (TREU strain 667) and *T. congolense* (TREU strain 623) in the presence of alimentary tract of *G. morsitans* pupae also using one of Trager's media.[29,30] The trypanosomes grown in this system resembled those observed in the midgut of infected fly vectors (procyclic trypomastigotes) and were noninfective for mammals.

Cunningham and Honigberg[31] improved the co-cultivation technique by cultivating tsetse fly-transmissible *T. b. brucei* (TREU strain 1193) in the presence of "head-salivary gland" explants obtained from pre-emergent adults of *G. m. morsitans* in a newly formulated medium[32] based partly on Grace's insect medium[33] and the amino acid compositions of tsetse fly hemolymph[34] and containing 20% fetal bovine serum. They first established procyclic form cultures of the trypanosomes following the procedure described by Trager.[24] The procyclic trypomastigotes which had been obtained by transformation in vitro and maintained in the culture medium for periods ranging from 10 days to 7 weeks were then each transferred into Leighton tubes or 25-cm^2 T-type plastic flasks containing 5 or 13-24 salivary gland explants, respectively. The trypanosomes propagated in this system reacquired their infectivity for mice. Few "metacyclic-like" forms were seen in the cultures of trypanosome populations which reacquired their animal infectivity. Although the system does not support the complete cyclic development in vitro, it reproduces to some extent the conditions prevailing in the tsetse fly vectors. Using the "salivary gland-trypanosome" system, Cunningham and Taylor further demonstrated that T. brucei (TREU strains 929 and 1275) grown in the presence of between 27 and 50 explants consistently produced infectivity in mice for a prolonged period (up to 88 days) whereas trypanosomes cultivated with 25 or fewer explants rarely infected mice.[35]

Steiger et al. utilized for the first time a continuous cell line of *G. morsitans* for the in vitro cultivation of *T. congolense* (TREU strain 1183) aimed at the cyclic development of the parasites through their insect vector stages.[36] Primary cultures of midmastigotes of these parasites were prepared at 27°C by seeding blood-agar plates with 0.1 mℓ of whole blood containing bloodstream forms following the procedure described by Tobie except using defibrinated bovine blood and different liquid overlays: HEPES-buffered TC-199 medium, Grace's insect tissue culture medium and Hank's balanced salt solution (BSS).[37] Midmastigotes induced in the blood-agar plates, or a small amount of fresh whole blood containing bloodstream forms were then transferred into tsetse fly cell cultures prepared according to the procedure by Schneider and maintained at 25°C.[16] Weekly subpassages were made in a modified Schneider's *Drosophila* medium. This vector cell system supported transformation of bloodstream trypomastigotes to procyclic trypomastigotes which continuously grew in vitro. Midgut trypomastigotes also differentiated reproducibly to epimastigotes. Although the parasites underwent partial cyclic development, metacyclic trypomastigotes were not produced in this system and all trypanosomes propagated were noninfective for mice and rats.

B. Hematozoic Trypomastigotes

All earlier attempts, using the systems described above (A), have failed to support the growth of hematozoic trypomastigotes for long periods of time without altering their biological, biochemical, physiological, and/or immunological properties. Unavailability of in vitro systems which support the continuous growth of bloodstream forms of salivarian trypanosomes had been considered to be one of the major obstructions for the rapid progress of fundamental studies of African trypanosomes in vitro. Therefore, other approaches have been made, aimed at the development of such in vitro methods (1) by cultivating bloodstream forms in the presence of bovine feeder layer cells, (2) by initiating cultures with insect vector organs excised from infective tsetse flies in the presence of bovine feeder layers, and (3) by explanting small pieces of the host animal skin removed from the sites of tsetse fly bites.

1. Bovine Feeder Layer Systems

Le Page earlier demonstrated that bloodstream trypomastigotes of *T. b. brucei* (Lister strain 427) could be maintained in vitro in the presence of mouse L cells at 37°C using NCTC109 medium (Difco) containing 10% (v/v) presuckling calf serum. The trypanosomes multiplied to a fivefold maximum within 66 hr, consistently retaining their infectivity for mice up to 36 hr and occasionally up to 70 hr.[38] Ward and West also demonstrated that the infectivity of *T. b. brucei* (EATRO-207) which was passaged through mice, rats, guinea pigs, and tsetse flies was retained up to 120 hr when they were maintained at 27°C in the presence of mouse fibroblast cells (L-929) with M-109, M-199, or NTC-135 medium supplemented with 5% fetal bovine serum.[39] Mendez and Honigberg[40] also attempted to cultivate *T. b. brucei* (EATRO 795 and 1587 and *T. b. rhodesiense* (EATRO 1985 and Bechuanaland 182A) in the the presence of mouse L cells at 27°C, using a modified Tobie's medium supplemented with 5% (v/v) inactivated, normal bovine serum and 10% (v/v) chick embryo extract.[41] Long slender bloodstream forms of the trypanosomes multiplied in this system, rapidly transforming to intermediate, short stumpy and procyclic forms, although *T. b. rhodesiense* (Bechuanaland 182A) was difficult to maintain in vitro. None of the populations of trypanosomes maintained in this system remained infective after 6 days of cultivation.

In these attempts, efforts were made to retain the infectivity for mammals during the course of in vitro cultivation. Although the earlier work using this approach failed to support the continuous growth of hematozoic trypomastigotes, one of the long standing requisites for the study of African trypanosomes in vitro has been recently achieved by the establishment of such an in vitro method that supports the growth of animal-infective bloodstream forms of *T. b. brucei* for prolonged periods of time (Hirumi et al.[42,43]). This method has been established by systematically testing a total of 96 different growth conditions for *T. b. brucei* (Stock 427): 12 different culture media, 4 different concentrations of fetal bovine serum (FBS), and the presence or absence of bovine feeder layer cells. Among these, the best growth rate was achieved by cultivating the bloodstream forms in HEPES (25 m*M*) buffered RPMI 1640 medium supplemented with 20% heat-inactived FBS at pH 7.2, osmolarity 290 mOs, at 37°C in the presence of bovine feeder layer cells. Bloodstream forms used to initiate the culture retained the morphological, as well as biological, properties of the original hematozoic trypomastigotes, including infectivity for mice and cattle. The method has been further standardized during the past three years (see details in Section III) and now hematozoic trypomastigotes of *T. b. brucei,* (Stock 427 and LUMP 227), can be cultivated under these conditions for an indefinite period of time (presently over 4 years).[8,9]

Using the same technique, Brun and his co-workers (personal communication) successfully cultivated the bloodstream forms of *T. b. brucei* (Stock 427 and Stock STIB 247). Hill et al. have also reproduced the same result by cultivating animal-infective *T. b. brucei* (Stock 427) in the presence of commercially available buffalo and Chinese hamster lung cells.[44] They further demonstrated that bloodstream forms of *T. b. rhodesiense* (EATRO 1895) and *T. b. brucei* (LUMP 1026 and EATRO 110) could also be cultivated in the presence of Chinese hamster lung cells under the same culture condition.[45,46] Utilization of the feeder layer system to propagate in vitro bloodstream forms of *T. congolense*[47] and *T. vivax*[48] has been made with some success. However, further extensive studies are still needed in order to establish a standardized method of cultivating the hematozoic trypomastigotes of these species.

Our recent preliminary studies indicated that hematozoic trypomastigotes of *T. congolense*, separated from infected rat blood, could continuously multiply in vitro in the presence of bovine epithelial cells either at 27 or 37°C using RPMI 1640 medium with 20% FBS. However, trypanosomes cultivated at both temperatures lost their surface coat and became noninfective for mice. To date, animal-infective bloodstream forms have been successfully propagated up to 35 days at 37°C in the presence of bovine embryonic endothelial cells using RPMI 1640 medium supplemented with 30% fetal bovine plasma (Hirumi, H. and Hirumi, K., unpublished data).

The role of mammalian feeder layer cells in supporting the growth of hematozoic trypomastigotes of African trypanosomes is still unknown. Nevertheless, this is the only presently available system to continuously propagate hematozoic trypomastigotes of *T. b. brucei* and *T. b. rhodesiense*.

2. Infected Vector Organs in Mammalian Feeder Layer Cultures

Using the feeder layer system described above, another approach has been recently reported. In this approach, mammalian cell cultures were inoculated with insect forms by placing vector organs obtained from infected tsetse flies in the cultures. Nyindo et al. fed tenereal *G. morsitans* on rats or mice infected with *T. b. brucei* (EATRO strain 999).[49] The intact alimentary tracts were then excised from the flies 3 to 96 hr after feeding and maintained at 29°C using RPMI 1640 medium supplemented with lactabumin hydrolysate and 20% FBS. By day 4 viable parasites were seen in the explants originating from flies fed 12-14 hours prior to dissection. On days 6 and 7 some explants were transferred to new flasks containing bovine embryonic spleen cells and maintained at either 28 or 39°C using the same medium. The trypanosomes grown under these conditions morphologically resembled insect midgut forms (procyclic trypomastigotes) with some epimastigotes present. Although electron microscopy examination revealed the absence of the surface coat, which is a basic characteristic of animal-infective bloodstream forms, 50% of rats inoculated with trypanosomes maintained in culture for 4 to 61 days reportedly became infected. Organisms cultivated in this system for 75 to 149 days were noninfective for animals. The results presented did not show if transformation of insect forms to complete hematozoic trypomastigotes occurred.

Brun and his co-workers recently demonstrated successful transformation of metacyclic trypomastigotes of pleomorphic *T. b. brucei* (STIB 247, 336A and 345A) and *T. b. rhodesiense* (STIB 350) to long slender bloodstream trypomastigotes in well cultures containing rabbit fibroblast-like cells.[50,51] Metacyclic form inoculum was prepared by letting infected *G. m. morsitans* salivate into a drop of warm medium. Each well was inoculated with 3,000 to 8,000 "salivary gland forms" and maintained at 36 to 37°C using Eagle's MEM modified by supplementing with HEPES (21 m*M*)-buffered glucose (1.5 g/ℓ), L-glutamine (0.3 g/ℓ, Na-pyruvate (1 m*M*), % MEM nonessential amino acid (100 ×) and 15% freshly prepared rabbit serum. All metacyclic trypomastigotes transformed to long slender bloodstream trypomastigotes within 24 hr and continued to multiply in the culture retaining their infectivity for mammals and original pleomorphism. The in vitro induced bloodstream trypomastigotes were also capable of transforming to procyclic trypomastigotes when transferred to SDM-79 medium[52] at 27°C.

This system, which supports partial cyclic development, would be of special interest for elucidating the mechanisms underlying (1) the transformation from metacyclic trypomastigotes to hematozoic trypomastigotes and (2) antigenic variation at its early stages.

3. Infected Host Skin Explant Culture

As briefly discussed above, reliable methods for the continuous cultivation of hematozoic trypomastigotes of *T. congolense, T. vivax,* and *T. b. gambiense* are still unavailable. In order to overcome the problem, a new approach has been recently made by Gray et al.[53] Based on the observation of the presence of extravascular *T. congolense* in local reaction sites of infective tsetse fly bites[54] and its close association with dermal collagen,[55] they suggested a possible use of the infected dermal regions as inocula for the establishment of long-term cultivation of this species in vitro.

In this approach, two strains of *G. morsitans* (TREU 1290 and TREU 1173) were fed on adult rabbits and 3 month-old bovine calves. Small areas of skin were then removed from the sites of tsetse fly bites on the animals after 8 days of postfeeding. Dermal collagen explants (approximately 5 × 10 mm) excised from the skin were placed in plastic 25 cm^2 T-type culture flasks containing 4 mℓ volumes of Eagle's MEM with Earle's salts and 25 m*M* HEPES or RPMI 1640 with 25 m*M* HEPES with added L-glutamine (292 μg/mℓ), both supplemented with 20% heat-inactivated FBS and antibiotics (penicillin, 100 IU/mℓ and streptomycin, 100 μg/mℓ) and incubated at 37°C. Some infected dermal explants were also placed in cultures containing fibroblasts derived from bovine embryonic skin, mouse (3T6 cell line, Flow) or buffalo (IMR-31 cell line, Flow).

In general, cultures initiated with bovine dermal explants yielded more trypanosomes at 24 hr than those prepared from rabbit explants. The number of trypanosomes declined after 15 days of cultivation. The organisms which survived up to 21 days retained the morphology of typical hematozoic trypomastigotes and were infective for mice. Although the preliminary results indicated that this system did not support the continuous growth of *T. congolense* bloodstream trypomastigotes, such an approach may be worth pursuing further in order to establish an in vitro system for the bloodstream form of this species.

C. Complete Cyclic Development

African trypanosomes undergo a complex cyclic development, involving mammalian hosts and tsetse fly vectors (see Introduction). Mechanisms underlying the complex life cycle have not yet been well understood. A number of in vitro systems described above support the growth of a certain specific stage of the trypanosomes or their partial cyclic development but none of them support the complete life cycle in vitro. However, a complete cyclic transformation of *T. b. brucei* (TC221, an in vitro derivative of Stock 427) has been recently demonstrated in vitro, initiating from the long slender hematozoic trypomastigotes through various insect stages, morphologically resembling midgut trypomastigotes, proventricular trypomastigotes, epimastigotes, and metacyclic trypanosomes, and subsequently back to the bloodstream forms.[8,9] The results presented may be summarized as follows:

Step I: hematozoic trypomastigote stage — Organisms at this stage could be maintained in vitro for an indefinite period of time in the presence of bovine feeder layer cells using HEPES-buffered RPNI 1640 supplemented with 20% (v/v) heat-inactivated FBS at 37°C. Such organisms when maintained under these conditions retained all the properties of the bloodstream form, including the infectivity for mammals and the glycoprotein surface coat.

Step II: procyclic trypomastigote stage (including the insect midgut form and proventricular form — When the bloodstream trypomastigotes propagated in vitro were transferred to 25°C and maintained under the same conditions, except for the temperature, the first appearance of short procyclic trypomastigotes (analogous to insect midgut forms) was seen on day 5 at 25°C. The procyclic trypomastigotes rapidly increased in number dominating the population (over 95%) by day 8-10. This stage could be maintained for long periods of time by replacing a portion (1/2 to 2/3) of culture medium with the same volume of fresh medium every 2 to 3 days and by subcultivating the trypanosomes into new feeder layer cultures every 10 to 14 days. The organisms at this stage were noninfective for mice, lacking the surface coat. When such cultures were maintained without changing the medium, short procyclic trypomastigotes transformed into long slender procyclic trypomastigotes (analogous to insect proventricular forms) within 5 to 10 days. Most of the trypanosome population (over 95%) consisted of these long slender forms by day 10 in such cultures. During this period many dividing forms were consistently seen, although the population density remained at more or less a constant level (1-5 × 10^6 trypanosomes/mℓ). This indicates that approximately the same number of trypanosomes, which are multiplying, also die during this period. This stage could be maintained for long periods under the same conditions by transferring the organisms to new feeder layer cell cultures every 10 to 14 days. These long slender proventricular forms were also noninfective for mice, lacking the surface coat. Short midgut forms in cultures, whose medium was unchanged but were opened to the air every 1 to 2 days, did not transform to long slender proventricular forms but changed to various abnormal forms.

Step III: epimastigote and metacyclic stages — Long slender proventricular forms, when the medium was again changed every 2 to 3 days, transformed to short procyclic forms, morphologically similar to those which appeared at Step II. Many trypanosomes changed to abnormal forms in such cultures. Among them, however, a small number (less than 1% of the population) of epimastigote-like forms were observed initially and subsequently metacyclic-like forms were seen. Populations containing such metacyclic-like forms reacquired their infectivity for mice. To date, populations consisting of only either epimastigotes or metacyclic forms have not been achieved.

Step IV: the 2nd hematozoic trypomastigote stage — When populations which contained the metacyclic-like forms were re-transferred to 37°C in new feeder layer cultures and maintained by changing the medium every 1 to 2 days, long slender hematozoic trypomastigotes reappeared in the cultures within 3 to 5 days and rapidly multiplied, completely dominating the population (over 99%) by day 8.

Complete cyclic development was achieved in over 40 populations of bloodstream trypomastigotes of *T. brucei* (TC221). Furthermore, a second cyclic development was also successfully demonstrated on all four populations tested after having undergone the first cyclic development. However, an in vitro induced clone (TC221-3)[56] as well as some populations derived from *T. brucei* (LUMP stock 227) failed to complete the cyclic development under these conditions.

It has been pointed out that the results presented above indicate "that (1) the complete cyclic development of *T. brucei* can be made in vitro and the presence of tsetse fly organs, tissues, and/or cells is not essential for cyclic development, (2) the transformation from bloodstream forms to insect midgut forms seems to be low temperature (25°C) dependent, (3) further transformation from midgut forms to proventricular forms requires additional chemical and/or physical changes in the growth environment, which are presently unidentified, (4) factor(s) promoting the development from proventricular forms to epimastigotes and subsequently to metacyclic forms cannot be identified at present either, and (5) the transformation of metacyclic forms to bloodstream forms seems to be also temperature (37°C) dependent".[9]

III. METHODS OF TRYPANOSOME CULTIVATION

Procyclic trypomastigotes of African trypanosomes can be readily cultivated in vitro by using various methods in many different media. Details of such widely used methods have been thoroughly reviewed previously.[7,17] On the contrary, the long-term cultivation of haematozoic trypomastigotes has been achieved only during the past 7 years. Since then attempts have been made to standardize the method in several laboratories and well standardized techniques for the cultivation of, at least, *T. b. brucei* and *T. b. rhodesiense* have recently been established. This section is, therefore, devoted to reviewing the standardized techniques for the long-term cultivation of hematozoic trypomastigotes.

Hematozoic trypomastigotes of salivarian trypanosomes appeared to be highly sensitive to even minor changes of their growth environments in vitro, considerably altering their various properties.[9] In order to obtain consistent yields of homogeneous populations and reproducible, experimental results, it is essential to standardize the following culture conditions and handling procedures: (1) preparation of trypanosome materials, (2) quality of culture medium and its serum supplement, (3) selection and preparation of mammalian feeder layer cell cultures, (4) initiation and maintenance of trypanosome cultures, and (5) cryopreservation of cultured trypanosome.

A. Preparation of Trypanosome Materials

1. Species and Stocks

Hematozoic trypomastigotes of the following stocks of *T. b. brucei* and *T. b. rhodesiense* have been successfully cultivated in vitro, initiating the cultures with either bloodstream trypomastigotes obtained from infected animal blood or "salivary gland forms" obtained from infective tsetse fly saliva.

- *T. b. brucei* (Stock 427) — This stock was originally isolated from sheep by Cunningham and Vickerman in 1962 and was maintained in laboratory animals by a number of syringe passages prior to the first in vitro cultivation in 1976.[57] Trypanosomes of this stock are known to be monomorphic and highly virulent. Cultures were initiated with bloodstream trypomastigotes.[42,45]
- *T. b. brucei* (Stock LUMP 227) — A derivative of Stock EATRO 889 which was originally isolated from cattle by Onyango et al. in 1966.[58] This stock had been also passaged a number of times in laboratory animals prior to the in vitro cultivation in 1978. Cultures were initiated with bloodstream trypomastigotes.[56]
- *T. b. brucei* (STIB 247) — A pleomorphic strain which was isolated from an infected hartebeest *(Alcelaphus buselaphus)* in the Serengeti National Park, Tanzania in 1971. The stock was cryopreserved after one rat passage. Prior to the in vitro cultivation, the organisms were cyclically passaged in tsetse fly vectors and laboratory animals for seven months. Cultures were initiated with "salivary gland forms."[50,51]
- *T. b. brucei* (STIB 345A) — A pleomorphic strain derived from Stock EATRO 1529 which was originally isolated from an infective *G. pallidipes* in Kiboko, Kenya, in 1969 and cryopreserved after six passages in mice. This material was stabilated after an additional five short passages in rats as STIB 345 in 1973. Cultures were initiated with "salivary gland forms."[50]
- *T. b. brucei* (STIB 336A) — A pleomorphic strain derived from Molteno Institute Stock (S42/030). The original stock was isolated from an infected warthog *(Phagochoerus aethiopicus)* in Tororo, Uganda, in 1966. Cultures were initiated with "salivary gland forms."[50]
- *T. b. brucei* (EATRO 110) and (LUMP 1026) — Cultures were initiated with bloodstream trypomastigotes.[46]
- *T. b. rhodesiense* (EATRO 1895) — A pleomorphic strain derived from a stock of the London School of Tropical Medicine and Hygiene. The strain was originally isolated from a naturally infected male patient in Lugala, Uganda. Cultures were initiated with bloodstream trypomastigotes.[45,46]
- *T. b. rhodesiense* (STIB 350) — A pleomorphic clone derived from Stock (STIB 24) which was isolated from a lion *(Panthera lea)* in the Serengeti National Park in 1971 and known to be infective for humans. Cultures were initiated with "salivary gland forms."[50]

Previously the history of trypanosome stocks (or strains) has been considered to be one of the most important factors for obtaining successful cultivation of salivarian trypanosomes in vitro.[7] It has been suggested that new isolates of trypanosomes are the most successful, whereas old laboratory strains which have been passaged in laboratory animals are very difficult or impossible to initiate in in vitro cultures. It has been recommended to use trypanosomes obtained from a donor animal one day after a peak of parasitemia to initiate the in vitro cultivation. Trypanosome populations comprising mostly long slender hematozoic trypomastigotes were found to give poor results in initiating the culture.[7] These observations now appear to be only applicable to the cultivation of noninfective procyclic forms.

On the contrary, bloodstream trypomastigotes of old laboratory strains, such as *T. b. brucei* (Stock 427), have been easily cultivated in vitro. Populations comprising mostly long slender bloodstream forms were found to give the best results in initiating the culture.[8,9,42,56] Hill and co-workers (unpublished data) have found that bloodstream forms of recently isolated strains of *T. b. brucei, T. b. gambiense,* and *T. b. rhodesiense* were difficult to grow in vitro. Bloodstream forms of certain pleomorphic strains have been found difficult to initiate in in vitro culture by Brun et al., but they solved this problem by initiating the culture with "salivary gland forms", presumably comprising many metacyclic forms.[50,51] We have also encountered some difficulties with a "slow" growing strain derived from *T. b. brucei* (Stock LUMB 227), although bloodstream forms of a "fast" growing strain derived from the same stock were cultivated without any difficulty.[9] This problem has been overcome by the selection of an appropriate feeder layer cell type (see Section III. C) for the "slow" growing strain. These results strongly suggest that the growth characteristics of each trypanosome may be more important factors than the history or the polymorphism of trypanosome strains. It would be, therefore, advisable to select the strain of trypanosomes and the type of feeder layer cells based on their growth characteristics which may be preexamined in laboratory animals and/or short-term cultures. The possibility of successful cultivation of hematozoic trypomastigotes should not be prejudged simply based on the history and/or polymorphism of the strains to be used.

2. *Trypanosome Inoculum*

In order to prepare an inoculum of bloodstream forms to give best results, two considerations have to be taken into account: (1) the stage of parasitemia in donor animals and (2) the method of trypanosome separation from the infected donor blood.

Stage of parasitemia — Trypanosome populations of both "monomorphic" and "pleomorphic" strains are comprised of long slender bloodstream forms during the early stage of parasitemia. Intermediate and stumpy forms appear in the middle (pleomorphic strain) or late (monomorphic strain) stages of the parasitemia, increasing in number through a peak of parasitemia. Since the best success rate with bloodstream form cultures has been achieved by initiating such cultures with an inoculum containing mostly long slender hematozoic trypomastigotes and many dividing forms, it is important to collect the infected blood at the logarithmic growth phase.[8,56] This can be done by inoculating lethally irradiated inbred mice with different doses (10^2-10^5 trypanosomes/mℓ) and examining the tail blood daily for trypanosome concentration and morphology.

Separation of bloodstream forms — Mechanical procedures, such as centrifugation, cellulose column separation, etc., often cause physical damage to the trypanosomes and give poor results in initiating the in vitro cultivation of hematozoic trypomastigotes. In earlier work bloodstream forms were separated from the infected blood by density gradient centrifugation or buffy coat separation techniques.[42,43] In order to minimize the possible physical damage, the following one step separation procedures have recently been recommended.[8] In one case infected blood (0.5 mℓ) was mixed with fresh medium (3.5 mℓ) and then centrifuged at low speed (150 g) for 5 min. The supernatant containing trypanosomes was used to initiate the culture. In the other case, a drop of tail blood from an infected mouse was suspended in 4 mℓ of fresh medium using a conical bottom centrifuge tube and kept at 37°C. After most of the red cells settled down, 1 mℓ of the trypanosome suspension was transferred into each culture flask containing 5 mℓ of fresh medium and mammalian feeder layer cells. A small number of red cells remaining in the final preparation did not affect the initial growth of trypanosomes.[8] Although these methods appear to be satisfactory for the separation of *T. b. brucei* bloodstream trypomastigotes, further improved techniques may be needed for the bloodstream forms of *T. congolense* and *T. vivax*.

B. Quality Control of Culture Medium and its Serum Supplement

1. Media

Although a number of culture media have been successfully used to cultivate procyclic trypomastigotes of African trypanosomes, none of them has supported the growth of hematozoic trypomastigotes in vitro.[6,7] However, it has been demonstrated that HEPES (25 m*M*)-buffered Rosewell Park Memorial Institute (RPMI) 1650 medium supplemented with fetal bovine serum (discussed below) could support the continuous growth of animal-infective bloodstream forms of *T. b. brucei* for long periods of time in the presence of bovine feeder layer cells.[42] Since then, this medium has been commonly used for the cultivation of hematozoic trypomastigotes of salivarian trypanosomes in many laboratories.

Our experience over the past years strongly indicates that the culture medium has to be highly standardized in order to obtain consistent yields of hematozoic trypomastigotes, since hematozoic trypomastigotes of African trypanosomes are more sensitive to minor changes in the culture medium than most mammalian cells. RPMI 1640 medium can be obtained from many commercial suppliers in either liquid or powder forms. The quality control tests which were made by the suppliers for certain mammalian cell types often appeared to be unsatisfactory for the bloodstream forms of salivarian trypanosomes. Trypanosomes grew differently in RPMI 1640 media which were (1) obtained from different suppliers, (2) prepared from different forms (liquid or powder), even those supplied by the same company, and (3) prepared from different batches of the medium obtained from the same supplier.[9] It has been, therefore, strongly recommended that (1) the medium be obtained from one supplier, (2) either liquid or powder form be used, whichever gives the best growth of the trypanosomes under individual laboratory conditions, and (3) tests on the growth efficiency of each batch of medium using one's own standard system be carried out prior to its use. It has been especially recommended to use highly purified water for the preparation of medium from its powder form.

2. Serum Supplement

It would be highly desirable to use a chemically defined medium for the cultivation of various protozoan parasites. Such a medium is, however, still unavailable for the cultivation of hematozoic trypomastigotes of African trypanosomes. Culture media used for the in vitro propagation of the bloodstream form of salivarian trypanosomes at present still require a serum supplement. The growth rate of bloodstream forms could be enhanced by supplementing RPMI 1640 medium with 20% (v/v) of heat-inactivated (56°C for 30 min) fetal bovine serum (HI-FBS).[42] However, qualities of FBS vary considerably among different batches regarding their enhancement and/or inhibition of cell growth in vitro. This variation is far greater than that of the culture media discussed above and hematozoic trypomastigotes are more sensitive to the variation than most of mammalian cells. The standardization of the quality of FBS is, therefore, essential. The following procedures have been recommended for the routine preparation and storage of FBS.[9]

1. Make a request for 2 to 3 samples in small volumes (usually 100 mℓ/sample) to commercial suppliers before the purchasing order is made.

Table 1
FEEDER LAYER CELLS AND CULTURE MEDIA USED FOR THE CULTIVATION OF HEMATOZOIC TRYPOMASTIGOTES OF AFRICAN TRYPANOSOMES

	Feeder layer cells		Culture media			
Trypanosomes	Type	Origin	Basic Medium	Supplement	(%)	Ref.
T. brucei	Fibroblast	African green monkey kidney[a]	RPMI 1640	HI-FBS	(20)	9
		Bovine adult blood	RPMI 1640	HI-FBS	(20)	9,42,50
		Bovine embryonic heart	RPMI 1640	HI-FBS	(20)	9
		Bovine embryonic lung	RPMI 1640	HI-FBS	(20)	9
		Bovine embryonic peritoneum	RPMI 1640	HI-FBS	(20)	9
		Bovine theileria-infected lung fluid	RPMI 1640	HI-FBS	(20)	9
		Buffalo lung[a]	RPMI 1640	HI-FBS	(20)	44
		Chinese hamster lung[a]	RPMI 1640	HI-FBS	(20)	44
		New Zealand white rabbit blood	Modified MEM	HI-RS	(15)	50
		Mouse connective tissue (L)[a]	RPMI 1640	HI-FBS	(20)	9
	Epithelial cell	Bovine embryonic pericardium	RPMI 1640	HI-FBS	(20)	**
		Bovine embryonic peritoneum	RPMI 1640	HI-FBS	(20)	9
		Dog kidney	RPMI 1640	HI-FBS	(20)	9
		New Zealand white rabbit kidney[a]	RPMI 1640	HI-FBS	(20)	9
T. b. rhodesiense	Fibroblast	Buffalo lung[a]	RPMI 1640	HI-FBS	(20)	45
		Chinese hamster lung[a]	RPMI 1640	HI-FBS	(20)	45
		New Zealand white rabbit blood	Modified MEM	HI-RS	(15)	50
T. congolense[b]	Endothelial cell	Bovine calf aorta	RPMI 1640	FBP	(30)	**

Note: HI-FBS: heat-inactivated fetal bovine serum; RS: rabbit serum; FBP: fetal bovine plasma.

[a] Commercially available cell lines.
[b] Hematozoic trypomastigotes have been cultivated up to 35 days.

2. Test the quality of the samples in comparison with a standardized reference medium using the cultured bloodstream form of *T. b. brucei* of which the growth rate in the medium has been previously established.
3. Place an order only for those samples which give a satisfactory result in the quality test.
4. Place frozen FBS in a freezer at −20°C immediately on arrival.
5. Defrost 1 to 2 ℓ of the frozen FBS at a time at 37°C using a water bath, 1 to 2 weeks prior to their use.
6. Remove any precipitated protein by centrifugation at 4°C for 20 min at 350 g.
7. Filter the supernatant aseptically by passing through 0.45 μm and subsequently 0.22 μm pore size Millipore (or similar) filters.
8. Aliquot the filtered FBS into small volumes (usually 100 mℓ/bottle) and store at −20°C until use. Avoid repeated freezing and thawing of FBS.

The ability of FBS to support the growth of bloodstream forms of *T. b. brucei* in vitro could be retained in this manner for at least up to 12 months.

A recent world-wide shortage of FBS has made the quality control of the serum rather difficult. Attempts have been made, therefore, to substitute FBS with other mammalian sera in various laboratories. Brun et al.[50] recently demonstrated that bloodstream forms of *T. b. brucei* and *T. b. rhodesiense* could be successfully cultivated by supplementing a modified Eagle's MEM with 15% of freshly prepared rabbit serum which could be stored at −20°C for several weeks. They also reported that rabbit sera obtained from five major commercial suppliers had given disappointing results.

C. Selection and Preparation of Feeder Layer Cells

Since the first long-term cultivation of bloodstream forms of *T. b. brucei* was achieved in the presence of bovine feeder layer cells, a number of different mammalian cells have been examined regarding their ability to support the growth of bloodstream trypomastigotes.[8,9,42,43,44,45,50] Cell lines examined to date can be grouped into (1) fibroblast type, (2) epithelial type, and (3) endothelial type cells, and have been summarized in Table 1. Although cells listed, regardless of their origin, support the continuous multiplication of hematozoic trypomastigotes of, at least, the *Brucei*- group, the growth characteristics of the organisms vary considerably depending on the type, growth rate, and density of the feeder layer cells. Therefore, feeder layer cell cultures have to be selected and prepared according to the objective and design of each experiment. The following general guidelines have been recommended based on extensive studies of the interaction between trypanosomes and feeder layer cells in vitro.[8,9]

1. Selection

Fibroblasts — Continuous lines of this type of cell can be initiated from various tissues of mammals and easily maintained in vitro following a number of standardized cell culture techniques. Technical details of one of such methods have been described earlier.[43,56] Fibroblasts develop loose multiple cell layers in culture flasks. Hematozoic trypomastigotes grow not only in suspension in the culture medium but also in intercellular spaces of these multiple layers. This type of feeder layer can be used for general purposes, such as the initiation and routine maintenance of bloodstream trypomastigote cultures which do not require accurate quantitative analyses. They may be also useful tools for the study of host cell-parasite relationship in vitro.

Epithelial cells — This type of cells forms a confluent monolayer in culture flasks. All hematozoic trypomastigotes grow in suspension and are unable to invade the feeder layer. This system, therefore, facilitates accurate quantitative analyses of trypanosome populations propagated in vitro. Continuous cell lines of this type of cells may also be initiated from various mammalian tissues. It is, however, rather difficult to maintain homogeneous populations of epithelial cells in vitro due to the frequent contamination with fibroblasts which easily overgrow the cultures. In our laboratory (ILRAD) the best epithelial cell cultures have been obtained from the bovine embryonic pericardium. Readers who are interested in the details of epithelial cell culture methods may consult recent technical articles on this topic.[59,60,61]

Endothelial cells — Our preliminary studies indicate that animal-infective bloodstream trypomastigotes of *T. congolense* are less adaptable to the growth environment in vitro in comparison with those of the *Brucei*-group and may require the presence of highly specific feeder layer cells (Hirumi, H. and Hirumi, K., unpublished data). The bloodstream forms derived from infected mouse blood continuously multiplied at 37°C, as well as at 27°C, in the presence of bovine epithelial cells for long periods of time (over 12 months) but rapidly became noninfective for mice losing their surface coats. To date, a limited success in cultivating animal-infective bloodstream forms (up to 35 days) has been obtained in the presence of bovine endothelial cells derived from the bovine calf aorta, using RPMI 1640 medium supplemented with 30% of fetal bovine plasma. The bloodstream forms grown in this system closely bind to the surface of the feeder layer cell in a similar manner observed in vivo on the surface of endothelial cells of small blood vessels.[5] These results suggest that in order to continuously synthesize the surface glycoprotein, hematozoic trypomastigotes of *T. congolense* may require some substances produced by the endothelial cells. Although further extensive studies are still needed, the endothelial feeder layer cell system is the only one presently available which supports, with certain limitations, the growth of hematozoic trypomastigotes of *T. congolense.* A factor which has been hindering the fast progress of this system is a difficulty in obtaining a continuous cell line of the endothelial cells.

General consideration — Growth patterns of various mammalian cells in vitro may be classified into four categories. When cultures are seeded with 10^5 cells/mℓ (6 mℓ/T-25 flask), (1) cells rapidly increase in number and reach their plateau at a high concentration ($1\text{-}2 \times 10^6$ cells/mℓ) within 4 to 6 days, (2) cells also grow rapidly but reach their plateau within 3 to 5 days and thereafter retain a low concentration ($4\text{-}5 \times 10^5$ cells/mℓ) for a long period of time, (3) cells slowly increase in number but reach a high concentration ($1\text{-}2 \times 10^6$ cells/mℓ) at their plateau within 7 to 8 days, and (4) cells grow slowly and reach their plateau at a low concentration ($4\text{-}5 \times 10^5$ cells/mℓ) within 7 to 8 days. The growth rates of various strains of trypanosomes also vary considerably. In addition to (1) and (2) mentioned above, the following considerations have been recommended:[9] (1) the fast growing cells which reach their plateau at low concentration can be generally used for cultivating fast growing trypanosomes, and (2) the slow growing cells which reach their plateau at a high concentration provide an excellent support for continuous propagation of both fast and slow growing trypanosomes. This type of feeder layer cell is also useful for initiating the cultures of trypanosomes whose growth rates are unknown.

2. Preparation

Old feeder layer cell cultures containing too high a concentration of cells (over 2×10^6 cells/mℓ), as well as young cultures containing too low concentrations (lower than 5×10^4 cells/mℓ), usually produce an unsatisfactory environment for the growth of hematozoic trypomastigotes.[9] In our laboratory (ILRAD), feeder layer cultures are routinely prepared as follows:

1. Select cell cultures which have reached their plateau a few days prior to the subcultivation and contain morphologically healthy cells with no cellular debris.
2. Collect the cells from the cultures by a standard cell culture technique and estimate the number of viable cells.
3. Set up feeder layer cultures by seeding culture flasks, or wells, with 10^4 cells/mℓ (fast growing cells), 5×10^4 cells/mℓ (medium growing cells) or 10^5 cells/mℓ (slow growing cells).
4. Use the feeder layer culture 3 to 5 days after subcultivation.
5. The feeder layer cultures of fast, medium and slow growing cells may be used for co-cultivating with trypanosomes for 5 to 7, 7 to 12, and 10 to 30 days, respectively. Thereafter, it is recommended the trypanosome populations be transferred to newly prepared, 3 to 5 days old feeder layer cultures.

D. Initiation, Maintenance and In Vitro Cloning of Bloodstream Form Culture

1. Initiation

From infected animal blood — The original method[42] in initiating bloodstream trypomastigote cultures of *T. b. brucei* has been further improved.[8,43] In general, the in vitro cultivation of trypanosomes whose growth rates are unknown can be initiated from infected animal blood following the procedures summarized below.

1. Select and prepare feeder layer cell cultures in either 25 mm^2 T-type culture flasks or 24-well (17 mm in diameter) tissue culture plates according to the suggestions and procedures mentioned above. It is recommended that 4 groups of 12 cultures in T-type flasks, or in 12-24 wells, at 5 to 7 day intervals be prepared.
2. Prepare a trypanosome inoculum minimizing possible damage due to physical treatment as described above, and make a series of tenfold dilutions at concentrations of 1-2 $\times 10^5$, $\times 10^4$, $\times 10^3$, and $\times 10^2$ trypanosomes/mℓ in fresh culture medium.
3. Place 1 mℓ of the final trypanosome suspension in each 3 to 5-day-old T-type flask cultures containing 5 mℓ fresh medium (3 flasks/dilution) or 0.2 mℓ in a well culture containing 0.8 mℓ fresh medium (3-6 wells/dilution) and incubate at 37°C. Well cultures should be incubated in a humidified CO_2 incubator (5% CO_2 in air).
4. Change a portion of culture medium (1/2-1/6, depending on the density of trypanosome population) every 24 hours and select the best culture during the first 7 days of cultivation.
5. Make subcultures from the best culture in new feeder layer cultures, 1:1, 1:2, 1:3, or 1:4 splits, depending on the density of the trypanosome populations. Keep the number of trypanosomes at a level of 1-5 $\times 10^5$/mℓ. Repeat this procedure 2 to 3 times every 7 to 10 days using new feeder layer cultures.
6. Establish long-term cultures from the trypanosome populations having consistent growth rates during the initial 30 to 40 days.

The selection of the best population may be made based on (1) the population density (select the populations of 5-8 $\times 10^5$ trypanosomes/mℓ), (2) the growth rate (select the populations having a consistent growth rate), (3) the morphology (select cultures in which long, slender bloodstream forms are predominant), (4) the motility (eliminate cultures containing many sluggish trypanosomes), (5) the number of dividing forms (select the populations containing many dividing forms) and (6) the size distribution (eliminate the populations consisting of a wide range of sizes).

Cultures of fast growing trypanosomes may be initiated by inoculating feeder layer cultures with low seeding densities (10^2 or 10^3 trypanosomes/mℓ) while those of slow growing trypanosomes should be initiated with high seeding densities (10^4 or 10^5 trypanosomes/mℓ).

From infective tsetse fly saliva — Brun et al. have demonstrated that bloodstream form cultures can also be initiated with metacyclic trypomastigotes derived from the salivary glands of infective tsetse flies as summarized below (see also Section II. B. 2).[50]

1. Infect teneral tsetse flies, *G. m. morsitans* with fly transmissible strains of *T. b. brucei* or *T. b. rhodesiense* according to the method of Jenni.[62]
2. Prepare feeder layer cell cultures in 24-well tissue culture plates (Coster, Cambridge, U.S.A.) by seeding the wells with fibroblast-like cells which are derived from blood of a young adult New Zealand white rabbit, following the procedure described by Hirumi et al.[43]
3. Collect metacyclic trypomastigotes from the salivary glands of infective flies by letting each salivate into one drop of warm culture medium (modified Eagle's MEM) and count the number of trypanosomes.
4. Inoculate each well containing 1 mℓ of medium with 3-8 $\times 10^3$ "salivary gland forms" exuded from one infective tsetse fly and incubate in 4% CO_2 in air at 36 to 37°C. After 24 hr, all metacyclic forms transform to long, slender bloodstream trypomastigotes.
5. After 48 hr, carefully remove 0.5 mℓ of the culture medium from each well and replace with the same volume of fresh medium. By this time, trypanosomes form clusters in intercellular spaces of the feeder layers.

6. After establishment of the intercellular cluster of bloodstream forms, change the culture medium completely every day.
7. Subcultivate in new feeder layer cultures about every 2nd week. Yields of 1-2 × 10^6 trypanosomes/mℓ can be achieved in this manner. (The original cultures may be maintained for 3 months by changing the medium every day).

2. Maintenance

It has been demonstrated that the growth rates of hematozoic trypomastigotes of *T. b. brucei* vary considerably depending on their population densities in vitro.[9] The trypanosomes which reach a maximum population density rapidly lose their growth potential in culture following subsequent dilution in fresh medium. The effects caused by overpopulation are so great that these may often lead to misinterpretation of experimental results. It is, therefore, essential to maintain the trypanosomes under optimum conditions in which they can produce their maximum growth (logarithmic growth phase) throughout the experiment. This is particularly important in any study which is dealing with accurate quantitative analyses. Bloodstream forms can be maintained at the logarithmic growth phase by the following procedure:

1. Select a culture containing mostly long, slender bloodstream forms at a concentration of 5 × 10^5-1 × 10^6 trypanosomes/mℓ and dilute the trypanosome population to 6 × 10^3/mℓ in fresh medium.
2. Place 1 mℓ of the final trypanosome suspension in each T-type flask (25 cm^2) containing feeder layer cells and 5 mℓ of fresh culture medium.
3. During the next 4 to 6 days, change a portion of culture medium (0.5 to 2.0 mℓ) with the same volume of fresh medium and estimate the population density every 24 hr.
4. When the population density reaches 5 × 10^5 trypanosomes/mℓ subculture the trypanosomes in new feeder layer cultures by inoculating them with 5 × 10^4 trypanosomes/mℓ. At this stage, the tryanosome populations consist mostly of long, slender bloodstream forms (over 99%) with many actively dividing forms.
5. Establish the maximum growth rate and population density of the organisms used during the next 48 hr without changing the medium.
6. Based on the maximum growth rate and population density established above, bloodstream trypomastigote cultures can be maintained at their logarithmic growth phase by diluting the populations every 24 hr so as to maintain their densities at a slightly lower level than the maximum density during every 24 hr of cultivation. For example, a bloodstream trypomastigote population whose maximum density is 2 × 10^6 trypanosomes/mℓ and the population doubling time is 6 hr may be diluted to 1/12 to 1/16 every 24 hr. In this way, the population can be maintained at a level of 0.9 to 1.0 × 10^6 trypanosomes/mℓ 24 hr after each dilution and consists of mostly long, slender bloodstream forms with a number of actively dividing trypanosomes. A population whose doubling time is 10 hr may be diluted to 1/4 to 1/5 every 24 hr.

During the long-term cultivation of hematozoic trypomastigotes, feeder layer cultures have to be also changed by subcultivating the trypanosome populations to new feeder layer cultures (3 to 5 days old) from time to time, depending on the type of feeder layer cells used (see Section III. C. 2.). It is thus necessary to have new feeder layer cultures always available for such subcultivation. One of the main factors which may account for the loss of long-term cultures is a lack of appropriate feeder layer cells at the time when the bloodstream forms need to be subcultured.

3. In Vitro Cloning

A prerequisite for certain investigations, such as antigenic variation of salivarian trypanosomes (see Section IV. 2.), is the availability of cloned populations of the trypanosomes in the absence of the host immune system. A technique to derive such populations in vitro has been recently established and the following step-by-step procedures have been recommended for future applications of the technique.[56]

1. Prepare (a) feeder layer cell cultures in a Microtest II Tissue Culture Plate (96 wells/plate, 0.39 mℓ/well size, Falcon: or similar one) and (b) 5-10 bloodstream form population at their logarithmic growth phase in 25 cm^2 T-type flasks containing feeder layer cells 3 days prior to in vitro cloning (see Section III. C. and D.).
2. On day 3, select the best population which mostly consists of long, slender bloodstream forms and many actively dividing forms and prepare a trypanosome suspension by 1-step dilution of the population to 3×10^3 trypanosomes/mℓ in RPMI 1640 medium supplemented with 50% FBS (isolation medium). Do not use populations containing many abnormal forms and/or sluggish trypanosomes.
3. Isolate a single organism from the final suspension by placing a small drop (approximately 0.3 μℓ) of the suspension in one well of a new Microtest Tissue Culture Plate without feeder layers. Place six drops in six wells at a time.
4. Promptly inspect the drops for the presence of a single trypanosome/well using an inverted phase-contrast microscope. Approximately 40 to 50% of drops are expected to contain a single organism.
5. Rapidly add 0.1 mℓ of standard culture medium containing 20% FBS to the wells containing a single organism. Take all possible precautions to prevent excessive drying of the isolation medium during steps 3, 4, and 5. The trypanosome suspension in the isolation medium should be replaced 30 min after its preparation.
6. Transfer the isolated single organism into a well of previously prepared (3-day-old) feeder layer cultures in the plate, and add 0.2 mℓ of fresh culture medium to each well. Place organisms in 6-24 wells following the same procedure and after replacing the lid incubate the plate in a humidified 5% CO_2 incubator at 37°C. Repeat steps 3, 4, 5, and 6 until sufficient numbers of single trypanososes have been isolated.
7. When the population density reaches over 10^2 trypanosomes/well, gently resuspend the trypanosomes and feeder layer cells in the well and transfer the mixed suspension into a T-type flask containing feeder layer cells and 6 mℓ of culture medium.
8. Replace a portion (0.5 to 2.0 mℓ) of old medium during the next 7 to 12 days. The volume to be changed and the time interval of medium change should be determined according to the number of trypanosomes and their growth rate in each culture.
9. When the population density reaches $1\text{-}5 \times 10^5$ trypanosomes/mℓ, subculture (1:2-1:6 split) into new feeder-layer cultures every 1 to 3 days during the next 10 to 14 days. Select cultures containing a number of actively dividing forms during this period. Discontinue cultures containing sluggish trypanosomes, many abnormal forms and/or much cellular debris.
10. After the cloned populations become stable (usually 14 to 21 days from cloning), the cultures can be maintained following the procedure described above (see Section III. D. 2). Approximately 50% of a cloning success rate may be expected using this technique.

Brun and Schönenberger[52] have also reported a similar technique (without feeder layer cells) for the in vitro cloning of procyclic forms of *T. b. brucei*, using a semidefined medium SDM-99 which has been modified from SDM-77.[23]

E. Large-Scale Cultivation

Certain experiments in various disciplines require large amounts of hematozoic trypomastigotes of African trypanosomes. However, large-scale culture techniques for the in vitro propagation of animal-infective bloodstream forms have not yet been developed. The maximum yields of the parasites in 25 cm^2 and 75 cm^2 T-type culture flasks by the presently available feeder layer cell system are 2.4×10^7 and 1.2×10^8 trypanosomes/flask, respectively. In order to meet this requirement a semilarge scale culture method has been developed.[8] Roller bottle (250 × 11 cm) cultures were seeded with 5×10^7 epithelial cells derived from bovine embryonic peritoneum 3 days prior to the cultivation of parasites. On day 3, each roller bottle (250 × 11 cm) was inoculated with 2×10^7 bloodstream forms of *T. b. brucei* (TC221) in 100 mℓ of RPMI 1640 medium with 20% FBS and incubated at 37°C using a Rollacell (New Brunswick Scientific, N.J.). A yield of 1.3×10^9 trypanosomes/bottle has been achieved in this manner within 48 hr of cultivation.

F. Cryopreservation

Hematozoic trypomastigotes propagated in vitro can be preserved in liquid nitrogen using either 7.5% dimethyl sulfoxide (DMSO) or 10% glycerol in RPMI 1640 medium with 20% FBS.[43] The following routine procedure has been successfully used in our laboratory (ILRAD) during the past 5 years.

1. Collect bloodstream forms by centrifugation at 1,000 g for 5 minutes at 4°C and discard the supernatant. (For example: 6 mℓ of culture medium containing 2×10^6 trypanosomes/mℓ from one 25 cm^2 T-type culture flask in a conical centrifuge, tube, 9 mℓ size).
2. Resuspend the trypanosomes in 7.5% DMSO in RPMI 1640 medium with 20% FBS (2 mℓ/tube) and transfer to a freezing tube.
3. Slowly freeze the trypanosome suspension in the vapor phase of liquid nitrogen for 2 hr, at the F position of BF-5 Biological Freezer on Liquid Nitrogen Refrigerator (Linde, Supper-30, Union Carbide) and then store in the liquid phase.
4. For recovery, rapidly defrost the frozen materials in a water bath at 37°C.
5. Immediately after thawing, transfer the trypanosome suspension into a centrifuge tube (9 mℓ conical) and centriguge at 1,000 g for 5 min.
6. After removing the supernatant, resuspend the trypanosomes in fresh culture medium (6 mℓ/tube), place in a 3 to 5-day-old feeder culture and maintain at 37°C following the procedures described above (see Section III. D. 2). Approximately 80% of the trypanosomes have been successfully recovered in this manner after 4 years of cryopreservation.

IV. RESEARCH APPLICATIONS AND FUTURE PROSPECTS

The procedures that have been used by Hill for growing *T. b. rhodesiense* are principally a modification of the initial studies by Hirumi et al.[43] Their studies were directed toward identifying established tissue culture cell lines that would support the growth of *T. b. rhodesiense* at high yield and developing simple procedures for establishing infective trypomastigotes in culture. In addition, they wanted to determine the biochemical similarities between *T. b. rhodesiense* cultured infective trypomastigotes maintained in culture and the slender bloodstream trypomastigotes isolated from a human. Successful cultures were established with a pleomorphic strain of *T. b. rhodesiense* (EATRO 1895) which was made available from the London School of Tropical Medicine and Hygiene. The strain was initially isolated from a naturally infected male patient from Lugala, Uganda.

The tissue culture cells that were found effective in growing *T. b. rhodesiense* in vitro include Chinese hamster *(Cricetulus griseus)* lung tissue culture cells (ATCC CCL 16 Don and embryonic bovine *(Bos taurus)* trachea tissue culture cells (ATCC CCL 44 EBTr). Both lines of tissue culture cells are available from the American Type Culture Collection, Rockville, Maryland. The tissue culture cells were maintained in RPMI 1640 (GIBCO) supplemented with 25 m*M* HEPES buffer, 20 m*M* sodium bicarbonate and 10 to 20% heat inactivated fetal bovine serum (Sterile Systems, Logan, Utah). In addition, the media contained per 100 mℓ, penicillin (10,000 U), fungizone (25 μg), streptomycin (10 mg), gentamycin (50 μg), and kanomycin (12.5 mg). The morphology of the tissue culture cell line was fibroblast-like. An inoculum of 1.8 - 2.0 × 10^6 viable tissue culture cells multiplied 1 to 3 times in 4 days in the described culture media at 37°. Carbon dioxide was not necessary for growth, the cells growing well in 100% air. The tissue culture cells were used just prior to becoming confluent, usually 3 to 4 days after a flask was inoculated. We have observed that either tissue culture cell line is able to support the growth of trypanosomes even after irradiation of the tissue culture cells at 5,000 rads.

A. Preparation of Trypanosomes

Successful cultures were established with a pleomorphic strain of *T. b. rhodesiense* from normal or lethally irradiated (800 rads) male rats weighing 200 to 300 g) that were inoculated intraperitoneally with 1-2 × 10^6 trypanosomes. When the parasitaemia reached a level of 1-5 × 10^8 trypanosomes/mℓ 10 to 12 days later, the rats were bled by cardiac puncture.

Approximately 90 to 95% of the trypanosomes present had the morphology of the slender trypomastigote, but some stumpy trypomastigotes were also observed. The trypanosomes were centrifuged at 1025 g for 7 min at 4°C and the buffy coat was removed. Special precautions were taken to prevent the removal of red blood cells below the buffy coat. In some experiments the trypanosomes were washed twice with sterile Hanks' balanced salt solution with 5% (w/v) sodium citrate. However, in other experiments, this step was not found to be desirous as the solution became quite viscous. In most experiments, the trypanosomes were removed from the buffy coat, resuspended in the plasma above the buffy coat and added to the flasks with the tissue culture cells. All the steps in the handling of the trypanosomes were done at 4°C under sterile conditions.

B. Inoculation of Tissue Flasks

The trypanosomes were inoculated into a total of 5.0 mℓ of the culture medium in T-25 Falcon flasks or in 15.0 mℓ in T-75 flasks. A typical experiment was performed by inoculating a T-25 flask containing a monolayer of Chinese hamster lung cells with an initial concentration of 1-2 × 10^5 bloodstream trypanosomes per mℓ. The culture was incubated at 37° in 100% air. Generally speaking, the trypanosome numbers remained low, less than 1 × 10^6 cells/mℓ for 5 to 7 days. However, the trypanosomes were very active in the media. In addition, numerous dividing forms were seen. The trypanosome population then began to increase and on day 12 to 14, reached 1 × 10^6 cells/mℓ. During the initial stages of the culture from day 0 to day 12 one third of the medium was changed when cell debris was evident in the media or when the pH dropped. After the trypanosomes reached a cell density of 1 × 10^6 cells/mℓ, one third of the media was changed daily. In addition, new flasks could be started at an initial concentration of 1-2 × 10^5 cells/mℓ. In subcultured flasks, the trypanosomes increased rapidly with the population doubling time of 9 to 11 hr, as noted in Table 2 by Hill et al.[45] By changing one third of the medium daily parasites have been maintained in vitro with a concentration of 3-5 × 10^6 cells/mℓ over Chinese hamster lung cells for 80 days. In addition, similar cultures have been established using embryonic bovine trachea tissue culture cells. In our routine procedures, the tissue culture cells are changed every 7 to 10 days.

On either of the tissue culture cell lines that we have used, the trypanosomes grow not only on the medium but also in spaces between the tissue culture cells. Before cell cultures become established, usually trypanosomes are seen initially among the tissue culture cells and then later in the supernatant. In addition, after the trypanosomes are established and

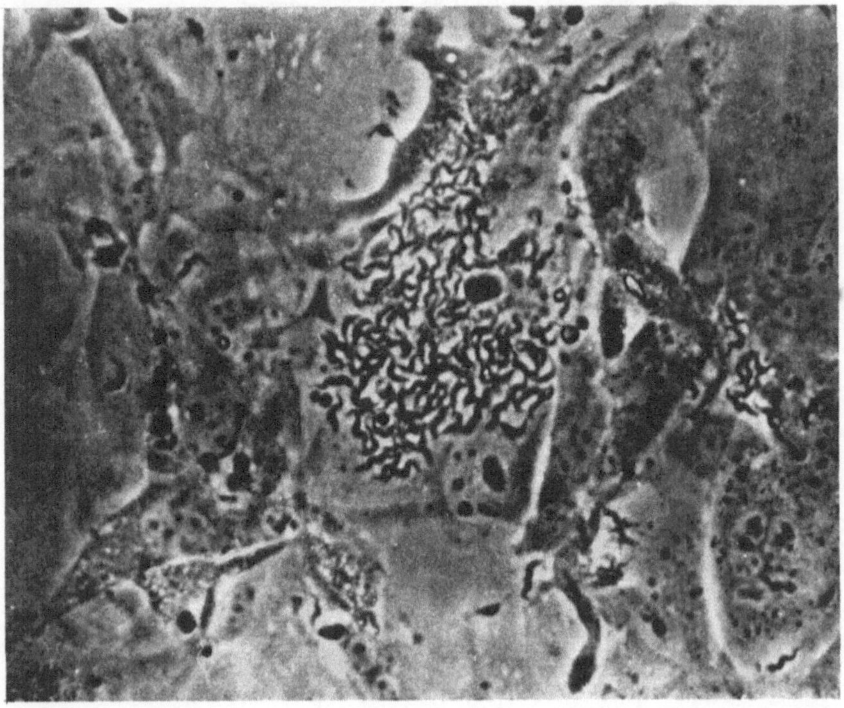

FIGURE 1. *T. rhodesiense* EATRO 1895 cultured infective trypomastigotes growing in the spaces between the bovine embryonic trachea tissue culture cells. The trypanosomes had been growing in culture for 50 days. ×200.

growing at a concentration above 2×10^6 cells/mℓ, trypanosomes are often observed in large clusters among the tissue culture cells. This can be seen in Figure 1. A similar observation was reported by Hirumi et al. for *T. b. brucei* growing on bovine fibroblast-like tissue culture cells.[43]

Giemsa-stained preparations of *T. b. rhodesiense* maintained in vitro in either tissue culture cell line revealed that the kinetoplast is subterminal, clearly distinguishing the trypanosomes as long, slender trypomastigotes. The ultrastructure of *T. b. rhodesiense* maintained on tissue culture cells has been previously published.[40] The surface coat is evident as well as the mitochondrion with few cristae, both features characteristic of slender, bloodstream trypomastigotes. Numerous electron micrographs of cultured trypomastigotes suggest that the thickness of the surface coat is significantly reduced from bloodstream trypomastigotes used as inocula for the tissue culture flasks. In addition, electron micrographs suggest very strongly that the trypanosomes in culture are attached to the tissue culture cells. Sections reveal a close proximity between the surface coat of the cultured trypomastigotes and the tissue culture cells. These results have been published for *T. b. brucei* as one can see in Figure 2, Hill et al.[44]

Experiments to test the infectivity of the culture were performed with stabilates of bloodstream trypomastigotes of *T. b. rhodesiense* as controls. Mice were inoculated with 10^5 and 10^6 cells. As previously reported, at early periods during cultivation, mice or rats inoculated with slender trypomastigotes from bloodstream stabilates required the same period before the death of the laboratory animals as cultured infective trypomastigotes.[45] However, the longer the trypanosomes remained in culture, the longer the period of time required before the trypanosomes killed the laboratory host. For example, cultures of *T. b. rhodesiense* which had been maintained for 80 days in vitro, required 25 to 30 days to kill mice, while

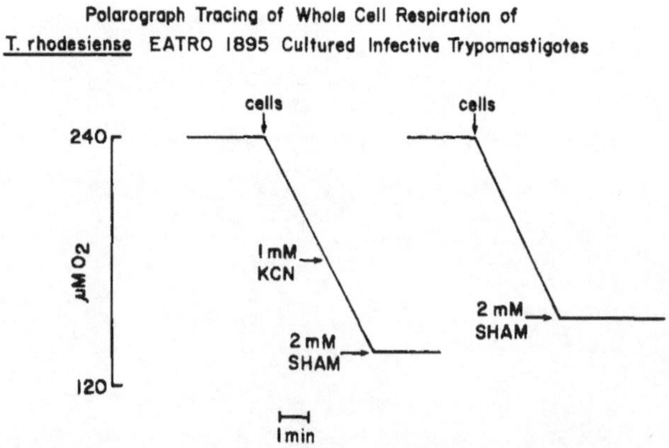

FIGURE 2a. Polarographic tracing of *T. rhodesiense* EATRO 1895 cultured infective trypomastigotes using the closed system. The cells had been growing in vitro at 37°C for 50 days. The concentration of cells was 2.0 $\times$ 10^8 trypanosomes/mℓ.

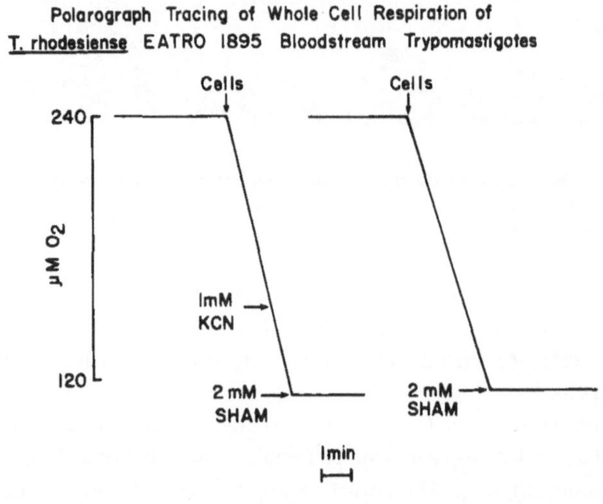

FIGURE 2b. Polarographic tracing of *T. rhodesiense* EATRO 1895 bloodstream trypomastigotes using the closed system. The concentration of cells was 2.0 $\times$ 10^8 trypanosomes/mℓ.

the same number of trypanosomes prepared from stabilates of *T. b. rhodesiense* required only 7 to 10 days. These results suggest that as the *T. b. rhodesiense* trypanosomes were maintained in vitro, the strain was becoming attenuated. Procyclic trypomastigotes grown in the absence of tissue culture cells at 37° were not infective for mice.

One of the important properties which has been considered is the electron transport system present in *T. b. rhodesiense* cultured infective trypomastigotes, bloodstream trypomastigotes, and procyclic trypomastigotes. In order to determine the terminal oxidases present in cultured infective trypomastigotes, we measured the steady state activity of the terminal oxidases present. Culture medium (90 mℓ) from three inoculated Chinese hamster cell cultures was removed at a concentration of 2.5 $\times$ 10^6 trypanosomes/mℓ. The organisms were collected by centrifugation and the steady state oxygen levels measured. A previously reported, salicylhydroxamic acid (SHAM) inhibited 90% of the cell respiration.[45] Neither azide nor cyanide had significant effect on cell respiration. These results are similar to our previously

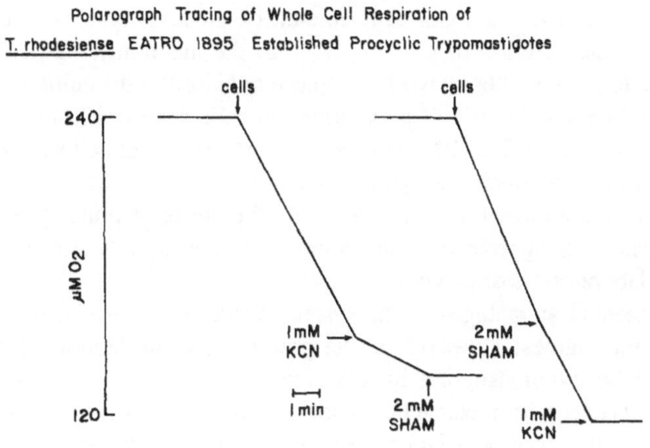

FIGURE 2c. Polarographic tracing of *T. rhodesiense* EATRO 1895 established procyclic trypomastigotes using the closed system. The concentration of cells was 2.0×10^8 trypanosomes/mℓ.

published results for bloodstream trypomastigotes of *T. b. rhodesiense* which demonstrated that these organisms are predominantly sensitive to SHAM.[63] These results clearly indicate that the SHAM-sensitive α-glycerophosphate oxidase is the predominant terminal oxidase in cultured infective trypomastigotes like *T. b. rhodesiense.* In contrast, procyclic trypomastigotes of *T. b. rhodesiense* have predominantly cytochrome aa_3 as the terminal oxidase present in its electron transport system. Results with the three different types of *T. rhodesiense* trypomastigotes can be seen in Figure 2a, 2b, and 2c, using the closed polarograph system.

C. Current Problems with In Vitro Techniques

While the tissue culture system as modified by Hirumi's procedures has proved an excellent means for obtaining cultured infective forms of *T. b. rhodesiense,* several problems remain. For example, it has been extremely difficult to grow and maintain recently isolated strains of *T. b. rhodesiense, T. b. brucei* or *T. b. gambiense*. Using the tissue culture cell lines that we have available, we have had little success in trying to maintain and grow recently isolated strains of *T. b. rhodesiense*. The organisms initially are maintained for 3 to 4 days in culture and then die out. This culture procedure seems to be extremely strain specific. For instance, a strain of *T. b. rhodesiense* which we have found most successful is *T. b. rhodesiense* EATRO 1895. In addition, we have had no difficulty in growing *T. b. brucei* EATRO 427, *T. b. brucei* LUMP 1026 and *T. b. brucei* EATRO 110. However, our results with other strains of *T. b. rhodesiense* have not been as successful.

One of the additional problems remaining is the presence of the tissue culture cells in the system. We have begun some studies using parabiotic chambers in order to try to separate the trypanosomes from the tissue culture cells and maintain their growth. The tissue culture cells could have several roles in this system. For example, the tissue culture cells could be providing some essential nutrient to the trypanosomes which is not in the RPMI medium. Alternatively, the tissue culture cells could be removing some growth inhibiting substances from the medium which the trypanosomes produce. Thirdly, it is possible that the tissue culture cells are absolutely required for initiation of the culture since very often the trypanosomes are seen among the tissue culture cells during the early stages of initiation of the cultures and before the cells appear in the supernatant. So far, it has not been possible to grow the trypanosome in the absence of tissue culture cells. Another technical problem at present is the requirement that one third of the medium must be removed from the culturing system daily. Preliminary studies have shown that 20 to 30% of the glucose in the medium is used in 24 hr. However, a significant amount of glucose remains. Studies are now in progress to see if the addition of glucose alone to the medium is sufficient to permit the cells to continue to grow without having to replace one third of the culture medium. Another concern, as noted earlier, is the cell population which is achieved in this culturing system. Maximum yield has been 3-5 × 10^6 trypanosomes/mℓ. This is quite low when one considers the growth that can be obtained from an inoculated rat (1-2 × 10^9 trypanosomes/mℓ blood). Clearly, there are some growth limiting factors in the present culture system which prevent the trypanosomes from approaching the number of cells usually obtained when they are infected into a laboratory host in vivo.

One of the potential advantages of the system would be its usefulness in the growth of metacyclic trypomastigotes. However, at the present time, no laboratory has been able to maintain metacyclic trypomastigotes in this system. Brun et al. have been able to initiate cultures of *T. b. brucei* with metacyclic trypomastigotes.[50] However, in their tissue culture system, the metacyclic trypomastigotes developed into bloodstream trypomastigotes. The growth of metacyclic trypomastigotes in vitro would be extremely helpful in being able to obtain large quantities of these important forms for molecular and biochemical studies.

D. Future Possibilities for Cultivation

The advantages of having the cultured infective trypomastigotes grown in vitro are numerous. One can use this culturing system in order to evaluate its potential for assaying for new trypanocidal drugs. The system would have the advantage that the drug could be added directly to a growing trypanosome culture and the effects noted on the parasites. Thus, one could do metabolic and ultrastructural studies as well as biochemical research without having to have the cultured infective trypanosomes put into the laboratory host system. However, a disadvantage of the system is that drugs that would be modified to their active form may not be active in this test system. However, it is possible that the presence of the tissue culture cells could ameliorate this disadvantage. The question of antigenic variation during the life cycle of trypanosomes is an important one. The advantage of the tissue culture system is that one can now consider the question of antigenic variation as a phenomena removed from the host system.

One of the areas of biochemical investigation which will be helped immensely by the availability of the tissue culture system is the question of differentiation during the life cycle of trypanosomes. With cultured infective forms of bloodstream trypomastigotes available, it will be possible to study differentiation totally in vitro. Thus, one is now able to take cultured infective trypomastigotes and examine differentiation of the trypanosomes in vitro. What biochemical changes are occurring as the trypanosomes go from bloodstream to procyclic trypomastigotes? What are the gene-regulating phenomena that are involved in this process? What are the environmental factors that control this differentiation? By removing the trypanosomes from the host, it will be possible to ask some of these questions in a markedly simplified system.

V. SUMMARY

While improvements clearly remain to be made in the culturing system for bloodstream trypomastigotes, the development of this system has led to the potential of studying the biochemistry of bloodstream trypomastigotes for the first time as they replicate in vitro. This system allows for the growth and maintenance of cultured infective trypomastigotes. This advantage is significant for individuals interested in the molecular biology and the biochemistry of bloodstream trypomastigotes. As the system continues to be improved, one can expect its application to numerous problems in the area of biochemistry and molecular biology of hemoflagellates.

ACKNOWLEDGMENTS

Some of the research discussed in this review was supported by an NIH Research Career Development Award (1-KO-4, AI-70813-05) to G.C.H., and NIH research Grant (1-AI-11622-08) and a contract from the U.S. Army Medical Research and Development Command (DAMD-17-74-4046). In addition, some investigations received financial support from the UNDP/World Bank/WHO Special Programme for Research and Training in Tropical Diseases.

REFERENCES

1. **Duggan, A. J.,** An historical perspective, in *The African Trypanosomes,* Mulligan, H. W., Ed., Wiley Interscience, New York, 1970, XIi.
2. **DeRaadt, P.,** African sleeping sickness today, *Trans. R. Soc. Trop. Med. Hyg.,* 70, 114, 1976.
3. **Hoare, C. A.,** *The Trypanosomes of Mammals,* Blackwell Scientific, Oxford, 1972, 51.
4. **Hoare, C. A.,** The mammalian trypanosomes of Africa, in *The African Trypanosomiases,* Mulligan, H. W., Ed., Wiley Interscience, New York, 1970, 3.
5. **Banks, K. L.,** Binding of *Trypanosoma congolense* to the walls of small blood vessels, *J. Protozool.,* 25, 241, 1978.
6. **Baker, K. L.,** Techniques for the detection of trypanosome infection, in *The African Trypanosomiases,* Mulligan, H. W., Ed., Wiley Interscience, New York, 1970, 67.
7. **Evans, D. A.,** Kinetoplastida, in *Methods in Cultivation Parasites In Vitro,* Taylor, A. E. R. and Baker, J. R., Ed., Academic Press, New York, 1978, 55.
8. **Hirumi, H.,** Cultivation of salivarian trypanosomes: application to *in vitro* studies of African trypanosomiases, in *Practical Tissue Culture Applications,* Maramorosch, K. and Hirumi, H., Eds., Academic Press, New York, 1979, 309.
9. **Hirumi, H., Hirumi, K., Nelson, R. T., and Bwayo, J. J.,** Present status of *in vitro* cultivation of animal-infective African trypanosomes, in *The In Vitro Cultivation of the Pathogens of Tropical Disease, WHO Trop. Dis. Res. Series 3,* Schwabe & Co. A. G., Basel, 1980, 165.
10. **Novy, F. G. and MacNeal, W. J.,** The cultivation of *Trypanosoma brucei.* A preliminary note, *JAMA,* 41, 1266, 1903.
11. **Novy, F. G. and MacNeal, W. J.,** On the cultivation of *Trypanosoma brucei, J. Infect. Dis.,* 1, 1, 1904.
12. **Baker, J. R.,** Techniques for the detection of trypanosome infection, in *The African Trypanosomiases,* Mulligan, H. W., Ed., Wiley Interscience, New York, 1970, 67.
13. **Brooks, M. A.,** Applications of insect tissue culture an elucidation of parasite transmission, in *Invertebrate Tissue Culture,* Maramorosch, K., Ed., Academic Press, New York, 1976, 181.
14. **Guttman, H. N. and Wallace, F. G.,** Nutrition and physiology of the Trypanosomatidae, in *Biochemistry and Physiology of Protozoa,* Vol. 3, Hutner, H., Ed., Academic Press, New York, 1964, 459.
15. **Lwoff, M.,** The nutrition of parasitic flagellates (Trypanosomidae, Trichomonadinae), in *Biochemistry and Physiology of Protozoa,* Vol. 1, Lwoff, A., Ed., Academic Press, New York, 1951, 129.
16. **Schneider, I.,** Tsetse fly tissue culture and its application to the propagation of African trypanosomes *in vitro,* in *Tissue Culture Application,* Maramorosch, K. and Hirumi, H., Eds., Academic Press, New York, 1979, 373.
17. **Tayler, A. E. R. and Baker, J. R.,** *The Cultivation of Parasites In Vitro,* Blackwell Scientific, Oxford, 1968, 29.
18. **Tobie, E. J.,** Cultivation of mammalian trypanosomes, *J. Protozool.,* 11, 418, 1964.
19. **Trager, W.,** Cultivation of parasites *in vitro, Am. J. Trop. Med. Hyg.,* 27, 216, 1978.
20. **Evans, D. A. and Brown, R. C.,** The utilization of glucose and proline by culture forms of *Trypanosoma brucei, J. Protozool.,* 19, 686, 1972.
21. **Cross, G. A. M. and Manning, J. C.,** Cultivation of *Trypanosoma brucei* sspp. in semi-defined media, *Parasitology,* 67, 315, 1973.
22. **Cross, G. A. M., Klein, R. A. and Linsteal, D. J.,** Utilization of amino acids by *Trypanosoma brucei* in culture: L-threonine as a precursor for acetate, *Parasitology,* 71, 311, 1975.
23. **Brun, R. and Jenni, L.,** A new semi-defined medium for *Trypanosoma brucei* sspp., *Acta Tropica,* 34, 21, 1977.
24. **Trager, W.,** Tsetse-fly tissue culture and the development of trypanosomes to the infective stage, *Ann. Trop. Med. Parasitol.,* 53, 473, 1959.

25. **Wyatt, S. S.,** Culture in vitro of tissue from the silkworm, *Bombyx mori L., J. Gen. Physiol.*, 39, 841, 1956.
26. **Trager, W.,** On the cultivation of *Trypanosoma vivax:* a tale of two visits in Nigeria, *J. Parasitol.*, 61, 3, 1956.
27. **Nicoli, J. and Vattier, G.,** Culture de *Trypanosoma rhodesiense* sur tissue de pupes de Glossines, *Bulletin de la Societe de Pathologie exotique (Paris)*, 57, 213, 1964.
28. **Amrein, Y., Geigy, R., and Kauffman, M.,** On the reacquisition of virulence in trypanosomes of the Brucei-group, *Acta Trop.*, 22, 193, 1965.
29. **Cunningham, I.,** *In vitro* cultivation of trypanosomes in association with tsetse fly tissues, Proc. 3rd International Colloq. on Invertebrate Tissue Culture, Smollenice near Bratislava, Czechoslovakia, 1971, 477.
30. **Cunningham, I.,** Quantative studies on trypanosomes in tsetse tissue culture, *Exp. Parasitol.*, 33, 34, 1973.
31. **Cunningham, I. and Honigberg, B. M.,** Infectivity reacquisition by *Trypanosoma brucei brucei* cultivated with tsetse salivary glands, *Science*, 197, 1279, 1977.
32. **Cunningham, I.,** New culture medium for maintenance of tsetse tissues and growth of trypanosomatids, *J. Protozool.*, 24, 325, 1977.
33. **Grace, T. D. C.,** Establishment of four strains of cells from insect tissues grown *in vitro, Nature (London)*, 195, 788, 1962.
34. **Cunningham, I. and Slater, J. S.,** Amino acid analyses of haemolymph of *Glossina morsitans morsitans* (Westwood), *Acta. Trop.*, 31, 83, 1974.
35. **Cunningham, I. and Taylor, A. M.,** Infectivity of *Trypanosoma brucei* cultivated at 28°C with tsetse fly salivary glands, *J. Protozool.*, 26, 428, 1979.
36. **Steiger, R. F., Steiger, E., Trager, W., and Schneider, I.,** *Trypanosoma congolense:* partial cyclic development in a Glossina cell system and oxygen consumption, *J. Parasitol.*, 63 (5), 861, 1977.
37. **Tobie, E. J.,** The cultivation of *Trypanosoma congolense in vitro, J. Parasitol.*, 44, 241, 1958.
38. **LePage, R. W. F.,** Short-term cultivation of *Trypanosoma brucei in vitro* at 37°C, *Nature (London)*, 216, 1141, 1967.
39. **Ward, R. A. and West, H. J. Jr.,** In vitro culture of *Trypanosoma brucei, Trans. R. Soc. Trop. Med. Hyg.*, 64, 159, 1970.
40. **Mendez, Y. and Honigberg, B. M.,** Infectivity of *Trypanosoma brucei*-subgroup flagellates maintained in culture, *J. Parasitol.*, 58(6), 1122, 1972.
41. **Wonde, T. and Honigberg, B. M.,** Morphology and infectivity of *Leishmania donovani* cultivated in nonliving media at elevated temperature, *Am. J. Trop. Med. Hyg.*, 20, 828, 1971.
42. **Hirumi, H., Doyle, J. J., and Hirumi, K.,** African trypanosomes: cultivation of animal-infective *Trypanosoma brucei* in vitro, *Science*, 196, 992, 1977.
43. **Hirumi, H., Doyle, J. J., and Hirumi, K.,** Cultivation of bloodstream *Trypanosoma brucei, Bull. W. H. O.*, 55, 405, 1977.
44. **Hill, G. C., Shimer, S., Caughey, B., and Sauer, S.,** Growth of infective forms of *Trypanosoma (T.) brucei* on buffalo lung tissue culture cells, *Acta Trop.*, 35, 201, 1978.
45. **Hill, G. C., Shimer, S. P., Caughey, B., and Sauer, L. S.,** Growth of infective forms of *Trypanosoma rhodesiense* in vitro, the causative agent of African trypanosomiasis, *Science*, 202, 763, 1978.
46. **Hill, G. C.,** Growth of infective forms of *Trypanosoma rhodensiense in vitro*, in *The In Vitro Cultivation of the Pathogens of Tropical Diseases, WHO Trop. Dis. Res. Series 3*, Schwabe & Co. AG, Basel, 1980, 202.
47. **El-on, J., Rosen, N. L. and Patton, L.,** *In vitro* growth of *Trypanosoma congolense* in the presence of mammalian tissue at 37°, Proc. of the 5th Internatl. Cong. of Protozool., Abstract No. 124, New York, 1977.
48. **Isoun, M. J.,** *In vitro* cultivation of blood stream forms of *Trypanosoma vivax*, in *The In Vitro Cultivation of the Pathogens of Tropical Diseases, WHO Trop. Dis. Res. Series 3*, Schwabe & Co. AG., Basel, 1980.
49. **Nyindo, M., Darji, N., Saminjolo, J. S., Jr., Chimtawi, M., Patel, N., and Onyango, P.,** *Trypanosoma brucei* — cultivation in vitro of infective forms derived from the midgut of *Glossina morsitans, J. Parasitol.*, 64, 469, 1978.
50. **Brun, R., Jenni, L., Tanner, M., Schoenenberger, M., and Schell, K. F.,** Cultivation of vertebrate infective forms derived from metacyclic forms of pleomorphic *Trypanosoma brucei* stocks, *Acta Trop.*, 36, 387, 1979.
51. **Brun, R., Jenni, L., and Tanner, M.,** Pleomorphic *Trypanosoma brucei* stains: *in vitro* transformation of metacyclic forms and subsequent cultivation of the resulting bloodstream forms, *The In Vitro Cultivation of the Pathogens of Tropical Diseases, WHO Trop. Dis. Res. Series 3*, Schwabe & Co. AG., Basel, 1980, 207.

52. **Brun, R. and Schönenberger, M.,** Cultivation and in vitro cloning of procyclic culture forms of *Trypanosoma brucei* in a semi-defined medium, *Acta Trop.*, 36, 289, 1979.
53. **Gray, M. A., Brown, C. G. D., Luckins, A. G., and Gray, A. R.,** Maintenance of infectivity of *Trypanosoma congolense* in vitro with explants of infected skin at 37°C, *Trans. R. Soc. Trop. Med. Hyg.*, 73, 406, 1979.
54. **Luckins, A. G. and Gray, A. R.,** An extravascular site of development of *Trypanosoma congolense*, *Nature (London)*, 272, 613, 1978.
55. **Roberts, C. J., Gray, M. A., and Gray, A. R.** Local skin reactions in cattle at the site of infection with *Trypanosoma congolense* by *Glossina morsitans* and *G. tachinoides*, *Trans. R. Soc. Trop. Med. Hyg.*, 63, 620, 1969.
56. **Hirumi, H., Hirumi, K., Doyle, J. J., and Cross, G. A. M.,** *In vitro* cloning of animal-infective bloodstream forms of *Trypanosoma brucei*, *Parasitology*, 80, 371, 1980.
57. **Cunningham, M. P. and Vickerman, K.,** Antigenic analysis in the *Trypanosoma brucei* group, using the agglutination reaction, *Trans. R. Soc. Trop. Med. Hyg.*, 56, 48, 1962.
58. **Onyango, R. J., van Hoeve, K., and de Raadt, P.,** The epidemiology of *Trypanosoma rhodesiense* sleeping sickness in Alego Location, Central Nyanza, Kenya. I. Evidence that cattle may act as reservoir hosts of trypanosomes infective to man, *Trans. R. Soc. Trop. Med. Hyg.*, 60, 175, 1966.
59. **Chopra, D. P., Yeh, K. Y., and Gordon, D. G.,** Isolation of epithelial cell types from the rat colon, *Tissue Culture Association Manual*, 6, 1165, 1979.
60. **Terracio, L. and Douglas, W. H. J.,** Primary culture of rat ventral prostate epithelial cells, *Tissue Culture Association Manual*, 6, 1169, 1979.
61. **Goldman, W. and Baseman, J. B.,** Selective isolation and culture of a proliferating cell population from the hamster trachea, *In Vitro*, 16, 313, 1980.
62. **Jenni, J.,** Comparisons of antigenic types of *Trypanosoma (T.) brucei* strains transmitted by *Glossina m. morsitans*, *Acta Trop. (Basel)*, 34, 35, 1977.
63. **Hill, G. C. and Degn, H.,** Characterization of the steady state oxygen kinetics of terminal oxidases in *Trypanosoma rhodesiense*, in *Biochemical and Clinical Aspects of Oxygen*, Caughey, W. S., Ed., Academic Press, New York, 1979, 405.

Chapter 6

CULTIVATION AND IN VITRO STUDIES OF *BABESIA**

Michael G. Levy and Miodrag Ristic

TABLE OF CONTENTS

* Supported in part by a research grant 1-44-44-30-377 from The Rockefeller Foundation, New York, NY.

I. INTRODUCTION

Babesia is a member of the suborder Piroplasmidea.[2] The suborder includes certain erythrocytic parasites of mammals belonging to the families Babesidae and Theileridae which do not produce the pigment hemozoin in erythrocytes following digestion of hemoglobin. *Babesia* is the most important genus in the family Babesidae. The presence of a complex organelle on the anterior extremity of *Babesia, Plasmodium, Toxoplasma, Sarcocystis, Besnoitia,* and the coccidia is the basis of the subphylum, Apicomplexa. This organelle is visible only with the electron microscope, and for this reason its significance was formerly unknown. The morphologic characteristics and dimensions of this organelle vary among different genera.

The term "babesia" was adopted to honor a Romanian scientist, Babes, who in 1888 was the first to describe the babesia parasite in blood of African cattle showing hemoglobinuria.[2] At approximately the same time, the classic work of Smith and Kilborne[3] which was underway in the U.S. signifying the beginning of the scientific history of babesiosis. It is through this study that an arthropod was recognized for the first time as a vector of a protozoan agent, *Babesia bigemina,* one of several *Babesia* spp. known to cause bovine babesiosis. For several decades thereafter, babesiosis caused by *B. bigemina,* widely known as Texas tick fever, was responsible for serious losses among the U.S. cattle population. The disease was gradually eliminated from the U.S. by eradication of the tick vector, *Boophilus annulatus.*

Important bovine *Babesia* species include *B. bigemina, B. bovis* (synonymous with *B. argentina), B. divergens,* and *B. major. Babesia bigemina* is a large piroplasm measuring approximately 4 to 5 μ long by 2 to 3μ wide. *Babesia bovis* is a small piroplasm measuring approximately 2.4 μ long and 1.5 μ wide. Both agents occur in Europe, Africa, Australia, Asia, and Central and South America. The disease caused by *B. bovis* is more severe and more difficult to control than that caused by *B. bigemina. Babesia bovis* frequently causes a cerebral form of the infection characterized by formation of thrombi and emboli in the brain capillaries. Such animals may develop clinical signs similar to rabies and usually die before any appreciable parasitemia is manifested in the peripheral blood (less than 2% infected erythrocytes). In general, and from the economic point of view, infections caused by *B. bovis* are considered the most important of all bovine babesial infections. *Babesia divergens* occurs more frequently in northern Europe and the United Kingdom; it is often confused with *B. bovis,* being a pyriform, paired, or club-shaped organism about 1.5μ long and 0.4 μ wide. *Babesia major,* found in many parts of the world, is similar to but smaller than *B. bigemina.*

II. PATHOLOGIC MECHANISMS OF BABESIOSIS

Two states of infection have been described as babesiosis and babesiasis. Babesiosis refers to the period when there is rapid growth and multiplication of the parasite and there are clinical signs of disease. The most common signs of acute babesiosis include fever, hemoglobinuria, anemia, and associated parasitemia. Babesiasis refers to the subclinical chronic infections observed in animals which have recovered from the clinical disease or were subjected to premunization.

Many of the pathologic effects of babesiosis are due to the indirect effects of babesial multiplication in the host. Immune complex disease and associated glomerulonephritis, anemia, and thrombocytopenia have been described in rodent babesiosis caused by *B. rodhaini.*[4] Circulating babesial antigens also appear to attach to parasitized and nonparasitized erythrocytes.[5] The formation of immune complexes on the erythrocyte surface may increase the permeability and fragility of circulating erythrocytes. Formation of such complexes in the plasma may serve as a means of evasion of parasite destruction by immune antibodies.

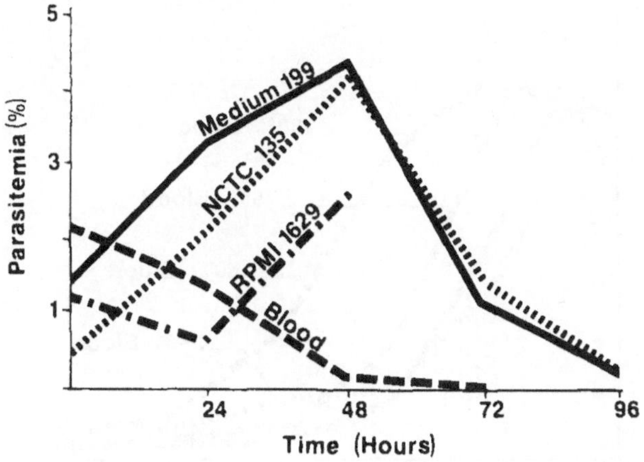

FIGURE 1. Growth of *Babesia bovis* in spinner flask cultures using various tissue-culture media according to the method of Erp et al. (1978).

During the exit of *B. bovis* and *B. bigemina* parasites from infected bovine erythrocytes, two or more parasite-associated proteolytic enzymes are released into the plasma. These enzymes and/or similar parasite metabolic products are believed to interact with blood components and may ultimately be responsible for several of the pathologic signs and symptoms associated with bovine babesiosis, including increased erythrocyte fragility, hypotensive shock, and disseminated intravascular coagulation.[6] Anemia is caused by intravascular destruction of erythrocytes by escaping babesias.[7] The reduction of hematocrits in acutely infected animals may also be associated with activation of kallikrein.

A peculiarity of babesiosis caused by certain species of babesia, including *B. bovis* and *B. canis*, is the clogging of small capillaries of the skin and brain with both parasitized and unparasitized erythrocytes.[8] It is believed that the coating of erythrocytes by parasite antigens neutralizes the normal surface charge of these cells, thus favoring erythrocyte autoagglutination. This effect is particularly striking in *B. bovis* infections.

Erythrocytes parasitized with *B. bovis* are more resistant to lysis than either normal erythrocytes from control animals or unparasitized erythrocytes from infected animals. On the other hand, erythrocytes parasitized with *B. bigemina* were larger and osmotically more fragile than normal erythrocytes from control animals.[6] There is an indication that *B. bigemina* preferentially invades young erythrocytes.[9]

III. CULTIVATION

Attempts to adapt the Trager and Jensen[10] methodology for the continuous cultivation of *Plasmodium falciparum* to the cultivation of *B. bovis* was unsuccessful.[11] Development of suspension cultures provided conditions which allowed short-term in vitro multiplication of *B. bovis*[11] and *B. canis*.[12] Minor modification of the Erp technology resulted in continuous in vitro growth of the important bovine parasite.[13] Although spinner flask (SF) cultures supported the growth of *B. bovis*, the large volumes and extensive manipulations required of these cultures, and the relatively slow growth rate achieved, limited utilization of this method. These objections have been overcome with development of the microaerophilous stationary phase culture system by Levy and Ristic.[14] High parasitemias, adaptation to microtiter culture plates, and growth characteristics similar to those observed in the bovine host make this an ideal system for experimental investigation.

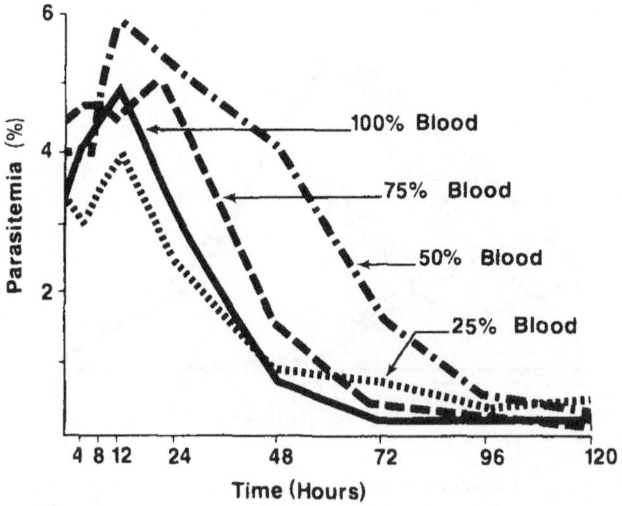

FIGURE 2. Growth of *Babesia bovis* in spinner flask cultures using varying ratios of blood and medium 199 according to the method of Erp et al. (1978).

A. Suspension Cultures of *Babesia bovis*

Erp and co-workers developed suspension cultures of *B. bovis* in an attempt to simulate in vivo conditions.[11,13] Infected and normal erythrocytes were suspended in bovine serum diluted with equal volumes of mammalian tissue culture media at physiological pH (7.4) under an atmosphere of 5% CO_2/air. Antibiotics and anticoagulants were not used, and cultures incubated at 38°C in 100 mℓ spinner flasks were stirred at approximately 100 rpm. Several tissue culture media were utilized as diluents and the results are presented in Figure 1. Dilution of infected blood with tissue culture media resulted in considerable multiplication of the parasites during the first 48 hr, whereas, the parasitemia in whole defibrinated blood declined throughout the incubation period. When infected blood was diluted with medium 199 in varying ratios (Figure 2), it appeared that equal volumes of blood and medium were optimal for short-term cultivation. However, parasite multiplication could not be sustained for greater than 48 hr. By changing the medium at 24-hr intervals some additional multiplication was achieved. To accomplish the medium change, cultures were transferred to centrifuge tubes and sedimented at 1000 × g for 10 min, after which the erythrocytes were returned to the culture flasks and the supernatant replaced with a mixture of 50% fresh normal bovine serum and 50% M199. This procedure resulted in a steadily increasing parasitemia during the first 72 hr in vitro, but the parasitemia failed to increase beyond 3 days. Subcultures with freshly collected bovine erythrocytes resulted in initiation of additional parasite growth; however, the rate of increase was less than the rate of erythrocyte dilution.

Utilizing short-term spinner flask cultures with daily medium changes as previously described, an attempt was made to determine what factors, if any, might bring about increased and sustained parasite multiplication. The acceptability of using heparin in blood collection was investigated as defibrination of freshly collected bovine blood is less convenient than using anticoagulants. Starting with an initial parasitemia of 1.8%, a fourfold increase was observed during 80 hr of in vitro cultivation when defibrinated serum was used. However, when heparinized plasma (25 units/mℓ) was substituted for serum, the parasitemia failed to increase. Other anticoagulants were considered undesirable, and defibrination with glass beads has remained the preferential method for blood collection.

In a further effort to simplify the culture method and improve pH control, the effect of HEPES (N-2-hydroxyethylpiperazine-N′-2-ethane sulfonic acid) as a supplemental buffer

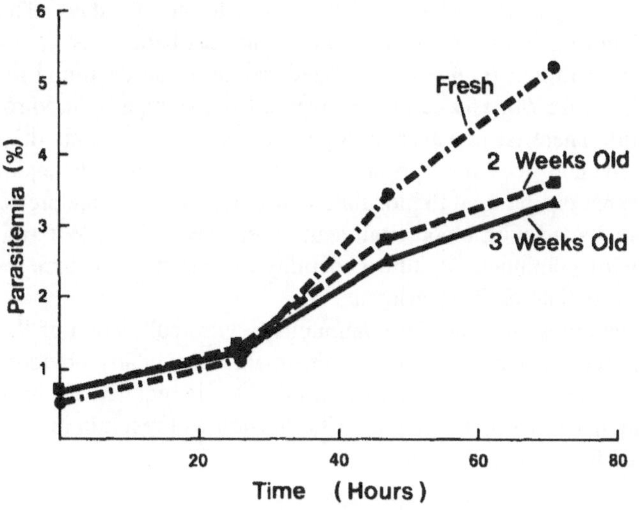

FIGURE 3. Growth of *Babesia bovis* in spinner flask culture according to the method of Erp et al. (1978) utilizing freshly collected and stored erythrocytes.

was evaluated. Three days of culture with daily medium change indicated that at a concentration of 25 m*M* this buffer failed to affect babesial growth. This suggested that it might be possible to dispense with supplemental CO_2 and rely entirely on the buffering capacity of the medium in order to maintain the desired pH. Attempts to cultivate *B. bovis* in the absence of CO_2 failed, indicating that CO_2 is a requirement for growth, and its concentration is an important culture parameter. It was later determined that CO_2 is necessary for infection of erythrocytes by merozoites (see below).

Addition of HEPES considerably improved pH control, and the effect of varying the pH was examined. The greatest multiplication was observed when the cultures were adjusted to pH 7.0. Maintaining the culture at pH 7.0 proved to be important in sustaining continuous in vitro growth. At this pH cultures were incubated at 38°C in 100 mℓ spinner flasks under an atmosphere of 5%CO_2/air. Subcultures were performed at 1 to 3-day intervals. Medium was changed daily, and a packed cell volume of 20 to 30% was maintained. During a 32-day period, during which 14 subcultures were performed, *B. bovis* increased by a factor of approximately 10^6-fold and the average generation time was 42 hr. This calculation fails to account for the substantial number of extra-erythrocytic *Babesia* which were discarded with the removed medium. Consequently, the actual rate of parasite multiplication was greater than recorded. Whether the extracellular merozoites are in some way defective or culture conditions are not optimal for infection is not known.

A very practical consideration for anyone using this culture methodology is the effect of storage on serum and erythrocytes. Figure 3 represents 3-day growth patterns of cultures containing fresh serum and serum from the same donor stored at 4°C for 2 and 3 weeks. The ability of serum to support babesial growth decreased with time, but storage for up to 1 week does not appear to be detrimental. The effect of storage on erythrocytes, however, appears to be more dramatic. Storage of defibrinated blood (pH corrected to 7.0) at 4°C for 6 days resulted in half the babesial growth as compared to freshly collected red cells from the same donor. As a result of these findings, erythrocytes were collected on the day needed and serum were stored for up to 1 week at 4°C.

Morphologically, *Babesia* which had been cultivated in vitro by the SF method continuously for 270 days could not be differentiated by light microscopy from organisms found in the blood of infected cattle. A fatal infection resulted from inoculation of a splenectomized calf with 5 × 10^9 organisms which had been in vitro for 85 days. This infection was clinically indistinguishable from tick or blood transmitted babesiosis.

The finding that cultured *B. bovis*-parasitized red cells can be stored in medium at 4°C for 1 month and that the *Babesia* can be recovered from storage by in vitro cultivation was particularly useful. There is, however, a large decrease in parasitemia during such storage but this culture method does not appear to have a seeding density requirement. *Babesia* stored in this manner can be used to inoculate new cultures. All of the previously mentioned experiments were started with blood from acutely infected calves. With this discovery and the development of continuous culture capability, it has become unnecessary to use an infected animal to initiate each experiment.

This culture method is adequate for continuous in vitro cultivation of *B. bovis*, but there are culture parameters which may require adjustment. Particularly obvious are oxygen and carbon dioxide tensions. We know that carbon dioxide is required for growth of *B. bovis*, so one might anticipate that optimizing the CO_2 tension will result in improved growth. The same may be true for oxygen tension.

B. Suspension Cultures of *Babesia canis*

Short-term in vitro growth of *B. canis* has been reported.[12] The methodology utilized by these investigators was similar to that reported by Erp et al.[11] Briefly, blood was collected when the parasitemia in peripheral blood of infected splenectomized dogs reached 3 to 4%. Following defibrination by shaking with glass beads, the blood was centrifuged at 400× g for 20 min at 4°C. After removal of the buffy coat cells, the remaining erythrocytes were washed three times in tissue culture medium. Washed erythrocytes were resuspended in four volumes of medium which consisted of RPMI 1640 pH 7.4, supplemented with 25 m*M* HEPES and 20% normal canine serum. Cultures were incubated at 38°C under an atmosphere of 5% CO_2/air with gentle stirring. After 48 hr of in vitro growth, the cultures were sedimented at 400× g for 10 min at 4°C, and the supernatant was replaced with fresh medium.

Under these conditions, the parasitemia increased from an initial 3.9% to 7.6% after 24 hr. By 48 hr the parasitemia decreased to 3.3%, and the cultures were discontinued at 96 hr, at which time few parasitized cells were observed. Similar to the report of Erp et al.,[11] large numbers of extra-erythrocytic forms were observed.

C. Stationary Culture of *Babesia bovis*

Cultivation of *B. bovis* in suspension cultures provided the first opportunity to examine this parasite under conditions that would support growth in vitro.[11,13] However, the relatively large volumes required for spinner flask cultures (100 mℓ), the requirement for daily centrifugation in order to provide for medium change, production of large numbers of extracellular forms incapable of invading new cells, and slow growth of the parasite under the described in vitro conditions limited the application of this methodology to experimental manipulation. Development of the microaerophilous stationary phase culture (MASP) system[14] has eliminated these objections. With this method, cultures are initiated using defibrinated peripheral blood of infected donor cattle when the parasitemia is between 0.1 and 2%. As with SF cultures, the use of standard anticoagulants, e.g., heparin, (ethylenedinitriol) tetraacetic acid (EDTA), or acid-citrate dextrose (ACD) in the collection of infected or normal blood is avoided. Infected erythrocytes are suspended to a final packed cell volume of 9.1% (5 to 10% is satisfactory) in a medium consisting of 60% medium 199 and 40% normal bovine serum supplemented with 15 m*M* HEPES, 100 mcg streptomycin, and 100 units penicillin G per mℓ. Serum obtained from mature *Bos taurus* cattle is preferable, as commercially obtained fetal calf, newborn calf, and bobby calf serum will not support growth. Additionally, the sera of calves (mixed dairy breeds under 3 months of age) obtained locally vary in their ability to support growth, with the sera of many failing to support parasite

Table 1
STATISTICAL DATA FOR MICROAEROPHILOUS STATIONARY PHASE (MASP) CULTURE OF *BABESIA BOVIS*

Days in culture	31
Number of subcultures	13
Parasitemia of 2-day subculture	19.9 ± 7.3%
Parasitemia of 3-day subculture	26.1 ± 7.1%
Maximum parasitemia	38.1 ± 3.0%
Cumulative dilution	3.15×10^{17}-fold
1 log increase[a]	42.3 hr

[a] Analysis of the exponential growth phase indicates a generation (doubling) time of 7-10 hr.

growth (see below). Following titration to pH 7 with 1 *N* HCl, culture suspensions are aliquoted into appropriate containers in a ratio of 0.62 mℓ/cm^2 and incubated at 37 to 38°C under an atmosphere of 5% CO/humidified air. At approximately 24-hr intervals, the overlying medium is removed and replaced with fresh babesia tissue culture medium #4 (BTCM4) consisting of 40% serum and 60% M199, supplemented with HEPES and antibiotics and titrated to pH7 as previously described. Following 48 to 72 hr, subcultures are accomplished by diluting old culture 3 to 25-fold with new culture (NC) containing freshly collected uninfected bovine erythrocytes. Parasite multiplication during each subculture does not persist beyond 3 to 4 days and maximum rates of growth are achieved when parasitemias are reduced to 0.5 to 1% at each pass. Infected cultures may be stored for up to 3 weeks at 4°C and should be diluted with freshly prepared NC in order to reinitiate active growth. Satisfactory results are obtained with NC stored for up to 7 days at 4°C providing the pH is readjusted to 7 before use.

For routine maintenance, flat-bottomed microtiter culture plates have been utilized. Each well has a diameter of 6.4 mm and receives 0.2 mℓ of culture suspension. Cultures have been grown in a variety of containers, and as long as this ratio of culture volume to area is maintained, there appears to be no limit to the volume of a single culture.

Under the described culture conditions, the layer of settled erythrocytes becomes dark-red to black, indicating a condition of low oxygen tension (deoxyhemoglobin). Resuspension of the erythrocytes in fresh oxygenated medium or lowering the temperature to 4°C results in a return to the bright-red appearance of oxygenated erythrocytes. The darkening of the erythrocytes cannot be explained by acidic conditions as the culture pH increases to approximately 7.2 during incubation. Generally, cultures containing less than 1% infected erythrocytes fail to darken. These observations indicate that the depletion of oxygen from the settled cell layer is a result of parasite metabolism. The column of medium above the parasitized cells acts as a barrier to oxygen exchange, and its depth is crucial to successful cultivation. Reducing the depth to 0.16 cm maintains the cultures in an oxygenated state and such cultures cannot be continuously maintained. Cultures which fail to achieve parasitemias great enough to darken the cells within 4 days produce aberrant and degenerate parasite forms (large rings with diffuse, poorly staining cytoplasm) and die out, indicating the requirement for a minimum seed density. The culture conditions allow the parasites to establish, as a result of their metabolic activity, a reduced oxygen environment; thereby preventing the toxic effect of high concentrations of this gas. It is for this reason that the term microaerophilous stationary phase (MASP) was used to describe this culture system. Numerous attempts to decrease or eliminate the seeding density requirement by placing cultures in candle jars similar to those used for the cultivation of *Plasmodium falciparum*[10]

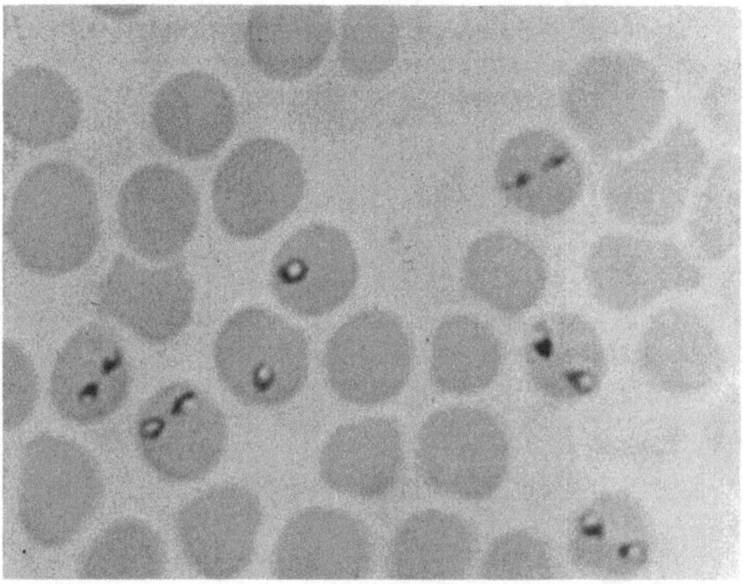

FIGURE 4. Giemsa-stained thin-blood film of *Babesia bovis* from microaerophilous stationary phase cultures following 105 days of continuous in vitro growth.

or under 5% CO_2, 5% O_2, 90% N have failed, as *B. bovis* did not grow under these atmospheric conditions.

The kinetics of growth during a primary culture are indistinguishable from those of later subcultures suggesting that a period of adaptation and/or selection does not occur. Statistical data for one isolate are presented in Table 1. Observed rates of parasite growth are comparable to those reported for infection in cattle.[15] Three months of cultivation with subcultures every 2 to 3 days produced *B. bovis* parasites which were indistinguishable from those obtained from infected cattle (Figure 4). Following 5 months in vitro, 8.29×10^6 parasitized erythrocytes were injected intravenously into a splenectomized Holstein-Friesian calf. An elevated temperature was first observed 5 days postinoculation, and the calf died on day 10 postinoculation from severe babesiosis. The maximum parasitemia observed in thin blood films of peripheral blood was 1.3%.

The unusual requirement for a relatively high serum concentration (i.e., 40%) is well documented (Table 2). However, even serum concentration which fail to support continuous growth, e.g., 80%, permitted increases of 15-fold during the first 48-hr subculture. This observation should lend caution to extrapolating the effects of altering culture conditions during brief time spans in determining the nutritional and other requirements of *Babesia*.

Whereas extracellular *Babesia* (merozoites) are commonly observed in SF cultures and short-term *B. microti* cultures,[16] they are rarely observed in exponential phase MASP cultures. This indicates that almost all the parasites are able to rapidly infect new erythrocytes. However, transferring MASP cultures to an incubator containing ambient air (CO_2 deprived) results in an accumulation of merozoites in the medium. Isolation of infectious *B. bovis* merozoites is accomplished by depriving MASP cultures of CO_2 for 4 to 6 hr. The cultures are then gently centrifuged (400 × g for 5 min) to sediment erythrocytes. The top two-thirds of the supernatant is sedimented twice at 750 × g for 10 min to bring down the remaining erythrocytes. The resulting supernatant is centrifuged at 1000 × g for 20 min in order to pellet the merozoites. All processing is at 4°C as spontaneous agglutination of this parasite form occurs at higher temperatures. Additionally, cultures should be prepared utilizing erythrocytes from cattle which have not been extensively bled, as reticulocytes will

Table 2
GROWTH OF *BABESIA BOVIS* IN MICROAEROPHILOUS STATIONARY PHASE (MASP) CULTURES WITH VARYING SERUM CONCENTRATION

% Serum	Initial % parasitemia (IP)	Final % parasitemia (FP)	Cumulative[a] increase (CI)	Morphology	% of 40%
		Experiment 1			
20	0.5	0.3	41	Degenerate	1
30	0.5	9.9	1451	Good	37
40	0.5	22.8	3958	Good	100
40 (frozen)[b]	0.5	26.4	4092	Good	103
		Experiment 2			
40	0.8	16.9	3177	Good	100
50	0.8	14.4	2689	Good	85
60	0.8	1.5	412	Degenerate	13
80	0.8	0.2	45	Degenerate	1
100	0.8	0.2	—	Degenerate	—

Note: Serum or erythrocytes (when required for subcultures) were collected on the day used and the results of three sequential subcultures are reported.

[a] CI = (FP/IP) CD when CD (cumulative dilution) is the factor by which the erythrocytes present at the start of the experiment are diluted with fresh erythrocytes during subcultures.

[b] Serum stored at −20°C for 2-4 weeks was used in preparation of medium and normal culture. As with other groups, freshly collected erythrocytes were used as host cells.

contaminate the merozoite pellet produced from such cultures. The kinetics of merozoite invasion are presented in Table 3. It is important to note that even after more than 16 hr at 4°C many merozoites remain infectious. Initiation of new cultures using these isolated erythrocyte-free merozoites is possible (Figure 5).

D. Cultivation of Rodent Babesias

Short-term cultivation of *B. rodhaini* was accomplished by Jack and Ward.[17,18] Their method consisted of collecting infected rat erythrocytes and diluting these cells with uninfected erythrocytes to yield a final parasitemia of 0.3 to 1.0%. The medium consisted of equal volumes of RPMI 1640 and serum (human or rat) with the final packed cell volume not reported. Two hundred microliters of this material was added to each well of a microtiter plate and incubated under an atmosphere of 5% CO_2 in humidified air. Medium was changed at 24-hr intervals and cultures were maintained for 3 days. A threefold increase in parasitemia occurred during the first 48 hr of in vitro cultivation.

Babesia microti has been cultivated for up to 4 days using modification of the Trager-Jensen method for cultivation of *P. falciparum*.[16,19] Their methodology differed from Trager-Jensen[10] in that fetal bovine serum was substituted for human serum and hamster erythrocytes were the host cell. Increases of up to threefold occurred during the first 48 to 72 hr in vitro. These investigators[16] were able to adapt *B. microti* to culture by alternate passage in culture (72 hr) and hamsters. Following ten such passages, a strain of *B. microti* was obtained which grew better in short-term cultures than the original strain, i.e., almost twice the maximum parasitemia as compared to the original isolate. The presence of immune hamster macro-

Table 3
KINETICS OF INVASION OF ISOLATED MEROZOITES OF *BABESIA BOVIS* IN MICROAEROPHILOUS STATIONARY PHASE CULTURES

Time (hr)	% Cells infected (± S.D.)
1.0	0.8 ± 0.2
2.0	1.0 ± 0.3
3.0	1.3 ± 0.4
4.3	1.5 ± 0.1
16.3	2.2 ± 0.6
16.3 (4 C)[a]	None seen

Note: Infection was accomplished by adding 0.020 mℓ of a merozoite suspension (parasites not counted) to wells containing 0.2 mℓ normal culture which was preincubated at 37°C under an atmosphere of 5% CO_2/air for 1 hr.

[a] Following additional incubation for 26.6 hr at 37°C, 0.4 ± 0.3% of erythrocytes become infected.

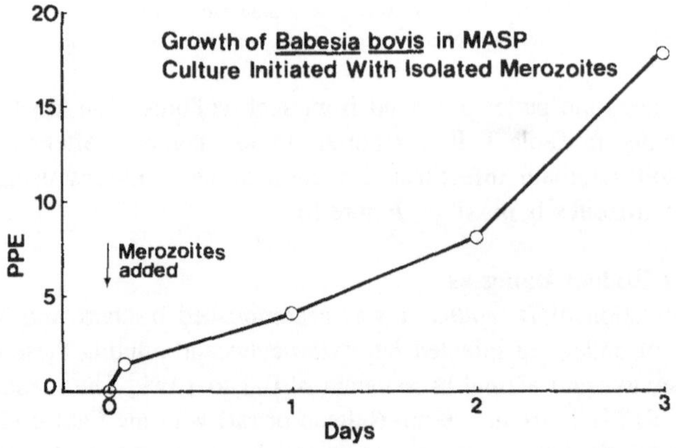

FIGURE 5. Growth of *Babesia bovis* in microaerophilous stationary phase cultures utilizing isolated erythrocyte-free merozoite to initiate cultures. PPE = percent parasitized erythrocytes.

phages or the supernatants from such macrophage cultures significantly suppressed growth of *B. microti* in vitro, demonstrating nonantibody mediated inhibition of *B. microti* growth.[16]

IV. APPLICATION OF CELL CULTURES TO IN VITRO STUDIES OF *BABESIA*

Numerous research achievements which follow the development of the MASP cultural method can be grouped into three categories: (1) Application of the cultural system to various in vitro studies of babesiosis. These include (a) use of the culture as an efficient in vitro

Table 4
COMPARISON OF INDIRECT FLUORESCENT ANTIBODY (IFA) TITERS AND MEROZOITE NEUTRALIZING (MN) ANTIBODY CAPACITY AGAINST *BABESIA BOVIS*

IFA titer 1171[a] postchallenge	MN[b] (%I)	IFA titer 1174[c] postchallenge	MN (%)
131,072	100	—	—
13,107	95	16,384	95
6,554	90	—	—
3,277	60	—	—
1,638	55	1,638	80
819	NS[d]	819	50
410	NS	410	NS
205	NS	205	NS
—	—	103	NS
—	—	51	NS
—	—	25	NS

[a] Immunized with soluble MASP antigen-challenged with 1×10^8 virulent *B. bovis*.
[b] MN titers are measured by comparing the percentage of infected erythrocytes following addition of merozoites to cultures containing normal (nonimmune) sera and similar percentage in the presence of immune sera (percent inhibition = %I).
[c] Preimmunized with attenuated *B. bovis* — challenged with 1×10^8 virulent *B. bovis*.
[d] NS = inhibition less than 50%.

system for detection and quantitation of protective anti-*B. bovis* antibodies, (b) evaluation of innate resistance factors governing babesial infection, (c) assay of drug sensitivity of *B. bovis*, and (d) interaction of *Babesia* spp. with complement; (2) adaptation of culture method to propagation of *B. bovis* in nonbovine erythrocyte-serum systems, and (3) the use of culture-derived antigens in development of diagnostic procedures and vaccines for bovine babesiosis. These recent findings, along with a review of previously acquired data are being presented below.

A. Merozoite Neutralization (MN) Test

Until the present time, the only method available to detect and measure the functional protective effect of antibabesia antibodies was by the use of passive immunization followed by challenge of susceptible cattle. Such a procedure was laborious, expensive, and not a precise means for quantitation and differentiation of protective immunoglobulins.

Availability of cell-free merozoites provided a tool to examine the protective role of antibabesia antibodies. A merozoite serum neutralization (MN) test was developed for that purpose. In the MN test MASP cultures containing uninfected bovine erythrocytes and varying dilutions of test sera (immune and normal) are allowed to preincubate in a CO_2 incubator at 37°C for 1 hr. Following this period, 20 $\mu\ell$ of a cell-free merozoite suspension is added to 200 $\mu\ell$ of the MASP cultures contained in microtiter plates. These are allowed to incubate for 1 to 16 hr after which thin blood films are prepared. The percent parasitemia is then determined by examination of Giemsa-stained blood films and the percent infection compared with that of control cultures containing nonimmune serum. The MN titers cor-

Table 5
GROWTH OF *BABESIA BOVIS* IN MASP CULTURES CONTAINING CALF OR ADULT ERYTHROCYTES AND/OR SERUM[a]

Erythrocytes	Serum	Cumulative increase[b]
Adult	Adult	5893-fold
Calf	Adult	1408-fold
Adult	Calf	423-fold
Calf	Calf	18-fold

[a] Calves less than 10 days old at the start of experiment. Blood collected on day of subculture.
[b] During three successive 2 or 3-day subcultures — the mean of 7 calves is reported.

related well with those of the indirect fluorescent antibody (IFA) test (Table 4). Furthermore, the specificity of the MN test was ascertained by preincubating immune sera with soluble antigen, which resulted in a loss of the ability of the antibody to prevent in vitro infection of erythrocytes by *B. bovis* merozoites. It was determined that the amount of antigen contained in 150 $\mu\ell$ of culture supernatant (average parasitemia was 16.4% following 3 days in vitro cultivation) will completely neutralize 80 $\mu\ell$ of immune serum (IFA titer 1:327,000). Dilution of antigen fivefold resulted in an 80% reduction antibody neutralization, i.e., an 80% increase in merozoite neutralization.[20]

B. Innate Resistance

Innate factors are possessed by the host as natural protective elements unrelated to a particular pathogen. This narrow host-specific resistance to babesia infection, however, must be reevaluated in view of recognized occurrence of babesiosis in human beings caused by various babesia strains.[21] Different breeds of cattle are known to differ in their susceptibility to infection and manifestation of clinical signs of babesiosis.[22] For example, *Bos indicus* cattle are more resistant to *B. bovis* than *Bos taurus*. Animals concurrently infected with other hemotropic parasites may show resistance to babesiosis. This resistance may be related to nonspecific stimulation of reticuloendothelial systems and possibly an alteration on the erythrocytes by a concurrent agent which makes these cells less suitable for babesia.

In general, field observations and experimental investigations have indicated that calves are clinically resistant to a number of hemotropic diseases, including anaplasmosis, theileriosis, and babesiosis. Although calves become infected, the course of the disease is generally mild and the animals are left as premunized carriers. In highly endemic areas, cattle herds are able to exist largely because the natural premunization of calves occurs when the disease is mild and provides protection from the severe disease they would suffer as susceptible adults. The reason young animals are resistant to severe disease is not known. However, speculation has been centered on colostral antibody providing passive protection, increased rate of erythrocyte removal by the spleens of young animals, and biochemical characteristics of calf erythrocytes (fetal hemoglobin).

Investigations into the causes of age resistance in living animals are difficult, if not impossible, as one cannot control the myriad of biochemical and physiological mechanisms which may be responsible for the phenomenon. Utilizing the MASP culture system for *B. bovis* in which all parameters can be quantitated and experimentally manipulated, we have examined the role of host blood elements in age resistance.[23] When varying mixtures of adult and calf serum were utilized to support parasite growth in adult erythrocytes, adult

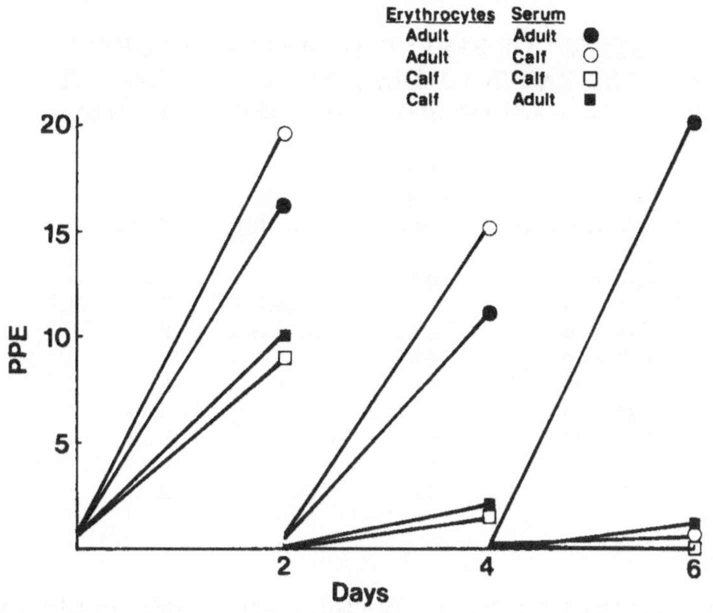

FIGURE 6. Growth of *Babesia bovis* in microaerophilous stationary phase cultures using various combinations of calf (8 days old) and adult bovine erythrocytes and serum. PPE = percent parasitized erythrocytes.

serum alone or equal volumes of the two sera (20% each) would support continuous growth. Calf sera alone (40%), 30% calf serum and 10% adult serum, or 36% calf serum and 4% adult serum would not support continuous growth.

It is evident that both erythrocyte and serum factors are involved in the mechanism of age resistance. Serum factors appear to be more important; greater multiplication is observed when parasites are grown in calf or adult erythrocytes with adult serum present (Table 5, Figure 6). Adult serum promoted almost 100-fold more growth in calf erythrocytes compared to the use of calf serum.

This investigation indicated that colostral antibody and phagocytic mechanisms need not be involved in age resistance in vivo. Because splenectomized calves support luxuriant babesia growth, the role of spleen-associated soluble factors is presently being investigated and will be reported at a later time. In addition, commercially available fetal calf, newborn calf, and calf sera would not support even limited parasite multiplication.

In collaboration with Dr. Anthony Cerami of the Rockefeller Foundation, we have utilized this methodology to determine whether porphyritic cattle which are resistant to trypanosomiasis may also exhibit increased resistance to babesiosis. In these cattle, a genetic metabolic defect leads to accumulation of porphyrin in the blood. In the presence of light, metabolically produced H_2O_2 releases oxygen which interacts with unsaturated fatty acids. This results in destruction of biological membranes in cells (parasite) lacking catalase activity. Preliminary investigation revealed that extracellular parasites (merozoites) were destroyed, whereas intracellular forms were unchanged.

The usefulness of MASP cultures in quantitating innate resistance to infection has been demonstrated. This methodology should be applied to the examination of other factors conferring resistance, e.g., breed resistance of *Bos indicus* cattle. Results of such studies may facilitate the selection of cattle breeds destined for endemic areas.

C. Drug Sensitivity

Screening of potential antibabesial drugs has been time consuming and costly due to the

Table 6
EFFECT OF BERENIL D1-(4-AMIDINOPHENYL)-TRIAZAN-(N-1,3) DIACETURATE ON GROWTH OF *BABESIA BOVIS* ON MASP CULTURES[a]

(Berenil) g/mℓ	Initial Parasitemia (48 hr)	Final parasitemia	Growth
Control	0.6	6.7	Black
1.37×10^{-4}	0.6	None seen	Red
1.37×10^{-5}	0.6	None seen	Red
1.37×10^{-6}	0.6	0.6	Red
1.37×10^{-7}	0.6	1.0	Red
1.37×10^{-8}	0.6	4.1	Black
1.37×10^{-9}	0.6	3.7	Black
1.37×10^{-10}	0.6	5.1	Black
1.37×10^{-11}	0.6	5.5	Black

[a] 3.5×10^{-6} g/mℓ is therapeutic dose in vivo.

necessity of screening drugs in animals. To this end, various models based on rodent babesias, e.g., *B. rodhaini*, have been developed. Drug screening results using rodent models have generally been applicable to bovine species; however, cost has greatly limited their further evaluation. An in vitro method for the screening of potential antibabesial drugs has been developed.[14] Cultures are initiated with 0.2 to 0.5% parasitemia in microtiter plates. The volume of each culture is 0.2 mℓ and 8 replicates are prepared at each dilution of the test substance. The color change of infected cultures from bright red (no infection) to dark red or black (heavy infection) offers a simple means to evaluate growth. This color change is evaluated following 24 to 48 hr in vitro maintenance. For example, addition of varying concentrations of the antibabesial drug D1-(4-amidinophenyl)-triazan-N-1,3) diaceturate (Berenil) to cultures allows one to titrate the drug's action because a lethal concentration (3.5×10^{-6}g) prevents culture darkening. A gradation of color change is observed between lethal and nonlethal dosages, and this color change was closely correlated with parasitemias. An example is given in Table 6.

Use of the above in vitro system for drug screening and potency evaluation would facilitate identification of new antibabesial drugs and thus substantially reduce development costs. It is anticipated that some variation in drug sensitivity may exist among babesias of various animal species. An adaptation of the respective babesia to growth in the MASP culture would easily eliminate this problem.

D. Interaction of *Babesia* with Complement

Babesia rodhaini, a parasite of rodents, is reported to require elements of the host's immune mechanism in order to effect attachment and penetration of host erythrocytes. Specifically, host cell penetration depends upon factors of the alternative complement pathway as well as ionic magnesium and the third component of complement.[17,18,24] Utilizing short-term cultures, these investigators demonstrated inhibition of erythrocyte invasion in complement-free serum. The MASP culture system was utilized in order to determine if an analogous situation was operational for *B. bovis*.

Although parasite multiplication was reduced when heat-inactivated serum (56°/30 min) replaced fresh serum in the culture medium, increases of 10 to 15-fold were routinely observed during 3-day incubation periods (Figure 7). This suggested loss of a heat-labile nutrient rather than a complement requirement. Additionally, zymosan treated serum (C_3 deficient) and complement deficient bovine erythrocytes (eluted at low pH) allowed normal

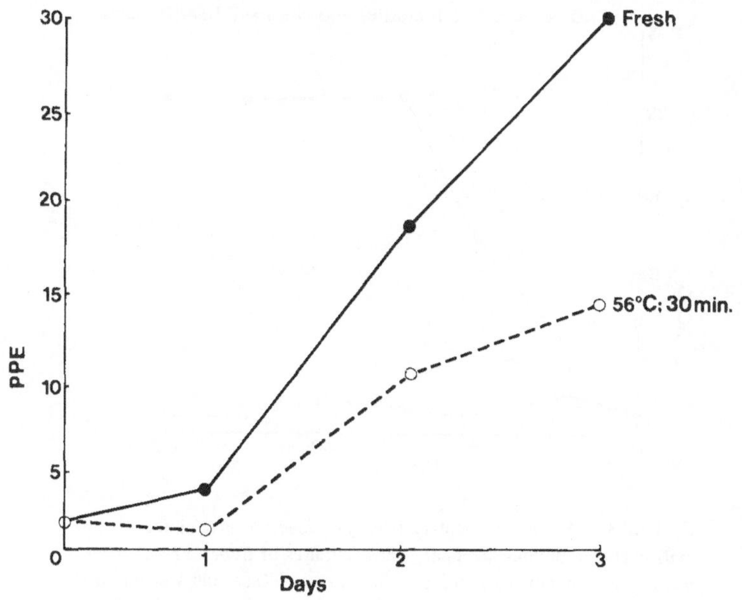

FIGURE 7. Growth of *Babesia bovis* in microaerophilous stationary phase cultures using fresh and heat-inactivated serum from the same donor cow. The 24-hr lag in growth has been demonstrated to be the result of centrifuging the seed culture material and did not not occur in washed but concentrifuged cultures (parasitized erythrocytes allowed to sediment at 1 × g during washing). PPE = percent parasitized erythrocytes.

growth of *B. bovis* in cultures. It appears, therefore, that unlike the rodent babesia *B. rodhaini*, *B. bovis* does not utilize complement during erythrocyte penetration.

In a variety of hemoparasitic infections, including malaria, trypanosomiasis, and babesiosis, hypocomplementempia is associated with the acute phase of parasitemia. Intact *B. rodhaini*, but not the supernatant from short-term cultures, activates complement.[17] Because we are able to obtain parasite metabolites unaltered by host biochemical and physiological mechanism in MASP cultures, we investigated the possibility that *B. bovis* produces anticomplementary materials. Analysis of culture supernatants revealed the presence of anticomplementary activity which was not present in the supernatants of noninfected cultures (Figure 8). This activity could not be accounted for by nonparasite products, e.g., erythrocyte stroma, hemoglobin, or normal serum. The level of residual complement in infected cultures decreased with time at rates exceeding those observed for noninfected cultures (Figure 9). This difference could be accounted for by the activity of supernatant anticomplement.[25]

E. Propagation of *B. bovis* in Nonbovine Cell Culture Systems

A traditional concept of host specificity for various babesia species was drastically changed in view of the reports of human babesiosis caused by bovine, equine, and rodent species of babesia. Moreover, splenectomized chimpanzees and rhesus monkeys have been experimentally infected with *B. divergens* and *B. bovis*. The findings suggest that under certain circumstances *Babesia* spp. are capable of transgressing natural defenses of a heterologous host, invading its erythrocytes, and pursuing successful growth and development in these cells. In the absence of interfering defense elements, the MASP culture system offered an opportunity to study the suitability of erythrocytes from nonbovine species to support growth of *B. bovis*.[26]

In the first series of experiments utilizing MASP culture system as described by Levy

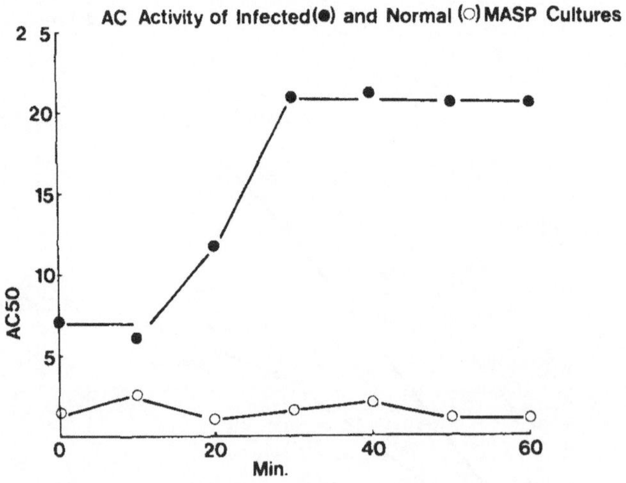

FIGURE 8. Anticomplementary activity of infected and noninfected (normal) microaerophilous stationary phase cultures of *Babesia bovis*. Complement determination according to the method of Kabat and Mayer (1961) in which culture supernatants were first heat-inactivated and then preincubated with fresh serum (complement source) before running assay. AC50 = anticomplement activity Min = time after replacing the supernatant of infected (4.2%) or noninfected cultures with freshly prepared medium containing fresh serum (1 hr post-collection from cow).

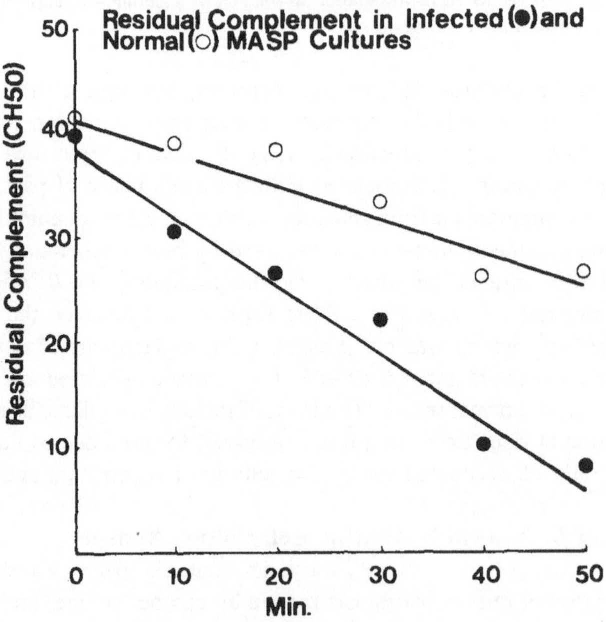

FIGURE 9. Residual complement in infected and noninfected microaerophilous stationary phase cultures. Assay of complement was according to the method of Kabat and Mayer[37] with the culture supernatant serving as the source of complement. Min = time in minutes after replacing the supernatant of infected (4.2%) or noninfected cultures with freshly prepared medium containing fresh serum (1 hr post-collection from cow).

Table 7
PROPAGATION OF *BABESIA BOVIS* IN BOVINE AND NONBOVINE ERYTHROCYTE-SERUM SYSTEMS USING MASP CULTURAL TECHNIQUE

Cells	Serum	Infection	Multiple continuously
Ovine	Bovine	Yes	No[a]
Ovine	Ovine	Yes	No
Goat	Bovine	Yes	No[a]
Goat	Goat	Yes	No
Horse	Bovine	Yes	No[a]
Horse	Horse	Yes	Yes
Rabbit	Bovine	Yes	Yes[a]
Rabbit	Rabbit	Yes	Yes

[a] Hemolysis.

and Ristic,[14] the serum of the sheep, goat, horse, and rabbit was used to gradually replace bovine serum until a 40% level of the heterologous species serum was reached in individual cultures. In a similar fashion, erythrocytes of the above animal species were used to completely replace bovine erythrocytes. The lapse between subcultures varied from 24 to 96 hr, depending on the multiplication rate of the parasites and the degree of hemolysis.

Results of these studies showed that erythrocytes of four animal species (sheep, goat, horse, rabbit) can be infected by *B. bovis*. Continuous propagation of the organism in a series of subcultures over a prolonged period of time, however, was accomplished with erythrocyte-serum systems of rabbit and horse only (Table 7). Major problems encountered in the process of development of the above study were agglutination and hemolysis during the period of co-existence in the culture of heterologous species sera and erythrocytes. This problem was resolved once both of these reagents in cultures were of a single animal species.

In both horse and rabbit cultures, multiple *B. bovis* organisms were noted in a single erythrocyte; moreover, the appearance of the organism in horse erythrocytes more closely resembled *Babesia equi* than *B. bovis*.

It is apparent from these studies that *B. bovis* is capable of transgressing host cell specificity levels by invading, developing, and multiplying in the erythrocytes of widely-separated, nonruminant animal species. The results further indicate that the potential host resistance to various babesias is not necessarily related to an erythrocyte receptor level or that such receptors and the essential metabolites needed for replication of *B. bovis* are not limited to bovine species only.

The traditional taxonomy of *Babesia* spp. was principally based upon the animal species in which these organisms are found and to a degree upon certain morphological features of these organisms. The above findings and additional studies yet to come may eventually reveal fundamental aspects of the nature of babesia which should serve as a more reliable basis for taxonomic affiliation of this organism.

There are other more practical advantages, which the above studies may offer. An adaptation and prolonged maintenance of *B. bovis* in cultures using heterologous species erythrocytes and serum may lead to selection of a mutant strain with potential immunogenic capabilities.

F. Slide Latex Agglutination Test (LAT)

Accurate identification of negative (highly susceptible) and positive (acute or carrier) animals is an essential initial step in controlling babesiosis. Whereas several diagnostic tests have been developed[27] none have proven useful under field conditions. Goodger and Mahoney[28] and Lopez and Todorovic[29] reported development of serodiagnostic tests for bovine babesiosis utilizing antigen sensitized latex particles. Antigens were obtained from infected bovine blood, were highly labile and contained host-tissue contaminants. The resulting unpredictability and inconsistency of the tests prevented widespread application under field conditions.

Availability of antigens generated in vitro and unaltered by host metabolic and immunogenic processes has resulted in development of a simple latex agglutination test (LAT) for *B. bovis*. Polystyrene latex particles 0.797μ suspended in 0.15 *M* glycine buffer pH 8.3 were treated at 56°C for 15 min with an optimal concentration of filtered soluble antigen derived from the supernatant of *B. bovis* MASP culture. These sensitized latex particles remain functional in the test for at least 2 months when stored at 4°C. The antigen and sera must be adjusted to room temperature prior to use in the test.[30] The LAT is performed on glass slides marked into squares. One drop of the sensitized latex particles is applied to each square and one drop of undiluted inactivated serum was added. This material is mixed by stirring with an applicator stick. The slide is then rotated for 1 to 2 min by hand and the suspension examined for agglutination over a dark background. Reactions were classified as strong agglutination where an abundance of clumps were observed, weak agglutination when there was a fine granular appearance, and negative when no visible reaction occurred. Both degrees of agglutination are considered as positive reactions. In each test, known positive and negative reference sera were used as controls. Inactivated normal bovine serum was used as diluent in the titration of the serum samples.

Thus far, more than 100 sera from cattle naturally infected with *B. bovis* and positive in the IFA test were examined by the LAT. There was 100% agreement between the two tests.

G. Vaccines from Cell Cultures

At least two types of inactivated vaccines can be prepared from the in vitro cultures. The first is a merozoite vaccine and the second is a parasite-free soluble merozoite surface coat vaccine. Both of these vaccines may be administered singly or in combination and can be preserved, stored, and used as stable lyophilized products.

Initially the antigens used for production of these vaccines were derived from spinner flask cell cultures which produced an average of 3% parasitemia.[31] They were admixed with incomplete Freund's adjuvant and administered subcutaneously in two doses at 2-week intervals. Four weeks after the initial inoculation, all principal and control animals were each challenged with 1000 *B. microplus* larvae from *B. bovis*-infected colonies.[32] Vaccinated animals were protected from death following challenge.[31]

In more recent studies, soluble *B. bovis* antigens were derived from the improved MASP culture which produced an average of 29% parasitemia. The antigens were concentrated approximately seven times by lyophilization, admixed with a new Saponin (Quil-A) adjuvant, and administered subcutaneously in two doses on days 1 and 14. Several small experiments were conducted, each consisting of three to four cattle. All inoculated animals showed a marked rise of IFA antibodies reaching a maximum titer of 1:20,000 at approximately 3 to 4 weeks after inoculation. The antibody titers persisted for 3 to 4 months and then subsided below detectable levels. Animals challenged at this time mounted a strong anamnestic antibody response which reached IFA titers of 1:327,680 at 24 days after challenge. In one case studied, no evidence of even subclinical infection was detected, indicating that the challenge infection was fully rejected (Figure 10). Conversely, all control animals developed clinical signs of the disease and died, or were treated and recovered. Strong anamnestic immune responses were also demonstrated in 60 mature mixed-breed animals maintained under field conditions in Mexico. These animals received a third (booster) vaccine dose at 7 1/2 months following primary vaccination. There was a prompt IFA antibody response in all animals reaching a level in excess of 1:40,000 within 3 weeks following the booster injection.

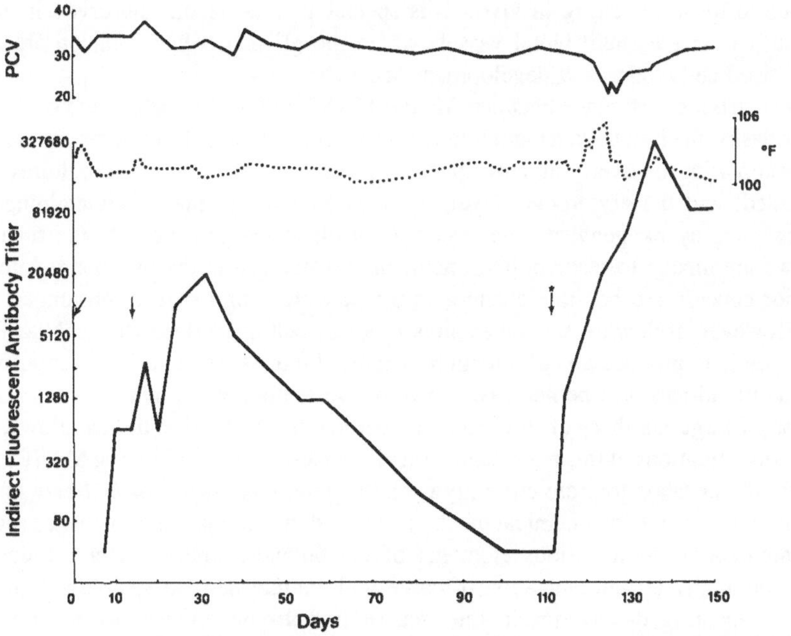

FIGURE 10. Sixteen-month-old Holstein cow vaccinated with soluble cell-culture derived antigen (culture supernatant) admixed with an adjutant (Quil-A) on days 0 and 14 (arrows). On day 112 this animal was challenged with 1 × 10^8 virulent *Babesia bovis* via the intramuscular route (starred arrow). A strong anamnestic response was measured (IFA titer 1:40 at time of challenge increasing to 1:1280 at 3 days postchallenge). In contrast control animal failed to demonstrate an antibody response (1:40 or above) prior to 10 days postchallenge and died on day 14 postchallenge. Fifty mℓ of blood from vaccinated animal was subinoculated into a susceptible splenectomized calf on day 28 postchallenge which failed to become infected indicating a sterile immunity. PCV = packed cell volume.

It is apparent from these and other observations that the soluble *B. bovis* vaccine which appears to be a glycoprotein is highly immunogenic and that primary stimulation by it results in prolonged maintenance of sensitized lymphocytes with a rapid cloning ability following restimulation by vaccination or challenge exposure. Translated into practical terms, this behavior of the immunogen should result in a prolonged period of protective immunity following vaccination.

V. CONCLUSIONS

Successful cultivation of *B. bovis* has been achieved utilizing contrasting methodologies. In an attempt to simulate the bovine circulatory system, Erp and co-workers developed a suspension culture which supported continuous growth, although at a rate much below that observed in the bovine host. Accumulation of large numbers of extracellular merozoites in the suspending medium rather than invasion of additional erythrocytes by these forms in a probable bottleneck in SF cultures. It was noted that *B. bovis* merozoites will undergo spontaneous agglutination at physiologic temperatures, and it is probable that such clumped merozoites are rendered noninfectious. Therefore, the process used to simulate in vivo conditions, i.e., stirring, may be responsible for the low multiplication rates observed, as this would increase the probability of merozoites contacting one another prior to attaching to and invading host cells. The stationary MASP culture more probably recreates the in vivo conditions required by *B. bovis*, an assumption that is supported by multiplication rates comparable to those occurring in vivo. This species is most readily observed in capillary blood, and it is in the small blood vessels where the pH and oxygen tension are reduced that parasitized cells lodge and development occurs.[8,33]

The most striking difference between SF and MASP cultures is erythrocyte oxygenation. In spinner flasks, the host cells remain bright red (oxygenated) and the cultures are maintained in equilibrium with the overlying atmosphere. In contrast, infected MASP cultures become deoxygenated, and the erythrocyte layer becomes dark-red to black. Maintaining MASP cultures in a highly oxygenated state results in death of the parasites. At the time of this writing, we are unable to resolve this fundamental difference in the two methodologies.

Of major concern are possible changes in the parasite resulting from prolonged in vitro growth. However, following several months in vitro, culture derived (SF or MASP) organisms were able to produce a fatal infection in cattle. Future studies utilizing larger numbers of intact adult animals will be necessary to fully substantiate this finding.

Isolation of large numbers of infectious merozoites from MASP cultures allows, for the first time, investigations of this parasite in an erythrocyte-free state. Utilizing MASP cultures, researchers in our laboratory are currently examining various aspects of *B. bovis*, including nutritional requirements, biochemical mechanisms, and the interaction of parasites (including isolated merozoites) with various elements of the bovine immune system. Induction of protective immunity by immunization of cattle with culture-derived soluble antigen is one of the most exciting discoveries in the area of hemotropic diseases to occur in recent years.[31,34-36] Moreover, the finding that *B. bovis* in MASP cultures will infect erythrocytes of species other than the natural host (e.g., ovine, caprine, equine, and rabbit), and preliminary indications that the organism may be continuously cultivated in erythrocytes and/or serum of the latter two species[26] raises new questions concerning host specificity and the relationship between various *Babesia* species.

It is obvious that development of both short-term and continuous methodologies for *Babesia* spp. has greatly facilitated our understanding of these important protozoan pathogens. Application of the described culture techniques to other species of *Babesia* and other intra-erythrocytic parasites remains largely untested. Development of continuous cultures for other important *Babesia* spp. is a major goal of researchers in the concern with hemotropic infections of animals.

REFERENCES

1. **Levine, N. D.,** Taxonomy of piroplasms, *Trans. Am. Microbiol. Soc.*, 90, 2, 1971.
2. **Babes, V.,** Sur L'hemoglobinuria bacterine de boeufs, Compte *Ren. Hebd. Sen. L'Acad. Sci. (Paris)*, 107, 692, 1888.
3. **Smith, T. and Kilbourne, F. L.,** Investigation into the nature, causation, and prevention of Texas or southern cattle fever, *U.S.D.A. Bur. An. Ind. Bull.*, 1, 1, 1893.
4. **Annable, C. R. and Ward, P. A.,** Immunopathology of the renal complications of babesiosis, *J. Immunol.*, 112, 1, 1974.
5. **Goodger, B. V.,** *Babesia argentina:* observations on the immunogenicity of the cryofibrinogen complex, *Zeit. Parasitenkd.*, 53, 47, 1977.
6. **Wright, I. G.,** Osmotic fragility of erythrocytes in actue *Babesia argentina* and *Babesia bigemina* infections in splenectomized *Bos taurus* calves, *Res. Vet. Sci.*, 15, 299, 1973.
7. **Callow, L. L. and Pepper, P. M.,** Measurement of and correlation between fever, changes in the packed cell volume and parasitemia in the evaluation of the susceptibility of cattle to infection with *Babesia argentina, Aust. Vet. J.*, 50, 1, 1974.
8. **Wright, I. G.,** An electron microscopic study of intravascular agglutination in the cerebral cortex due to *Babesia argentina* infection, *Int. J. Parasitol.*, 2, 209, 1972.
9. **Wright, I. G. and Kerr, J. D.,** The preferential invasion of young erythrocytes in acute *Babesia bigemina* infections of splenectomized calves, *Zeit. Parasitenkd.*, 43, 63, 1974.

10. **Trager, W. and Jensen, J. B.,** Human malaria parasites in continuous culture, *Science*, 193, 673, 1976.
11. **Erp, E. E., Gravely, S. M., Smith, R. D., Ristic, M., Osorno, B. M., and Carson, C. A.,** Growth of *Babesia bovis* in bovine erythrocyte cultures, *Am. J. Trop. Med. Hyg.*, 27, 1061, 1978.
12. **Molinar, E., Kakoma, I., Aikawa, M., and Ristic, M.,** Development and ultrastructure of *Babesia canis* in primary cell culture, manuscript in preparation.
13. **Erp, E. E., Smith, R. D., Ristic, M., and Osorno, M. B.,** Continuous *in vitro* cultivation of *Babesia bovis*, *Am. J. Vet. Res.*, 41, 1141, 1980.
14. **Levy, M. G. and Ristic, M.,** *Babesia bovis:* continuous cultivation in a microaerophilous stationary phase culture, *Science*, 207, 1218, 1980.
15. **Mahoney, D. F., Kerr, J. D., Goodger, B. V., and Wright, I. G.,** The immune response of cattle to *Babesia bovis* (syn. *B. argentina).* Studies on the nature and specificity of protection, *Int. J. Parasitol.*, 9, 297, 1979.
16. **Bautista, C. R. and Kreier, J. P.,** The action of macrophages and immune serum on growth of *Babesia microti* in short-term cultures, *Tropenmed. Parasitol.*, in press, 1980.
17. **Jack, R. M. and Ward, P.,** *Babesia rodhaini* interactions with complement: relationship to parasitic entry into red cells, *J. Immunol.*, 124, 1566, 1980.
18. **Jack, R. M. and Ward, P.,** The role *in vivo* of C3 and C3b receptor in babesial infection in the rat, *J. Immunol.*, 124, 1574, 1980a.
19. **Bautista, C. R. and Kreier, J. P.,** Effect of immune serum on the growth of *Babesia microti* in hamster erythrocytes in short-term culture, *Infect. Immunol.*, 25, 470, 1979.
20. **Levy, M. G. and Ristic, M.,** Merozoite neutralization test in bovine babesiosis, manuscript in preparation.
21. **Ristic, M. and Healy, G. R.,** Babesiosis, in *The Parasitic Zoonosis — I. Protozoan Zoonosis*, Steele, J. H. and Arambulo, P., Eds., CRC Press, Boca Raton, Fla., 1980.
22. **Aragon, R. S.,** Bovine babesiasis: a review, *Vet. Bull.*, 46, 903, 1976.
23. **Levy, M. G., Clabaugh, G., and Ristic, M.,** Age resistance in bovine babesiosis: the role of blood factors in resistance to *Babesia bovis*, *Infect. Immunol.*, 37, 1127, 1982.
24. **Chapman, W. E. and Ward, P. A.,** *Babesia rodhaini:* requirement of complement for penetration of human erythrocytes, *Science*, 196, 67, 1977.
25. **Levy, M. G., Kakoma, I., and Ristic, M.,** Interaction of *Babesia bovis* with complement *in vitro*, manuscript in preparation.
26. **Esparaza, H., Levy, M. G., and Ristic, M.,** Propagation of *Babesia bovis* in non-bovine erythrocytes, manuscript in preparation.
27. **Todorovic, R. A.,** Serological diagnosis of babesiosis: a review, *Trop. An. Heal. Prod.*, 7, 1, 1975.
28. **Goodger, B. V. and Mahoney, D. F.,** A rapid slide agglutination test for herd diagnosis of *Babesia argentina* infections, *Aust. Vet. J.*, 50, 250, 1974.
29. **Lopez, G. V. and Todorovic, R. A.,** Rapid latex agglutination (RLA) test for the diagnosis of *Babesia argentina*, *Vet. Parasitol.*, 4, 1, 1978.
30. **Montenegro, S., James, M. A., Levy, M. G., Preston, M. D., Esparsa, A., and Ristic, M.,** Utilization of culture-derived soluble antigen in the latex agglutination test for bovine babesiosis and anaplasmos, *Vet. Parisitol.*, 8, 291, 1981.
31. **Smith, R. D., Carpenter, J., Cabrera, A., Gravely, S. M., Erp, E. E., Osorno, M. B., and Ristic, M.,** Bovine babesiosis. Vaccination against tickborne challenge exposure with culture-derived *Babesia bovis* immunogens, *Am. J. Vet. Res.*, 40, 1678, 1979.
32. **Smith, R. D., Osorno, M. B., Brewer, J., de LaRosa, R., and Ristic, M.,** Bovine babesiosis: severity and reproducibility of *Babesia bovis* infections induced by *Boophilus microplus* under laboratory conditions, *Res. Vet. Sci.*, 24, 287, 1978.
33. **Potgieter, F. T. and Els, H. J.,** An electron microscopic study of intra-erythrocytic stages of *Babesia bovis* in the brain capillaries of infected splenectomized calves, *Onderstepoot J. Vet. Res.*, 46, 41, 1979.
34. **Ristic, M. and Levy, M. G.,** A new era of research toward solution of bovine babesiosis, in *Proceedings International Conference Malaria and Babesiosis*, Academic Press, New York, 1980.
35. **Smith, R. D. and Ristic, M.,** Immunization against bovine babesiosis with cell culture-derived antigens, *Proc. Int. Conf. Malaria and Babesiosis*, Ristic, M. and Kreier, J. P., Eds., Academic Press, New York, 1980.
36. **Ristic, M.,** Babesiosis, in *In Diseases of Cattle in the Tropics*, Ristic, M. and MacIntyre, I., Eds., Junk Press, The Hague, Holland, in press.
37. **Kabat, E. A. and Mayer, M. M.,** *Experimental Immunochemistry*, Charles C Thomas, Springfield, Ill., 1961, 133.

Chapter 7

THEILERIA

C. G. D. Brown

TABLE OF CONTENTS

I. INTRODUCTION

Pathogenic species of *Theileria* cause tick-borne protozoan diseases of cattle, sheep, and goats known as theilerioses. These organisms also infect and may cause disease in wild bovids, notably buffalo and eland, while many member species of the family Bovidae may be asymptomatic carriers of *Theileria spp.*

Theileria are transmitted trans-stadially but not transovarially by 2- and 3-host ixodid ticks, from larva to nymph or nymph to adult. In the vertebrate host *Theileria* undergo an exoerythrocytic schizogonous cycle in lymphoid cells. These schizonts release merozoites which invade erythrocytes, becoming piroplasms. The diseases produced by these parasites are highly pathogenic and often fatal lymphoproliferative disorders which may cause leukopenia and anemia.

The genus *Theileria* is controversially regarded as being the solitary genus within the family Theileriidae, one of two families in the order Piroplasmida. The *Theileria* are thus closely related to another family of parasites of veterinary importance — the *Babesia*. Though the classification of these parasites, along with other sporozoan protozoa, has been modified considerably of recent years, the latest consensus on the phylum Apicomplexa[1] has been generally accepted by protozoologists and is given here in Table 1. Within the genus *Theileria*, however, the taxonomic situation is not quite so clear. This is exemplified by the comprehensive review of Uilenberg[2] which was followed shortly afterwards by a reappraisal of the classification and taxonomic position of three theilerial species by Morel and Uilenberg.[3] A comparatively simple summary of the main features of the pathogenic *Theileria spp.* is given in Table 2. The most important of these species are

- *Theileria parva* — an organism transmitted by *Rhipicephalus appendiculatus*, causing East Coast fever (ECF) of cattle in East and Central Africa. The organisms *T. parva lawrencei* and *T. p. bovis* are part of the *T. parva* complex of parasites, while *T. mutans* and *T. taurotragi* may contribute to the ECF syndrome, to which some 50 million cattle are at risk.
- *Theileria annulata* — transmitted by ticks of the genus *Hyalomma*, causing tropical theileriosis of cattle and possibly domestic buffalo *(Bubalus bubalis)* from Morocco to China. *T. sergenti* of cattle in Japan, Korea, and the Eastern seaboard of Asia may overlap with *T. annulata*. An estimated 200 million cattle are at risk to infection with *T. annulata*.
- *Theileria hirci* — transmitted by *Hyalomma anatolicum* and probably other ticks of the genus *Hyalomma*, causing malignant ovine theileriosis of sheep (and goats) in the Middle East and Indian subcontinent where some 150 million animals are at risk.

Nowhere has the seriousness of these protozoan parasites been more evident than in the catastrophic epizootic of ECF in Southern Africa at the turn of the century. This followed the introduction into Mozambique, Rhodesia (now Zimbabwe), and South Africa of infected cattle from German East Africa (now Tanzania) and has been described in graphic terms by Henning[4] and Gutsche[5] in her biography of Sir Arnold Theiler after whom the genus was named.[6] At the same time as ECF was devastating Rhodesia and South Africa, another, yet similar disease manifested itself in Transcaucasia, (now Southern USSR).[7] This eventually proved to be tropical theileriosis or Mediterranean Coast fever, caused by *T. annulata*.

When first recorded in German East Africa in 1897[8] the pathogenic form of *T. parva* was thought to be intraerythrocytic and the disease was thus initially confused with redwater or Texas fever. At an early stage in the intensive investigation into the disease, however, Lounsbury,[9] in an elegant piece of work, showed the parasite was transmitted trans-stadially but not transovarially by *Rhipicephalus appendiculatus;* Hutcheon[10] differentiated the pi-

Table 1
CLASSIFICATION OF THEILERIA

Subkingdom		Protozoa
Phylum		Apicomplexa
Class		Sporozoea
Subclass		Piroplasmia
Order		Piroplasmida
Family	Babesiidae	Theileriidae
Genus	*Babesia*	*Theileria*

roplasm of ECF from that of redwater and Koch[11] demonstrated the intra-lymphocytic form of the parasite — the macroschizont or "Koch's blue body". Gonder,[12] a few years later, related these two forms and thus illustrated at an early stage of the work on the *Theileria* almost all that is known of the cycle of the parasite in its vertebrate host.

Less was known of the life cycle of the parasites in the vector ticks until 1932 when Cowdry and Ham,[13] using classical histological techniques, defined the whole cycle of *T. parva* in *R. appendiculatus*. This piece of work was regrettably so far ahead of its time that the results could not be reproduced and the significance of the information was not appreciated until Cowdry and Ham's interpretations were vindicated by Schein and colleagues working in Berlin with *T. annulata* in *H. anatolicum excavatum*[14] and *T. parva* in *R. appendiculatus*.[15]

At the present time it is accepted that the piroplasm is ingested by an immature stage (larva or nymph) of the arthropod host as it engorges on the blood of an infected bovid. In the gut of the engorged tick, micro- and macro-gametes are released, zygotes are formed in cells of the gut wall, and at the time when the tick moults, kinetes migrate in the hemolymph from gut to salivary glands. There they commence sporogony, multiplying to form large numbers of sporozoites when the next stage of the tick commences to feed.

In the vertebrate host the sporozoite inoculated by the infective tick (usually after 1 to 5 days' attachment) when feeding enters a lymphoid cell. The intracytoplasmic initial body develops to become a macroschizont which multiplies synchronously and interdependently with its host cell. Later, merozoites are budded off the macroschizont nuclei and, while still intracellular this collection of transitional merozoites is known as a microschizont. These merozoites are then released from the host cell cytoplasm, invading erythrocytes where they are termed piroplasms. The piroplasms multiply in the erythrocytes to varying degrees and possibly differentiate into gametocytes, the forms infective for ticks. The cycle of development of *Theileria* can thus be summarized as:

- Schizogony — in mammalian lymphocytes
- Gametogony — in erythrocytes and the tick gut
- Sporogony — in the tick salivary gland

In susceptible animals theileriosis is manifested by massive hyperplasia of the lymphoid organ and metastatic proliferation of parasitized lymphoblasts. There may be leukopenia and thrombocytopenia, as in ECF. In those diseases where there is rapid proliferation of intraerythrocytic piroplasms as in tropical theileriosis, there is frequently a catastrophic anemia. In exotic cattle *(Bos taurus)* the mortality from ECF may be 95 to 100%. Mortality from tropical theileriosis in similar susceptible cattle is lower, but 40 to 60% is common. In indigenous Zebu cattle *(Bos indicus)* reared in endemic areas, mortality from either disease is significantly lower and confined mainly to calves.

Cattle recovering from these diseases develop a significant degree of immunity to challenge with the same strain of parasite. For *T. parva* infection this immunity may be sterile and long lasting (more than 3 years).[16] In *T. annulata* infection a carrier status, with persistence

Table 2
PATHOGENIC THEILERIA OF DOMESTIC RUMINANTS

Species	Host	Vector	Disease and distribution	Serotype	Schizonts	Piroplasms
Theileria parva parva	Domestic cattle (buffalo)	*Rhipicephalus appendiculatus*	East Coast fever (ECF). East & Central Africa.	*T. parva*	+ + +	+ + +
T. parva bovis	Cattle (buffalo)	*R. appendiculatus*	January disease. Central Africa.	*T. parva*	+	+
T. parva lawrencei	Buffalo, cattle	*R. appendiculatus*	Buffalo disease, Corridor disease. East, Central and Southern Africa.	*T. parva*	+	±
T. taurotragi	Eland, cattle	*R. appendiculatus*	Benign theileriosis (? turning sickness). East, Central and Southern Africa.	*T. taurotragi*	±	+
T. mutans	Cattle	*Ambylomma spp.*	Benign theileriosis, turning sickness. Africa.	*T. mutans*	± (morphology distinct)	+ + (+)
T. velifera	Cattle buffalo	*Ambylomma spp.*	Africa South of the Sahara	?	?	+ (distinct veil)
T. annulata	Cattle, swamp buffalo	*Hyalomma spp.*	Mediterranean coast fever, tropical theileriosis. Morocco to China.	*T. annulata*	+ + +	+ + +
T. sergenti	Cattle	*Haemaphysalis longicornis* + *Haemaphysalis spp.*	Theileriosis sergenti. Japan, Korea. (Worldwide.)	*T. sergenti*	?	+ +
T. hirci	Sheep & goats	*Hyalomma spp.*	Malignant ovine theileriosis. North Africa, Middle East to India.	*T. hirci*	+ + +	+ + +
T. ovis	Sheep & goats	? *R. bursa*	Benign ovine theileriosis Worldwide	*T. ovis*	?	+ +
Others: *T. separata* *T. spp.*	Sheep & deer	?	Germany, U.S.A. and elsewhere	?	?	+ (may be characteristic)

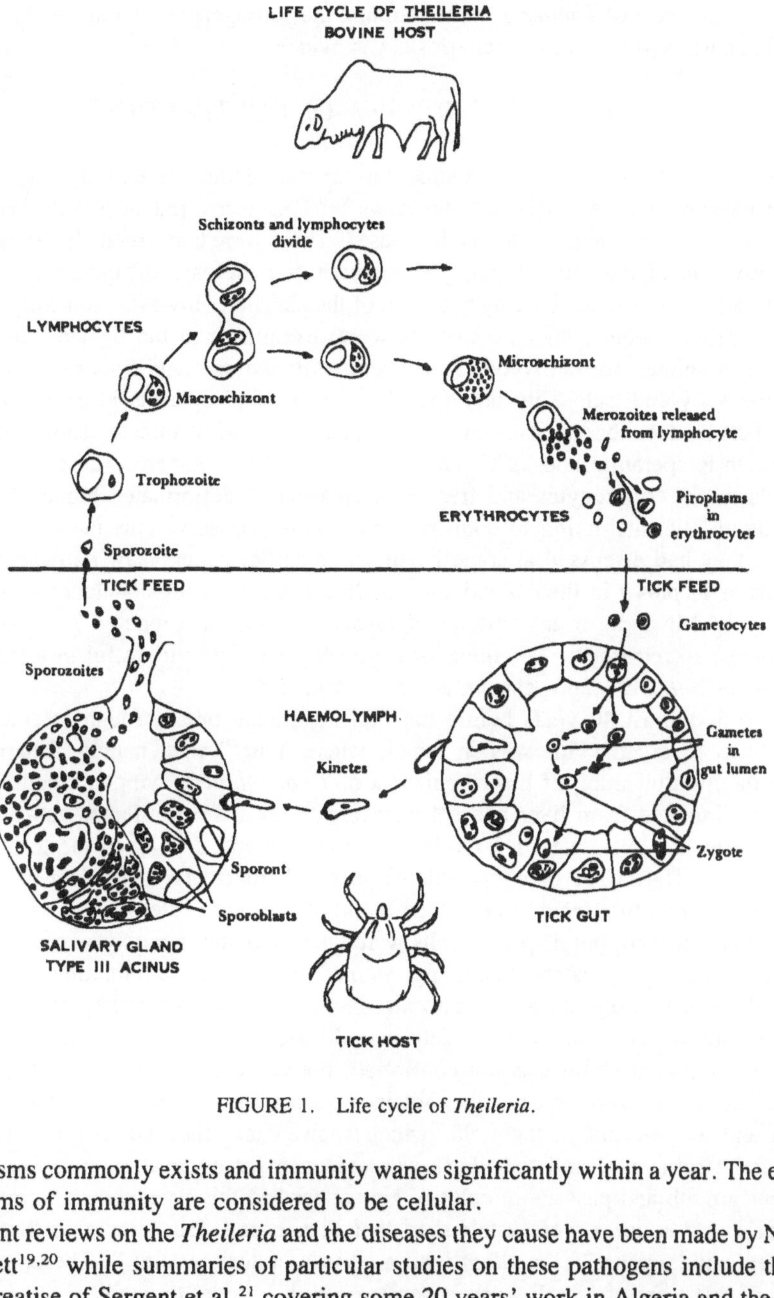

FIGURE 1. Life cycle of *Theileria*.

of piroplasms commonly exists and immunity wanes significantly within a year. The effective mechanisms of immunity are considered to be cellular.

Excellent reviews on the *Theileria* and the diseases they cause have been made by Neitz[17,18] and Barnett[19,20] while summaries of particular studies on these pathogens include the monumental treatise of Sergent et al.[21] covering some 20 years' work in Algeria and the reviews of Wilde[22] and Purnell.[23] This last, which summarizes the work conducted by the FAO Tickborne Diseases Project and their collaborators at Muguga in Kenya is complemented by FAO Animal Production & Health Paper No. 19,[24] a mammoth tome encompassing 142 papers published on ECF and related diseases.

Theileria and the theilerioses were considered in depth at a recent tickborne diseases meeting in Edinburgh[25] and have been the subject of two workshops[26,27] held in association with the International Laboratory for Research on Animal Diseases (ILRAD), Nairobi, Kenya. In the first of these the importance of theileriosis as a limiting factor on the development of the cattle industry in Africa, the Indian subcontinent and the Near East was emphasized. At the most recent meeting the role cell culture has played and will play in studying the life cycle of *Theileria*, understanding the pathogenesis and in developing methods for diagnosis and control of theileriosis was evident.

II. HISTORICAL REVIEW OF CULTIVATION

In vitro cultivation of *Theileria* is a small but far from exclusive area of study within the field of protozoan diseases. Three recent reviews[28,29,30] have covered most of the brief history of the technical developments in the methods used. These were concerned almost exclusively in the cultivation of the intralymphocytic schizont and the establishment of transformed lymphoblastoid cell lines infected by this form of the parasite. However, some of the earliest attempts to grow *Theileria* in vitro may be worth examining in the light of recent developments in techniques for cultivating *Plasmodium falciparum*[31] and *Babesia bovis*.[32]

Dschunkowsky and Luhs,[7] in their first description of piroplasms in cattle, maintained what can best be described as hanging drop cultures of blood of infected cattle for 20 to 25 days at room temperature and 38°C. During this time they claimed to observe motility of the piroplasms in erythrocytes and free in the plasma of defibrinated blood. Nuttall and Graham-Smith[33] in attempting to confirm some startling observations made by Miyajima and which they had discussed at some length in an earlier paper[34] saw no development of piroplasms of *T. parva* in their blood and bouillon cultures. They concluded that the flagellates seen by Miyajima in his cultures of Japanese *Theileria*, probably *T. sergenti*, were trypanosomes. In retrospect, Miyajima was probably the first successful in vitro isolation of *Trypanosoma theileri*, an organism as yet unidentified.

It was to be almost 40 years before the first significant breakthrough in cultivation of *Theileria* was recorded. This was in Israel, where Tsur,[35] then named Tchernomoretz, observed the multiplication of macroschizonts of *T. annulata* in lymphoid cells in plasma clot cultures initiated from liver and spleen explants taken from infected cattle. Tsur,[36,37] alone and with his colleagues in Israel[38-43] followed up and developed the technology he had introduced. They were regularly able to grow schizonts of *T. annulata* in monolayer cultures, established first from tissues, then blood of infected cattle. Cultures initially were of the plasma clot type, but dispersed cells were later used and monolayers of infected cells were readily and rapidly established using, as medium, a basal salt solution supplemented with lactalbumin hydrolysate and yeast extract. In one experiment they claimed to have effected transfer of schizonts to rodent cells co-cultivated with the infected bovine lymphoid cells. This significant claim was not confirmed, nor could it be repeated. Of greater significance, however, were the results, obtained as a group working in almost complete isolation and as pioneers in the field, which showed that they could infect cattle with cryopreserved schizonts, and that there was evidence for attenuation of *T. annulata* on continuous growth and passage in culture. No sooner had this been shown, some 20 years after Tsur's first cultures were established that Pipano and Tsur[44] reported that they had immunized cattle against *T. annulata* with a "tissue culture vaccine". The pragmatic application of this team of Israeli workers must serve as a model to us all.

In seeking to take the message abroad, Tsur et al.[45] had found in South Africa that *T. parva* was not as easy to cultivate as *T. annulata*. Using a development of the original plasma clot technique they claimed to have maintained intact, apparently viable schizonts in vitro for 2 weeks at 37°C. Brocklesby and Hawking[46] working in Kenya, in comparative trials using similar techniques found that *T. parva* would not grow in conditions in which *T. annulata* proliferated. In 1963, Hulliger and colleagues, also in Kenya and having little success in their attempts to cultivate *T. parva*, obtained the Tova strain of *T. annulata* from Professor Tsur to use as a model on which to develop techniques. They were immediately able to establish cultures of *T. annulata* as schizont-infected lymphoid cell lines and in these record the observation which put the cultivation of *Theileria* in perspective with other advances being made in the field of cell culture. This description[47] of the mode of multiplication of *Theileria* in association with the lymphoid cells which it infects illustrated perhaps a unique example of cell/parasite interdependence and emphasized the value of the *Theileria* culture system as a model for comparative studies with other lymphoblastoid cell lines. The schizont was shown to divide in association with the host cell, giving rise to daughter cells which are almost invariably parasitized (see Figure 6). The presence of the schizont within the lymphoid cell appears to be a prerequisite for its continued replication. Like the Epstein - Barr virus[48] or the virus of Marek's disease[49] the theilerial schizont appears to confer on an infected cell the power of unlimited replication.

Hulliger introduced two significant modifications to the Israeli technology in her attempts to grow *T. parva.*[50] These were the use of complex synthetic media, notably Eagle's MEM and feeder layers, in this case the fibroblastic baby hamster kidney line BHK21. Using these techniques with the Tova strain of *T. annulata* Hulliger appeared to cultivate both *T. parva* and cattle-adapted strains of *T. lawrencei* successfully. The most significant difference from *T. annulata* was the difficulty experienced in initiating the cultures and the time taken to start transforming. However, when such cultures were inoculated into cattle they induced a disease syndrome resembling that induced by the Tova strain of *T. annulata.* Cattle inoculated with what were presumed to be cell cultures of *T. parva* and *T. lawrencei* were repeatedly shown to resist challenge with *T. annulata* blood or cell culture schizonts yet to be fully susceptible to ECF on challenge with ticks infected with *T. parva.*[22]

These results cast doubt on the validity of Hulliger's apparently successful cultivation of *T. parva* and added significance to the successful establishment of a *T. parva*-infected and transformed cell line by Malmquist et al.[51] This cell line was proved serologically and by xenodiagnosis to be *T. parva,* inducing ECF in cattle and completing its life cycle in vivo to infect susceptible *R. appendiculatus* ticks which, on moulting, induced fatal ECF in susceptible calves.[52] This breakthrough of Malmquist et al.[51] could perhaps be attributed chiefly to two factors: first the use of fetal calf serum (FCS) at high concentrations (20%) in the medium and secondly the patience to wait for the cells to transform.

From these exciting developments came an increased interest in theilerial cultivation in vitro and the applications which such culture systems have. Moulton et al.[53,54] studied *T. parva* cell line first isolated by Malmquist. In recording the transformation phenomenon they considered that this was dependent on the isolation of infected reticulum cells in the culture and that it was these cells which "transformed" to give rise to infected lymphoblastoid cell lines. Doubts were cast on this hypothesis by Martini and Moulton[55] who, in histological and cytological studies on the pathogenesis of ECF, failed to demonstrate any parasitized reticulum cells in cattle patently demonstrating macroschizonts in the lymphoid organ, or in the prepatent period of the disease. Developments of culture techniques, notably improved methods of establishing cultures infected with *T. parva,* provide evidence that there is no need to hypothesize an intermediate "preschizogonous" phase of the life cycle in a cell such as a reticulum cell. Malmquist and Brown[56] clearly illustrated the benefits of feeder layers when establishing cultures from lymph node biopsy material while Brown et al.[57] were able to initiate cultures by infection and transformation in vitro of lymphoid cells by sporozoites of *T. parva* harvested from infective *R. appendiculatus.*

This particular technical breakthrough, which took some years to achieve, depending as it did on the ability to produce large numbers of infective sporozoites from ticks and to inoculate them into cultures without causing heavy contamination, has perhaps lead to the most exciting developments in the technology of theilerial culture. The method has recently been improved considerably by a number of workers who have been able to infect peripheral blood lymphocytes (PBL) of cattle with theilerial sporozoites harvested from ticks. Brown[58] established lymphoblastoid cell lines infected with both *T. parva* and *T. annulata,* while Kurtti et al.[59,60] and Stagg et al.[61] both achieved excellent results in infecting and transforming cells with *T. parva* sporozoites. These successes may be attributed to improved techniques for separating PBL using density gradients and for handling defined cell populations in semimicro systems in vitro, to improvements in methods for examining and monitoring infection and transformation in vitro using a cytocentrifuge,[62] to the use of suitable feeder layers to offer a stable culture environment and, perhaps most important, to the provision of heavily infected tick batches from which to harvest the sporozoites.

While the attention on *T. parva* has been focused on the use of the in vitro infection system, elsewhere cultures have been established by more orthodox methods. *T. annulata*-infected cell lines have now been established in many parts of the world. In the USSR, Zablotsky,[63,64] Konyukhov,[65,66] and, more recently, Mutuzkina[67,68] and many other workers have isolated cultures from lymph node, spleen, liver, and peripheral blood lymphocytes using recognized techniques. In Iran, Hooshmand Rad and Hashemi Fesharki[69] isolated a number of strains of *T. annulata*, finding increasing difficulty in establishing cell lines from the more virulent strains of the parasite. Van den Ende and Edlinger,[70] in Tunisia, isolated and grew a local strain of *T. annulata* in suspension, introducing for the first time RPMI 1640 as a medium for the maintenance of a bovine lymphoblastoid cell line. Hooshmand Rad,[71] in consequence, was subsequently able to grow a number of Iranian and Iraqi strains of *T. annulata* in suspension culture. Further afield, a team from Gansu,[72] in China and Gill et al.[73] in India were the first to establish *T. annulata*-infected lymphoid cell lines in their countries, while more recently, Ozkoc and Pipano[74] report the isolation of a Turkish strain from near Ankara, using the techniques described and developed by Pipano.[75]

Using modifications of the technique developed for the establishment of *T. parva* cell lines, Stagg et al. established *T. lawrencei*-infected cell lines[76] from buffalo *(Syncerus caffer)* and two types of cell line infected with *T. taurotragi*[77] from an eland, *Taurotragus oryx*. One of these cell lines consisted of large, often multinucleate, adherent macrophage-like cells, invariably infected with many schizonts, giving the impression of cytomere formation. A number of these cells would, in certain culture conditions, detach from the monolayer and go into suspension, at which time the macroschizonts transformed into microschizonts, releasing merozoites. This phenomenon was unique to that culture and all subsequent cell lines isolated from that or other eland have been lymphoid, nonadherent suspension cultures.

In Iraq the parasite of sheep and goats, *T. hirci*, has been isolated in culture by Hooshmand Rad and Hawa.[78] Methods used for its isolation and maintenance, are similar to those of transformed cell lines infected with *T. annulata*. The cells of sheep, buffalo, or eland infected with these different theilerial species, grow in suspension culture in a very similar manner to that of *T. parva* or *T. annulata*-infected bovine cells.

III. CULTURE TECHNIQUES

Techniques used for the initiation, establishment and maintenance of cell cultures infected with *Theileria* around the world are many and varied. However all established techniques relate to the growth of schizont-infected lymphoid cell lines. No reproducible methods for the culture of other stages of the theilerial life cycle have yet been recorded. For simplicity, in this chapter, the techniques described are those used and developed in Edinburgh for culturing the exoerythrocytic vegetative form of *Theileria*, the intralymphocytic schizont. These techniques have been developed primarily for use with the bovine pathogens *T. annulata* and *T. parva* but in the experience of the author they are applicable with slight modification to any other *Theileria spp.* or host cell for which culture techniques have been established to date. Few papers have been written describing specifically the techniques used in the cultivation of *Theileria*. Since the first such review of the relevant technology[28] recent papers by Kurtti et al.[60] and Stagg et al.[61] have recorded developments of methods for the establishment of cultures infected with *T. parva* which have given excellent results. The techniques given here are thus intended as a guide to a technology which is still an art, not a science. The principles of cell culture methodology apply here to all the techniques described. For background reading on the subject, where indicated, excellent reviews and descriptions relevant to these methods can be found in a number of volumes.[79-81]

A. Isolation of Bovine Peripheral Blood Lymphocytes (PBL)

1. Clip and scrub jugular furrow with 70% alcohol. Collect 100 mℓ blood by jugular puncture (14 gauge needle) into sterile screwtop bottle containing 20 mℓ ACD* or 1 mℓ Heparin (sodium salt, 1,000 μ/mℓ) and mix well by shaking bottle. Alternatively, collect in 150 mℓ Erlenmeyer flask and defibrinate with three wooden bacterial swab sticks.
2. In laboratory pipette 50 mℓ aliquots of blood into 50 mℓ centriguge tubes and spin at 2800 × g, 5°C for 20 min (4000 rpm MSE Coolspin). (After removing tubes, run centrifuge empty at 15°C in preparation for 6.)
3. Remove most of the supernatant plasma with a 10 mℓ pipette and discard. Using a Pasteur pipete remove the remaining plasma to within 1 to 2 mm of the buffy coat.
4. With a 5 mℓ pipette, recover the buffy coat in a total volume of less than 3 mℓ and transfer to a 25 mℓ screwtop Universal bottle containing 9 mℓ of PBS (20°C Dulbecco A, pH 7.3). Mix well with pipette.
5. Carefully layer this buffy coat/PBS suspension (12 mℓ) onto 8 mℓ of Ficoll-paque** in a Universal bottle.
6. Centrifuge at 800 × g, 15°C, for 35 min (2800 rpm).
7. With a 5 mℓ pipette discard the supernatant above the interface.
8. Recover the mononuclear cells at and below the interface using a 5 mℓ pipette. Take care to pick up cells from just above the interface moving down. Take up to 5 mℓ in total.
9. Add these cells in Ficoll-paque to 20 mℓ of PBS in a Universal bottle and centrifuge at 300 × g (1500 rpm) 15°C for 10 min.
10. Pour off supernatant PBS, add 20 mℓ fresh PBS (20°C) and recentrifuge at 300 × g, 15°C, for 5 min (1500 rpm).
11. Discard supernatant PBS and resuspend cells in 10 mℓ cold, complete culture medium. Count and determine viability of cells, adjust to required cell concentration (10^6 - 10^7/mℓ) and prepare cytocentrifuge smears, stained with Giemsa.

* Acid citrate dextrose
Each 100 mℓ solution contains:

Citric acid BP	0.8 g
Sodium citrate BP	2.2 g
Anhydrous dextrose PhEur	2.24 g

In distilled water. Autoclave or filter 0.22 μm MF filter (Millipore Corporation).

** Ficoll-paque (Pharmacia-Fine Chemicals AB, Uppsala, Sweden). Sterile solution, sp. gr. 1.077 Nonpyrogenic. Each 100 mℓ solution contains:

Ficoll 400	5.7 g
Diatrizoate sodium	9.0 g
Edetate calcium sodium	

B. Separation of Bovine Buffy Coat Cells by Ammonium Chloride Lysis (A Modification of the Shortman et al. Technique* Applied to Packed WBC)

1. Bleed 50 mℓ blood in anticoagulant (Heparin 20 iu/mℓ) or ACD 1 in 6 or defibrinate with swab sticks**.

2. Spin at 2000 × g 20 min 5°C in conical graduated 50 mℓ centrifuge tube (400 rpm MSE Chilspin).
3. Remove plasma aseptically and keep.
4. Take off buffy coat in not more than 2 mℓ and transfer to 4 mℓ *cold* PBS in plastic universal. Mix well by pipetting.
5. Add 3 volumes (not more than 18 mℓ) cold (+ 4°C) 0.17 *M* NH_4Cl, mix gently and stand for 10 min at 4°C.
6. Layer 5 mℓ autologous plasma *beneath* this lysed suspension *carefully* and centrifuge at 400 × g for 7 min (1500 rpm MSE Chilspin).
7. Remove supernatant and resuspend cells in the requisite cold complete culture medium to 10 mℓ. Count, adjust cell concentration as required and prepare cytospin smears for differentials.
8. Set up cultures as required: these *should* be sufficient cells for 50 mℓ culture suspension at 2-4 × 10^6/mℓ.

* Shortman, Williams, and Adams, *J. Immunol. Methods,* 1, 273, 1972.

** The use of anticoagulant will imply the presence of both granulocytes and platelets. To remove platelets, wash 2 × in PBS, 300 × g 10 min as for PBL separations on Ficoll-paque. Defibrinated blood gives PBL by this technique without further washing.

C. Isolation of Bovine PBL from Small Volumes of Blood

1. Bleed into 10 mℓ vacutainer* containing lithium heparin then separate
   a. using Ficoll-paque as per method 1. except:
      i. Mix 10 mℓ blood with 10 mℓ PBS and layer direct onto 8 mℓ Ficoll-paque sp. gr. 1.077.
      ii. Centrifuge 800 × g, 30 min, 15°C and continue as for method 1.
   (Expected yield from this method will be 2-4 × 10^6 PBL).
   b. using NH_4Cl as per method 2. except:
      i. Centrifuge blood in vacutainer 2000 × g, 20 min, 5°C to obtain buffy coat.
      ii. Remove buffy coat in 2 mℓ and mix with 4 mℓ PBS.
      iii. Add and mix with 18 mℓ 0.17 *M* NH_4Cl and continue as per method 2.

* Vacutainer 606484 10 mℓ Lithium Heparin 143 U.S.P. Units (Becton Dickinson).

D. Preparation of Sterile Ground Up Tick Supernate (GUTS)

1. Adult *Hyalomma anatolicum* ticks infected with *Theileria annulata* and fed for 3 days on rabbit ears, or *Rhipicephalus appendiculatus* infected with *Theileria parva* fed for 4 days.
2. Wash with 1% benzalkonium chloride* × 1. (Roccal - Winthrop Laboratories, Surrey, England.)
3. Wash with 70% ethanol × 3.
4. Wash with Eagles MEM with Hanks salts and antibiotics at double concentration (penicillin, 200 iu/mℓ, streptomycin 200 μg/mℓ, nystatin, 100 μg/mℓ) × 3.
5. Leave 10 min in 4th wash of Eagles MEM/ antibiotics as above.
6. Discard medium, add 2 to 5 mℓ MEM/BPA 3.5% (Bovine Plasma Albumin-Armour Fraction V) containing antibiotics as above and transfer ticks to a sterile mortar.
7. Grind thoroughly with pestle (3 to 4 rounds) and remove supernatant suspension repeatedly, each time adding and recovering a recorded volume of medium e.g., 3 to 5 mℓ MEM/BPA 3.5% as above. Repeat till a final concentration of 4 tick equivalents (t.e.) per mℓ is reached.
8. Centrifuge at 100 × g for 5 min (750 rpm MSE Coolspin). Aseptically recover supernate containing sporozoites.

9. Regrind debris in Griffith's tube with 3 to 5 mℓ extra MEM/BPA, recentrifuge as above, collect supernate and add to the first harvest of sporozoites.
10. Filter through sterile 25 or 47 mm Millipore Swinnex filter, with AP prefilter and 8μ MF filter (Millipore Corporation) using a 20 to 50 mℓ syringe.
11. Cytocentrifuge 50 to 100 μℓ filtered GUTS at 1000 rpm for 10 min, stain with Giemsa and check for presence of sporozoites in GUTS filtrate.

* Roccal antiseptic (Winthrop Laboratories, Surrey, England) contains 1% benzalkonium chloride.

E. In Vitro Infection of Bovine PBL with *Theileria* Sporozoites in Sterile GUTS

According to the amount of material available and the purpose for which the cultures are intended, this can be done with varying numbers of PBL and volumes of GUTS. Cultures can thus be initiated in microtiter plates with wells of total volume of 0.2 mℓ, wells of 2 cm^2 base surface area and a volume of 2.5 mℓ, or 25 cm^2 flasks which will normally hold up to 10 mℓ culture medium when horizontal. For clarity this technique is given for the initiation of cultures in 2 cm^2 wells of a multiwell plate (Costar).

1. Seed the required number of wells with 0.25 mℓ bovine PBL suspension prepared as techniques 1, 2, or 3 containing 2×10^6 cells in complete culture medium. For *T. parva* the wells should have been seeded the previous day with a suitable feeder layer.*
2. Add an equal volume (0.25 mℓ) of sterile 8 μ GUTS filtrate prepared as in technique 4 of the optimal concentration to each of 3 to 6 wells to give a series of replicate cultures. The concentration of sporozoites and contaminant debris in the GUTS filtrate may be ascertained on Giemsa-stained cytocentrifuge smears. It may be necessary to prepare serial dilutions of the concentrated sporozoite suspension, a 1:10 or 1:100 dilution in MEM/BPA sometimes proving less toxic to the cells than undiluted suspension at 4 t.e./mℓ.
3. Place the multiwell plate in a humidified incubator gassed with 5% CO_2 in air and incubate at 37°C. Alternatively place in a sterilized box which can be sealed and which has two vent holes on opposite sides which in turn can be closed. Humidify the box using a petri dish containing 10 mℓ sterile deionized water. Gas this box with 5% CO_2 in air, seal the vents and incubate at 37°C. (Gassing is not necessary for media not containing $NaHCO_3$ e.g., Leibovitz L15, and some HEPES-buffered media with minimal bicarbonate levels.)
4. On day 1 of culture, i.e., after 16 to 24 hr incubation, add 0.5 mℓ complete culture medium to each well, thus doubling the original volume.
5. On day 2, double the medium volume again by adding 1 mℓ.
6. From day 3 onwards the culture should be handled at least 3 times weekly as indicated by its rate of metabolism and appearance of cells in cytocentrifuge smears (which can be prepared from 50 μℓ aliquots taken from a well in which the cells have been thoroughly resuspended). Medium change involves careful removal of supernatant medium and its replacement with an equal or lesser volume of fresh medium. The culture volume in each well will thus be maintained between 0.5 and 2 mℓ.
7. Once infected and transformed lymphoblasts are detected in a well it will probably be found that its metabolism rate will increase in parallel with the degree of transformation. This will necessitate more frequent changing of larger volumes of medium. When

* Feeder layers are used for fastidious parasites with a low cloning efficiency. They improve the 'take' of *T. parva* and *T. lawrencei* cell lines immeasurably. Those most frequently used have been slow growing fibroblastic, epithelial, or endothelial monolayers of bovine origin. These may be primary cultures such as bovine embryo spleen (BESP) or bovine embryo thymus (BETH), cultures of limited growth such as bovine embryo skin (BES), bovine aortic endothelium (BAE), or such established cell lines as buffalo lung (IMR 31).

some 50% of cells are infected it will be necessary to transfer the culture to a 25 cm^2 flask (Nunc). One or more replicate cultures can be transferred together to a flask, again with an equal volume of fresh medium to give a total volume of 4 to 6 mℓ in a flask incubated horizontally. Gas flask with 5% CO_2 in air.

8. Once this first passage or sub-culture has been made the culture should metabolize well and grow vigorously. There may, however, be a check at this stage (and this is common with fastidious strains of *T. parva* or *T. lawrencei)* and the culture will need to be handled as if it were a primary culture again. In certain instances it may be necessary to seed the cells back onto a feeder layer once more to give them a suitable environment in which to reestablish. Once a growth rate of 10 times in 3 days has been obtained and there is a cell viability of >85%, the culture is ready to subculture normally and will behave as an established line.

F. Establishment of *Theileria* in Cell Culture from Infected Animals

This is the most usual way of initiating schizont-infect lymphoblastoid cell lines. Any source of lymphoid material is satisfactory, ideally taken from an animal patently exhibiting a schizont parasitosis. It is, however, frequently possible to initiate such cell lines from blood or tissues of animals which may be persistent schizont carriers yet in which schizonts are not detectable by standard cytological techniques. In the living animal, blood is perhaps the most easily accessible source of lymphocytes. Techniques 1, 2, and 3 are suitable for isolating PBL from infected cattle and are particularly relevant to the provision of material for initiating cell cultures of *T. annulata*. For *T. parva*, however, where the leukopenia associated with the clinical reaction reduces the number of PBL which can be harvested from a volume of blood, it is sometimes necessary to obtain schizont-infected lymphocytes from other organs. The most simple tissue to biopsy is superficial lymph node (parotid or prescapular) and this is the technique given here.

Postmortem, depending on the environmental temperature, tissues will yield viable schizont-infected cells for at least 2 hr. Any tissue containing reasonable numbers of lymphocytes (lymph node, spleen, thymus, bone marrow, lung, liver, or kidney have all been used) which can be taken aseptically may be used in the manner described for the biopsy here.

1. Place 5 to 10 mℓ complete medium containing preservative-free heparin 10 μ/mℓ in a sterile universal bottle.
2. Draw up 5 mℓ heparinized medium into a 10-mℓ syringe via a 1$^1/_2$ in. 14 g biopsy needle.
3. Clip, shave, and surface-sterilize the skin over the lymph node which is immobilized with one hand.
4. Biopsy the lymph node by penetrating rapidly with the needle attached to the syringe and probing while applying negative pressure with the syringe plunger.
5. Still applying negative pressure and squeezing the lymph node, withdraw the needle and express biopsy material and medium into the universal bottle. Keep cool before processing.
6. In the laboratory either pipette the tissue to break up the clumps or break up firm tissue by applying pressure (the conical plunger of a disposable syringe in the conical base of a disposable universal bottle does this well).
7. Centrifuge the cell suspension for 5 min at 400 × g, discard supernate and resuspend in 5 to 10 mℓ growth medium.
8. Seed, with or without feeder layers as indicated (1 of each if sufficient material has been obtained) 2 flasks of 25 cm^2 or an equivalent number of 2 cm^2 wells. (Due to the residual clumps of cells which are important to the establishment of the culture there is little point in making cell counts at this stage.)

9. Infected cultures initiated in this way are then handled in the same way as those infected in vitro described in technique 5. Supernatant medium should be removed and replaced with fresh medium at frequent intervals. This should be done without disturbing the cells if possible (as it can easily with wells in a plate). Cell free supernates are not readily removed from flasks maintained horizontally, however. If necessary, the supernatant medium removed can be centrifuged (400 × g, 5 min), spent medium discarded, the cells resuspended in fresh medium and returned to the flask. Not more than two-thirds of the medium should be changed at any one time.
10. Transformation can be ascertained both by regular examination on an inverted microscope and of cytocentrifuge smears, and by detection of increased metabolic rate of the culture. Continue to feed the culture with increasing frequency until 50% or more cells are parasitized, then subculture *with care* as described in technique 5. Gas the fresh flask with 5% CO_2 in air.

G. Maintenance of Established Cell Lines

1. Cultures are routinely incubated at 37°C in 25 cm^2 plastic disposable flasks (Nunc) held horizontally and containing 5 to 10 mℓ cell suspension in growth medium.
2. Though sealed, if using a medium with high bicarbonate levels, it is advisable to gas the cultures with 5% CO_2 in air when subculturing.
3. Cultures will grow rapidly if healthy, increasing Log_{10} every 48 to 72 hr. They thus require subculture every 2 to 3 days if in complete growth medium. The seeding rate at subculture is usually 1-2 × 10^5/mℓ. This is usually simply affected by seeding at 1:10, i.e., 1 mℓ into 9 mℓ fresh medium or 0.5 mℓ to 4.5 mℓ. This 'subculture' need not necessarily be to a new flask provided the cells are fully in suspension. Where adherent they can usually be removed by vigorous pipetting or the use of a silicone rubber policeman on a glass rod, resuspended and seeded as for cells in suspension.
4. Should larger volumes be required they can be obtained to a limited degree in static culture using standard disposable tissue culture flasks. Held horizontal, a 75 cm^2 flask will hold 50 to 100 mℓ cells in suspension and a 175 cm^2 flask 100 to 200 mℓ. Volumes as great as 1 to 2 ℓ can be obtained from static cultures grown in Roux flasks and Thompson or Pavitsky bottles. In static culture, however, depth of medium is an important factor in determining oxygen tension at the cell interface. In consequence, at the larger volumes quoted here, growth rates and cell viabilities decline if cultures are maintained too long.

H. Suspension Cultures

Theileria-infected lymphoblastoid cell lines grow readily to suspension culture in suitable medium, e.g., RPM1 1640 supplemented with serum at 5 to 10% and in flasks designed for the purpose, e.g., Wheaton Celstir. Cultures of 25 mℓ to 10 ℓ volumes can be initiated directly from static cultures adapted to the same medium. Even larger volumes, up to 20 ℓ have been grown in comparatively simple apparatus, comprising a glass flask with a wide-mouthed top, one or more side arms for introduction of media and sampling of culture, and a magnetic follower serving as an impeller. Cultures in such flasks may be maintained at 37°C in a hot room or incubator, be water jacketed or on a heated stand. The impeller or stirrer arm, a Teflon®-coated magnet which maintains the cells in suspension, is rotated at approximately 100 rpm by placing the whole unit on a magnetic stirrer. This should be of heavy duty type capable of running under load for weeks on end in an environment at 37°C.

1. Seed cultures heavily, at approximately 5 × 10^5 cells/mℓ, from a 2-day-old static culture into RPM1 1640/10% FCS. The volume of cell suspension is determined by the amount needed to cover the impeller arms, e.g., for a 1 ℓ stirrer flask, where 250

mℓ is sufficient, seed 125 mℓ static culture at 10^6 cells/mℓ and add 125 mℓ fresh medium.

2. Double volume of medium daily to maintain cells in exponential growth phase.
3. Harvest 2 days after final medium addition, anticipating an optimal yield of 1.5 to 3 $\times$ 10^6 cells/mℓ (at lower serum concentrations this yield will be lower).
4. For regular supplies of large volumes of cell suspension small stirrer flasks can be maintained to provide seed material for the larger production flasks. Alternatively, with care and attention to sterility, a large suspension culture can be maintained for a long time by repeatedly harvesting 75 to 90% of cell suspension, then bulking up again by daily doubling of medium volume. The need to maintain sterility in such culture conditions cannot be emphasized enough. The commitment made in time and materials to the production of such large volumes of culture means that contamination at this stage can prove particularly costly. If long-term maintenance is required some form of sterile vacuum system is needed to enable large volumes to be withdrawn speedily and aseptically. If in any doubts about the efficiency of your sterile system it is safer to initiate separate individual bulk cultures of this kind to avoid reliance on long-term sterility in bulk culture. There is every indication that really large volumes of *Theileria*-infected lymphoblastoid cell lines could be regularly produced by growing in specially designed fermentor or continuous culture systems (e.g., New Brunswick Scientific) providing hundreds of liters of suspension and kilograms of cells.

I. Assessment of Cultures

Cell growth and behavior of *Theileria*-infected cells need to be assessed routinely to ensure that cultures are behaving normally. Cultures may be monitored intermittently, at each subculture, or, as indicated for isolates of new strains, daily over three or more subcultures. Two sets of observation are indicated:

1. Determination of cell number and viability. Cell concentration in a culture can be assessed by doing direct hemocytometer counts or using an electronic particle counter (Coulter Electronics). With experience the health and viability of cells can readily be evaluated when counting the cells by the former method. If using a particle counter, however, great care must be taken in determining the level of the lower threshold to avoid counting particles (schizonts or nuclei) released from broken cells. Moreover, at lower cell concentrations, special dilution techniques are necessary to ensure accuracy. For specific counts of viable cells, routine dye exclusion tests may be used. Suitable dyes are 0.45% nigrosin or 0.3% trypan blue prepared in normal saline and distributed in sterile aliquots in Bijoux. To count viable cells, mix 0.1 mℓ cell suspension with 0.1 mℓ dye, place in hemocytometer and count the percentage of live cells (those which do not take up dye).
2. Cell morphology and characteristics. The character of cells and schizonts within a culture is best assessed on Giemsa-stained smears of thin air-dried preparations. Such preparations can be made by spinning 50 μℓ aliquots of culture suspension onto clean glass microscope slides at 600 to 1000 rpm for 5 min in a cytocentrifuge (Cytospin, Shandon Southern Instruments). Note that for optimal preparations it may be desirable to pre-wet the slides by doing a preliminary centrifugation of tissue culture medium onto the slide. Cytocentrifuge smears made in this way are air dried, fixed methanol and stained for 40 min with 5% Giemsa buffered at pH 7.2. For optimal visualization of schizont nuclei the smears can be hydrolyzed after fixation by before staining by immersing in 1N HCl for 6 min then washing in buffer solution. Observations should include counts to give: (1) percentage infected cells (PIC); (2) mean schizont nuclear number (MSN); (3) multinucleate cells (MNC); (4) mitotic index (MI).

J. Cryopreservation of Cell Cultures Using DMSO

Freezing

Materials

1. Suspension cultures: 100 mℓ of a day 2 culture in an 80 cm^2 flask at approximately 10^6 cells/mℓ.
2. Monolayer cultures: harvest cells as soon as a confluent monolayer is formed on the surface of a 175 cm^2 flask after seeding with the contents of a 25 cm^2 flask. (Appendix a).

Method

1. Approximately 2 hr prior to cryopreservation prepare the following: (a) for each flask of culture to be cryopreserved put 6 × 2 mℓ vials into compartmented box at −20°C; (b) 5 mℓ complete medium (RPMI 1640/20% FCS — Appendix c) put in ice bucket; (c) 5 mℓ of 20% dimethyl sulfoxide (DMSO, Appendix d) in complete medium put into ice bucket.
2. Centrifuge cells at 300 × g for 5 min. Discard supernatant. Have media, vials and ice bucket ready. Make sure that there will be no impediments on the journey to dry ice container or −70°C Cliffco.
3. Add 5 mℓ ice cold medium to cell pellet and resuspend gently. Add 5 mℓ ice cold 20% DMSO dropwise and mix by gentle pipetting. *Keep on ice.* Add approximately 1.5 mℓ by pasteur pipette to each vial. Tighten caps fully. Replace in box and *rush* to dry ice or Cliffco.
4. Transfer to gas phase liquid nitrogen bank (Appendix e) the following day.

Appendix

a. To harvest monolayer cells by (1) physical methods and (2) chemical method: (i) scrape off cells with a silicone rubber-encased glass rod (policeman) and break up by pipetting up and down vigorously. (ii) Remove cells with versene (diamino-ethane-tetra-acetic acid). Wash surface of flask once with PBS. Add sufficient 0.02% versene in PBS to cover monolayer. Incubate at 37°C until cells have just detached. Pipette vigorously and transfer to a centrifuge tube. Wash out flask with complete medium, 2-3 × the volume of versene used and add to centrifuge tube. Spin at 1000 g for 5 min.

b. Cryopreservation vials. 2 mℓ glass vials. Wheaton Scientific, Millville, New Jersey 08332, U.S.A.

c. Culture medium. RPMI 1640 with 25 m*M* HEPES and 20% fetal calf serum. Gibco Europe Ltd., Paisley, PA3 4EP.

d. Dimethyl Sulfoxide. BDH Chemicals Ltd., Poole, BH12 4NN.

e. LR40 II, BOC Ltd., 150 Polmadie Road, Glasgow, G5 OHN.

Resuscitation of Cultures Cryopreserved in DMSO

1. Prepare a 10 mℓ centrifuge tube with 8 mℓ of warm (37°C) medium. (RPMI 1640/20% FCS).
2. Thaw vial rapidly in a 37°C water bath. Add contents of vial to the centrifuge tube and mix gently.
3. Centrifuge at 300 × g for 5 min.
4. Resuspend pellet in 5 mℓ medium.
5. Add cell suspension to flask (25 cm^2, Nunc) and gas well with 5% CO_2 in air.
6. Depending on cell concentration, add 5 mℓ medium the same evening or following day.
7. 24 hr later, spin down cells gently (200 × g) and resuspend in fresh medium.

K. Cryopreservation of Sporozoites Using 7.5% Glycerol: Preparation of Tick Stabilates

Freezing

Materials

1. Ground up tick supernate (GUTS) at 4-5 tick equivalents per mℓ prepared in Eagle's minimum essential medium with 3.5% bovine plasma albumin (MEM/BPA) as described in Technique 4.
2. 'Analar' grade glycerol, sterilized by autoclaving at 15 p.s.i., 20 min, as the cryoprotectant.
3. Either a 'Cliffco' mechanical deep freeze at −70°C or dry ice cabinet, and a liquid nitrogen container — gas phase LR 40 used for routine storage.

Methods

1. Add pre-warmed (37°C) glycerol to MEM/BPA to make a solution of 15% glycerinated MEM/BPA cryoprotectant medium. Pipette thoroughly and repeatedly.
2. Add an equal volume of 15% glycerinated MEM/BPA dropwise with gentle shaking to the GUTS. Pipette gently to mix thoroughly.
3. Allow to stand at room temperature (20°C) for approximately 30 min to equilibrate.
4. Distribute 1 to 2 mℓ aliquots of the 7.5% glycerinated GUTS into 2 mℓ glass vials (see technique 5) with repeated mixing of the GUTS suspension.
5. After a total of 30 to 45 min equilibration, pack the vials, appropriately labelled with zinc oxide tape and indelible ink — parasite strain, stabilate number — in a compartmented box, *vertically*, and place at −70°C overnight.
6. The following day, transfer to the liquid nitrogen bank. Record stabilate in stabilate record, cross-indexed for parasite, stabilate no. and bank position.

Resuscitation of GUTS Stabilate Cryopreserved in Glycerol

1. Thaw vial rapidly in a 37°C water bath. Remove once thawed and allow to equilibrate at room temperature (20°C) for 15 to 45 min.
2. If to be used to infect cattle in vivo: (a) pool contents of all vials resuscitated and take relevant aliquots from pool for inoculation; (b) if to be titrated, dilute, after 15 to 30 min equilibration, in MEM/BPA with 7.5% glycerol; (c) aim to inoculate recipients *subcutaneously*, in a site drained by a specific peripheral lymph node, e.g., in front of the ear — parotid, L. N., in front of shoulder — prescapular L. N.

If to be Used to Infect Culture In Vitro

1. As in 2a above
2. If to be titrated, dilute gently, stepwise, with twofold dilution steps in MEM/BPA 1:2, 1:4, to 1:10, at 15-min intervals, allowing equilibration between each step, then subsequent dilutions can be handled normally. This is done to reduce effect of hypotonic shock.
3. Inoculate stabilate and diluted aliquots of same, to cultures in an *equal volume of medium*, e.g., 1 mℓ + 1 mℓ in 25 cm^2 flasks, 0.25 mℓ + 0.25 mℓ in 2 cm^2 wells.
4. Cultures should ideally be infected 30 to 60 min after resuscitation of stabilate.
5. Feed cultures by addition of an equal volume of medium on day 1 and 2 after infection, then as indicated by metabolism.

L. Karyotyping/Chromosome Preparations

1. Into 10 mℓ suspension culture in a flask (25 cm^2) add 0.2 mℓ colcemid (10 mcg/mℓ — Gibco lyophilized, cat. no. 120-5211 — reconstituted in distilled water), to give 0.2 mcg/mℓ final concentration (2×10^{-7}).
2. Incubate culture at 37°C for 90 to 120 min. (Longer for slower-growing cells — the shorter the better preparation.)

3. Spin down at 1500 MSE Chilspin (300 × g) for 5 min.
4. Discard supernatant and resuspend pellet in 5 mℓ freshly prepared 0.75% sodium citrate (held at 37°C) by sucking up and down gently 3 to 4 times.
5. Incubate at 37°C (water bath) for 35 min*
6. Spin down at 1500 rpm (300 × g) for 5 min.
7. Discard supernate and add *fresh* ice cold acetic alcohol (1 part glacial acetic acid: 3 parts methanol) — 5 mℓ and resuspend by pipetting up and down *gently* 3-4 times.
8. Hold at 4°C for 10 min.
9. Repeat 6, 7, and 8.
10. Spin down finally at 1500 rpm, discard supernatant and resuspend cell pellet in a volume of fresh acetic alcohol (ice cold) equal to the cell pellet volume.
11. With pasteur pipette, gently break up the cell clumps and drop, from a height of approximately 1 ft onto *very clean,* ice cold, wet slides.
12. Dry above a bunsen flame.
13. Stain with Giemsa pH 7.2 - 1 hr 3%. (There may be some merit in keeping unstained slides in a drawer for days/weeks/months before staining — this may improve the preparations.)

M. Appendix to Culture Techniques

Culture Vessels

Theileria cultures are particularly fastidious and sensitive to environmental factors. It is thus essential that all materials used are of the highest standard and maintained to that standard. Glassware used should be borosilicate or Pyrex, washed with nonresidue-forming detergents specifically designed for cell culture and sterilized by dry heat or autoclaving. Should it not be possible to obtain good glassware and keep it clean and sterile it will be necessary to use disposable plastic tissue culture ware. It is in any event advisable to maintain some cultures in disposable flasks — for safety, control of sterility and chemical or organic contamination, and for the excellent optical properties they offer.

In this laboratory a compromise is reached with all pipettes, centrifuge tubes, blood and media storage bottles being reusable glassware, while cultures are grown in disposable plastic ware — initiated in plates containing 24 × 2 cm^2 wells maintained routinely in 25 cm^2 flasks and bulked in 75 cm^2 or 125 cm^2 flasks (Costar or Nunc).

Media

Lymphoid cell lines infected and transformed by *Theileria* have been and can be grown in a wide variety of media. The most common currently used synthetic media are those intended primarily for human diploid cultures and lymphoblastoid cell lines — RPMI 1640 and Eagle's Minimum Essential Medium (MEM). If intended for use in open culture systems it is advisable to use a medium buffered with the zwitterionic HEPES buffer at approximately 25 m*M*, supplementing the existing bicarbonate buffer systems incorporated in the standard formulation. Alternatively, for this purpose, Leibovitz L15 has proved to be satisfactory, requiring no atmospheric CO_2 since it has a bicarbonate-free buffering system.

The media used can be commercially-prepared 1 × media containing all buffers and supplements. Alternatively they may be bought as 10 × concentrated media, diluted with sterile distilled, deionized water of the highest purity and then supplemented as required, or purchased as dry, powdered media, reconstituted to 1 × strength in deionized distilled water and sterilized by membrane filtration (Millipore). These three alternatives, if carefully controlled, provide a range of options suitable for all situations. The first is most expensive but requires neither sterilization nor water of the highest grade of purity. The latter two

* *All* glassware used to be siliconized beforehand, (except slides).

require some degree of capital outlay on deionization and filtration equipment but offer medium which is significantly less expensive and perhaps more reproducible — a single batch can be tested, bought and stored for repeated use.

The media currently used in this laboratory are 1 × liquid media (Gibco - Europe) as follows:

- RPMI 1640 with 25 m*M* HEPES - 041 - 2400
- MEM with Earle's salts and 25 m*M* HEPES - 041 - 2360
- L-15 (Leibovitz) Medium - 041 - 1415

They have osmolarities ranging between 270 - 310 (RPMI 1640), 260 - 330 (MEM) and 290 - 340 (L-15). All media, whether bought with glutamine present or not, have glutamine added to a concentration of 2 m*M*. The glutamine is stored as a 200 m*M* solution at −20°C, thawed and added at 1 mℓ per 100 mℓ medium.

While the growth of *Theileria* in vitro does not appear particularly sensitive to the synthetic medium used, it may be sensitive to that all-important component of cell culture media — serum. *T. parva* and *T. lawrencei* are particularly so and it is difficult, if not impossible, to establish cell lines infected with these organisms without supplementing the medium with fetal calf serum (FCS) to a concentration of 20%. This FCS may, but need not, be heat-inactivated (56°C, 30 min) but should be selected carefully. Many commercially available batches are toxic or will not support the growth of *T. parva*. Samples should be pretested by growing three successive subcultures of a fastidious *T. parva* cell line in them before selecting for purchase. Alternatively FCS should be prepared from embryos obtained at the local abattoir, the essential point of preparation being to remove the serum on the day of bleeding with minimal haemolysis, and to have the fewest possible freeze-thaw cycles, certainly no more than one before filtration.

Once established, *T. parva* cultures may be maintained in medium with 10% FCS or even weaned onto normal bovine serum (NBS). *T. annulata* can be established and maintained in medium supplemented with NBS - 20% to initiate and 5 to 10% to maintain, though growth rates and yields are lower than those of similar cultures supplemented with FCS. Horse serum can be used as a partial or complete replacement for FCS in suspension cultures of *T. parva* while FCS, NBS, or normal sheep serum all appear satisfactory for the maintenance of *T. hirci* in ovine lymphocytes.

In this laboratory, for ease of administration, a standard serum supplementation is used for all cultures, arthropod or vertebrate, initiation or maintenance, *T. parva* or *T. annulata*. To a single 500 mℓ bottle of 1 × medium is added 100 mℓ heat inactivated FCS (Gibco), giving a final serum concentration of 16.6%.

Other nutritional additives to the media in addition to glutamine and serum are

- Tryptose phosphate broth (Gibco 043 - 8050) at 10% to L-15
- MEM nonessential amino acids (Gibco 043 - 1140) to MEM used for the culture of cell lines
- Bovine plasma albumin (BPA - Sigma, Armour Fraction V) added to MEM at 3.5% and membrane filtered — used for preparing GUTS

In medium used for cell cultures the antibiotics penicillin (100 μ/mℓ) and streptomycin (100 μg/mℓ) are standard additives. If mycoplasma contamination is a problem or expected, Kanamycin (100 μg/mℓ) may be added. Fungistats are not used routinely but nystatin (Mycostatin) is added both to the medium used to wash ticks and to the MEM/BPA in which they are ground at a concentration of 100 μ/mℓ.

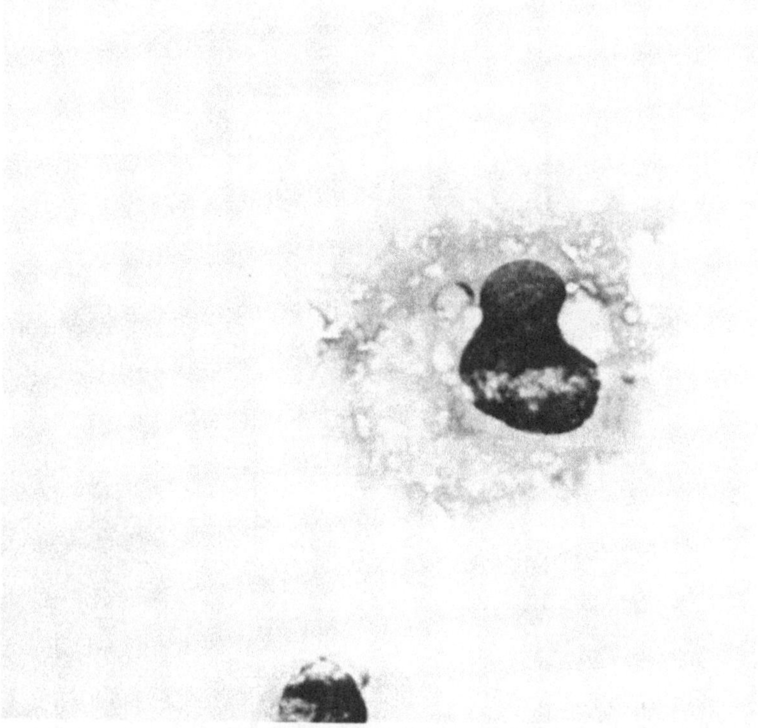

FIGURE 2. Photomicrographs (Figures 2-7) of Giemsa-stained cytocentrifuge smears of cultures of bovine lymphoid cells infected and transformed in vitro by sporozoites of *Theileria annulata* (all × 1280). Large basophilic lymphoid cell with uninucleate trophozoite, 16 hr after infection.

IV. CULTURES AS A RESEARCH TOOL

The culture system described here in which lymphocytes are infected and transformed in vitro by theilerial sporozoites has proved to be ideal for cytological and ultrastructural studies of the mechanisms of sporozoite invasion and the subsequent schizont/cell interrelationship. A series of photomicrographs (Figures 2 to 7) illustrate the development of *T. annulata* in such a system. The rapid penetration of target cells by sporozoites has been described and emphasized by Brown[57] and Jura[82] and clearly illustrated by Stagg et al.[61] who showed, using both living and fixed material, the sequence of events which occurs on co-cultivation of heavy concentrations of sporozoites with bovine peripheral blood lymphocytes. These results were particularly gratifying, depending as they did on the development of a technique for serial centrifugation of suspensions of ground up infective ticks (GUTS) to concentrate the sporozoites of *T. parva*.

In his study, Jura[82] was able to examine the invasion and transformation of lymphocytes at an ultrastructural level. A series of electron micrographs from this investigation, again showing sporozoite invasion, dedifferentiation to give a uninucleate trophozoite, subsequent division of trophozoite nuclei to form a schizont and the relationship between schizont and host cell at interphase and in division is shown here (Figures 8 to 15). This association of schizont and host cell, with involvement of annulate lamellae and the spindle microtubules of the host cell has also been described by Stagg et al.[83] using a cell line of *T. taurotragi*[77] and Musisi et al.,[84] using cell lines infected with an Iranian strain of *T. annulata*.[69] In their work they not only confirmed the observations of Hulliger et al.[47] and Hulliger[50] who had

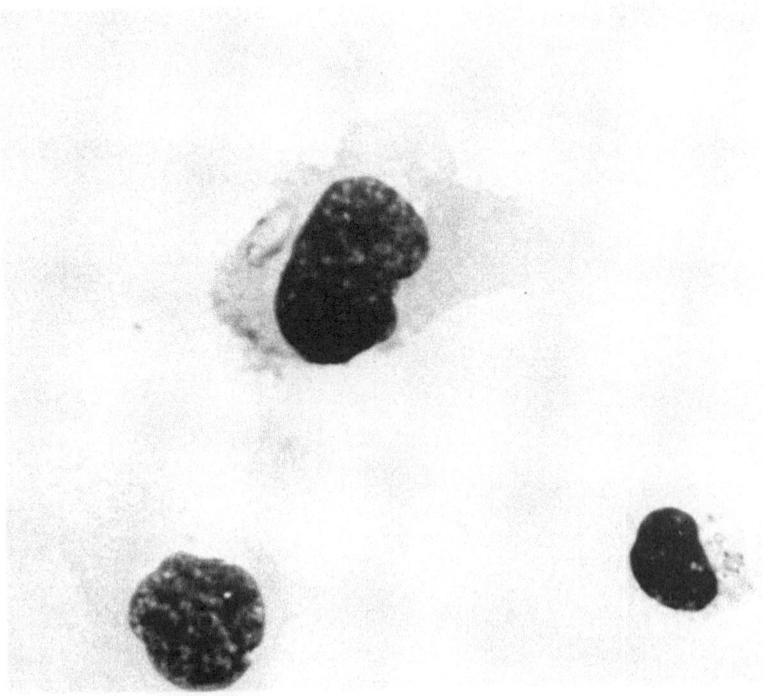

FIGURE 3. Pyriform trophozoite at point of first nuclear division, 24 hr after infection.

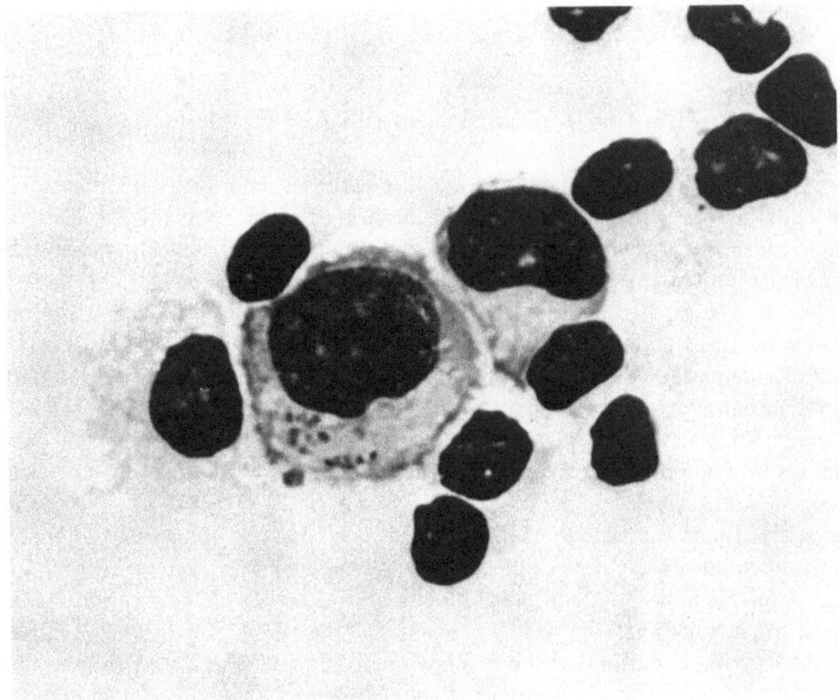

FIGURE 4. Schizont-infected, transformed lymphoblast surrounded by small, uninfected lymphocytes and two adjacent cells apparently transforming, 3 days after infection.

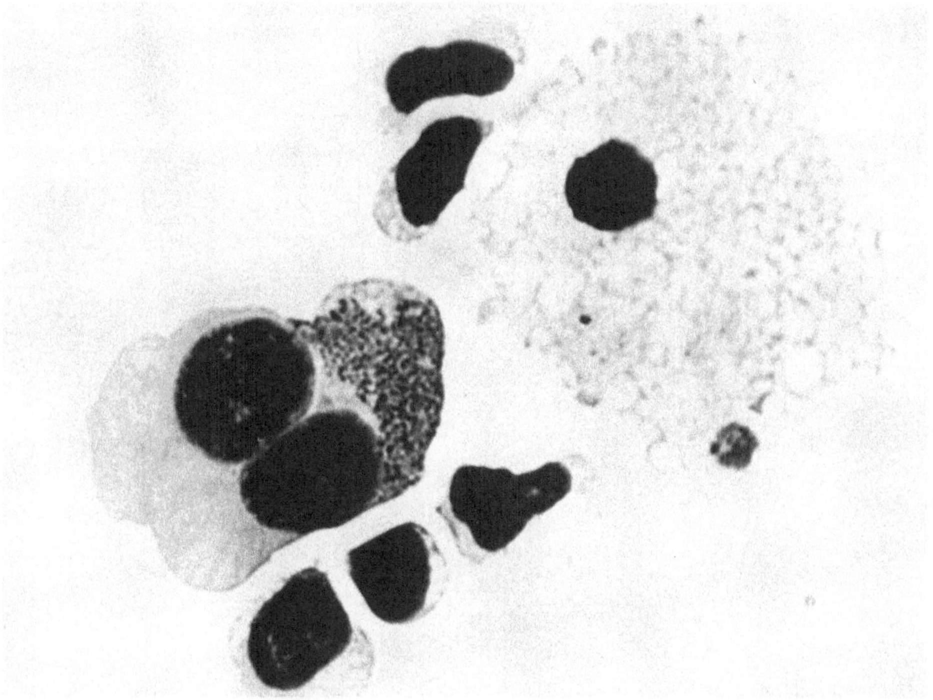

FIGURE 5. Binucleate lymphoblast and intracellular microschizont with uninfected macrophage and small lymphocytes, after 7 days in culture.

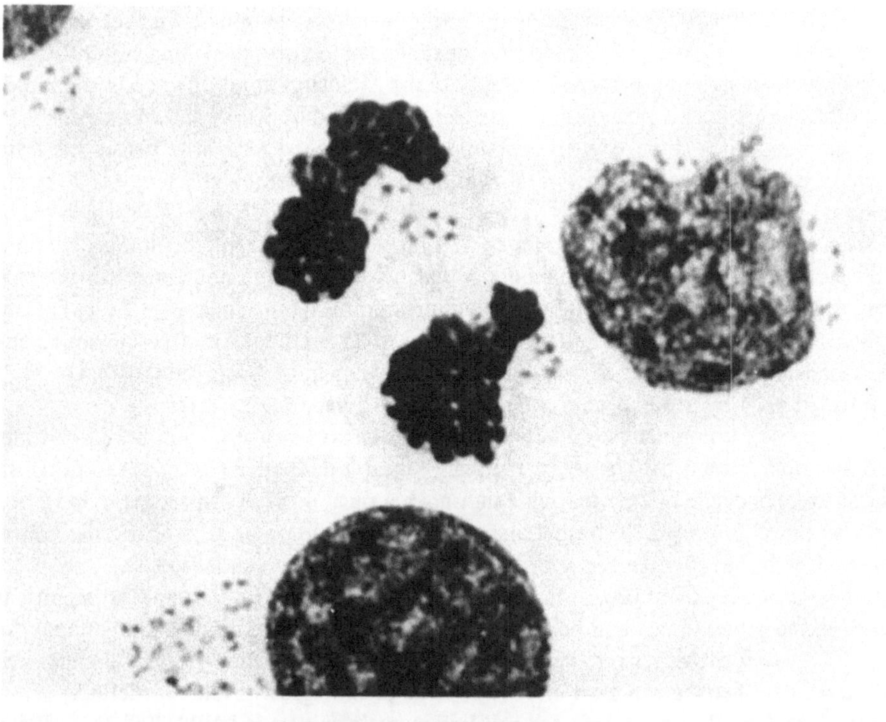

FIGURE 6. Infected, transformed lymphoblasts showing a pair of cells in reconstruction, each with daughter schizonts and streaming of spindle microtubules: 14 days after infection.

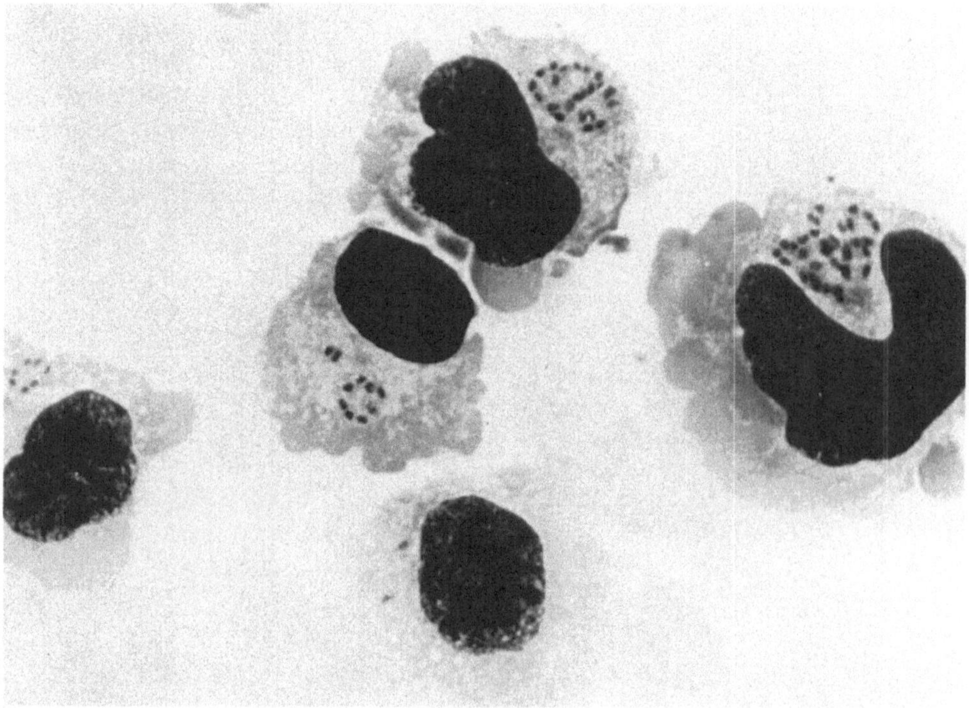

FIGURE 7. Transformed lymphoblastoid cell line, 98% of cells containing *T. annulata* schizonts: 3 weeks.

described the schizonts' evident close association with the golgi apparatus of the host cell, they also showed how the schizonts are apparently treated as if they were chromosomes at metaphase in cell division and pulled apart during anaphase, telophase and reconstruction of the daughter cells (Figures 14, 15 and 6). Once infected in vitro, providing culture systems are optimal, cells containing parasites multiply to the exclusion of all other cells and rapidly take over the cultures. This phenomenon has been illustrated by Kurtti and Munderloh[59] and Kurtti et al.[60] and goes some way to explaining the prolonged lag phase before the transformation of early cultures of *T. parva*. It is evident that the cloning efficiency of *T. parva*, as illustrated by Malmquist et al.[51] is very different from that of *T. annulata* which can be grown from a single cell[85] given the right environment. The use of microcultures containing 5 to 100 $\mu\ell$ of optimally buffered medium in balance with the gaseous phase of the environment is ideally necessary for initiating cultures from single cells. For *T. parva*, the addition of a feeder layer to this system has proved to be essential for effecting single cell-derived clones.[86] Malmquist et al.[51] claimed an improvement of 4 $\log_{10}$ in the "take" of *T. parva*-infected cells by seeding onto a bovine embryo spleen feeder layer.

There have been a number of studies defining the behavior of lymphoblastoid cell lines infected and transformed by *Theileria*. Hulliger[50] used a mathematical model to prove the intrinsic interdependence of cell and schizont on what must now be considered to have been the Tova strain of *T. annulata*. Malmquist et al.[51] and Moulton et al.,[53] using the C2 cell line infected with the Muguga strain of *T. parva* recorded very similar results.

In studies involving over 100 isolates of *T. parva*, Brown et al.[87] found no significant difference in the growth rates and behavior of cell lines infected with different strains of *T. parva* or *T. lawrencei*. Moreover, buffalo cells infected with *T. lawrencei*, and eland cells with *T. taurotragi* behaved in a similar way. Examples of growth rates in defined culture media are given for a *T. parva*- infected cell line in static culture in MEM/20% FCS (Figure 16) and in suspension culture in a spinner flask in RPMI 1640/10% FCS (Figure 17). Growth

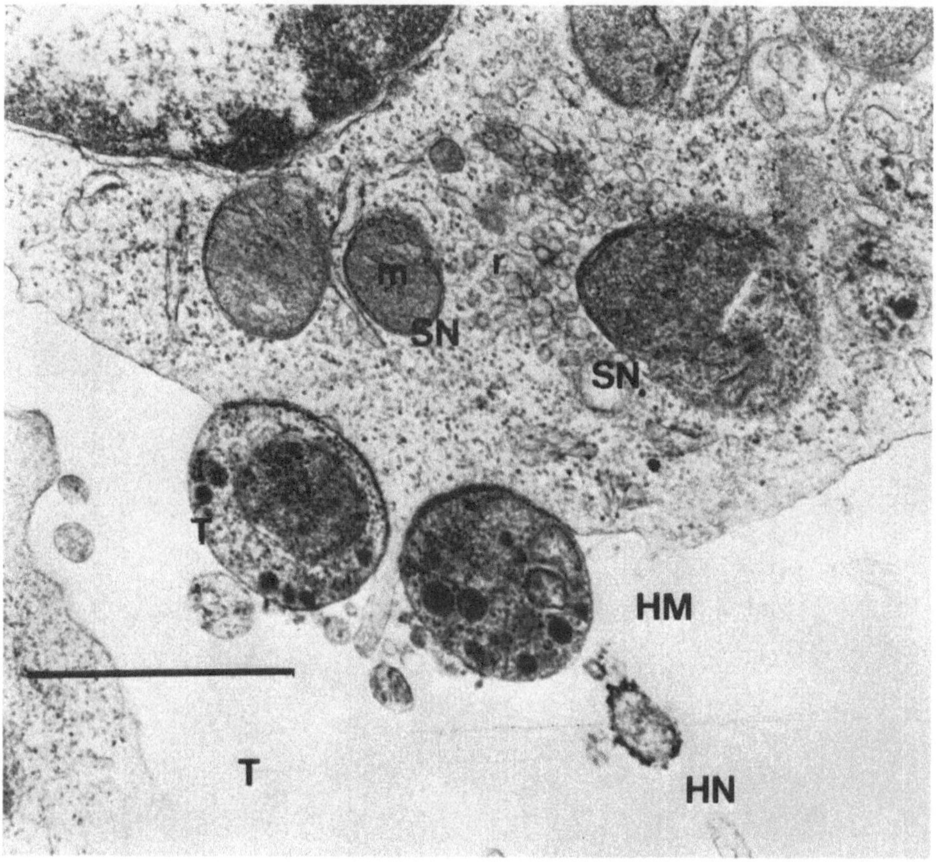

FIGURE 8. Electron micrographs (Figures 8-15) of cultures of bovine lymphoid cells prepared at intervals after infection in vitro with sporozoites of *Theileria annulata*. Bar = 1 μm. 1 hr after infection, showing multiple attachment and penetration. Each of two sporozoites has made contact with the cell plasma membrane by the aspect of its pellicle adjacent to the nucleus (SN). The apex of the sporozoite containing the rhoptries (r) and mitochondria (m) is distant from the lymphocyte plasmalemma. Within the cell cytoplasm are two trophozoites. (Host mitochondria — HM, magnification × 43,000.) (Courtesy of Dr. H. G. Z. O. Jura.)

parameters for *T. annulata* cell lines are perhaps marginally better but this probably relates to their cloning efficiency and their ability to maintain exponential growth despite the trauma of medium and flask changing. In optimal situations, all these infected cell lines have doubling times between 18 and 21 hr over a 3-day period. Synchronization of cultures would indicate that 11 hr is perhaps the fastest doubling time within that period. This means a Log_{10} increase in cell concentration over 3 days (as low as 2 days for *T. annulata)* in cultures maintaining a cell viability of >95%, with 98% of cells containing schizonts. From suspension cultures, massive cell yields are obtainable. Taking a conservative estimates of 1-2 × 10^6 cells/mℓ, 10^9 or more cells (1 to 2 g) can be harvested from 1 ℓ, providing a ready source of material for biochemical or immunological studies.

Despite this source of material, however, few studies on the biochemistry of the parasite in vitro have been published. This is perhaps due to the fact that separation of the schizont from its host cell have proved to be particularly difficult. Hirumi et al.[88] and Nyormoi and Bwayo[89] claimed some success, using a Stansted cell disrupter, but the results of their isoenzyme studies showed the presence of contaminating host cell material with concomitant loss of parasite-specific enzyme activity in the purified preparations. Melrose et al.[90] and Musisi et al.[91] showed that, provided uninfected lymphocytes and lymphoid cell lines could

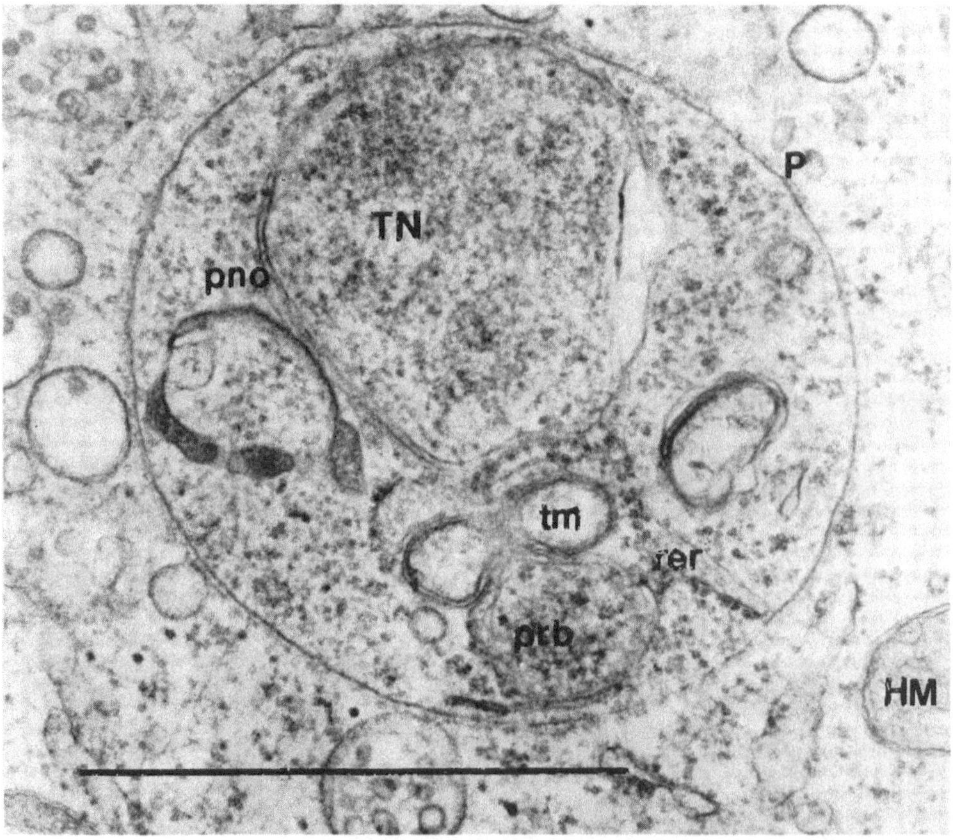

FIGURE 9. 18 hr after infection. An intracellular trophozoite showing its round form, nucleus (TN), tubular, acristate mitochondria (tm), profiles of rough endoplasmic reticulum (rer), many polyribosomes (prb), a unit pellicle (P) in direct contact with the cytoplasm, and a perinuclear organelle (pno). A portion of cristate host mitrochondria (HM) is seen in close proximity to the trophozoite. (Magnification × 92,000.)

be incorporated as controls, they were able to define parasite-specific isoenzyme patterns within the cultures examined. Strain-specific patterns for glucosephosphate isomerase (GPI) were detected in both *T. parva* and *T. annulata* cell lines, but GPI was the only enzyme of 27 examined which gave interpretable results. The results were, however, sufficiently encouraging for this technique to be considered of merit in attempting to characterize field isolates of *Theileria* in epidemiological studies (Figure 18).

Pharmacological investigations with *Theileria*-infected cell lines are similarly in their infancy. However, in recent years, these have proved to be particularly rewarding. The first, inauspicious study, was that of Hawking,[92] who examined the effects of 40 different chemotherapeutic agents, mostly antimalarials or antibiotics, on a *T. annulata* culture without observing any specific antiparasitic activity. Nearly 20 years later Irvin and Stagg[93] reported on an elegant piece of work investigating purine and pyrimidine metabolism of cells infected with *T. parva*. This work clearly illustrated the specific anti-theilerial activity of the antifolates aminopterin and amethopterin, lending support to the in vivo results obtained with pyrimethamine.[22] Shortly after this, however, long overdue developments in the field of theilerial chemotherapy occurred. These were initiated primarily by McHardy[94] and McHardy et al.[95] who, using a cell line infected with *T. parva* (Muguga), showed first that the naphthoquinone, menoctone, and subsequently a series of analogues of this antimalarial, were highly effective in selectively killing *T. parva* schizonts within lymphoid cells in vitro.

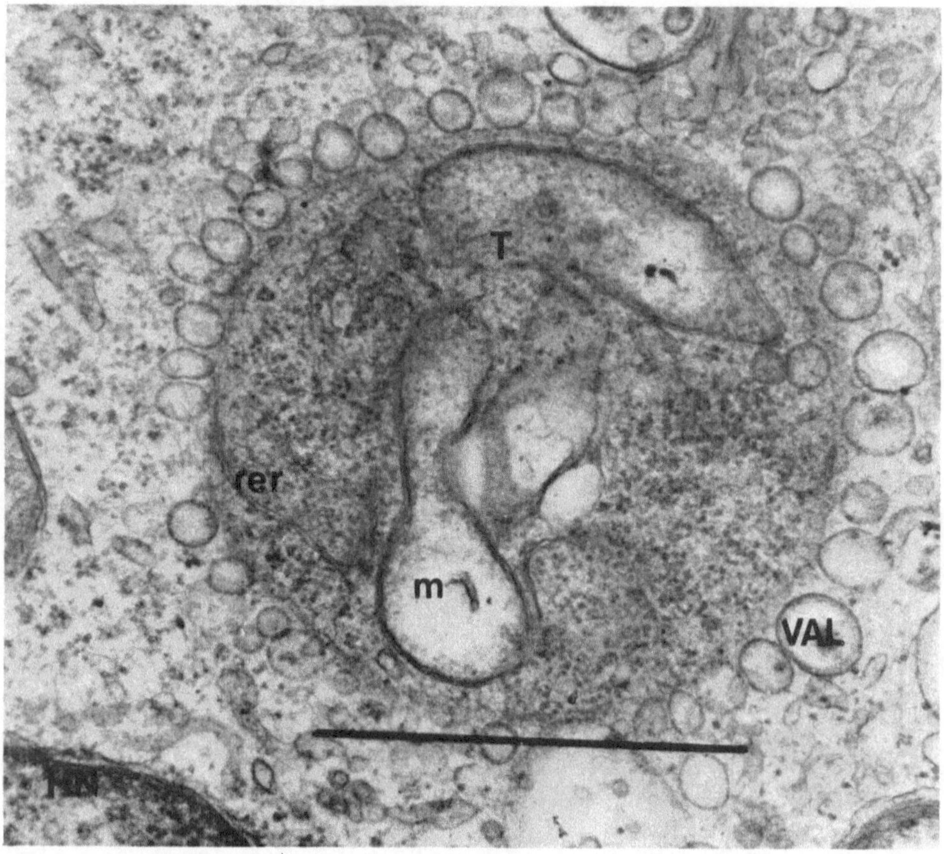

FIGURE 10. 18 hr after infection. A trophozoite (T) in the 'nuclear hof' region of the parasitized host lymphocyte is tightly surrounded by vesiculate annulate lamellae (VAL) and manifests mitochondria (m) and well-developed rough endoplasmic reticulum (rer) continuous with the pellicle. The parasitized lymphocyte is still in interphase as shown by its nucleus (HN). (Magnification × 71,000.)

This result was confirmed in studies on cattle conducted by Dolan and McHardy[96] who were able to compare in vitro efficacy of menoctone and amethopterin and obtain gratifying correlation of their results between the two systems.

A second group of compounds, the quinazilonones, represented by the coccidiostat halofuginone, a synthetic febrifugine were examined for chemotherapeutic activity against *T. parva* and *T. annulata* infections of cattle. Halofuginone, a coccidiostat active against *Leucocytozoon* infections was shown by Schein and Voigt[97] to be highly effective in curing clinical theileriosis. The drug, when examined in vitro by Uilenberg et al.[98] and McHardy and Morgan[99] was shown to exert maximum activity at the dose of 0.015 mg/ℓ, similar to that of 993C, the most active analogue of menoctone. The role played by theilerial cell cultures in the identification of novel chemotherapeutic agents for the treatment of theileriosis has been comprehensively emphasized in the review by Dolan.[100] Activity against *T. parva* and *T. annulata* has now been identified in vitro for three series of compounds — the quinones, including menoctone and 993C, the febrifuginones, notably halofuginone, and the ionophores, monensin and lasalocid.

For many years workers have been attempting to produce a small animal model for the study of theileriosis. In the course of this work they have attempted to infect laboratory animals with *Theileria*. While the development of a cell culture system has to some extent reduced the need for a small animal model it also provided a source of material with which

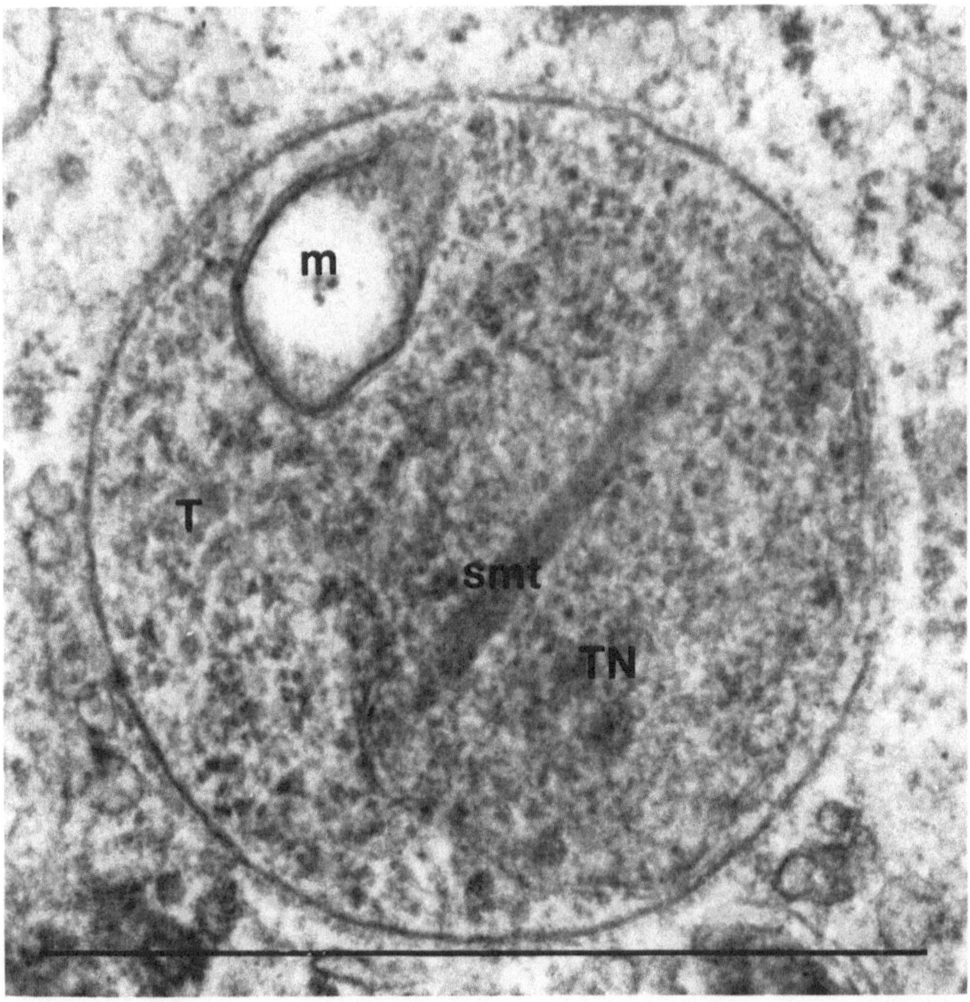

FIGURE 11. Trophozoite (T) after 24 hr incubation is in the process of developing into a schizont, as depicted by the development within its nucleus (TN) of anisotropic spindle microtubules (smt). (Magnification × 120,000, m = mitochondrion.)

to create an in vivo laboratory model. Irvin et al.[101] showed that culture-derived *T. parva*-infected bovine lymphoid cells are tumorigenic in immunodeficient mice. Lesions produced by the subcutaneous inoculation of 5 × 10^7 such cells in the optimal system of athymic nude mice irradiated with 800 rad were more invasive than those produced by inoculation of the same number of cells from a bovine lymphosarcoma cell line.[102] However, there was no evidence for the transfer of schizonts from the inoculated cells to those of the host. In an attempt to introduce *T. parva* schizonts into immune cells Irvin et al.[103,104] attempted to fuse cells in vitro using Sendai virus. They were able to induce heterokaryon formation between bovine cells infected with *T. parva* and mouse and hamster cells but were not able to select and grow infected synkaryons.

More recently Brown and Gray[105] have been able to infect bovine fibroblastic and other nonlymphoid cells with *Theileria* and to establish and maintain such infected cell lines in vitro. Using phenotypically different cells of this type they have been able to effect and study the transfer of schizonts from cell to cell in a culture system. It is hoped that this technique will answer a number of questions concerning the use of *Theileria*-infected cell cultures to immunize cattle against theileriosis. At present it is believed that, in order to

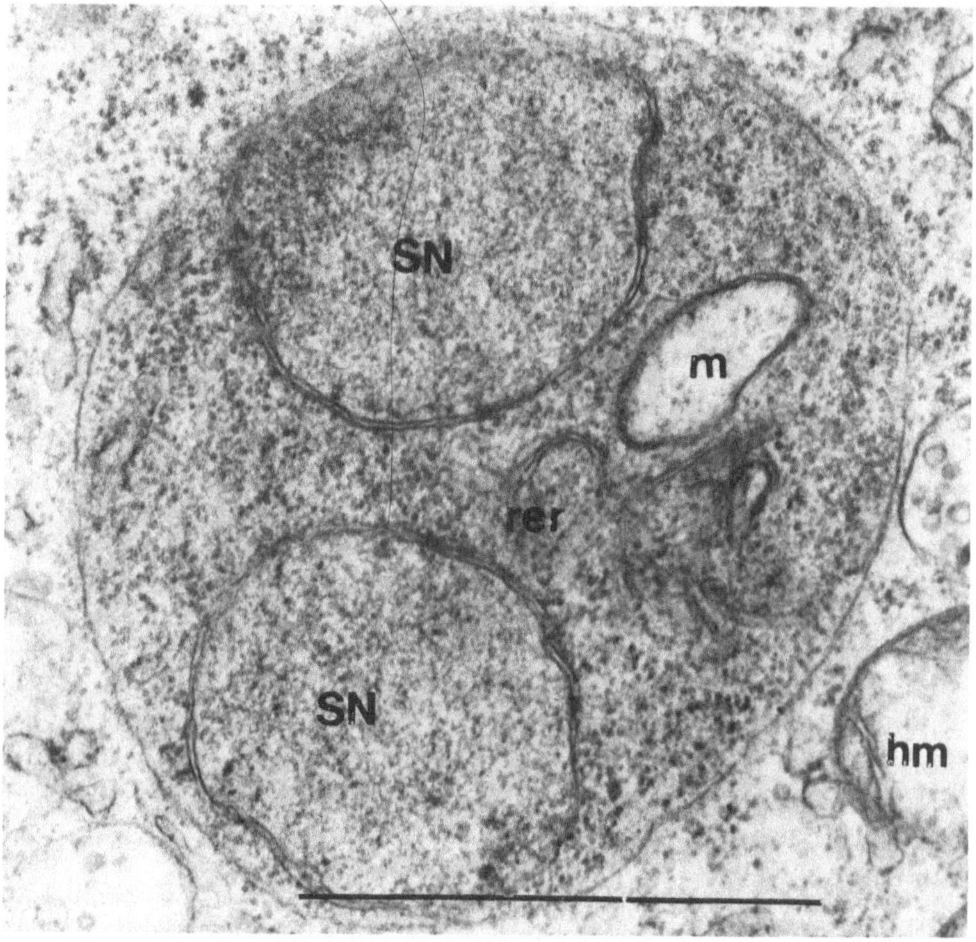

FIGURE 12. Specimen fixed and processed after 24 hr incubation showing a binucleate schizont resulting from the closed mitosis shown in Figure 11. The schizont at this stage is just slightly larger than a trophozoite and possesses nuclei (SN), rough endoplasmic reticulum (rer) and mitochondria (m). (Magnification × 71,000, hm = host mitochondrion.)

immunize cattle, it is necessary for the parasite to establish within the cells of the host.[87,106] Inoculating viable, infective cell cultures into cattle and using karyotypic analysis to monitor schizont transfer, it has been seen that, while schizonts of *T. annulata* transfer within hours from inoculated male cells to the female cells of the recipient,[22] *T. parva* schizont transfer is very inefficient and may take many days.[52] It was postulated that, with *T. parva* at least, acceptance of the parasite was dependent on its presentation and the histocompatibility antigens of the cells in which it was invested.[87,106] It has now been shown that lines of bovine lymphoid cells infected with either *T. parva* or *T. annulata* manifest histocompatibility BoLA antigens on their surface.[107] Recently, Morrison et al.[108] showed that cattle could be infected and immunized (it was assumed that one was the result of the other) by as few as 10^2 autologous *T. parva*-infected cells from a cell line produced by in vitro infection and transformation of bovine lymphocytes.[57] To achieve the same effect with allogeneic cell lines these workers needed 10^7 cells. Such a phenomenon has not been reported for *T. annulata* where there is no evidence that the parasite invested in allogeneic cells is any less infective than in a cell line autologous to the recipient. These results have provided one reason for the limited success in developing a cell culture vaccine against ECF whereas this is the standard method of controlling tropical theileriosis.

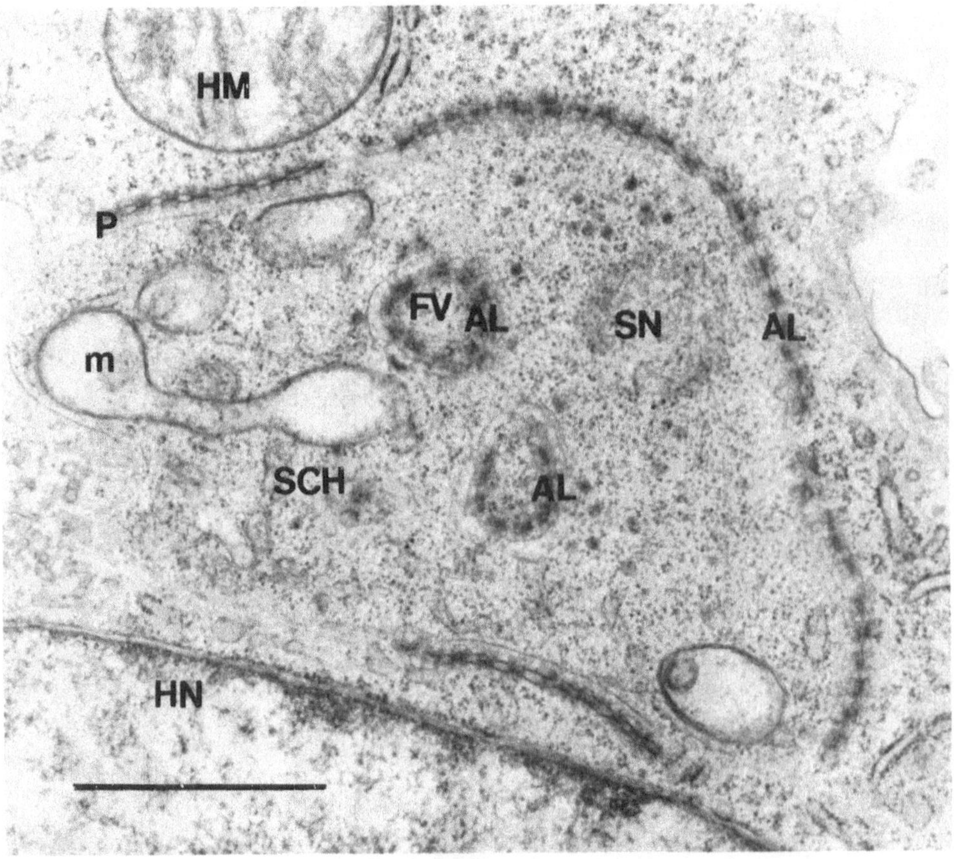

FIGURE 13. Infected and transformed lymphocyte after 6 days in culture with interphase nucleus (HN), showing a well developed schizont (SCH) almost completely surrounded by annulate lamellae (AL) closely applied to the schizont pellicle (P). The chunks of annulate lamellae taken in by phagotrophy are seen in the food vacuole (FV) of the schizont. (Magnification × 43,000. SN = schizont nucleus, m = schizont mitochondrion, HM = host mitochondrion.)

The role of the bovine MHC in recognition of theilerial antigens has been the subject of recent reviews of the immunological research on theileriosis.[109,110,111] The work reported in these reviews has confirmed the prediction by Hulliger et al.[112] that immunity to theileriosis is cell-mediated. Cell cultures have been used extensively in studies of both humoral and cellular mechanisms of immunity to theileriosis, the work developing from the pragmatic use of cultured schizont antigens for sero-diagnostic purposes to the development of in vitro correlates of cell-mediated immunity (CMI). Burridge and Kimber[113] have described the development of an indirect fluorescent antibody test (IFAT) using *T. parva* schizont cultures as antigen. Subsequently *T. lawrencei* antigen was produced from cell lines of this parasite grown in lymphoid cultures established from *Bos taurus* and *Syncerus caffer*.[114] The IFAT was also used to examine cross-reactions between cattle infected with *T. parva* and *T. annulata* using antigens of both organisms grown in suspension culture.[115] Similar bulk cultures of infected cells have been a source of masses of material for antigenic analysis.[116,117]

In studies of the effector mechanisms of immunity, and particularly CMI, cell culture systems have played a major role. While early examinations of this system involved the production of lymphokines, inhibition of migration and a study of antibody-dependent cell cytotoxicity,[118-121] an elegant series of experiments by workers at ILRAD has concentrated on the induction and specificity of cytotoxic T cells in *T. parva* infections. This work

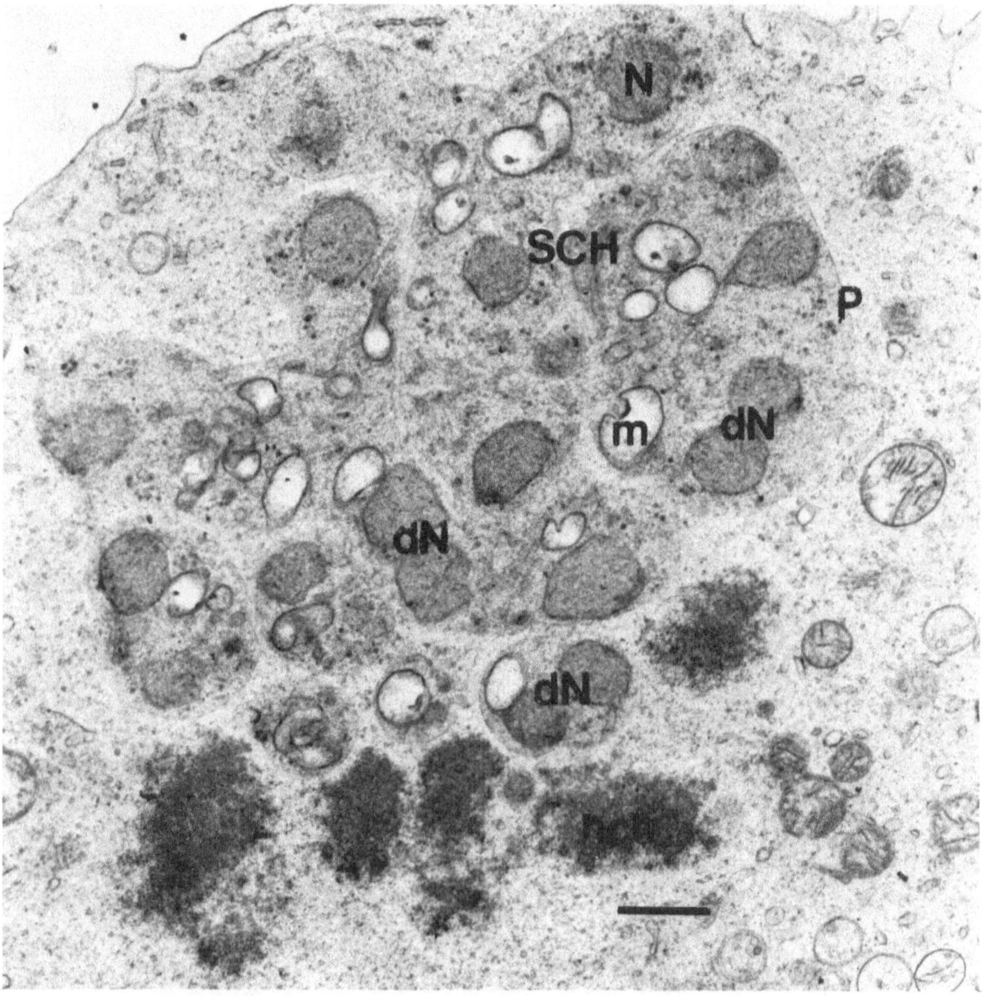

FIGURE 14. Established culture showing *T. annulata*-infected and transformed lymphocyte in prometaphase as depicted by condensed masses of chromosomal material (hch). At this stage many of the schizont nuclei are in an advanced stage of division (dN) without cleavage of the schizont pellicle (P). (Magnification × 12,000.)

depended on the technique of establishing autologous infected cell lines by in vitro infection to act as targets for cytotoxicity studies. Two outstanding investigations[122,123] have indicated the course for future studies. The first showed there are parasite-specific antigens on the surface of infected, transformed lymphoid cells which will induce a mixed lymphocyte reaction with autologous lymphocytes from peripheral blood and that this will induce the production of cytotoxic T cells if the donor animal is immune to ECF. The second produced conclusive evidence for genetic restriction in immunity to ECF. A series of papers by members of the ILRAD[27] group on the immunology of theileriosis has involved pioneering work on effector mechanisms of pathogenesis and immunity and the definition of parasite and cell surface antigens using monoclonal antibodies raised against cell cultures. It appears that parasite-specific monoclonal antibodies to *T. parva*[124] and *T. annulata*[125] may perhaps be used to distinguish strains of each of these organisms, proving invaluable in the selection of the antigenic mosaic necessary for immunization against theileriosis.

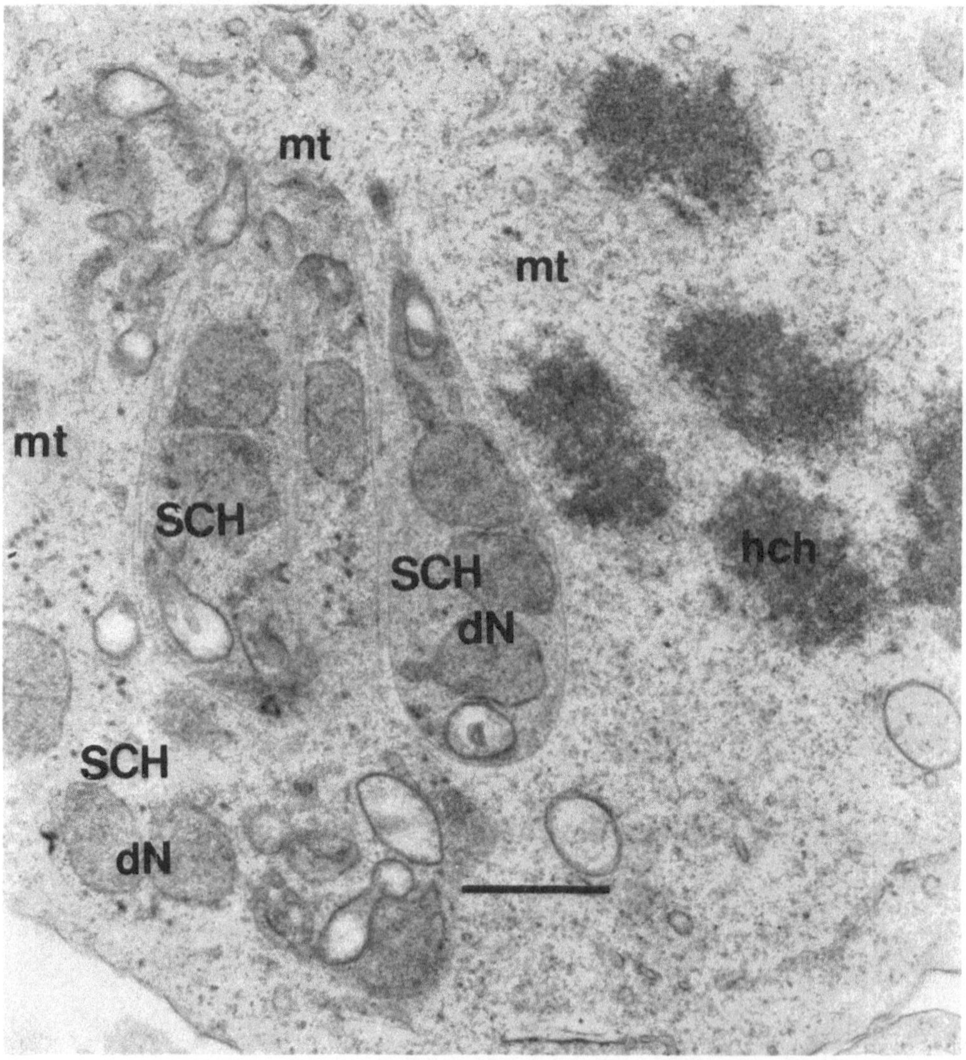

FIGURE 15. *T. annulata*-infected and transformed lymphocyte in metaphase. The condensed host cell chromosomal material (hch) and the pellicles of the individual schizonts (SCH) are already associated with spindle microtubules of the lymphocyte (mt). A number of parasite nuclei are in an advanced stage of division (dN). (Magnification × 20,000.)

V. UNSOLVED PROBLEMS AND THE PROSPECTUS

While there has been gratifying progress in the application of theilerial cell cultures in recent years, perhaps the development of culture technology *per se* has not moved at the same speed. To date the systems which have been developed are restricted most severely by the universal limiting factor of the need to incorporate serum and, for *T. parva* at least, fetal calf serum in the medium. John Hanks[126] has all too pointedly reminded us that the best culture media developed to date are merely supplements for fetal bovine serum. The development of truly synthetic, serum-free media such as those presently being studied by Iscove and Melchers[127] would enable long overdue critical biochemical work to be started.

A second significant limitation to the use and application of cell cultures to the study of theileriosis has been that, to date, only the schizonts have been cultivated. Moreover, with

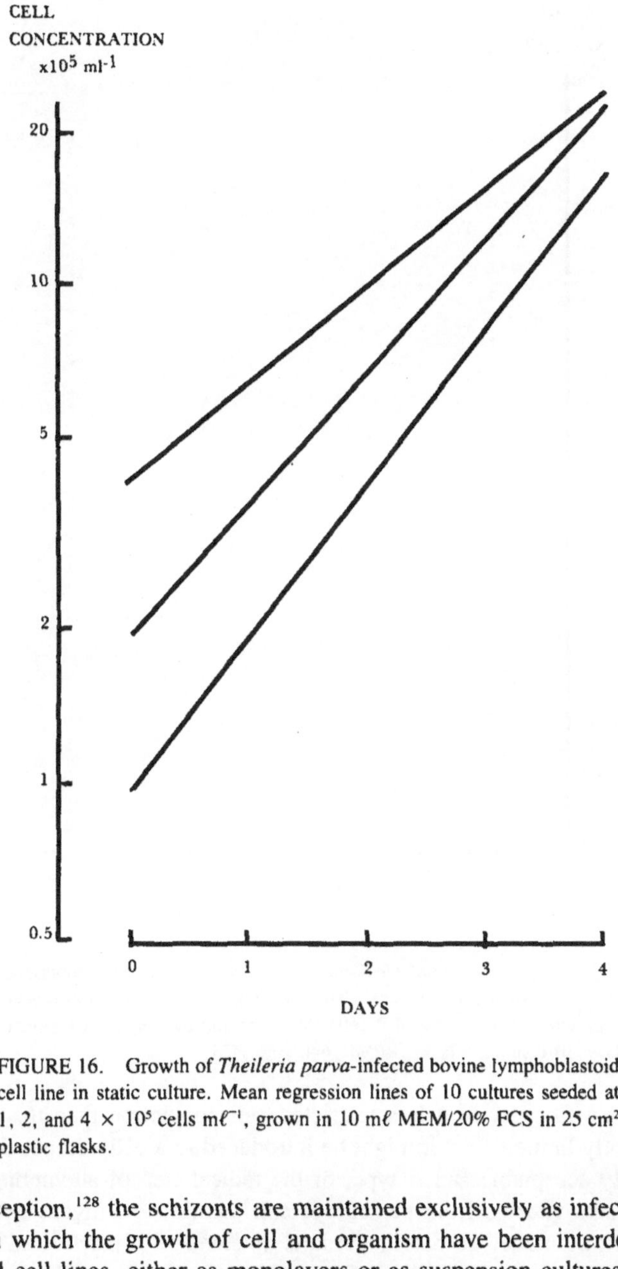

FIGURE 16. Growth of *Theileria parva*-infected bovine lymphoblastoid cell line in static culture. Mean regression lines of 10 cultures seeded at 1, 2, and 4 × 10^5 cells $m\ell^{-1}$, grown in 10 mℓ MEM/20% FCS in 25 cm^2 plastic flasks.

one reported exception,[128] the schizonts are maintained exclusively as infected lymphoblastoid cell lines in which the growth of cell and organism have been interdependent. These schizont-infected cell lines, either as monolayers or as suspension cultures, do have many applications however, the most notable of these being their use as vaccines, particularly against *T. annulata* infections. The work in this area, which has been comprehensively reviewed by Pipano[75,129,130,131] and Hooshmand Rad,[132] is extensive and laboratory and field trials of *T. annulata* cell culture vaccines have now been conducted in Tunisia,[70] Turkey,[74,133] Iraq,[132] Iran,[134] India,[73] USSR,[135] and China[72] in addition to Israel where the pioneering work was done. The vaccines are now applied in the field in a number of these countries. A preliminary trial with a *T. hirci* culture vaccine has also produced promising results in protecting sheep against malignant ovine theileriosis.[136] While vaccines against ECF using *T. parva*-infected cell lines have not proved successful[30,87] due, it has been inferred, to the need for the parasite to be present in an identifiable package, i.e., in autologous cells,[108] it seems feasible that this problem may be circumvented in one of three ways. The parasite

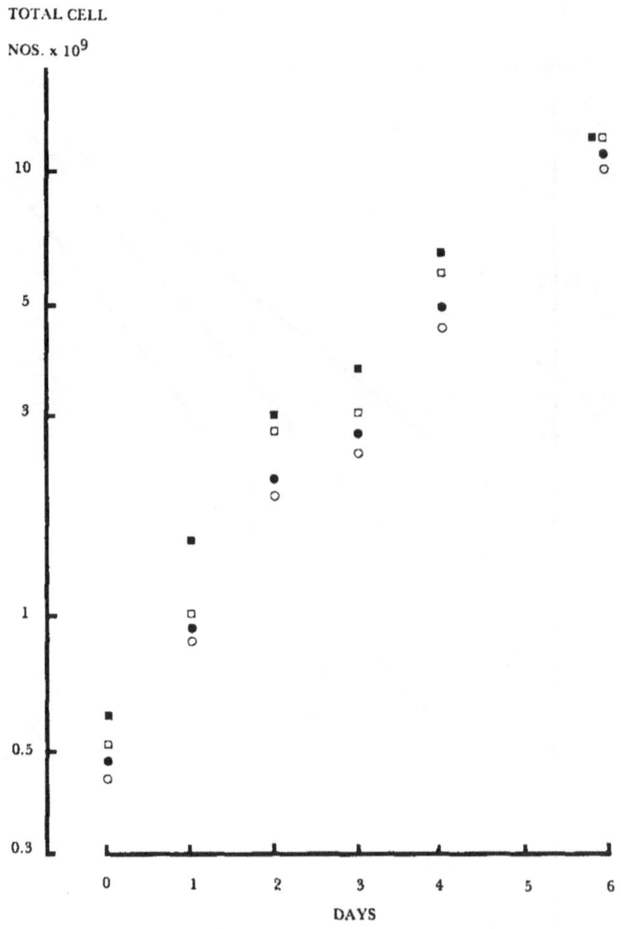

FIGURE 17. Growth of *Theileria parva*-infected bovine lymphoblastoid cell line in suspension culture. Cell yield from individual cultures seeded with approximately 5×10^8 cells (500 mℓ) and bulked to 10 ℓ over 6 days: 10 ℓ spinner flasks, RPMI 1640/10% FCS.

might be persuaded to grow in a less restricted cell population than the T cell subset to which it is presently limited[137,138] it might be introduced to a cell without BoLA antigens[107] or into universally acceptable BoLA type; or the radical step of attempting to produce an inactivated protozoal vaccine from parasite or cell membrane antigens derived from such cell cultures might be made. To date the results obtained with inactivated antigens administered with adjuvants have been particularly disappointing for both *T. parva*[139] and *T. annulata*.[140] While in both systems high antibody titres were induced, no effective protection against tick-derived challenge was observed. Moreover, with *T. annulata* the administration of such inactivated antigens effectively 'blocked' immunization with living cell culture. There is thus a considerable way to go before defined antigens can be identified and selected, perhaps with monoclonal antibodies and used as inactivated vaccines. If protection could be induced with such antigens there would perhaps be an indication for recombinant DNA technology at some time in the future.

For immunological studies on theileriosis the development of such a system in which lymphoid cells could be infected and transformed by sporozoites in vitro has perhaps indicated an approach which is already being examined by workers in other fields. The need to return continually to the tick and the sporozoite of *T. parva* was forced on those researchers studying

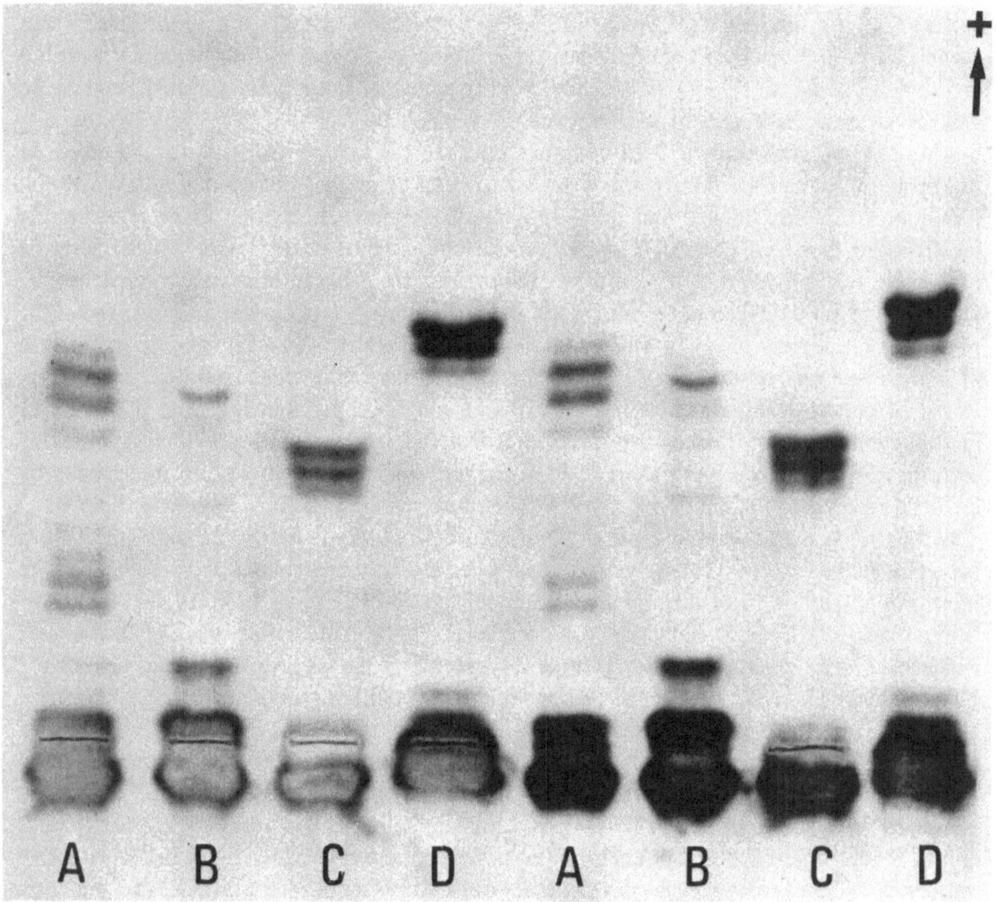

FIGURE 18. Comparison of the GPI isoenzyme pattern in bovine lymphoblastoid cells infected with four stocks of *T. annulata* from different geographical areas. (A) *T. annulata* (Ankara) Turkey. (B) *T. annulata* (Hissar) India. (C) *T. annulata* (Soba) Sudan. (D) *T. annulata* (Razi S3) Iran. (Courtesy of Mr. T. R. Melrose, 22.5.81.)

ECF since they were not able to passage the disease mechanically. This meant that this parasite was always studied in its natural biological system, the infective form of the parasite from the vector tick being of paramount importance.[141] Malariologists are now using a similar system, returning to the sporozoite from the mosquito as their fundamental antigen,[142] immunogenic[143] and infective material.[144] As a result, the cultivation of exoerythrocytic forms of the mammalian plasmodium has once more become a tangible prospect.[145] An in vitro system such as that would provide the opportunity for intensive evaluation of antisporozoite agents, in which field malariologists have perhaps shown the theilerial workers the way ahead, with irradiated sporozoite vaccines,[143] monoclonal antibodies against sporozoites[142] and merozoites[146] and real prospects of a genetically engineered vaccine.[147]

A fascinating product of the in vitro infection technique has been the recent study of the development of *Babesia equi*.[148] In this, sporozoites of *B. equi*, harvested from the tick *H. anatolicum excavatum*, were shown to have an exoerythrocytic cycle, infecting equine lymphocytes in vitro before undergoing a schizogonous cycle which results in *Theileria*-like transformation of the infected cells culminating in the release of merozoites which invade erythrocytes. This not only throws open once more the contention over the classification of the piroplasms but provides support for the argument that such a system lends itself to fundamental study of many vector-borne organisms. It also exemplifies two further limitations of the *Theileria* culture systems; namely that many *Theileria spp.* have yet to be cultured in vitro, notably *T. mutans* and *T. sergenti*, not to mention some of the contentious organisms in limbo somewhere between the *Theileria* and the *Babesia* (including *Nuttallia = Babesia danii, Babesia microti*, and others); and that cultivation systems need to be developed for many stages of the life cycle of *Theileria*.

A number of workers have tried without success to culture in vitro stages of the theilerial life cycle other than the intralymphocytic schizont. Attempts to induce merogony with the formation of microschizonts and release of merozoites have frequently given encouraging results which then have not proved to be reproducible. Hulliger et al.[149] raised the temperature of their cultivation system to trigger the transformation from macroschizont to microschizont. Danskin and Wilde[150] added bovine lymph to their cultures. Jongejan et al.[151] increased the frequency of medium changes with the same objective in mind. In each instance the release of merozoites was observed but of a frequency less than that often recorded in the early stages of culture transformation before schizont and cell have synchronized their rate of multiplication. In no instance have merozoites from schizont cultures convincingly infected erythrocytes to produce piroplasms in vitro, neither has it been possible to infect ticks by artificial feeding or inoculation[151,152] of such culture material.

Not only has the vertebrate cycle of *Theileria* not yet been completed in vitro, there have been few studies of the complex yet potentially rewarding development cycle of the parasite in arthropod culture. The technology of tick cell culture has recently been reviewed most comprehensively.[153,154] Systems for growing tick cells are in their infancy but good techniques now exist for producing cell lines from embryonic[155,156] and moulting nymphal ticks.[157,158] Unfortunately, though a number of *R. appendiculatus* cell lines have now been established, *Hyalomma spp.* have proved very difficult to culture and, to date, have only been maintained as prolonged primary cultures (L. Bell, M. Samish personal communication). However, in the short-term at least, it seems likely that cell lines may not be the answer to the production of invertebrate stages of *Theileria* in vitro. Hadani[159] from his studies on *N. danii*[160] has argued that an organ culture system might possibly be more rewarding, since the invertebrate protozoan cycle is complex and development occurs in very different tissues, ranging from gut to salivary gland. Coincidentally, such a system has been used for the production of metacyclic trypanosomes in this laboratory, using organ cultures of tsetse flies to initiate development of the protozoan. Massive numbers of trypanosomes infective for the vertebrate host have been produced from cultures which were started from infected *Glossina* salivary glands — for *T. brucei*[161] — and proboscides — for *T. congolense*.[162] Development of tick culture has not yet proved so fruitful but, in a tick organ culture system, Bell[163] has been able to harvest kinetes of *Theileria* migrating in supernatant culture medium from gut to salivary gland. Such a system provides the potential for cloning these parasites, a technique essential for the recombinant studies which are fundamental to the examination of the genetics of *Theileria* and the immunology of the diseases they cause.[164] Moreover, the production of massive numbers of sporozoites from arthropod cultures would make the development of a sporozite vaccine more feasible. It is already possible to demonstrate in vitro the effect of irradiation on the infectivity of sporozoites for target lymphocytes.[30] Given an arthropod culture system it would then be possible to complete the life cycle of these complex and fascinating organisms in vitro.

VI. APPENDIX: SUPPLIERS OF TISSUE CULTURE MATERIALS AND EQUIPMENT

- Baird & Tatlock Ltd., P. O. Box 1, Romford, RM1 1HA, England.
  LTE and BTL incubators.
- Becton-Dickinson UK Ltd., York Hse., Empire Way, Wembley, Mddx., England.
  Plastipak disposable syringes.

- Bellco Glass Inc., Vineland, New Jersey, 08360, U.S.A.
  Glassware for T.C., pipettes, centrifuge tubes, culture flasks, etc., apparatus including laminar flow hoods, magnetic stirrers.
- British Oxygen Company Ltd., Seafield Road, Edinburgh, Scotland.
  Compressed gases, medical grade and liquid nitrogen.
- Camlab Ltd., Nuffield Road, Cambridge, England.
  Micro-osmometer.
- Costar, 205 Broadway, Cambridge, Massachusetts, U.S.A.
  Tissue culture, multiwell plates and Cuprak cloning dishes.
- Coulter Electronics Ltd., Harpenden, Herts., England.
  Coulter, ZB1 electronic particle counter.
- Decon Labs. Ltd., Ellen St., Portslade, Brighton BN4 1EQ, England.
  Decon 75 and 90 detergents.
- Evans Medical Ltd., Speke, Liverpool, England.
  Pularin Heparin B.P. (mucous) without preservative.
- Falcon Plastics, c/o Becton-Dickinson, Empire Way, Wembley, England.
  Disposable tissue culture flasks.
- Fisons Scientific Apparatus, Bishop Meadow Rd., Loughborough, Leics., LE11 ORG, England.
  Water stills Fi-streem 8 and Fi-streem 4BD.
- Gibco Europe, 3 Washington Rd., Paisley, PA3 4EP, Scotland.
  Tissue culture media, sera, antibiotics, colcemid, agents for Nunc disposable flasks.
- E. Leitz Wetzlar, GmbH, Postfach 2020, D-6330 Wetzlar, W. Germany.
  Leitz Diavert inverted microscope, Orthoplan microscope.
- E. Merck, Darmstadt, W. Germany.
  Giemsa and Azure 11 stains.
- Microflow Ltd., Fleet Mill, Minley Rd., Fleet, Aldershot, Hants., GU13 8RD, England.
  Laminar flow cabinet - Microflow Laminar Flow Work Station.
- Millipore Corporation, Bedford, Mass., 01730, U.S.A.
  Filters, sterilizing filtration systems, water deionizers Milli-Q, Milli-RO.
- M.S.E. Scientific Instruments, Manor Royal, Crawley, Sussex, England.
  Centrifuges and centrifuge tubes.
- New Brunswick Scientific Co. Inc., 44 Talmadge Rd., Edison, N.J., 08817, U.S.A.
  Fermentors, suspension culture systems.
- Nunc, c/o Gibco-Biocult in U.K.
  Plastic disposable flasks, tubes, plates, pipettes, etc.
- Pharmacia, Uppsala, Sweden.
  Ficoll paque, percoll for differential centrifugation of cells.
- Shandon Southern Instruments Ltd., Chadwick Rd., Astmoor, Runcorn, Cheshire WA7 1PR, England.
  Shandon cytospin cytocentrifuge.
- Sherwood Medical Industries Inc., Deland, Fla. 32720, U.S.A.
  Disposable syringes and needles.
- Sigma Chemical Co., Box 14508, St. Louis, Mo. 63178, U.S.A.
  Fine chemicals, amino acids, bovine plasma albumin.
- E. R. Squibb and Sons Inc., New York (N.Y.) 10022, U.S.A.
  Nystatic (Mycostatin) fungistat.
- Sterilin Ltd., 43-45 Broad St., Teddington TW11 8QZ, England.
  Disposable plastic ware - centrifuge tubes, universals, Bijoux.
- TAAB Lab. Equip. Ltd., 40 Grovelands Rd., Reading, Berks., England.
  Materials for electron microscopy.

- Union Carbide U.K. Ltd., Cryogenics Dept., Redworth Way, Aycliffe Industrial Estate, Darlington, Durham, England.
  Cryogenic containers LR40, LR19S.
- Whatman Ltd., Springfield Mill, Maidstone, Kent, ME14 2LE, England.
  Cellulose filters, in-line air filters.
- Wheaton Scientific, Millville, N.J. 08332, U.S.A.
  Cryogenic Cryule ampoules and vials, Celstir spinner flasks for suspension culture.
- Winthrop Labs., Surbiton, Surrey, England.
  Kannusyn - kanamycin.

REFERENCES

1. **Levine, N. D., et al.,** A newly revised classification of the Protozoa, *J. Protozool.*, 27, 37, 1980.
2. **Uilenberg, G.,** Theilerial species of domestic livestock, in *Advances in the Control of Theileriosis*, Irvin, A. D., Cunningham, M. P., and Young, A. S., Eds., Martinus Nijhoff Publishers, The Hague, 1981, 4.
3. **Morel, P. C. and Uilenberg, G.,** The nomenclature of some *Theileria* species (Sporozoa, Babesioidea) of domestic ruminants, *Rev. Ele. Med. Vet. Pays Trop.*, 34, 139, 1981.
4. **Henning, M. W.,** *Animal diseases in South Africa*, 3rd ed., Central News Agency, Cape Town, 1956, 593.
5. **Gutsche, T.,** *There Was Man. The Life and Times of Sir Arnold Theiler K.C.M.G. of Onderstepoort*, Howard Timmins, Cape Town, 1979.
6. **Bettencourt, A., Franca, C., and Borges, J.,** Un cas de piroplasmose bacilliforme chez le daim, *Arq. Inst. Bacteriol. Camara Pestana (Lisbon)*, 1, 341, 1907.
7. **Dschunkowsky, E. and Luhs, J.,** Die piroplasmosen der Rinder, *Zentralblatt Bakteriol. Parasitkd. Infek.*, 35, 486, 1904.
8. **Koch, R.,** Weiterer Bericht uber das Texasfieber, *Centralblatt Bakteriol. Parasit. Infekt.*, 1, 24, 202, 1898.
9. **Lounsbury, C. P.,** Transmission of African Coast fever, *Agric. J. Cape Good Hope*, 24, 428, 1904.
10. **Hutcheon, D.,** The new form of red-water in the Transvaal, *Transvaal Agric. J.*, 1(3), 45, 1903.
11. **Koch, R.,** Beitrage zur Entwickelungsgeschichte der Piroplasmen, *Z. Hyg. Infekt.*, 54, 1, 1906.
12. **Gonder, R.,** The life cycle of *Theileria parva*, the cause of East Coast fever in cattle in South Africa, *J. Comp. Pathol. Ther.*, 23, 329, 1910.
13. **Cowdry, E. V. and Ham, A. W.,** Studies on East Coast fever. I. Life cycle of the parasite in ticks, *Parasitology*, 24, 1, 1932.
14. **Schein, E.,** On the life cycle of Theileria annulata (Dschunkowsky and Luhs, 1904) in the mid-gut and haemolymph of *Hyalomma anatolicum* excavatum (Koch, 1844), Z. *Parasitenk.*, 47, 165, 1975.
15. **Schein, E., Warnecke, M., and Kirmse, P.,** Development of *Theileria parva* (Theileri, 1904) in the gut of *Rhipicephalus appendiculatus* (Neumann, 1901), *Parasitology*, 75, 309, 1977.
16. **Burridge, M. J., Morzaria, S. P., Cunningham, M. P., and Brown, C. G. D.,** Duration of immunity to East Coast fever *(Theileria parva)* infection of cattle, *Parasitology*, 64, 511, 1972.
17. **Neitz, W. O.,** Theileriosis, gonderiosis and cytauxzoonosis: a review, *Onderstepoort J. Vet. Res.*, 27, 275, 1957.
18. **Neitz, W. O.,** Theilerioses, in *Advances in Veterinary Science*, Brandly, C. A. and Jungherr, E. L., Eds., Vol. 5, Academic Press, New York, 1959, 241.
19. **Barnett, S. F.,** Theileriasis, in *Infectious Blood Diseases of Man and Animals*, Weinman, D. and Ristic, M., Eds., Vol. 2, Academic Press, New York, 1968, 269.
20. **Barnett, S. F.,** *Theileria*, in *Parasitic Protozoa*, Kreier, J. P., Ed., Vol. 4, Academic Press, New York, 1977, 77.
21. **Sergent, E., Donatien, A., Parrot, L., and Lestoquard, F.,** *Études sur les piroplasmoses bovines*, Institut Pasteur d'Algerie, Algiers, 1945.
22. **Wilde, J. K. H.,** East Coast fever, in *Advances in Veterinary Science*, Brandly, C. A. and Cornelius, C., Eds., Vol. 11, Academic Press, New York, 1967, 207.
23. **Purnell, R. E.,** East Coast fever: some recent research in East Africa, *Adv. Parasitol.*, 15, 83, 1977.
24. F.A.O., East Coast fever and related tick-borne diseases, *F.A.O. Animal Production and Health Paper 19*, Food and Agriculture Organization of the United Nations, Rome, 1980.
25. **Wilde, J. K. H., Ed.,** *Tick-Borne Diseases and their Vectors*, Centre for Tropical Veterinary Medicine, University of Edinburgh, 1978.

26. **Henson, J. B. and Campbell, M., Eds.,** *Theileriosis*, ILRAD/IDRC, International Development Research Centre, Ottawa, 1977.
27. **Irvin, A. D., Cunningham, M. P., and Young, A. S.,** *Advances in the Control of Theileriosis*, Martinus Nijhoff Publishers, The Hague, 1981.
28. **Brown, C. G. D.,** Propagation of *Theileria*, in *Practical Tissue Culture Applications*, Maramorosch, K. and Hirumi, H., Eds., Academic Press, New York, 1979, 223.
29. **Brown, C. G. D.,** *In vitro* cultivation of *Theileria*, in *The in vitro Cultivation of the Pathogens of Tropical Diseases*, Rowe, D. S. and Hirumi, H., Eds., Schwabe and Co. AG, Basel, 1980, 127.
30. **Brown, C. G. D.,** Application of *in vitro* techniques to vaccination against Theileriosis, in *Advances in the Control of Theileriosis*, Irvin, A. D., Cunningham, M. P., and Young, A. S., Eds., Martinus Nijhoff Publishers, The Hague, 1981, 104.
31. **Trager, W. and Jensen, J. B.,** Cultivation of malaria parasites, *Nature (London)*, 273, 621, 1978.
32. **Levy, M. G. and Ristic, M.,** *Babesia bovis:* continuous cultivation in a microaerophilous stationary phase culture, *Science*, 207, 1218, 1980.
33. **Nuttall, G. H. F. and Graham-Smith, G. S.,** *Theileria parva:* Attempts at cultivation, *Parasitology*, 2, 208, 1909.
34. **Nuttall, G. H. F. and Graham-Smith, G. S.,** The development of *Piroplasma canis* in culture, *Parasitology*, 1, 243, 1908.
35. **Tsur (Tchernomoretz), I.,** Multiplication *in vitro* of Koch bodies of *Theileria annulata, Nature (London)*, 156, 391, 1945.
36. **Tsur, I.,** Multiplication *in vitro* of Koch bodies of *Theileria annulata, Refu. Vet.*, 4, 86, 1947.
37. **Tsur, I.,** *Theileria annulata* et *Leishmania* en culture de tissu, 15th Int. Veterinary Congr., Stockholm, 1, 26, 1953.
38. **Tsur, I. and Adler, S.,** Cultivation of *Theileria annulata* schizonts in monolayer tissue cultures, *Refu. Vet.*, 19, 225, 1962.
39. **Tsur, I. and Adler, S.,** The cultivation of lymphoid cells and *Theileria annulata* schizonts from infected bovine blood, *Refu. Vet.*, 22, 62, 1965.
40. **Tsur, I., Adler, S., Pipano, E., and Senft, Z.,** Continuous growth of *Theileria annulata* schizonts in monolayer tissue cultures, Proc. 1st Int. Congr. Parasitology, Rome, 1, 266, 1964.
41. **Tsur, I. and Pipano, E.,** Growth and multiplication of Koch bodies of *Theileria annulata* in rodents' spleen tissue culture, *Refu. Vet.*, 16, 51, 1959.
42. **Tsur, I. and Pipano, E.,** The successful infection of cattle with *Theileria annulata* parasites preserved in the frozen state, *Refu. Vet.*, 19, 110, 1962.
43. **Tsur, I. and Pipano, E.,** Attenuation of *Theileria annulata* by growth and passage through tissue culture, *J. Protozool.*, 13, 33, 1966.
44. **Pipano, E. and Tsur, I.,** Experimental immunization against *Theileria annulata* with a tissue culture vaccine. I. Laboratory trials, *Refu. Vet.*, 23, 186, 1966.
45. **Tsur, I., Neitz, W. D., and Pols, J. W.,** The development of Koch bodies of *Theileria parva* in tissue culture, *Refu. Vet.*, 14, 53, 1957.
46. **Brocklesby, D. W. and Hawking, F.,** Growth of *Theileria annulata* and *T. parva* in tissue culture, *Trans. R. Soc. Trop. Med. Hyg.*, 52, 414, 1958.
47. **Hulliger, L., Wilde, J. K. H., Brown, C. G. D., and Turner, L.,** Mode of multiplication of *Theileria* in culture of bovine lymphocytic cells, *Nature (London)*, 203, 728, 1964.
48. **Henle, G. and Henle, W.,** Immunofluorescence, interference and complement fixation techniques in the detection of herpes-type virus in Burkitt tumor cell lines, *Cancer Res.*, 27, 2442, 1967.
49. **Churchill, A. E. and Biggs, P. M.,** Agent of Marek's disease in tissue culture, *Nature (London)*, 215, 528, 1967.
50. **Hulliger, L.,** Cultivation of three species of *Theileria* in lymphoid cells *in vitro, J. Protozool.*, 12, 649, 1965.
51. **Malmquist, W. A., Nyindo, M. B. A., and Brown, C. G. D.,** East Coast fever: cultivation *in vitro* of bovine spleen cell lines infected and transformed by *Theileria parva, Trop. Anim. Health Prod.*, 2, 139, 1970.
52. **Brown, C. G. D., Malmquist, W. A., Cunningham, M. P., Radley, D. E., and Burridge, M. J.,** Immunization against East Coast fever. Inoculation of cattle with *Theileria parva* schizonts grown in cell culture, *J. Parasitol.*, 57(4), 59, 1971.
53. **Moulton, J. E., Krauss, H. H., and Malmquist, W. A.,** Growth characteristics of *Theileria parva* — infected bovine lymphoblast cultures, *Am. J. Vet. Res.*, 32, 1365, 1971.
54. **Moulton, J. E., Krauss, H. H., and Malmquist, W. A.,** Transformation of reticulum cells to lymphoblasts in cultures of bovine spleen infected with *Theileria parva, Lab. Invest.*, 24, 187, 1971.
55. **De Martini, J. C. and Moulton, J. E.,** Responses of the bovine lymphatic system to infection by *Theileria parva*. I. Histology and ultrastructure of lymph nodes in experimentally infected calves, *J. Comp. Pathol.*, 83, 281, 1973.

56. **Malmquist, W. A. and Brown, C. G. D.,** Establishment of *Theileria parva*-infected lymphoblastoid cell lines using homologous feeder layers, *Res. Vet. Sci.*, 16, 134, 1974.
57. **Brown, C. G. D., Stagg, D. A., Purnell, R. E., Kanhai, G. K., and Payne, R. C.,** Infection and transformation of bovine lymphoid cells *in vitro* by infective particles of *Theileria parva*, *Nature (London)*, 245, 101, 1973.
58. **Brown, C. G. D.,** *In vitro* infection and transformation of lymphoid cells by sporozoites of *Theileria parva* and *T. annulata*, *J. S. Afr. Vet. Association*, 50, 345, 1979.
59. **Kurtti, T. J. and Munderloh, U. G.,** Factors influencing the *in vitro* establishment of bovine lymphoblastoid cells infected with *Theileria parva*, in *Proc. XI Int. Congr. Diseases of Cattle*, Mayer, E., Ed., Tel Aviv, 1980, 684.
60. **Kurtti, T. J., Munderloh, U. G., Irvin, A. D., and Buscher, G.,** *Theileria parva:* early events in the development of bovine lymphoblastoid cell lines persistently infected with macroschizonts, *Exp. Parasitol.*, 52, 280, 1981.
61. **Stagg, D. A., Dolan, T. T., Leitch, B. L., and Young, A. S.,** The initial stages of infection of cattle cells with *Theileria parva* sporozoites *in vitro*, *Parasitology*, 83, 191, 1981.
62. **Watson, P.,** A slide centrifuge: an apparatus for concentrating cells in suspension onto a microscope slide, *J. Lab. Clin. Med.*, 68, 494, 1966.
63. **Zablotsky, V. T.,** The biological properties of the pathogen of theileriasis of cattle grown in cell culture, in *Progress in Protozoology*, Nauka, Leningrad, 1969, 277.
64. **Zablotsky, V. T.,** Specific prophylaxis of bovine theileriosis, *(Theileria annulata* infection), in *Arachnid Infestations and Protozoal Diseases of Farm Animals*, Stepanova, N. I., Ed., (in Russian), Usesoyuznyi Institu Eksperimental'noi Veterinarii, Moscow, 1977, 121.
65. **Konyukhov, M. P.,** The interrelation of *Theileria* and lymphoid cells of lymph nodes in tissue culture, in *Progress in Protozoology*, Nauka, Leningrad, 1969, 284.
66. **Konyukhov, M. P.,** Experience in cultivating *Theileria annulata* in cell culture of lymphoid tissue from lymph nodes, in *Progress in Protozoology*, Nauka, Leningrad, 1969, 284.
67. **Mutuzkina, Z. P.,** Cultivation of *Theileria annulata* in tissue cultures, *Veterinariya*, 4, 56, 1975.
68. **Mutuzkina, Z. P.,** Culture and morphological and cytogenetic characteristics of cells infected with *Theileria annulata*. Antigenic properties of the parasite in the slow complement fixation test, *Byull. Vses. Inst. Eksp. Vet.*, 31, 21, 1977.
69. **Hooshmand Rad, P. and Hashemi Fesharki, R.,** The effect of virulence on cultivation of *Theileria annulata* strains in lymphoid cells which have been cultured in suspension, *Arch. Inst. Razi*, 20, 85, 1968.
70. **van den Ende, M. and Edlinger, E.,** Culture des lignées lymphocytaires bovines infectées par *Theileria annulata*, *Arch. Inst. Pasteur Tunis*, 1-2, 45, 1971.
71. **Hooshmand Rad, P.,** The growth of *Theileria annulata* infected cells in suspension culture, *Trop. Anim. Health Prod.*, 7, 23, 1975.
72. Gansu Provincial Institute of Veterinary Medicine. Studies of the cell culture and the immunogen of schizonts of *Theileria annulata*, *Acta Zool. Sinica*, 21, 243, 1975.
73. **Gill, B. S., Bhattacharyulu, Y., Kaur, D., and Singh, A.,** Vaccination against bovine tropical theileriasis *(Theileria annulata)*, *Nature (London)*, 264, 355, 1976.
74. **Ozkoc, E. and Pipano, E.,** Trials with cell culture vaccine against theileriosis in Turkey, in *Advances in the Control of Theileriosis*, Irvin, A. D., Cunningham, M. P., and Young, A. S., Eds., Martinus Nijhoff Publishers, The Hague, 1981, 256.
75. **Pipano, E.,** Basic principles of *Theileria annulata* control, in *Theileriosis*, Henson, J. B. and Campbell, M., Eds., I.D.R.C., Ottawa, 1977, 55.
76. **Stagg, D. A., Brown, C. G. D., Crawford, J. G., Kanhai, G. K., and Young, A. S.,** *In vitro* cultivation of *Theileria lawrencei*-infected lymphoblastoid cell lines derived from a buffalo *(Syncerus caffer)*, *Res. Vet. Sci.*, 16, 125, 1974.
77. **Stagg, D. A., Kanhai, G. K., Young, A. S., and Brown, C. G. D.,** The establishment of *Theileria*-infected cell lines from an eland *(Taurotragus oryx* Lydekker, 1906), *Res. Vet. Sci.*, 20, 122, 1976.
78. **Hooshmand-Rad, P. and Hawa, N. J.,** Cultivation of *Theileria hirci* in sheep lymphoid cells, *Trop. Anim. Health Prod.*, 7, 121, 1975.
79. **Paul, J.,** *Cell and Tissue Culture*, 5th ed., Churchill Livingstone, Edinburgh, 1975.
80. **Jakoby, W. B. and Pastan, I. H., Eds.,** *Methods in Enzymology: Vol. 58, Cell Culture*, Colowick, S. P. and Kaplan, N. O., Eds., Academic Press, New York, 1979.
81. **Maramorosch, K. and Hirumi, H., Eds.,** *Practical Tissue Culture Applications*, Academic Press, New York, 1979.
82. **Jura, W. G. Z. O.,** Light and electron microscopic study of bovine lymphocyte penetration by *Theileria annulata* sporozoites *in vitro*, *Trans. R. Soc. Trop. Med. Hyg.*, 75, 898, 1981.
83. **Stagg, D. A., Chasey, D., Young, A. S., Morzaria, S. P., and Dolan, T. T.,** Synchronization of the division of *Theileria* macroschizonts and their mammalian host cells, *Ann. Trop. Med. Parasitol.*, 74, 263, 1980.

84. **Musisi, F. L., Bird, R. G., Brown, C. G. D., and Smith, M.,** The fine structural relationship between *Theileria* schizonts and infected bovine lymphoblasts in culture, *Z. Parasitenk.*, 65, 31, 1981.
85. **Sharma, R. D. and Brown, C. G. D.,** *In vitro* studies on 2 strains of *Theileria annulata*, in *Advances in the Control of Theileriosis*, Irvin, A. D., Cunningham, M. P., and Young, A. S., Eds., Martinus Nijhoff Publishers, The Hague, 1981, 140.
86. **Nelson, R. T. and Hirumi, H.,** *In vitro* cloning of *Theileria*-infected lymphoblastoid cells: standardization and characterization, in *Advances in the Control of Theileriosis*, Irvin, A. D., Cunningham, M. P., and Young, A. S., Eds., Martinus Nijhoff Publishers, The Hague, 1981, 120.
87. **Brown, C. G. D., Crawford, J. G., Kanhai, G. K., Njuguna, L. N., and Stagg, D. A.,** Immunization of cattle against East Coast fever with lymphoblastoid cell lines infected and transformed by *Theileria parva*, in *Tick-Borne Diseases and their Vectors*, Wilde, J. K. H., Ed., Centre for Tropical Veterinary Medicine, University of Edinburgh, 1978, 331.
88. **Hirumi, H., Nyormoi, O., Nelson, R., and Simmons, C.,** Semi-large scale cultivation and isolation of macroschizonts of *Theileria parva*, in *The In Vitro Cultivation of the Pathogens of Tropical Diseases*, Rowe, D. S. and Hirumi, H., Eds., Schwabe and Co. AG, Basel, 1980, 153.
89. **Nyormoi, O. and Bwayo, J. B.,** Isoenzyme patterns of *Theileria parva*-infected bovine lymphoblastoid cells and purified *Theileria* macroschizonts, in *Advances in the Control of Theileriosis*, Irvin, A. D., Cunningham, M. P., and Young, A. S., Eds., Martinus Nijhoff Publishers, The Hague, 1981, 383.
90. **Melrose, T. R., Brown, C. G. D., and Sharma, R. D.,** Glucosephosphate isomerase isoenzyme patterns in bovine lymphoblastoid cell lines infected with *Theileria annulata* and *T. parva* with an improved enzyme visualisation method using Meldola blue, *Res. Vet. Sci.*, 29, 298, 1980.
91. **Musisi, F. L., Kilgour, V., Brown, C. G. D., and Morzaria, S. P.,** Preliminary investigations on isoenzyme variants of lymphoblastoid cell lines infected with *Theileria* species, *Res. Vet. Sci.*, 30, 38, 1981.
92. **Hawking, F.,** Chemotherapeutic screening of compounds against *Theileria annulata* in tissue culture, *Br. J. Pharmacol.*, 13, 458, 1958.
93. **Irvin, A. D. and Stagg, D. A.,** *Theileria parva:* purine and pyrimidine metabolism and the action of folate antagonists in parasitised bovine lymphoid cells, *Exp. Parasitol.*, 41, 172, 1977.
94. **McHardy, N.,** *In vitro* studies on the action of menoctone and other compounds on *Theileria parva* and *Theileria annulata*, *Ann. Trop. Med. Parasitol.*, 72, 501, 1978.
95. **McHardy, N., Haigh, A. J. B., and Dolan, T. T.,** Chemotherapy of *Theileria parva* infection, *Nature (London)*, 261, 698, 1976.
96. **Dolan, T. T. and McHardy, N.,** The chemotherapy of experimental *Theileria parva* infection, in *Tick-Borne Diseases and their Vectors*, Wilde, J. K. H., Ed., Centre for Tropical Veterinary Medicine, University of Edinburgh, 1978, 318.
97. **Schein, E. and Voigt, W. P.,** Chemotherapy of bovine theileriosis with halofuginone, *Acta Tropica*, 36, 391, 1979.
98. **Uilenberg, G., Jongejan, F., Perie, N. M., and Franssen, F. F. J.,** Chemotherapy of cattle theileriosis with halofuginone, an anticoccidian. Preliminary note, *Rev. Ele. Med. Vet. Pays Trop.*, 33, 33, 1980.
99. **McHardy, N. and Morgan, D. W. T.,** Studies *in vivo* and *in vitro* with 993C and halofuginone, in *Advances in the Control of Theileriosis*, Irvin, A. D., Cunningham, M. P., and Young, A. S., Eds., Martinus Nijhoff Publishers, The Hague, 1981, 209.
100. **Dolan, T. T.,** Progress in the chemotherapy of theileriosis, in *Advances in the Control of Theileriosis*, Irvin, A. D., Cunningham, M. P., and Young, A. S., Eds., Martinus Nijhoff Publishers, The Hague, 1981, 186.
101. **Irvin, A. D., Brown, C. G. D., Kanhai, G. K., and Stagg, D. A.,** Establishment of *Theileria parva*-infected bovine tissue culture in Swiss and athymic (nude) mice, *Vet. Parasitol.*, 3, 141, 1977.
102. **Irvin, A. D., Brown, C. G. D., Kanhai, G. K., and Stagg, D. A.,** Transplantation of bovine lymphosarcoma cells to athymic (nude) mice, *Res. Vet. Sci.*, 22, 53, 1977.
103. **Irvin, A. D., Brown, C. G. D., Kanhai, G. K., Stagg, D. A., and Rowe, L. W.,** Cell fusion, using Sendai virus, to effect interspecies transfer of a cell-associated parasite *(Theileria parva)*, *Int. J. Parasitol.*, 4, 519, 1974.
104. **Irvin, A. D., Brown, C. G. D., Stagg, D. A., Kanhai, G. K., and Rowe, L. W.,** Hybrid cells, infected with *Theileria parva*, formed by fusion of hamster and mouse cells with parasitized bovine lymphoid cells, *Res. Vet. Sci.*, 19, 142, 1975.
105. **Brown, C. G. D. and Gray, M. A.,** Infection of bovine fibroblastic cell lines with *Theileria parva* and *T. annulata*, *Trans. R. Soc. Trop. Med. Hyg.*, 75, 323, 1981.
106. **Brown, C. G. D., Cunningham, M. P., Joyner, L. P., Purnell, R. E., Branagan, D., Corry, G. L., and Bailey, K. P.,** *Theileria parva:* significance of leukocytes for infecting cattle, *Exp. Parasitol.*, 45, 55, 1978.

107. **Spooner, R. L. and Brown, C. G. D.,** Bovine lymphocyte antigens (BoLA) of bovine lymphocytes and derived lymphoblastoid lines transformed by *Theileria parva* and *Theileria annulata, Parasite Immunol.*, 2, 163, 1980.
108. **Morrison, W. I., Buscher, G., Emery, D. L., Nelson, R. T., and Murray, M.,** The kinetics of infection with *Theileria parva* in cattle and the relevance to the development of immunity, in *Advances in the Control of Theileriosis,* Irvin, A. D., Cunningham, M. P., and Young, A. S., Eds., Martinus Nijhoff Publishers, The Hague, 1981, 311.
109. **Allison, A. C. and Eugui, E. M.,** Theileriosis. Cell-mediated and humoral immunity, *Am. J. Pathol.*, 102, 114, 1980.
110. **Allison, A. C. and Eugui, E. M.,** Approaches to vaccines against protozoan diseases of cattle, in *The Ruminant Immune System, Advances in Experimental Medicine and Biology,* Vol. 137, Butler, J. E., Ed., Plenum Press, New York, 1981, 225.
111. **Allison, A. C.,** Cellular immune responses in theileriosis, in *Advances in the Control of Theileriosis,* Irvin, A. D., Cunningham, M. P., and Young, A. S., Eds., Martinus Nijhoff Publishers, The Hague, 1981, 273.
112. **Hulliger, L., Brown, C. G. D., and Wilde, J. K. H.,** Theileriosis *(T. parva)* immune mechanism investigated *in vitro*, in *Progress in Protozoology, Proceedings of the 2nd International Congress on Protozoology, London,* Excerpta Medical Foundation, Amsterdam, 1965, 37.
113. **Burridge, M. J. and Kimber, C. D.,** The indirect fluorescent antibody test for experimental East Coast fever *(Theileria parva* infection of cattle). Evaluation of a cell culture schizont antigen, *Res. Vet. Sci.*, 13, 451, 1972.
114. **Burridge, M. J., Young, A. S., Stagg, D. A., Kanhai, G. K., and Kimber, C. D.,** *Theileria lawrencei* infection of cattle and African buffalo: evaluation of a buffalo-derived cell culture schizont antigen for the indirect fluorescent antibody test, *Res. Vet. Sci.*, 17, 285, 1974.
115. **Burridge, M. J., Brown, C. G. D., and Kimber, C. D.,** *Theileria annulata:* cross-reactions between a cell culture schizont antigen and antigens of East African *Theileria* species in the indirect fluorescent antibody test, *Exp. Parasitol.*, 35, 374, 1974.
116. **Wagner, G. G., Brown, C. G. D., Duffus, W. P. H., Kimber, C. D., Crawford, J. G., and Lule, M.,** Immunochemical studies on East Coast fever. I. Partial segregation and characterization of the *Theileria parva* schizont antigen, *J. Parasitol.*, 60, 848, 1974.
117. **Allsop, B. A., Wagner, G. G., Matthews, K. P., and Kariavu, C. G.,** Purification and characterisation of precipitating antigens from *Theileria parva, J. Gen. Microbiol.*, 100, 319, 1977.
118. **Wagner, G. G. and Duffus, W. P. H.,** Anti-lymphocyte antibody response in cattle inoculated with *Theileria parva*-infected lymphoblastoid cell lines, in *Parasitic Zoonoses, Clinical and Experimental Studies,* Soulsby, E. J. L., Ed., 1974, 97.
119. **Duffus, W. P. H., Wagner, G. G., and Preston, J. M.,** Initial studies on the properties of a bovine lymphoid cell culture line infected with *Theileria parva, Clin. Exp. Immunol.*, 34, 347, 1978.
120. **Muhammed, S. I.,** The effect of leucocytes from cattle immunized against East Coast fever on the migration of bovine lymphoblasts infected with *Theileria parva, Zentralblatt Vet.*, 22, 455, 1975.
121. **Duffus, W. P. H., Butterworth, A. E., Wagner, G. G., Preston, J. M., and Franks, D.,** Antibody-dependent cell-mediated cytotoxicity in cattle: activity against ^{51}Cr-labeled chicken erythrocytes coated with protozoal antigens, *Infect. Immun.*, 22, 492, 1978.
122. **Pearson, T. W., Lundin, L. B., Dolan, T. T., and Stagg, D. A.,** Cell-mediated immunity to *Theileria*-transformed cell lines, *Nature (London),* 281, 678, 1979.
123. **Eugui, E. M. and Emery, D. L.,** Genetically-restricted cell-mediated cytotoxicity in cattle immune to *Theileria parva, Nature (London),* 290, 251, 1981.
124. **Pinder, M. and Hewett, R. S.,** Monoclonal antibodies detect antigenic diversity in *Theileria parva* parasites, *J. Immunol.*, 124, 100, 1980.
125. **Roelants, G. E. and Pinder, M.,** Applications of monoclonal antibody technology to parasite immunology, in *The Immune System,* Vol. 2, Steinberg, C. M. and Lefkovits, I., Eds., S. Karger, Basel, 1981, 446.
126. **Hanks, J. H.,** Balance salt solutions and general purpose culture media, in *Practical Tissue Culture Applications,* Maramorosch, K. and Hirumi, H., Eds., Academic Press, New York, 1979, 3.
127. **Iscove, N. N. and Melchers, F.,** Complete replacement of serum by albumin, transferrin and soybean lipid in cultures of lipopolysaccharide-reactive B lymphocytes, *J. Exp. Med.*, 147, 923, 1978.
128. **Nyindo, M. B. A., Kaminjolo, J. S., Wagner, G. G., and Lule, M.,** East Coast fever: cultivation *in vitro* of cell-free schizonts and merozoites of *Theileria parva* and their immunogenicity in cattle, *Am. J. Vet. Res.*, 39, 37, 1978.
129. **Pipano, E.,** Immunological aspects of *Theileria annulata* infection, *Bull. Office Int. Epizoot.*, 81, 139, 1974.
130. **Pipano, E.,** Virulence and immunity of cultured *Theileria annulata* schizonts, *J. S. Afr. Vet. Assoc.*, 50, 332, 1979.
131. **Pipano, E.,** Immunization against intracellular blood protozoans of cattle, in *Progress in Clinical and Biological Research,* Vol. 47, Alan R. Liss Inc., New York, 1980, 301.

132. **Hooshmand Rad, P.,** Some Studies on Theileria Annulata with Special Reference to an Attenuated Vaccine, Ph.D. thesis, University of Edinburgh, 1973.
133. **Pipano, E. and Shkap, V.,** Attenuation of two Turkish strains of *Theileria annulata, J. Protozool.*, 26, 80A, 1979.
134. **Hashemi-Fesharki, R.,** Quantitative studies of three different strains of *Theileria annulata* in experimental calves, in *Tick-Borne Diseases and their Vectors,* Wilde, J. K. H., Ed., Centre for Tropical Veterinary Medicine, Edinburgh, 1978, 357.
135. **Stepanova, N., Zablotskii, V., Mutuzkina, Z., Rasulov, I., Umarov, I., and Tuhtaev, B.,** Live cell culture vaccine against *Theileria annulata* infection, *Veterinariya,* 3, 69, 1977.
136. **Hawa, N. J., Latif, B. M. A., and Ali, S. R.,** Immunization of sheep against *Theileria hirci* infection with schizonts propagated in tissue culture, *Vet. Parasitol.*, 9, 91, 1981.
137. **Pinder, M., Withey, K. S., and Roelants, G. E.,** *Theileria parva* parasites transform a subpopulation of T lymphocytes, *J. Immunol.*, 127, 389, 1981.
138. **Black, S. J., Jack, R., Lalor, P., and Newson, J.,** Analyses of *Theileria*-infected cell surface antigens with monoclonal antibodies, in *Advances in the Control of Theileriosis,* Irvin, A. D., Cunningham, M. P., and Young, A. S., Eds., Martinus Nijhoff Publishers, The Hague, 1981, 327.
139. **Wagner, G. G., Duffus, W. P. H., and Burridge, M. J.,** The specific immunoglobulin response in cattle immunized with isolated *Theileria parva* antigens, *Parasitology,* 69, 43, 1974.
140. **Pipano, E., Goldman, M., Samish, M., and Friedhoff, K. T.,** Immunization of cattle against *Theileria annulata* using killed schizont vaccine, *Vet. Parasitol.*, 3, 11, 1977.
141. **Theiler, A. and du Toit, P. J.,** The transmission of tick-borne disease by the intrajugular injection of the emulsified intermediary host itself, 13th and 14th Reports of the Director of Veterinary Education and Research, Union of South Africa, 17, 1928.
142. **Yoshida, N., Nussenzweig, R. S., Potocnjak, P., Nussenzweig, V., and Aikawa, M.,** Hybridoma produces protective antibodies directed against the sporozoite stage of malaria parasite, *Science,* 207, 71, 1980.
143. **Nussenzweig, R. S.,** Immunoprophylaxis of malaria: sporozoite-induced immunity, *Adv. Exp. Med. Biol.*, 93, 75, 1977.
144. **Strome, C. P. A., De Santis, P. L., Leef, J. L., and Beaudoin, R. L.,** Large scale isolation of *Plasmodium berghei* sporozoites from infected mosquitoes; *J. Tissue Culture Methods,* 6, 13, 1980.
145. **Strome, C. P. A., De Santis, P. L., and Beaudoin, R. L.,** The cultivation of the exoerythrocytic stages of *Plasmodium berghei* from sporozoites, *In Vitro,* 15, 531, 1979.
146. **Freeman, R. R., Trejdosiewicz, A. J., and Cross, G. A. M.,** Protective monoclonal antibodies recognizing stage-specific merozoite antigens of a rodent malaria parasite, *Nature (London),* 284, 366, 1980.
147. **Kilejian, A.,** Prospects and problems in the use of recombinant DNA for the production of a malaria vaccine, *Am. J. Trop. Med. Hyg.*, 29 (5), 1125, 1980.
148. **Schein, E., Rehbein, G., Voigt, W. P., and Zweygarth, E.,** *Babesia equi* (Laveran 1901). I. Development in horses and in lymphocyte culture, *Tropenmed. Parasitol.*, 32, 223, 1981.
149. **Hulliger, L., Brown, C. G. D., and Wilde, J. K. H.,** Transition of developmental stages of *Theileria parva in vitro* at high temperature, *Nature (London),* 211, 328, 1966.
150. **Danskin, D. and Wilde, J. K. H.,** The effect of calf lymph and bovine red blood cells on *in vitro* cultivation of *Theileria parva*-infected lymphoid cells, *Trop. Anim. Health Prod.*, 8, 175, 1976.
151. **Jongejan, F., Perié, N. M., Franssen, F. F. J., and Uilenberg, G.,** Artificial infection of *Rhipicephalus appendiculatus* with *Theileria parva* by percutaneous injection, *Res. Vet. Sci.*, 29, 320, 1980.
152. **Schreuder, B. E. C. and Uilenberg, G.,** Studies on Theileriidae (Sporozoa) in Tanzania. V. Preliminary experiments on a new method for infecting ticks with *Theileria parva* and *Theileria mutans, Tropenmed. Parasitol.*, 27, 422, 1976.
153. **Kurtii, T. J. and Buscher, G.,** Trends in tick cell culture, in *Practical Tissue Culture Applications,* Maramorosch, K. and Hirumi, H., Eds., Academic Press, New York, 1979, 351.
154. **Kurtii, T. J. and Munderloh, U. G.,** Tick cell culture: characteristics, growth requirements and applications to parasitology, in *Invertebrate Cell Culture Applications,* Maramorosch, K. and Mitsuhashi, J., Eds., Academic Press, New York, in press.
155. **Pudney, M., Varma, M. G. R., and Leake, C. J.,** Culture of embryonic cells from the tick *Boophilus microplus* (Ixodidae), *J. Med. Entomol.*, 10, 493, 1973.
156. **Yunker, C. E. and Meibos, H.,** Culture of embryonic tick cells (Acari: Ixodidae), *Tissue Culture Association Manual,* 5, 1015, 1979.
157. **Varma, M. G. R., Pudney, M., and Leake, C. J.,** The establishment of three cell lines from the tick *Rhipicephalus appendiculatus* (Acari: Ixodidae) and their infection with some arboviruses, *J. Med. Entomol.*, 11, 698, 1975.
158. **Pudney, M., Varma, M. G. R., and Leake, C. J.,** Establishment of cell lines from ixodid ticks, *Tissue Culture Association Manual,* 5, 1003, 1979.

159. **Hadani, A.,** In vitro development of *Babesia* and *Theileria* parasites in organs and tissues of ticks, in *Babesiosis,* Ristic, M. and Kreier, J. P., Eds., Academic Press, New York, 1981, 225.
160. **Hadani, A., Kaufman, W., and Barnett, S. F.,** Development of *Nuttallia danii* (Protozoa: Babesidae) within tick salivary glands cultured, *in vitro, J. Parasitol.,* 64, 501, 1978.
161. **Cunningham, I. and Taylor, A. M.,** Infectivity of *Trypanosoma brucei* cultivated at 28°C with tsetse fly salivary glands, *J. Protozool.,* 26, 428, 1979.
162. **Gray, M. A., Cunningham, I., Gardiner, P., Taylor, A. M., and Luckins, A. G.,** Cultivation of infective forms of *Trypanosoma congolense* from trypanosomes in the proboscis of *Glossina morsitans, Parasitology,* 82, 81, 1981.
163. **Bell, L. J.,** Organ culture of *Rhipicephalus appendiculatus* with maturation of *Theileria parva* in tick salivary glands *in vitro, Acta Tropica,* 37, 319, 1980.
164. **Irvin, A. D. and Boarer, C. D. H.,** Some implications of a sexual cycle in *Theileria, Parasitology,* 80, 571, 1980.

INDEX

A

B

C

D

E

F

G

H

I

J

K

L

M

N

O

P

Q

R

S

U

V

W

X

Y

Z